Studies in Computational Intelligence

Volume 669

Series editor

Janusz Kacprzyk, Polish Academy of Sciences, Warsaw, Poland
e-mail: kacprzyk@ibspan.waw.pl

About this Series

The series "Studies in Computational Intelligence" (SCI) publishes new developments and advances in the various areas of computational intelligence—quickly and with a high quality. The intent is to cover the theory, applications, and design methods of computational intelligence, as embedded in the fields of engineering, computer science, physics and life sciences, as well as the methodologies behind them. The series contains monographs, lecture notes and edited volumes in computational intelligence spanning the areas of neural networks, connectionist systems, genetic algorithms, evolutionary computation, artificial intelligence, cellular automata, self-organizing systems, soft computing, fuzzy systems, and hybrid intelligent systems. Of particular value to both the contributors and the readership are the short publication timeframe and the worldwide distribution, which enable both wide and rapid dissemination of research output.

More information about this series at http://www.springer.com/series/7092

Juan Julián Merelo · Agostinho Rosa
José M. Cadenas · António Dourado Correia
Kurosh Madani · António Ruano
Joaquim Filipe
Editors

Computational Intelligence

International Joint Conference, IJCCI 2015
Lisbon, Portugal, November 12–14, 2015,
Revised Selected Papers

 Springer

Editors
Juan Julián Merelo
Computer Architecture and Computer
  Technology
Universidad de Granada
Granada
Spain

Agostinho Rosa
aSEEB-ISR-IST
Technical University of Lisbon (IST)
Lisbon
Portugal

José M. Cadenas
Facultad de Informática
University of Murcia
Murcia
Spain

António Dourado Correia
Departamento de Engenharia Informatica
University of Coimbra
Coimbra
Portugal

Kurosh Madani
Images, Signals and Intelligence Systems
  Laboratory
University PARIS-EST Creteil (UPEC)
Créteil
France

António Ruano
Campus de Gambelas
Universidade do Algarve
Faro
Portugal

Joaquim Filipe
Escola Superior de Tecnologia de Setúbal
Polytechnic Institute of Setúbal/INSTICC
Setúbal
Portugal

ISSN 1860-949X   ISSN 1860-9503   (electronic)
Studies in Computational Intelligence
ISBN 978-3-319-83957-8   ISBN 978-3-319-48506-5   (eBook)
DOI 10.1007/978-3-319-48506-5

Printed on acid-free paper

This Springer imprint is published by Springer Nature
The registered company is Springer International Publishing AG
The registered company address is: Gewerbestrasse 11, 6330 Cham, Switzerland

Preface

The present book includes extended and revised versions of a set of selected papers from the Seventh International Joint Conference on Computational Intelligence (IJCCI 2015). IJCCI was sponsored by the Institute for Systems and Technologies of Information, Control and Communication (INSTICC). This conference was held in Lisbon, Portugal, from November 12 to 14, 2015.

IJCCI was technically co-sponsored by IEEE Systems, Man, and Cybernetics Society and by International Federation of Automatic Control (IFAC). It was held in cooperation with the ACM SIGAI—ACM Special Interest Group on Artificial Intelligence, AI*IA Associazione Italiana per l'Intelligenza Artificiale, INNS—International Neural Network Society, AAAI—Association for the Advancement of Artificial Intelligence, EUSFLAT—European Society for Fuzzy Logic and Technology, APPIA Associação Portuguesa para a Inteligência Artificial, IFSA—International Fuzzy Systems Association and APNNA—Asia Pacific Neural Network Assembly. Since its first edition in 2009, the purpose of the International Joint Conference on Computational Intelligence (IJCCI) has been to bring together researchers, engineers and practitioners in computational technologies, especially those related to the areas of fuzzy computation, evolutionary computation, and neural computation. IJCCI is composed of three co-located conferences, each one specialized in one of the aforementioned areas, namely:

– International Conference on Evolutionary Computation Theory and Applications (ECTA)
– International Conference on Fuzzy Computation Theory and Applications (FCTA)
– International Conference on Neural Computation Theory and Applications (NCTA)

Their aim is to provide major forums for scientists, engineers and practitioners interested in the study, analysis, design, and application of these techniques to all fields of human activity. In ECTA, evolutionary computation is associated with systems that use computational models of evolutionary processes as the key elements in design and implementation, i.e., computational techniques which are based to some degree on the evolution of biological life in the natural world. A number of evolutionary computational models have been proposed, including evolutionary algorithms, genetic algorithms, evolution strategy, evolutionary programming, and swarm intelligence. These techniques form the basis of several disciplines such as artificial life and evolutionary robotics. FCTA is concerned with modeling and implementation of fuzzy systems, in a broad range of fields. Fuzzy computation is a field that encompasses the theory and application of fuzzy sets and fuzzy logic to the solution of information processing, system analysis and decision problems. Supported by the information technology developments, fuzzy computation has grown continuously during the last decades, and actually leads to major applications in many fields such as medical diagnosis, machine learning, image understanding, automation, and process control. NCTA is focused on modeling and implementation of artificial neural networks (ANN) and neural computing architectures. Neural computation and ANN have seen an explosion of interest over the last few decades, and are being successfully applied across an extraordinarily wide range of problems and domains, in areas as diverse as finance, medicine, engineering, geology, and physics, in problems of complex dynamics and complex behaviour prediction, classification or control. Various structural designs, learning strategies, and algorithms have been introduced in this highly dynamic field in the last couple of decades.

The joint conference IJCCI received 127 paper submissions from 45 countries, of which 20 % were presented as full papers. The high quality of the papers received imposed difficult choices in the review process. To evaluate each submission, a double-blind paper evaluation method was used: each paper was reviewed by at least two experts from the independent international Program Committee, in a double-blind review process, and most papers had three reviews or more. This book includes revised and extended versions of a strict selection of the best papers presented at the conference.

On behalf of the Conference Organizing Committee, we would like to thank all participants. First of all to the authors, whose quality work is the essence of the conference, and to the members of the Program Committee, who helped us with their expertise and diligence in reviewing the papers. As we all know, producing a post-conference book, within the high technical level exigency, requires the effort of many individuals. We wish to thank also all the members of our Organizing Committee, whose work and commitment were invaluable.

November 2015   Juan Julián Merelo
                                                                                                             Agostinho Rosa
                                                                                                          José M. Cadenas
                                                                                                           António Ruano
                                                                                                          Kurosh Madani
                                                                                                        António Dourado
                                                                                                            Joaquim Filipe

Organization

Conference Chair

Joaquim Filipe   Polytechnic Institute of Setúbal/INSTICC, Portugal

Program Co-chairs

ECTA

Agostinho Rosa   ISR - Instituto de Sistemas e Robótica, Portugal
Juan Julian Merelo   University of Granada, Spain

FCTA

António Dourado   University of Coimbra, Portugal
José M. Cadenas   University of Murcia, Spain

NCTA

Kurosh Madani   University of Paris-EST Créteil (UPEC), France
António Ruano   University of Algarve, Portugal

ECTA Program Committee

Chang Wook Ahn	Sungkyunkwan University, Korea, Republic of
Richard Allmendinger	University of Manchester, UK
Keijiro Araki	Kyushu University, Japan
Thomas Baeck	Leiden University, Netherlands
Tim Blackwell	University of London, UK
Alan Blair	University of New South Wales, Australia
Terry Bossomaier	Charles Sturt University, Australia
William R. Buckley	California Evolution Institute, USA
Pedro A. Castillo	University of Granada, Spain
Sung-Bae Cho	Yonsei University, Korea, Republic of
Chi-Yin Chow	City University of Hong Kong, Hong Kong
Antonio Della Cioppa	University of Salerno, Italy
Ernesto Costa	Universidade De Coimbra, Portugal
John Drake	University of Nottingham, China
Peter Duerr	Sony Corporation, Japan
Marc Ebner	Ernst-Moritz-Arndt-Universität Greifswald, Germany
Andries Engelbrecht	University of Pretoria, South Africa
Fabio Fassetti	DIMES, University of Calabria, Italy
Carlos M. Fernandes	University of Lisbon, Portugal
Stefka Fidanova	Bulgarian Academy of Sciences, Bulgaria
Bogdan Filipic	Jozef Stefan Institute, Slovenia
Valeria Fionda	University of Calabria, Italy
Dalila B.M.M. Fontes	Faculdade de Economia and LIAAD-INESC TEC, Universidade do Porto, Portugal
Marcus Gallagher	The University of Queensland, Australia
Crina Grosan	Brunel University London, UK
Steven Guan	Xian Jiaotong-Liverpool University, China
Jörg Hähner	Universität Augsburg, Germany
Lutz Hamel	University of Rhode Island, USA
Julia Handl	University of Manchester, UK
Thomas Hanne	University of Applied Arts and Sciences Northwestern Switzerland, Switzerland
Mohsin Bilal Hashmi	Umm Al-Qura University, Saudi Arabia
Wei-Chiang Hong	Oriental Institute of Technology, Taiwan
Jeffrey Horn	Northern Michigan University, USA
Takashi Ikegami	University of Tokyo, Japan
Katsunobu Imai	Hiroshima University, Japan
Seiya Imoto	University of Tokyo, Japan
Hesam Izakian	Alberta Centre for Child, Family & Community Research, Canada
Yaochu Jin	University of Surrey, UK

Colin Johnson	University of Kent, UK
Ed Keedwell	University of Exeter, UK
Wali Khan	Kohat University of Science & Technology (KUST), Kohat, Pakistan
Mario Köppen	Kyushu Institute of Technology, Japan
Ondrej Krejcar	University of Hradec Kralove, Czech Republic
Jiri Kubalik	Czech Technical University, Czech Republic
Halina Kwasnicka	Wroclaw University of Technology, Poland
Dario Landa-Silva	University of Nottingham, UK
Per Kristian Lehre	University of Nottingham, UK
Piotr Lipinski	University of Wroclaw, Poland
Shih-Hsi Liu	California State University, Fresno, USA
Wenjian Luo	University of Science and Technology of China, China
Penousal Machado	University of Coimbra, Portugal
Stephen Majercik	Bowdoin College, USA
Rainer Malaka	Bremen University, Germany
Davide Marocco	Plymouth University, UK
Euan William McGookin	University of Glasgow, UK
Jörn Mehnen	Cranfield University, UK
Fernando Melício	Instituto Superior de Engenharia de Lisboa, Portugal
Juan Julian Merelo	University of Granada, Spain
Daniel Merkle	University of Southern Denmark, Denmark
Marjan Mernik	University of Maribor, Slovenia
Konstantinos Michail	Cyprus University of Technology, Cyprus
Chilukuri Mohan	Syracuse University, USA
Ambra Molesini	Alma Mater Studiorum—Università di Bologna, Italy
Antonio Mora	University of Granada, Spain
Luiza de Macedo Mourelle	State University of Rio de Janeiro, Brazil
Pawel B. Myszkowski	Wroclaw University of Technology, Poland
Tomoharu Nakashima	Osaka Prefecture University, Japan
Andrzej Obuchowicz	University of Zielona Góra, Poland
Kei Ohnishi	Kyushu Institute of Technology, Japan
Schütze Oliver	CINVESTAV-IPN, Mexico
Yukiko Orito	Hiroshima University, Japan
Ender Özcan	University of Nottingham, UK
Gary B. Parker	Connecticut College, USA
Nelishia Pillay	University of KwaZulu-Natal, South Africa
Walter Potter	University of Georgia, USA
Aurora Pozo	Federal University of Parana, Brazil
Joaquim Reis	ISCTE, Portugal
Robert Reynolds	Wayne State University, USA
Mateen Rizki	Wright State University, USA

Katya Rodriguez   Instituto de Investigaciones en Matemáticas
                                 Aplicadas y en Sistemas (IIMAS), Mexico
Olympia Roeva   Institute of Biophysics and Biomedical
                                 Engineering, Bulgarian Academy of Sciences,
                                 Bulgaria
Peter Ross   Napier University, UK
Suman Roychoudhury   Tata Consultancy Services, India
Filipe Azinhais Santos   ISCTE-IUL, Portugal
Miguel A. Sanz-Bobi   Comillas Pontifical University, Spain
Emmanuel Sapin   University of Exeter, UK
Robert Schaefer   AGH University of Science and Technology,
                                 Poland
Patrick Siarry   University Paris 12 (LiSSi), France
Jim Smith   The University of the West of England, UK
Giovanni Stracquadanio   University of Oxford, UK
Antonio J. Tallón-Ballesteros   Universidade de Lisboa, Portugal
Jonathan Thompson   Cardiff University, UK
Krzysztof Trojanowski   Cardinal Stefan Wyszynski University, Poland
Elio Tuci   Aberystwyth University, UK
Lucia Vacariu   Technical University of Cluj Napoca, Romania
Neal Wagner   Massachusetts Institute of Technology Lincoln
                                 Laboratory, USA
Lusheng Wang   Hefei University of Technology, China
Peter Whigham   University of Otago, New Zealand
Li-Pei Wong   Universiti Sains Malaysia, Malaysia
Gary Yen   Oklahoma State University, USA
Alexandru Ciprian Zavoianu   Johannes Kepler University Linz, Austria

ECTA Additional Reviewers

Tarek R. Besold   University of Osnabrueck, Germany
Sagar Sunkle   Tata Consultancy Services, India

FCTA Program Committee

Rafael Alcala   University of Granada, Spain
Adel M. Alimi   University of Sfax, Tunisia
Ismail H. Altas   Karadeniz Technical University, Turkey
Michela Antonelli   University of Pisa, Italy
Sansanee Auephanwiriyakul   Chiang Mai University, Thailand
Bernard De Baets   Ghent University, Belgium

Ahmad Lotfi   Nottingham Trent University, UK
Edwin Lughofer   Johannes Kepler University, Austria
Francesco Marcelloni   University of Pisa, Italy
Corrado Mencar   University of Bari, Italy
José M. Merigó   University of Chile, Chile
Ludmil Mikhailov   University of Manchester, UK
Valeri Mladenov   Technical University Sofia, Bulgaria
Javier Montero   Complutense University of Madrid, Spain
Hiroshi Nakajima   Omron Corporation, Japan
Yusuke Nojima   Osaka Prefecture University, Japan
Vilém Novák   University of Ostrava, Czech Republic
Parag Pendharkar   Pennsylvania State University, USA
Irina Perfilieva   University of Ostrava, Czech Republic
Radu-Emil Precup   Politehnica University of Timisoara, Romania
Daowen Qiu   Sun Yat-sen University, China
Jordi Recasens   Universitat Politècnica de Catalunya, Spain
Antonello Rizzi   University of Rome "La Sapienza", Italy
Alessandra Russo   Imperial College London, UK
Indrajit Saha   National Institute of Technical Teachers' Training
                                       & Research, India
Jurek Sasiadek   Carleton University, Canada
Steven Schockaert   Cardiff University, UK
Daniel Schwartz   Florida State University, USA
Hirosato Seki   Kwansei Gakuin University, Japan
Dipti Srinivasan   NUS, Singapore
Hooman Tahayori   Ryerson University, Canada
Vicenc Torra   University of Skövde, Sweden, Sweden
Dat Tran   University of Canberra, Australia
Junzo Watada   Waseda University, Japan
Chung-Hsing Yeh   Monash University, Australia
Jianqiang Yi   Institute of Automation, Chinese Academy
                                       of Sciences, China
Slawomir Zadrozny   Polish Academy of Sciences, Poland
Hans-Jürgen Zimmermann   ELITE (European Laboratory for Intelligent
                                       Techniques Engineering), Germany

NCTA Program Committee

Francisco Martínez Álvarez   Pablo de Olavide University of Seville, Spain
Veronique Amarger   University PARIS-EST Creteil (UPEC), France
Vijayan Asari   University of Dayton, USA
Amir Atiya   Cairo University, Faculty of Engineering, Egypt

Mourad Oussalah	University of Birmingham, UK
Parag Pendharkar	Pennsylvania State University, USA
Neil Rowe	Naval Postgraduate School, USA
Christophe Sabourin	IUT Sénart, University Paris-Est Creteil (UPEC), France
Carlo Sansone	University of Naples, Italy
Jurek Sasiadek	Carleton University, Canada
Gerald Schaefer	Loughborough University, UK
Alon Schclar	Academic College of Tel-Aviv Yaffo, Israel
Christoph Schommer	University Luxembourg, Campus Kirchberg, Luxembourg
Ivan Nunes Da Silva	University of São Paulo, Brazil
Lambert Spaanenburg	Lund University Lund Institute of Technology, Sweden
Jochen Steil	Bielefeld University, Germany
Ruedi Stoop	Universität Zürich/ETH Zürich, Switzerland
Mu-Chun Su	National Central University, Taiwan
Norikazu Takahashi	Okayama University, Japan
Antonio J. Tallón-Ballesteros	Universidade de Lisboa, Portugal
Juan-Manuel Torres-Moreno	Universite d'Avignon et des Pays de Vaucluse, France
Carlos M. Travieso	University of Las Palmas de Gran Canaria, Spain
Alfredo Vellido	Universitat Politècnica de Catalunya, Spain
Brijesh Verma	Central Queensland University, Australia
Ricardo Vigário	Aalto University School of Science, Finland
Shuai Wan	Northwestern Polytechnical University, China
Weiwei Yu	Northwestern Polytechnical University, China
Wenwu Yu	Southeast University, China
Cleber Zanchettin	Federal University of Pernambuco, Brazil
Michalis Zervakis	Technical University of Chania, Greece
Huiyu Zhou	Queen's University Belfast, UK

Invited Speakers

Julian Togelius	New York University, USA
Yaochu Jin	University of Surrey, UK
Anna Esparcia-Alcázar	Universitat Politècnica de València, Spain
Edwin Lughofer	Johannes Kepler University, Austria

Contents

Invited Paper

AI Researchers, Video Games Are Your Friends!

Julian Togelius[(✉)]

New York University, New York, USA
julian@togelius.com

Abstract. If you are an artificial intelligence researcher, you should look to video games as ideal testbeds for the work you do. If you are a video game developer, you should look to AI for the technology that makes completely new types of games possible. This chapter lays out the case for both of these propositions. It asks the question "what can video games do for AI", and discusses how in particular general video game playing is the ideal testbed for artificial general intelligence research. It then asks the question "what can AI do for video games", and lays out a vision for what video games might look like if we had significantly more advanced AI at our disposal. The chapter is based on my keynote at IJCCI 2015, and is written in an attempt to be accessible to a broad audience.

Keywords: Artificial intelligence · Games · Artificial general intelligence

1 Introduction

Video games and artificial intelligence are two of my favorite topics. Both as work and hobby. The great thing is that they go together so well: there is a great need for video games in artificial intelligence and for artificial intelligence in video games. In this chapter, I discuss what video games can do for AI and what AI can do for video games.

In the first part, I discuss the need for benchmarks in AI research and how games have historically been used as AI benchmarks. I then argue the advantages of video games over classic board games as AI benchmarks, and in particular the advantages of *general* video game playing. I present the general video game playing competition and benchmark, and the vision of having games both generated and played automatically. I discuss how this fits into the idea of artificial general intelligence, the idea of developing AI that is good not only at a single things but at all things, or at least most of them.

In the second part of the chapter, I discuss what AI can do in and for games. Lots of things, it turns out—playing them is what most people think of first, and it is true that there is a need for skilled and interesting adversaries and other non-player characters in many games—but perhaps even more exciting is all the possibilities that AI offers for modeling players, generating levels and perhaps even whole games, adapting games to suit players, and assisting game designers. The second section is structured as a vision of what playing an open-world game

© Springer International Publishing AG 2017
J.J. Merelo et al. (eds.), *Computational Intelligence*, Studies in Computational Intelligence 669,
DOI 10.1007/978-3-319-48506-5_1

might be like in a future where we have the AI technologies to truly make the game we like, followed by a brief description of some of the research challenges involved in getting there.

It is important to note that this paper does not go into any technical depth on any particular topic, nor is it a comprehensive survey of the field. It is instead meant as an accessible, informal and inspirational introduction as well as a long-form argument. It is equal parts propaganda and science fiction. However, throughout the text I provide a number of references for further reading if you are interested in knowing the technical details or the full state of the field.

2   What Video Games Can Do for AI

The most important thing for humanity to do right now is to invent true artificial intelligence (AI): machines or software that can think and act independently in a wide variety of situations. Once we have artificial intelligence, it can help us solve all manner of other problems.

Luckily, thousands of researchers around work on inventing artificial intelligence. While most of them work on ways of using known AI algorithms to solve new or existing problems, some work on the overarching problem of artificial general intelligence. I do both. As I see it, addressing applied problems spur the invention of new algorithms, and the availability of new algorithms make it possible to address new problems. Having concrete problems to try to solve with AI is necessary in order to make progress; if you try to invent AI without having something to use it for, you will not know where to start. My chosen domain is games, and I will explain why this is the most relevant domain to work on if you are serious about AI.

But first, let us acknowledge that AI has gotten a lot of attention recently. In particular work on "deep learning" is being discussed in mainstream press as well as turned into startups that get bought by giant companies for bizarre amounts of money. There have been some very impressive advances during the past few years in identifying objects in images, understanding speech, matching names to faces, translating text and other such tasks. By some measures, the winner of the recent ImageNet contest is better than humans at correctly naming things in images [1,2]; sometimes I think Facebook's algorithms are better than I am at recognizing the faces of my acquaintances [3].

With few exceptions, the tasks that deep neural networks have excelled at are what are called pattern recognition problems [4]. Basically, take some large amount of data (an image, a song, a text) and output some other (typically smaller) data, such as a name, a category, another image or a text in another language. To learn to do this, they look at tons of data to find patterns. In other words, the neural networks are learning to do the same work as our brain's sensory systems: sight, hearing, touch and so on. To a lesser extent they can also do some of the job of our brain's language centra.

However, this is not all that intelligence is. We humans don't just sit around and watch things all day. We do things: solve problems by taking decisions and carrying them out. We move about and we manipulate our surroundings. (Sure, some days we stay in bed almost all day, but most of the rest of the time we are active in one way or another.) Our intelligence evolved to help us survive in a hostile environment, and doing that meant both reacting to the world and planning complicated sequences of actions, as well as adapting to changing circumstances [5,6]. Pattern recognition - identifying objects and faces, understanding speech and so on - is an important component of intelligence, but should really be thought of as one part of a complete system which is constantly working on figuring out what to do next. Trying to invent artificial intelligence while only focusing on pattern recognition is like trying to invent the car while only focusing on the wheels.

2.1   The Need for AI Benchmarks

In order to build a complete artificial intelligence we therefore need to build a system that takes actions in some kind of environment. How can we do this? Perhaps the most obvious idea is to embody artificial intelligence in robots. And indeed, robotics has shown us how even the most mundane tasks, such as walking in terrain or grabbing strangely shaped objects, are really rather hard to accomplish for robots [7]. In the eighties, robotics research largely refocused on these kind of"simple" problems, which led to progress in applications as well as a better understanding of what intelligence is all about [8]. The last few decades of progress in robotics has fed into the development of self-driving cars, which is likely to become one of the areas where AI technology will revolutionize society in the near future.

Now, working with robots clearly has its downsides. Robots are expensive, complex and slow. When I started my PhD, my plan was to build robot software that would learn evolutionarily from its mistakes in order to develop increasingly complex and general intelligence—this undertaking generally goes by the name "evolutionary robotics" [9]. But I soon realized that in order for my robots to learn from their experiences, they would have to attempt each task thousands of times, with each attempt maybe taking a few minutes. This meant that even a simple experiment would take several days - even if the robot would not break down (it usually would) or start behaving differently as the batteries depleted or motors warmed up. In order to learn any more complex intelligence I would have to build an excessively complex (and expensive) robot with advanced sensors and actuators, further increasing the risk of breakdown. I also would have to develop some very complex environments where complex skills could be learned. This all adds up, and quickly becomes unmanageable. Problems such as these is why the field of evolutionary robotics has not scaled up to evolve more complex intelligence.

I was too ambitious and impatient for that. I wanted to create complex intelligence that could learn from experience. So I turned to video games.

2.2   Games as AI Benchmarks

Games and artificial intelligence have a long history together. Even since before artificial intelligence was recognized as a field, early pioneers of computer science wrote game-playing programs because they wanted to test whether computers could solve tasks that seemed to require "intelligence". Alan Turing, arguably the principal inventor of computer science, (re)invented the Minimax algorithm and used it to play Chess [10]. (As no computer had been built yet, he performed the calculations himself using pen and paper.) Chess was for a long time one of the most important AI benchmarks [11]. Arthur Samuel was the first to invent the form of machine learning that is now called reinforcement learning; he used it in a program that learned to play Checkers by playing against itself [12]. Much later, IBM's Deep Blue computer famously won against the reigning grandmaster of Chess, Gary Kasparov, in a much-publicized 1997 event [13, 14]. Currently, many researchers around the world work on developing better software for playing the board game Go; up until recently, the best software is still no match for good human players [15, 16]. Between the first and the second revision of this chapter, Google DeepMind (Google's primary AI research division) announced in *Nature* that their *AlphaGo* Go-playing program had beaten the European champion at this game [17].

Classic board game such as Chess, Checkers and Go are nice and easy to work with as they are very simple to model in code and can be simulated extremely fast - you could easily make millions of moves per second on a modern computer - which is indispensable for many AI techniques. Also, they seem to require thinking to play well. Many classib both depth and accessibility, meaning that they take "a minute to learn, but a lifetime to master". It is indeed the case that games have a lot to do with learning, and good games are able to constantly teach us more about how to play them. Indeed, to some extent the fun in playing a game consists in learning them and when there is nothing more to learn we largely stop enjoying them. This suggests that better-designed games are also better benchmarks for artificial intelligence. However, judging from the fact that now have (relatively simple) computer programs that can play Chess better than any human, it is clear that you don't need to be truly, generally intelligent to play such games well. When you think about it, they exercise only a very narrow range of human thinking skills; it's all about turn-based movements on a discrete grid of a few pieces with very well-defined, deterministic behavior.

But, despite what your grandfather might want you to believe, there's more to games than classical board games. In addition to all kinds of modern boundary-pushing board games, card games and role-playing games, there's also video games. Video games owe their massive popularity at least partly to that they engage multiple senses and multiple cognitive skills. Take a game such as Super Mario Bros. It requires you not only to have quick reactions, visual understanding and motoric coordination, but also to plan a path through the level, decide about tradeoffs between various path options (which include different risks and rewards), predict the future position and state of enemies and other characters of the level, predict the physics of your own running and jumping, and balance the demands of limited time to finish the level with multiple goals. Other games introduce demands of handling incomplete information (e.g. *StarCraft*), understanding narrative (e.g. *Skyrim*), or very long-term planning (e.g. *Civilization*).

On top of this, video games run inside controllable environments in a computer and many (though not all) video games can be sped up to many times the original speed. It is simple and cheap to get started, and experiments can be run many thousands of times in quick succession, allowing the use of learning algorithms.

So it is not surprising that AI researchers are increasingly turning to video games as benchmarks. Researchers such as myself have adapted a number of video games to function as AI benchmarks. To make it easier to participate in this field and to provide common challenges for researchers to work on, we have organized competitions where researchers can submit their best game-playing AIs and test them against the best that other researchers can produce. Having recurring competitions based on the same game allows competitors to refine their approaches and methods, hoping to win next year. Games for which we have run such competitions include *Super Mario Bros* [18,19], *StarCraft* [20], the *TORCS* racing game [21], *Ms. Pac-Man* [22], a generic *Street Fighter*-style fighting game [23], Angry Birds [24] and several others. In most of these competitions, we have seen performance of the winning AI player improve every time the competition is run. These competitions play an important role in catalyzing research in the community, and every year many papers are published where the competition software is used for benchmarking some new AI method. There are by now a set of best practices for how to organize such competition so as to maximize research value [25]. Thus, we advance AI through game-based competitions.

2.3　Artificial General Intelligence and General Game Playing

There's a problem with the picture I just painted. Can you spot it?

That's right. Game specificity. The problem is that improving how well an artificial intelligence plays a particular game is not necessarily helping us improve artificial intelligence in general. It's true that in most of the game-based competitions mentioned above we have seen the submitted AIs get better every time the competition ran. But in most cases, the improvements were not because of better AI algorithms, but because of even more ingenious ways of using these algorithms for the particular problems. Sometimes this meant relegating the AI to a more peripheral role. For example, in the car racing competition the first years were dominated by AIs that used evolutionary algorithms to train a neural network to keep the car on the track. In later years, most of the best submissions used hand-crafted "dumb" methods to keep the car on the track, but used learning algorithms to learn the shape of the track to adapt the driving [21]. This is a clever solution to a very specific engineering problem but says very little about intelligence in general.

In order to make sure that what such a competition measures is anything approaching actual intelligence, we need to recast the problem. To do this, it's a great idea to define what it is we want to measure: general intelligence. Shane Legg and Marcus Hutter have proposed a very useful definition of intelligence, which is roughly the average performance of an agent on all possible problems [26]. (In their original formulation, each problem's contribution to the average is weighed by its simplicity, but let's disregard that for now.) Obviously, testing an AI on all possible problems is not an option, as there are infinitely many problems. But maybe we could test our AI on just a sizable number of diverse problems? For example on a number of different video games [27]?

The first thing that comes to mind here is to just to take a bunch of existing games for some game console, preferably one that could be easily emulated and sped up to many times real time speed, and build an AI benchmark on them. This is what the Arcade Learning Environment (ALE) does [28]. ALE lets you test your AI on more than a hundred games released for 70 s vintage *Atari 2600* console. The AI agents get feeds of the screen at pixel level, and have to respond with a joystick command. ALE has been used in a number of experiments, including those by the original developers of the framework. Perhaps most famously, Google Deep Mind published a paper in *Nature* last year showing how they could learn to play several of the games with superhuman skill using deep learning (Q-learning on a deep convolutional network) [29].

ALE is an excellent AI benchmark, but has a key limitation. The problem with using Atari 2600 games is that there is only a finite number of them, and developing new games is a tricky process. The Atari 2600 is notoriously hard to program, and the hardware limitations of the console tightly constrain what sort of games can be implemented. More importantly, all of the existing games are known and available to everyone. This makes it possible to tune your AI to each particular game. Not only to train your AI for each game (DeepMind's results depend on playing each individual game millions of times to train on it) but to tune your whole system to work better on the games you know you will train on.

Can we do better than this? Yes we can! If we want to approximate testing our AI on all possible problems, the best we can do is to test it on a number of unseen problems. That is, the designer of the AI should not know which problems it is being tested on before the test. At least, this was our reasoning when we designed the *General Video Game Playing Competition*.

2.4   General Video Game Playing

The General Video Game Playing Competition (GVGAI) allows anyone to submit their best AI players to a special server, which will then use them to play ten games that no-one (except the competition organizers) have seen before [30,31]. These games are of the type that you could find on home computers or in arcades in the early eighties; some of them are based on existing games such as *Boulder Dash*, *Pac-Man*, *Space Invaders*, *Sokoban* and *Missile Command*. The winner of the competition is the AI that plays these unseen games best. Therefore, it is impossible for the creator of the AI to tune their software to any particular game. Around 60 games are currently available for training your AI on and 20 unseen games are available to test on; every iteration of the competition increases this number as the testing games from the previous iteration become available to train on, and new testing games are created.

Now, 60 games is not such a large number; where do we get new games from? To start with, all the games are programmed in something called the *Video Game Description Language* (VGDL) [32,33]. This is a simple language we designed to be able to write games in a compact and human-readable way, a bit like how HTML is used to write web pages. The language is designed explicitly to be able to encode classical arcade games; this means that the games are all based on the movement of and interaction between sprites in two dimensions. This is how essentially all video games were designed before *Wolfenstein 3D*, and quite a few games are still designed that way. In any case, the simplicity of this language makes it very easy to write new games, either from scratch or as variations on existing games. (Incidentally, as an offshoot of this project we are exploring the use of VGDL in a prototyping tool for game developers.)

2.5   General Video Game Generation

Even if it's simple to write new games, that doesn't solve the fundamental problem that someone has to write them, and design them first. For the GVG-AI competition to reach its full potential as a test of general AI, we need an endless supply of new games. For this, we need to generate them. We need software that can produce new games at the press of a button, and these need to be good games that are not only playable but also require genuine skill to win. (As a side effect, such games are likely to be enjoyable for humans.)

I know, designing software that can design complete new games (that are also good in some sense) sounds quite hard. And it is. However, I and a couple of others have been working on this problem on and off for a couple of years, and I'm firmly convinced it is doable. Cameron Browne has already managed to build a complete generator for playable (and enjoyable) board games [34], and several people including myself have attempted to automatically generate video games using different methods [35–38], or just generating interesting variations of existing video games [39]. Some of our recent work has focused on generating simple VGDL games, and though we've had some success there is much left to do [40,41]. Also, it is clearly possible to generate parts of games, such as game levels; there has been plenty of research within the last five years on procedural content generation - the automatic generation of game content [42]. Researchers have demonstrated that methods such as evolutionary algorithms, planning and answer set programming can automatically create levels, maps, stories, items and geometry, and basically any other content type for games [43,44]. Now, the research challenges are to make these methods general (so that they work for all games, not just for a particular game) and more comprehensive, so that they can generate all aspects of a game including the rules. Most of the generative methods include some form of simulation of the games that are being generated, suggesting that the problems of game playing and game generation are intricately connected and should be considered together whenever possible.

Once we have extended the General Video Game Playing Competition with automated game generation, we have a much better way of testing generic game-playing ability than we have ever had before. The software can of course also be used outside of the competition, providing a way to easily test the general intelligence of game-playing AI.

2.6   What Kind of AI Will We Need?

So far we have only talked about how to best test or evaluate the general intelligence of a computer program, not how to best create one. Well, this post is about why video games are essential for inventing AI, and I think that I have explained that pretty well: they can be used to fairly and accurately benchmark AI. But for completeness, let us consider which are the most promising methods for creating AIs of this kind. As mentioned above, (deep) neural networks have recently attracted lots of attention because of some spectacular results in pattern recognition. I believe neural networks and similar pattern recognition methods will have an important role to play for evaluating game states and suggesting actions in various game states. In many cases, evolutionary algorithms are more suitable than gradient-based methods when training neural networks for games.

But intelligence can not only be pattern recognition. (This is for the same reason that behaviorism is not a complete account of human behavior: people don't just map stimuli to responses, sometimes they also think.) Intelligence must also incorporate some aspect of planning, where future sequences of actions can be played out in simulation before deciding what to do. Recently an algorithm called Monte Carlo Tree Search, which simulates the consequences of long sequences of actions by doing statistics of random actions, has worked wonders on the board game Go [45]. It has also done very well on GVGAI. Another family of algorithms that has recently shown great promise on game planning tasks is rolling horizon evolution [46]. Here, evolutionary algorithms are used not for long-term learning, but for short-term action planning.

I think the next wave of advances in general video game-playing AIs will come from ingenious combinations of neural networks, evolution and tree search. (Case in point: Google's recent success on the game of Go stemmed from a combination of Monte Carlo Tree Search and two different types of neural networks [17].) And from algorithms inspired by these methods. The important thing is that both pattern recognition and planning will be necessary in various different capacities. Of course, we cannot predict what will work well in the future (otherwise it wouldn't be called research), but I bet that exploring various combinations of these method will inspire the invention of the next generation of AI algorithms.

2.7   The Even Bigger Picture

Now, you might object that this is a very limited view of intelligence and AI. What about text recognition, listening comprehension, storytelling, bodily coordination, irony and romance? Our game-playing AIs can't do any of this, no matter if it can play all the arcade games in the world perfectly. To this I say:

Patience! One day. None of these things are required for playing early arcade games, that is true. But as we master these games and move on to include other genres of games in our benchmark, such as role-playing games, adventure games, simulation games and social network games, many of these skills will be required to play well. As we gradually increase the diversity of games we include in our benchmark, we will also gradually increase the breadth of cognitive skills necessary to play well. Of course, our game-playing AIs will have to get more advanced to cope. Understanding language, images, stories, facial expression and humor will be necessary. And don't forget that closely coupled with the challenge of general video game playing is the challenge of general video game generation, where plenty of other types of intelligence will be necessary. I am convinced that video games (in general) challenges all forms of intelligence except perhaps those closely related to bodily movement, and therefore that video games (in general) are the best testbed for artificial intelligence. An AI that can play almost any video game and create a wide variety of video games is, by any reasonable standard, intelligent.

"But why, then, are not most AI researchers working on general video game playing and generation?"

To this I say: *Patience! One day.*

This argument has become rather long and winding. Let me sum it up in a handy paragraph, so you remember what this was all about:

It is crucial for artificial intelligence research to have good testbeds. Games are excellent AI testbeds because they pose a wide variety of challenges and are highly engaging. But they are also simpler, cheaper and faster than robots, permitting a lot of research that is not practically possible with robotics. Board games have been used in AI research since the field started, but in the last decade more and more researchers have moved to video games because they offer more diverse and relevant challenges. (They are also more fun.) Competitions play a big role in this. But putting too much effort into AI for a single game has limited value for AI in general. Therefore we created the General Video Game Playing Competition and its associated software framework. This is meant to be the most complete game-based benchmark for general intelligence. AIs are evaluated on playing not a single video game, but on multiple games which the AI designer has not seen before. It is likely that the next breakthroughs in general video game playing will come from a combination of neural networks, evolutionary algorithms and Monte Carlo Tree Search. Coupled with the challenge of playing these games is the challenge of generating new games and new game content. The plan is to have an infinite supply of games to test AIs on. While playing and generating simple arcade games tests a large variety of cognitive capacities - more diverse than any other AI benchmark - we are not yet at the stage where we test all of intelligence. But there is no reason to think we would not get there, given the wide variety of intelligence that is needed to play and design modern video games.

It is now time to turn the perspective around a full radian, and ask not what video games can do for AI, but what AI can do for video games.

3   What AI Can Do for Video Games

Let's start in the here and now. The phrase "game AI" is usually understood as the artificial intelligence you find inside a video game, for example for controlling various non-player characters (NPCs). But is there really any AI in a typical video game? Depends on what you mean. The kind of AI that goes into most video games deals with pathfinding and expressing behaviors that were designed by human designers. The sort of AI that we work on in university research labs is often trying to achieve more ambitious goals, and therefore often not yet mature enough to use in an actual game. Alex Champandard, a prominent developer/researcher at the interface between academic and game-industrial AI, suggests that the "next giant leap of game AI is actually artificial intelligence" [47]. And there's indeed lots of things we could do in games if we only had the AI techniques to do it.

So let's step into the future, and assume that many of the various AI techniques we are working on at the moment have reached perfection, and we could make games that use them. In other words, let's imagine what games would be like if we had good enough AI for anything we wanted to do with AI in games. Imagine that you are playing a game of the future.

You are playing an "open world" game, something like *Grand Theft Auto V* or *Skyrim*. Instead of going straight to the next mission objective in the city you are in, you decide to drive (or ride) five hours in some randomly chosen direction. The game makes up the landscape as you go along, and you end up in a new city that no human player has visited before. In this city, you can enter any house (though you might have to pick a few locks), talk to everyone you meet, and involve yourself in a completely new set of intrigues and carry out new missions. If you would have gone in a different direction, you would have reached a different city with different architecture, different people and different missions. Or a huge forest with realistic animals and eremites, or a secret research lab, or whatever the game comes up with.

Talking to these people you find in the new city is as easy as just talking to the screen. The characters respond to you in natural language that takes into account what you just said. These lines are not read by an actor but generated in real-time by the game. You could also communicate with the game though waving your hands around, dancing or using other exotic modalities for expressing emotions and intentions. Of course, in many (most?) cases you are still pushing buttons on a keyboard or controller, as that is often the most efficient way of telling the game what you want to do.

Perhaps needless to say, but all the non-player characters (NPCs) navigate and generally behave in a thoroughly believable way. For example, they will not get stuck running into walls or repeat the same sentence over and over (well, not more than an ordinary human would). This also means that you have interesting adversaries and collaborators to play any game with, without having to resort either to waiting for your friends to come online or to being matched with annoying thirteen year-olds.

Within the open world game, there are other games to play, for example by accessing virtual game consoles within the game or proposing to play a game with some NPC. These NPCs are capable of playing the various sub-games at whatever level of proficiency that fits with the game fiction, and they play with human-like playing styles. It is also possible to play the core game at different resolutions, for example as a management game or as a game involving the control of individual body parts, by zooming in or out. Whatever rules, mechanics and content are necessary to play these sub-games or derived games are invented by the game engine on the spot. Any of these games can be lifted out of the main game and played on its own.

The game senses how you feel while playing the game, and figures out which aspects of it you are good at as well as which parts you like (and conversely, which parts you suck at and despise). Based on this, the game constantly adapts itself to be more to your liking, for example by giving you more story, challenges and experiences that you will like in that new city which you reached by driving five hours in a randomly chosen direction. Or perhaps by changing its own rules. It's not just that the game is giving you more of what you already liked and mastered. Rather more sophisticatedly, the game models what you preferred in the past, and creates new content that answers to your evolving skills and preferences as you keep playing.

Although the game you are playing is endless, of infinite resolution and continuously adapts to your changing tastes and capabilities, you might still want to play something else at some point. So why not design and make your own game? Maybe because it's hard and requires lots of work? Sure, it's true that back in 2016 it required hundreds of people working for years to make a high profile game, and it required at least a handful of highly skilled professionals to make any notable game at all, even if small. But now that it's the future and we have advanced AI, this can be used not only inside of the game but also in the game design and development and process. So you simply switch the game engine to edit mode and start sketching a game idea. A bit of a storyline here, a character there, some mechanics over here and a set piece on top of it. The game engine immediately fills in the missing parts and provides you with a complete, playable game. Some of it is suggestions: if you have sketched an in-game economy but have no money sink, the game engine will suggest one for you, and if you have designed gaps that the player character can not jump over, the game engine will suggest changes to the gaps or to the jump mechanic. You can continue sketching, and the game engine will convert your sketches into details, or jump right in and start modifying the details of the game; whatever you do, the game engine will work with you to flesh out your ideas into a complete game with art, levels and characters. At any time you can jump in and play the game yourself, and you can also watch a number of artificial players play various parts of the game, including players that play like you would have played the game or like your friends (with different tastes and skills) would have played it.

If you ask me, I'd say that this is a rather enticing vision of the future. I'll certainly play a lot of games if this is what games will look like in a decade or so.

But will they? Will we have the AI techniques to make all this possible? Well, me and a bunch of other people in the CI/AI in Games research community are certainly working on it. (Whether that means that progress is more or less likely to happen is another question...) My team and I are in some form working on all of the things discussed above, except the natural interaction parts (talking to the game etc.).

Let's start with the goal of generating complete games [36,41,48,49]. This requires generating a large number of different aspects of the game, including levels, rules, items, quests, textures etc. The generation of various types of game content is commonly referred to as *procedural content generation* [42,50]. We work mainly within the search-based procedural content generation paradigm [43], where evolutionary algorithms are used to generate content; often, this takes the form of searching for game content that, according to a player model, creates some particular type of player experience [51]. This of course requires us to have models of player experience and player behavior [52–55], so we can predict what players will do when faced with a particular type of game content and how they will experience it. Given that we for the foreseeable future will not be able to completely automate all parts of the game creation process we need to find ways to involve humans inside the game and content generation process; we need *mixed-initiative* tools that combine the best of human and machine creativity [56–59]. In order to assess the quality of games and game content we need to be able to playtest them. Therefore we need strong AI capable of playing any game—which, not coincidentally, is what the first part of this chapter focuses on. Once you have a strong game-playing AI, you might also need to restrict it or otherwise modify it so that it plays the game in a human-like manner; it is common that strong AI players play in a somewhat "machine-like way" [60–62].

By now you probably see how it all fits together. In order to generate games you need to generate various types of content, and in order to do that you need good player models and good artificial players to play the games in a human-like manners. But in order to develop good game-playing AI you need to test your players on multiple games, and in order to do so you need to automatically generate games and game content of high quality [63]. It's like a web, where every part is dependent on every other part. Games are essential to furthering AI, but AI also has a lot to give games. This chapter has tried to explain some of the various ways in which these research questions interact.

This chapter is also an invitation to you to start working within the field of AI in games, and address some of its many fascinating questions. If you are already an AI researcher, you should consider working on games. If you are a researcher in a different field interested in games, consider taking the artificial intelligence perspective on the research problems associated with games. There is a lot of work to do, and you are welcome to join our research community.

References

1. Deng, J., Dong, W., Socher, R., Li, L.J., Li, K., Fei-Fei, L.: ImageNet: a large-scale hierarchical image database. In: IEEE Conference on Computer Vision and Pattern Recognition, CVPR 2009, pp. 248–255. IEEE (2009)
2. Krizhevsky, A., Sutskever, I., Hinton, G.E.: ImageNet classification with deep convolutional neural networks. In: Advances in Neural Information Processing Systems, pp. 1097–1105 (2012)
3. Taigman, Y., Yang, M., Ranzato, M., Wolf, L.: Deepface: closing the gap to human-level performance in face verification. In: Proceedings of the IEEE Conference on Computer Vision and Pattern Recognition, pp. 1701–1708 (2014)
4. Duda, R.O., Hart, P.E., Stork, D.G.: Pattern Classification, 2nd edn. Wiley, New York (2001)
5. Barkow, J.H., Cosmides, L., Tooby, J.: The Adapted Mind: Evolutionary Psychology and the Generation of Culture. Oxford University Press, New York (1996)
6. Buss, D.: Evolutionary Psychology: The New Science of the Mind. Psychology Press, Hove (2015)
7. Arkin, R.: Behavior-Based Robotics. The MIT Press, Cambridge (1998)
8. Brooks, R.: Intelligence without representation. Artif. Intell. **47**, 139–159 (1991)
9. Nolfi, S., Floreano, D.: Evolutionary Robotics. MIT Press, Cambridge (2000)
10. Turing, A.M., Bates, M., Bowden, B., Strachey, C.: Digital computers applied to games. Faster than Thought **101**, 390 (1953)
11. Newell, A., Shaw, J.C., Simon, H.A.: Chess-playing programs and the problem of complexity. IBM J. Res. Dev. **2**, 320–335 (1958)
12. Samuel, A.: Some studies in machine learning using the game of checkers. IBM J. Res. Dev. **3**, 210–229 (1959)
13. Campbell, M., Hoane, A.J., Hsu, F.-H.: Deep blue. Artif. Intell. **134**, 57–83 (2002)
14. Newborn, M.: Kasparov vs. Deep Blue. Computer Chess Comes of Age. Springer, New York (1997)
15. Lee, C.S., Wang, M.H., Chaslot, G., Hoock, J.B., Rimmel, A., Teytaud, O., Tsai, S.R., Hsu, S.C., Hong, T.P.: The computational intelligence of MoGo revealed in Taiwan's computer go tournaments. IEEE Trans. Comput. Intell. AI Games **1**, 73–89 (2009)
16. Müller, M.: Computer go. Artif. Intell. **134**, 145–179 (2002)
17 Silver, D., Huang, A., Maddison, C.J., Guez, A., Sifre, L., van den Driessche, G., Schrittwieser, J., Antonoglou, I., Panneershelvam, V., Lanctot, M., et al.: Mastering the game of go with deep neural networks and tree search. Nature **529**, 484–489 (2016)
18. Karakovskiy, S., Togelius, J.: The Mario UI benchmark and competitions. IEEE Trans. Comput. Intell. AI Games **4**, 55–67 (2012)
19. Togelius, J., Shaker, N., Karakovskiy, S., Yannakakis, G.N.: The Mario AI championship 2009–2012. AI Mag. **34**, 89–92 (2013)
20. Ontanón, S., Synnaeve, G., Uriarte, A., Richoux, F., Churchill, D., Preuss, M.: A survey of real-time strategy game AI research and competition in starcraft. IEEE Trans. Comput. Intell. AI Games **5**, 293–311 (2013)
21. Loiacono, D., Lanzi, P.L., Togelius, J., Onieva, E., Pelta, D.A., Butz, M.V., Lonneker, T.D., Cardamone, L., Perez, D., Sáez, Y., et al.: The 2009 simulated car racing championship. IEEE Trans. Comput. Intell. AI Games **2**, 131–147 (2010)
22. Rohlfshagen, P., Lucas, S.M.: Ms Pac-Man versus ghost team CEC 2011 competition. In: 2011 IEEE Congress on Evolutionary Computation (CEC), pp. 70–77. IEEE (2011)

23. Lu, F., Yamamoto, K., Nomura, L.H., Mizuno, S., Lee, Y., Thawonmas, R.: Fighting game artificial intelligence competition platform. In: 2013 IEEE 2nd Global Conference on Consumer Electronics (GCCE), pp. 320–323. IEEE (2013)
24. Renz, J.: AIBIRDS: the angry birds artificial intelligence competition. In: AAAI, pp. 4326–4327 (2015)
25. Togelius, J.: How to run a successful game-based AI competition. IEEE Trans. Comput. Intell. AI Games 8(1), 95–100 (2014)
26. Legg, S., Hutter, M.: Universal intelligence: a definition of machine intelligence. Minds Mach. 17, 391–444 (2007)
27. Schaul, T., Togelius, J., Schmidhuber, J.: Measuring intelligence through games. Arxiv preprint arXiv:1109.1314 (2011)
28. Bellemare, M., Naddaf, Y., Veness, J., Bowling, M.: The arcade learning environment: an evaluation platform for general agents. Arxiv preprint arXiv:1207.4708 (2012)
29. Mnih, V., Kavukcuoglu, K., Silver, D., Rusu, A.A., Veness, J., Bellemare, M.G., Graves, A., Riedmiller, M., Fidjeland, A.K., Ostrovski, G., et al.: Human-level control through deep reinforcement learning. Nature 518, 529–533 (2015)
30. Perez, D., Samothrakis, S., Togelius, J., Schaul, T., Lucas, S., Couëtoux, A., Lee, J., Lim, C.U., Thompson, T.: The 2014 general video game playing competition. IEEE Trans. Comput. Intell. AI Games (2015)
31. Perez-Liebana, D., Samothrakis, S., Togelius, J., Schaul, T., Lucas, S.M.: General video game AI: Competition, challenges and opportunities. In: AAAI (2016)
32. Ebner, M., Levine, J., Lucas, S.M., Schaul, T., Thompson, T., Togelius, J.: Towards a video game description language (2013)
33. Schaul, T.: A video game description language for model-based or interactive learning. In: Proceedings of the IEEE Conference on Computational Intelligence in Games, Niagara Falls. IEEE Press (2013)
34. Browne, C., Maire, F.: Evolutionary game design. IEEE Trans. Comput. Intell. AI Games 2, 1–16 (2010)
35. Nelson, M., Mateas, M.: Towards automated game design. In: Procedings of the 10th Congress of the Italian Association for Artificial Intelligence (2007)
36. Togelius, J., Schmidhuber, J.: An experiment in automatic game design. In: Proceedings of the IEEE Symposium on Computational Intelligence and Games (2008)
37. Cook, M., Colton, S.: Multi-faceted evolution of simple arcade games. In: IEEE Conference on Computational Intelligence in Games, pp. 289–296 (2011)
38. Zook, A., Riedl, M.O.: Automatic game design via mechanic generation. In: AAAI, pp. 530–537 (2014)
39. Isaksen, A., Gopstein, D., Togelius, J., Nealen, A.: Discovering unique game variants. In: Computational Creativity and Games Workshop at the 2015 International Conference on Computational Creativity (2015)
40. Nielsen, T.S., Barros, G.A.B., Togelius, J., Nelson, M.J.: General video game evaluation using relative algorithm performance profiles. In: Mora, A.M., Squillero, G. (eds.) EvoApplications 2015. LNCS, vol. 9028, pp. 369–380. Springer, Heidelberg (2015). doi:10.1007/978-3-319-16549-3_30
41. Nielsen, T.S., Barros, G.A., Togelius, J., Nelson, M.J.: Towards generating arcade game rules with VGDL. In: IEEE Conference on Computational Intelligence in Games (2015)

42. Shaker, N., Togelius, J., Nelson, M.J.: Procedural content generation in games: a textbook and an overview of current research. In: A Textbook and an Overview of Current Research, Procedural Content Generation in Games (2015)
43. Togelius, J., Yannakakis, G., Stanley, K., Browne, C.: Search-based procedural content generation: a taxonomy and survey. IEEE Trans. Comput. Intell. AI Games (2011)
44. Smith, A.M., Mateas, M.: Answer set programming for procedural content generation: a design space approach. IEEE Trans. Comput. Intell. AI Games **3**, 187–200 (2011)
45. Browne, C., Powley, E., Whitehouse, D., Lucas, S., Cowling, P., Rohlfshagen, P., Tavener, S., Perez, D., Samothrakis, S., Colton, S.: A survey of Monte Carlo tree search methods. IEEE Trans. Comput. Intell. AI Games **4**(1), 1–43 (2012)
46. Perez, D., Samothrakis, S., Lucas, S., Rohlfshagen, P.: Rolling horizon evolution versus tree search for navigation in single-player real-time games. In: Proceedings of the 15th Annual Conference on Genetic and Evolutionary Computation, pp. 351–358. ACM (2013)
47. Graft, K.: When artificial intelligence in video games becomes artificially intelligent. Gamasutra (2015)
48. Togelius, J., Nelson, M.J., Liapis, A.: Characteristics of generatable games. In: Foundations of Digital Games, vol. 9, p. 20 (2014)
49. Font Fernández, J.M., Manrique Gamo, D., Mahlmann, T., Togelius, J.: Towards the automatic generation of card games through grammar-guided genetic programming. In: EvoApps (2013)
50. Togelius, J., Champandard, A.J., Lanzi, P.L., Mateas, M., Paiva, A., Preuss, M., Stanley, K.O.: Procedural content generation: goals, challenges and actionable steps. Dagstuhl Follow-Ups **6** (2013)
51. Yannakakis, G.N., Togelius, J.: Experience-driven procedural content generation. IEEE Trans. Affect. Comput. **2**, 147–161 (2011)
52. Yannakakis, G.N., Spronck, P., Loiacono, D., André, E.: Player modeling. Dagstuhl Follow-Ups **6** (2013)
53. Smith, A.M., Lewis, C., Hullett, K., Smith, G., Sullivan, A.: An inclusive taxonomy of player modeling. Technical report, UCSC-SOE-11-13, University of California, Santa Cruz (2011)
54. Pedersen, C., Togelius, J., Yannakakis, G.N.: Modeling player experience for content creation. IEEE Trans. Comput. Intell. AI Games **2**, 54–67 (2010)
55. Mahlmann, T., Drachen, A., Togelius, J., Canossa, A., Yannakakis, G.N.: Predicting player behavior in tomb raider: underworld. In: 2010 IEEE Symposium on Computational Intelligence and Games (CIG), pp. 178–185. IEEE (2010)
56. Liapis, A., Yannakakis, G.N., Togelius, J.: Sentient sketchbook: computer-aided game level authoring. In: FDG, pp. 213–220 (2013)
57. Yannakakis, G.N., Liapis, A., Alexopoulos, C.: Mixed-initiative co-creativity. In: Proceedings of the 9th Conference on the Foundations of Digital Games (2014)
58. Shaker, N., Shaker, M., Togelius, J.: Evolving playable content for cut the rope through a simulation-based approach. In: AIIDE (2013)
59. Shaker, N., Shaker, M., Togelius, J.: Ropossum: an authoring tool for designing, optimizing and solving cut the rope levels. In: AIIDE (2013)
60. Hingston, P.: Believable bots. In: Hingston, P. (ed.) Can Computers Play Like People? Springer, Heidelberg (2012)

61. Shaker, N., Togelius, J., Yannakakis, G.N., Poovanna, L., Ethiraj, V.S., Johansson, S.J., Reynolds, R.G., Heether, L.K., Schumann, T., Gallagher, M.: The turing test track of the 2012 Mario AI championship: entries and evaluation. In: IEEE Conference on Computational Intelligence in Games (CIG), pp. 1–8. IEEE (2013)
62. Ortega, J., Shaker, N., Togelius, J., Yannakakis, G.N.: Imitating human playing styles in super Mario Bros. Entertainment Comput. **4**, 93–104 (2013)
63. Yannakakis, G.N., Togelius, J.: A panorama of artificial and computational intelligence in games. IEEE Trans. Comput. Intell. AI Games **7**(4), 317–335 (2014)

Evolutionary Computation Theory and Applications

Evolution of Cellular Automata-Based Replicating Structures Exhibiting Unconventional Features

Michal Bidlo$^{(\boxtimes)}$

Faculty of Information Technology, IT4Innovations Centre of Excellence,
Brno University of Technology, Božetěchova 2, 61266 Brno, Czech Republic
bidlom@fit.vutbr.cz
http://www.fit.vutbr.cz/~bidlom

Abstract. Replicating loops represent a class of benchmarks, which is commonly studied in relation with cellular automata. Most of the known loops, for which replication rules exist in two-dimensional cellular space, create the copies of themselves using a certain construction algorithm that is common for all the emerging replicas. In such cases, the replication starts from a single instance of the loop (represented as the initial state of the cellular automaton) and is controlled by the transition function of the automaton according to which the copies of the loop are developed. Despite the fact that universal replicators in cellular automata are possible (for example, von Neumann's Universal Constructor), the process of replication of the loops is usually specific to the shape of the loop and the replication rules given by the transition function. This work presents a method for the automatic evolutionary design of cellular automata, which allows us to design transition functions for various structures that are able to replicate according to a given specification. It will be shown that new replicating loops can be discovered that exhibit some unconventional features in comparison with the known solutions. In particular, several scenarios will be presented which can, in addition to the replication from the initial loop, autonomously develop the given loop from a seed, with the ability of the loop to subsequently produce its replicas according to the given specification. Moreover, a parallel replicator will be shown that is able to develop the replicas to several directions using different replication algorithms.

Keywords: Genetic algorithm · Cellular automaton · Transition function · Conditional rule · Replicating loop

1 Introduction

Since the introduction of cellular automata (CA) in [20], researchers have dealt, among others, how to effectively design a cellular automaton (and its transition function in particular) to solve various problems. For example, cellular automata have been studied for their ability to perform computations, e.g. using principles

© Springer International Publishing AG 2017
J.J. Merelo et al. (eds.), *Computational Intelligence*, Studies in Computational Intelligence 669,
DOI 10.1007/978-3-319-48506-5_2

from the famous Conway's Game of Life [1] or by simulating elementary logic functions in non-uniform cellular matrix [15].

One of the topics widely studied in the area of artificial life is the problem of (self-)replicating loops. Since the introduction of probably the most known loop by Langton [8], which is able to replicate in 151 steps in a CA working with 8 states, some other researchers have dealt with this topic trying to simplify the replication process or enhance the abilities of the loop during replication. For example, Byl introduced a smaller loop that is able to replicate in 25 steps using a CA that works with 6 cell states [4]. Later, several unsheathed loops were proposed by Reggia et al. from which the simplest loop consists of 6 cells only and is able to replicate using 8-state CA in 14 steps [12]. On the other hand, Tempesti studied a possibility to introduce construction capabilities into the loops and proposed a 10-state CA that allows to generate patters inside the replicating structures [19]. Perrier et al. created a "self-reproducing universal computer" using 64-state CA by "attaching" executable programs (Turing Machines) on the loops [11]. Although the aforementioned solutions were achieved using analytic methods, the process of determining suitable transition rules for a given problem represents a difficult task and requires an experienced designer (the process of "programming" the CA is not intuitive). As the number of cell states increases, the process of the CA design becomes challenging due to a significant increase of the solution space. Moreover, for some problems no analytic approach has yet been known to the design of the transition rules. In such cases various unconventional techniques have been applied including Genetic Algorithm (GA) [7], possibly in combination with other heuristics.

For example, Mitchell et al. investigated a problem of performing computations in cellular automata using GA [9]. Their work contains a comparison with the original results obtained by Packard in [10] which can be considered as a milestone in applying evolutionary algorithms (EA) to the design and optimisation of cellular automata. In particular, the authors in [9] claim: "Our experiment produced quite different results, and we suggest that the interpretation of the original results is not correct." It may indicate that the research of cellular automata (and their typical features like emergent behaviour or cooperative cell signalling by means of local rules) using various computing techniques can provide valuable information for advanced studies and applications in this area. Note that Mitchell et al. considered binary (i.e. 2-state) 1D cellular automata only which represent a fundamental concept for advanced models. Sipper proposed a technique called Cellular Programming (a spatially distributed and locally interacting GA) that allows for the automatic design of non-uniform CA that are well suited to various problems [16]. Sapin et al. introduced a GA-based approach to the design of gliders and glider guns in 2D cellular automata [13,14]. It was shown that a spontaneous emergence of glider guns in CA can occur with a significant number of new gun-based and glider structures discovered by EA. The aim of the glider research was to construct a system for collision-based computationally universal cellular automata that are able to simulate Turing machines [14]. In recent years, several solutions emerged that aim to optimize the CA design by introducing various evolution-based and soft-computing techniques in combination with suitable representations of the transition functions. For example, Elmenreich et al. proposed an original technique for the calculation of the transition function using neural networks (NN) [5]. The goal was to train the NN by means of Evolutionary Programming [6] in order to develop self-organising structures in the CA. A novel technique for encoding the transition functions of CA, called Conditionally Matching Rules, was introduced in [3], and some applications in binary CA with advantages over the conventional (table-based) encoding were presented in [2].

Whilst the most of the aforementioned studies considered binary CA (i.e. those working with two cell states only), which may be suitable for straightforward hardware implementations (e.g. Sipper's Firefly machine [17]), multi-state CA can provide a more efficient way for the representation and processing of information in CA thanks to the ability of individual cells to work with more than two states. This feature is important for studying complex systems that are in most cases described by integer (or real-valued) variables. In addition, the introduction of more than two states per cell in the CA may allow to reduce the resources needed to solve a given problem (e.g. the size of the cellular array or dimension of the automaton). For example, Yunès studied computational universality in multi-state one-dimensional cellular automata [21]. A technique for the construction of computing systems in 2D CA was demonstrated by Stefano and Navarra in [18] using rules of a simple game called Scintillae working with 6 cell states. Their approach allows to design components (building blocks) for the construction of bigger systems, e.g. on the basis of gate-level circuits.

The goal of this study is to demonstrate the evolutionary design of 2D cellular automata, using the concept of conditionally matching rules to encode the transition functions, which are able to replicate the given structures with respect to a given arrangement in the cellular array. In particular, uniform, multi-state cellular automata will be treated, the cells of which work with 8 and 10 states. The GA will be applied in order to design suitable transition rules that perform replication of the given structure according to the designer's specification. It will be shown that novel replication scenarios can be found in CA that can copy the given loop not only from its initial instance but also, from a seed the loop can autonomously grow. Moreover, a parallel replication scheme will be presented, the objective of which is to speed-up the replication process by allowing the structures to replicate to more directions in the 2D CA. The results will demonstrate the ability of the GA to discover different replication scenarios for the replicas developing in parallel in the cellular automaton, which will be encoded in a single evolved transition function.

2   Cellular Automata

The original concept of cellular automaton, introduced in [20], which will be considered in this study, assumes a 2D matrix of cells, each of which at a given moment acquires a state from a finite set of states. The development of the CA is performed synchronously in discrete iterations (time steps) by updating the cell states according to local transition functions of the cells. Uniform cellular automata will be investigated in which the local transition function is identical for all cells and hence it can be considered as a transition function of the CA. The next state of each cell depends on the combination of states in its neighbourhood. In this work, von Neumann neighbourhood will be assumed that includes a given (*C*entral) cell to be updated and its immediate neighbours in the *N*orth, *S*outh, *E*ast and *W*est direction (i.e. it is a case of a 5-cell neighbourhood).

Since the CA behaviour can practically be evaluated in the cellular array of a finite size, boundary conditions need to be specified in order to correctly determine cell states at the edge of the array. Cyclic boundary conditions will be implemented which means that cells at an edge of the CA are "connected" to the appropriate cells on the opposite edge (i.e. these cells are considered as neighbours) in each dimension. In case of the 2D CA the shape of such cellular array can be viewed as a toroid.

The transition function is usually defined as a mapping that for all possible combinations of states in the cellular neighbourhood determines a new state. This mapping can be represented as a set of rules of the form $N_t \ W_t \ C_t \ E_t \ S_t \rightarrow C_{t+1}$ where N_t, W_t, C_t, E_t and S_t denote cell states in the defined neighbourhood at a time t and C_{t+1} is the new state of the cell to be updated. It means that for every possible combination of states $N_t \ W_t \ C_t \ E_t \ S_t$ a new state C_{t+1} needs to be specified. However, if the number of cell states increases, the number of possible transition rules grows significantly which is inconvenient for efficient CA design. Of course, not all transition rules need to be specified explicitly but the problem is how to choose the rules which modify the central cell in the neighbourhood. Therefore, an advanced representation of the transition rules was proposed and denominated as Conditionally Matching Rules [3]. Conditionally matching rules allows us to reduce the size of representation of the transition functions especially with respect to the evolutionary design of cellular automata.

3   Conditionally Matching Rules

The concept of conditionally matching rules (CMR) showed as a very promising technique in comparison with the conventional (table-based) approach considering various experiments with binary cellular automata (e.g. pattern development task [3] or binary multiplication in 2D CA [2]). In this work, evolutionary design of the CMR-based representation will be investigated in order to design cellular automata with up to 10 cell states that support replication of a given structure.

A conditionally matching rule represents a generalised rule of a transition function for determining a new cell state. Whilst the common approach specifies a new state for every given combination of states in the cellular neighbourhood, the CMR-based approach allows to encode a wider range of combinations into a single rule. A CMR is composed of two parts: a condition part and a new state. The number of items (size) of the condition part corresponds to the number of cells in the cellular neighbourhood. Let us define a condition item as an ordered pair consisting of a condition function and a state value. The condition function is typically expressed as a function whose result can be interpreted as either true or false. The condition function evaluates the state value in the condition item with respect to the state of the appropriate cell in the cellular neighbourhood. In particular, each item of the condition part is associated with a cell in the neighbourhood with respect to which the condition is evaluated. If the result of such evaluation is true, then the condition item is said to match with the state of the appropriate cell in the neighbourhood. In order to determine a new cell state according to a given CMR, all its condition items must match (in such case the CMR is said to match).

The following condition functions will be considered: $== 0, \neq 0, \leq, \geq$. Note that this condition set represents a result of our long-term experimentation and experience with the CMR approach and will be used for all the experiments in this study. The condition $== 0$, respective $\neq 0$, evaluates whether the corresponding cell state is equal to 0 (i.e. a "dead" state), respective whether it is different from state 0. Note that the state value of the condition item for $== 0$ and $\neq 0$ is considered implicitly within the condition itself. The conditions $\leq$ and $\geq$ represent relational operators "less or equal" and "greater or equal" respectively for which the state value of the condition item must be explicitly specified.

Figure 1 shows an example of conditionally matching rules defined for a 2D CA with the 5-cell neighbourhood together with the illustration of cells the condition items are related to. CMR (A) is a matching CMR since all the conditions of its condition part are evaluated as true with respect to the sample neighbourhood shown in the left part of Fig. 1. On the other hand, CMR (B) does not match because the second condition item $! = 2$ evaluates as false with respect to the west cell that possesses state 2. Similarly, the third condition $== 0$ of CMR (B) is not true as the central cell is in state 2.

A CMR-based transition function can be specified as a finite (ordered) sequence of conditionally matching rules. The following algorithm will be applied to determine a new state of a cell. The CMRs are evaluated sequentially one by one. The first matching CMR in the sequence is used to determine the new state. If no of the CMRs matches, then the cell keeps its current state. These conventions for evaluating and applying the CMRs ensure that the process of calculating the new state is deterministic (it is assumed that the condition functions are deterministic too). Therefore, it is possible to convert the CMR-based transition function to a corresponding table-based representation which preserves the fundamental concept of cellular automata. Moreover, every condition set that includes relation $==$ allows to formulate transition rules for specific combinations of states if needed (by specifying $==$ for all condition items of the CMR).

In order to obtain the conventional (table-based) representation of the transition rules from an evolved CMR-based solution, the following algorithm is applied using the same CA that was considered during evolution. Let C_t and C_{t+1} denote states of a cell in two successive steps of the CA at time t and $t+1$

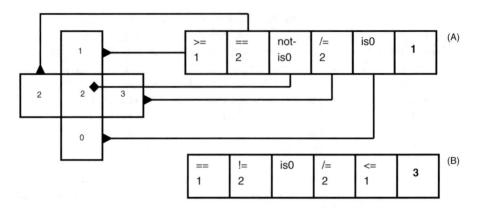

Fig. 1. Example of a conditionally matching rule specified for 5-cell neighbourhood. The value of the new state is written in bold. (A) example of a matching CMR, (B) example of a CMR that does not match – the second and third condition is evaluated as false.

CMR 1	CMR 2	...	CMR n
cn sn | cw sw | cc sc | ce se | cs ss | ns	cn sn | cw sw | cc sc | ce se | cs ss | ns		cn sn | ... | ns

Fig. 2. Structure of a chromosome for genetic algorithm encoding a CMR-based transition function. cx denote a condition for the cell at position x in the neighbourhood, sx represents the state value to be investigated using the appropriate condition with respect to the state of cell at position x, ns specifies the next state for a given CMR. All the conditions and state values are represented by integer numbers.

respectively. A transition rule of the form $N_t\ W_t\ C_t\ E_t\ S_t \rightarrow C_{t+1}$ is generated for the combination of states in the cellular neighbourhood if $C_t \neq C_{t+1}$. This process is performed after each step and for each cell until the CA reaches a stable or periodic state. The set of rules obtained from this process represents the corresponding conventional prescription of the transition function. Note that only the rules that modify the cell state are generated, all the other rules are implicitly considered to preserve the current state.

4   Evolutionary System Setup

A genetic algorithm is utilized for the evolution of CMR-based transition functions in order to achieve the given behaviour in cellular automata. Each chromosome of the GA represents a candidate transition function encoded as a finite sequence of CMRs. The chromosome is implemented as a vector of integers in which the condition items and next states of the CMRs are encoded. Note that the population consists of chromosomes of a uniform length (given by the number of CMRs) which is specified as a parameter for a specific experiment. The structure of a chromosome is depicted in Fig. 2.

The population of the GA consists of 8 chromosomes that are initialised randomly at the beginning of the evolutionary process. In each generation, four individuals are selected randomly from the current population, the best one of which is considered as a parent. In order to generate an offspring, the parent undergoes a process of mutation as follows. A random integer M in range from 0 to 2 is generated. Then M random positions in the parent chromosome are selected. The offspring is created by replacing the original integers at these positions by new valid randomly generated values. If M equals 0, then no mutation is performed and the offspring is identical to the parent. The process of selection and mutation is repeated until the entire new population is created. Crossover is not applied because no benefit of this operator was observed during the initial experiments. Note that the same GA has successfully been applied since the introduction of CMRs in various case studies [2,3]. Although no optimal (evolutionary) approach has yet been known for uniform CA, our experiments indicate that small-population EA (i.e. less than 10 individuals) with a simple mutation operator may represent a suitable class of algorithms to obtain working solutions with a reasonable success rate and computational effort. However, the detailed analysis and wider comparison of different techniques is not a subject of this study.

For each experiment, the GA is executed for 3 million generations. If no correct solution is found within this limit, the evolution is terminated. The evaluation of the chromosomes (i.e. the fitness function) and details regarding various experimental settings are described in the next section.

5   Experimental Results

This section summarises statistics of the evolutionary experiments performed and presents some results together with a more detailed analysis. Two sets of experiments are considered, the goal of each is to design CA that is able to replicate the given loop. The first set works with a *big loop* (the denomination is chosen for the purposes of this work with respect to the loop in the second set of experiments), the objective is to design transition rules that are able to develop a single replica of the loop in a given arrangement against the initial loop. In the second set, a simpler, *small loop* is treated, the goal is to find replication rules for the development of two independent replicas in parallel on the left and right side of the initial loop. Note that the loops consist of cells in 7 different states (including state 0). In both sets of experiments, the CA working with 8 and 10 cell states are investigated. Moreover, different numbers of CMRs (varying from 20 to 50) encoded in the GA chromosomes are considered. For each setup, 100 independent evolutionary runs are executed. The experiments were executed using the Anselm cluster[1], the time of a single run (3 million generations) is approximately 12 h.

---

[1] https://docs.it4i.cz/anselm-cluster-documentation/hardware-overview.

5.1   Replication Evolution of the Big Loop

A big loop is considered for the replication in the first set of experiments, the structure of which is shown in Fig. 3a. The genetic algorithm is applied to design the transition rules for the CA, which perform the replication of the loop in a maximum of 30 steps. The required CA state, that contains the replica, is depicted in Fig. 3b. The following algorithm is applied to the evaluation of the candidate solutions during evolution and the calculation of the fitness function. A partial fitness function is evaluated after each CA step as the number of cells in correct states with respect to Fig. 3b. The final fitness value of a given candidate solution is defined as the maximum of the partial fitness values. It this case the replication can be considered as a pattern transformation problem from a single (initial) loop onto two loops in a given arrangement. However, the loop is expected to replicate again and again during the subsequent CA development, which will be validated for the results obtained from the evolution. Moreover, an assumption is considered that each newly created loop is shifted by two cells down with respect to its predecessor (as shown in Fig. 3b). Therefore, the solutions obtained are further investigated using a visual software simulator developed by the author of this work in order to check that. The goal of this approach is to determine whether the GA is able to discover various new general replication scenarios. Note that, for the purposes of this study, the term "general" means the ability of a solution to repeatedly produce more replicas of the given loop, not an ability to replicate arbitrary loops.

Table 1 summarises the results of experiments with the big loop and provides an overview of some basic parameters of the CA that can be observed during its development using the evolved transition functions. As evident, the maximum success rate achieved during the experiments is only 12 % which is not very high. Note, however, that the replication of the proposed loop represents a problem for which no working solution was found during our previous experiments using the table-based transition functions.

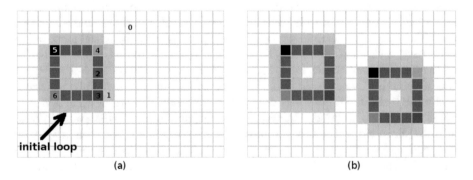

Fig. 3. The structure of the big loop in the cellular automaton that was evaluated during evolution: (a) the initial CA state containing the loop to be replicated, (b) the target state specifying the replica arrangement.

Table 1. Results of the evolutionary experiments considering the design of transition functions for the replication of the loop from Fig. 3a. Success rate – the number of successful experiments out of 100 independent experiments performed that has met the fitness specification in a limit of 3 million generations, Replicates repeatedly – the number of results from the successful experiments that are able to produce more replicas during the subsequent CA development, Min. steps – the minimal number of steps of the CA needed to create the replica (i.e. the lowest value of this parameter from the group of "Replicates repeatedly" solutions, Min. rules – the minimal number of table-based transition rules obtained (i.e. the lowest value of this parameter from the group of "Replicates repeatedly" solutions.

Num. of CMRs	CA with 8 cell states				CA with 10 cell states			
	Success rate [%]	Replicates repeatedly	Min. steps	Min. rules	Success rate	Replicates repeatedly	Min. steps	Min. rules
20	0	-	-	-	1	0	-	-
30	10	6	19	84	12	9	21	146
40	9	4	20	139	12	6	16	186
50	10	6	18	130	12	6	21	177

In addition to the results obtained for the CA working with 8 cell states, some successful solutions have even been obtained for 10 cell states which indicates that the CMRs are an efficient encoding of the transition rules that allows for the design of more complex multi-state CA. The solutions obtained in this work demonstrate a wide range of various replication schemes that can be performed using CA. For example, a solution was found that is able to replicate the loop in 16 steps (the best result achieved for this loop) whilst some CA require 30 steps (the maximal allowed number of steps) in order to finish the replication. Similarly, the number of transition rules generated from the CMRs varies from 84 to more than 1500 rules. These results indicate that cellular automata can in some cases exhibit behaviour that has not yet been discovered which may be beneficial not only for the area of CA but also, for the study of complex systems in general.

Figure 4 shows a CA development performed by one of the successful transition functions obtained for the replication of the given loop. It is one of the best solutions discovered in this work with respect to the number of steps needed to create a copy of the loop. The transition function was found with 30 CMRs in the GA chromosomes and the corresponding conventional representation contains 238 transition rules. If the development of the initial loop is considered (see the upper parts of each step in Fig. 4), the CA needs 21 steps to create a complete replica. As shown by the last step, more replicas can be created in the same way according to the original specification if the CA development continues. However, a more detailed investigation of this result showed that the complete initial loop is not strictly needed in order to successfully perform the replication. For example, the loop is able to emerge even from a single seed – the lower parts of each step presented in Fig. 4 shows a development of the loop from

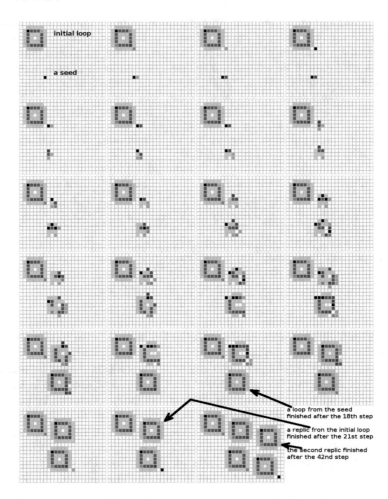

Fig. 4. Develpment of a CA performing replication of the loop from Fig. 3a. The sequence of steps reads from left to right and top to bottom. The upper part of each step of the CA illustrates the replication of the initial loop. The bottom part demonstrates a seed represented by a cell in state 5. Note that after the loop is finished, its replication continues in the same way as from the initial instance (shown by the last CA state).

a single initial cell (a seed) in state 5. As marked by the up-most black arrow a complete loop is developed from the seed after 18 steps which is by 3 steps faster compared to the development from the initial loop. This behaviour is caused by a need of the initial loop to generate a cell in state 5 (i.e. the same state as the seed) from which the replica can be developed (it takes 3 steps – see the top-right CA state in Fig. 4). The process of finishing the replica is identical with the development from the seed. Note that the ability of the transition function to develop and replicate the loop from a seed was not explicitly required in the

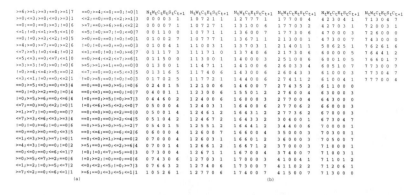

Fig. 5. Transition function for the CA in Figs. 6 and 7: (a) the evolved representation with 50 CMRs, (b) the corresponding conventional representation consisting of 130 rules. This result represents one of the best solutions discovered for the replication of the big loop.

fitness evaluation. Hence it can be considered as an additional, unconventional feature of this solution.

Another result is presented in the form of an evolved transition function (Fig. 5) and the appropriate CA development (Figs. 6 and 7). This cellular automaton demonstrates a development process from a seed that at first creates rather a chaotic structure even larger than the required loop itself. A "mature" loop is developed from this structure during the subsequent CA development that is able to replicate itself. Whilst the replication of the initial loop takes 25 steps (marked by the black arrow in Fig. 6), the development of the chaotic structure needs 36 steps. Starting by step 37 (Fig. 7) the loop is developed from that structure in the same way as from the initial loop. It was verified that the loops are able to replicate repeatedly if the CA development continues.

For both the presented solutions the transition function was identified as redundant (i.e. not all the conventional transition rules generated from the CMR representation are needed for the replication of the initial loop required by the fitness function). A more detailed analysis showed that this redundancy is caused by the finite CA size with cyclic boundary conditions and by generating the transition rules from the CMRs until the CA reaches a stable or periodic state. Although this approach leads to more complex table-based transition functions, in this case it showed as very beneficial for achieving some additional features that were not required during evolution (especially the ability to develop the loops from a seed). Advanced experiments with the resulting CA showed that if the transition functions are optimized (i.e. only the rules for the development of a single replica from the initial loop are considered), the CA in most cases loose the ability of the development from the seed. It was also determined that the seed-based development does not work in case of the known replicating loops (e.g. Langton's or Byl's loop). In the future, this ability may be beneficial for the advanced study of complex systems in which a given (complex) configuration

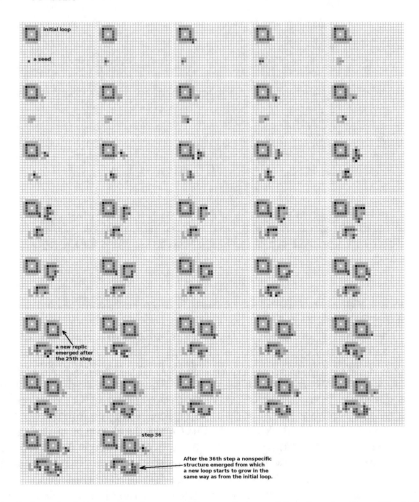

Fig. 6. Part 1 of the replication according to the transition function from Fig. 5. The sequence of steps reads from left to right and top to bottom. The development shows a replication of the initial loop (the upper part of each step) and a growth of a non-specific structure from a seed allowing to create the loop autonomously (the lower part of each step). The seed is represented by a cell in state 7.

needs to be achieved—distributed—from a single cell or a simple initial configuration. In addition to the results presented herein, various other solutions were found that are able to replicate a given structure. It indicates that the replication in CA is not limited to known schemes only but can be performed in many different ways.

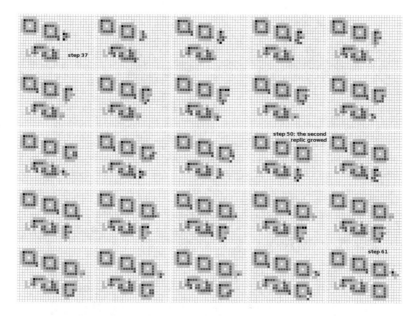

Fig. 7. Part 2 of the replication according to the transition function from Fig. 5. The sequence of steps reads from left to right and top to bottom. The development shows an autonomous growth of the loop from a non-specific structure that emerged in the last step of Fig. 6 (the bottom part of each step). It was verified that the loop is able to replicate in the same way as the initial loop during the subsequent CA development.

5.2   Parallel Replication of the Small Loop

The second set of experiments presents the evolution of parallel replication techniques of a small loop with its structure shown in Fig. 8a. As with the evolution of the big loop, the CA behaviour is evaluated for 30 steps using the partial fitness calculated after each step with respect to the target arrangement of the replicas shown in Fig. 8b, and the final fitness value is given by the maximum of the partial fitness values. In this case, however, two replicas are required with the arrangement on the left and right side of the original loop. The hypothesis evaluated herein is that if suitable transition functions exist for the development of the replicas, then at least a subset of the results will produce the replicas repeatedly in the given directions during the subsequent CA development (i.e. for the purposes of this study, such the solutions will be considered as general). Since the loop is not fully symmetric with respect to the cell states on the sides of the loop, it is expected that different replication algorithms (i.e. sequences of the CA steps) need to be designed to produce the replicas.

Table 2 summarises the results of experiments with the small loop and provides an overview of some basic parameters of the CA that can be observed during its development using the evolved transition functions. Although the shape of the small loop is simpler than the big loop, the requirement of two

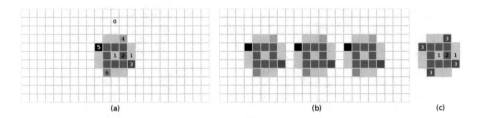

Fig. 8. The structure of the small loop in the cellular automaton that was evaluated during evolution: (a) the initial CA state containing the loop to be replicated, (b) the target state specifying the replicas arrangement, (c) example of a symmetric loop.

Table 2. Results of the evolutionary experiments considering the design of transition functions for the replication of the loop from Fig. 8a. The success rate, the number of general solutions, the minimal number of transition rules and the minimal number of CA steps needed to create the replicas were evaluated.

Num. of CMRs	CA with 8 cell states				CA with 10 cell states			
	Success rate [%]	Replicates repeatedly	Min. steps	Min. rules	Success rate	Replicates repeatedly	Min. steps	Min. rules
20	0	-	-	-	2	0	-	-
30	9	5	18	134	9	3	17	120
40	7	6	17	123	9	4	17	134
50	8	4	18	157	12	7	17	219

independent replicas increases the overall complexity of this task, the maximum success rate achieved does not exceed 12 %. Despite this fact, the evolution provided some solutions that perfectly fulfil the target specification and, in addition, also exhibit the capability of the seed-based development which was not explicitly required.

Figure 9 shows a CA that performs a successful parallel replication of the small loop. The CA works with 8 cell states and, in addition to the replication of the initial loop, is also able to perform the development and replication of the loop from a seed. This is one of the most efficient and compact solution obtained in this study regarding the number of CA steps and the number of transition rules. The corresponding table-based transition function consists of 154 rules as shown in Fig. 10. The CA needs to perform 23 steps in order to finish the replicas of the initial loop. However, if a cell is initialised as a seed by one of the states 1, 3, 5, 6, or 7, the small loop autonomously grows into its full shape and subsequently is able to replicate according to the original specification. The analysis of the seed-based development showed that the loop needs 19 steps to fully develop from state 1, 18 steps from states 3, 6 and 7, and 24 steps from state 5. An interesting behaviour of the CA can be observed after finishing the seed development when the loop ought to be replicated. In particular, the loop replicates according to the given specification from states

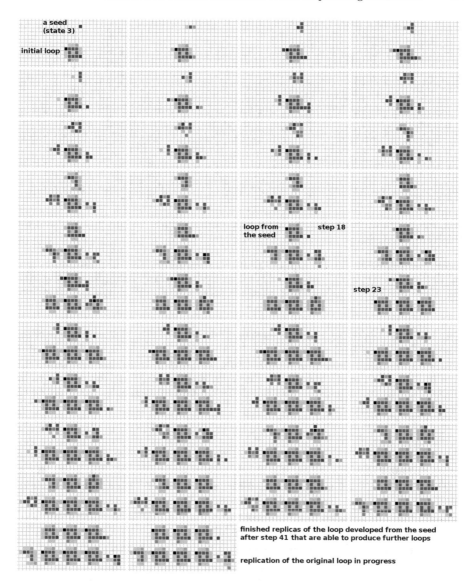

Fig. 9. A sequence of CA steps demonstrating the parallel replication of the small loop according to the evolved transition function from Fig. 10. The states are ordered from left to right and top to bottom. The bottom part of each state shows the replication from the initial loop, the top part of each state demostrates the development and replication of the loop from a seed.

1, 3, 6, and 7. However, the state-5 seed creates an undesirable structure that prevents the loop replication to the left side, i.e. the loop developed from state 5 can replicate to the right side only (see Fig. 11). This indicates that a wide

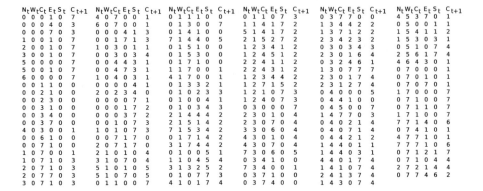

N_t	W_t	C_t	E_t	S_t	C_{t+1}	N_t	W_t	C_t	E_t	S_t	C_{t+1}	N_t	W_t	C_t	E_t	S_t	C_{t+1}	N_t	W_t	C_t	E_t	S_t	C_{t+1}	N_t	W_t	C_t	E_t	S_t	C_{t+1}	N_t	W_t	C_t	E_t	S_t	C_{t+1}
0	0	0	1	0	7	4	0	7	0	0	1	0	1	1	1	0	0	0	1	1	0	7	3	0	3	7	7	0	0	4	5	3	7	0	1
0	0	0	4	0	3	6	0	7	0	0	1	0	1	3	0	0	7	1	1	4	1	7	2	1	3	4	4	2	2	0	5	0	0	1	1
0	0	0	7	0	3	0	0	0	4	1	3	0	1	4	1	0	0	5	1	4	1	7	2	1	3	7	1	2	2	1	5	4	1	1	2
1	0	0	1	0	7	0	0	1	7	1	3	7	1	4	4	0	5	2	1	5	2	7	2	2	3	4	2	3	2	1	5	3	0	3	1
2	0	0	1	0	7	1	0	3	0	1	1	0	1	5	1	0	0	1	2	3	4	1	2	0	3	0	3	4	3	0	5	1	0	7	4
3	0	0	1	0	7	0	0	3	0	3	4	0	1	5	3	0	0	1	2	4	5	1	2	2	3	0	1	6	4	2	5	6	1	7	4
5	0	0	0	0	7	0	0	4	4	3	1	0	1	7	1	0	0	2	2	4	1	1	2	0	3	2	4	6	1	4	6	4	3	0	1
5	0	0	1	0	7	0	0	4	7	3	1	1	1	7	0	0	1	2	2	4	3	1	2	1	3	0	7	7	7	0	7	0	0	0	1
6	0	0	0	0	7	3	1	4	0	3	1	4	1	7	0	0	1	1	2	3	4	4	2	2	3	0	1	7	4	0	7	0	1	0	1
0	0	1	1	0	0	0	0	0	0	4	1	0	1	3	3	2	1	1	2	7	1	5	2	2	3	1	2	7	4	0	7	0	7	0	1
0	0	2	1	0	0	2	0	2	3	4	0	0	1	0	2	3	3	1	2	1	0	7	3	0	4	0	0	0	5	1	7	0	0	0	7
0	0	2	3	0	0	0	0	0	0	7	1	0	1	0	0	4	1	1	2	4	0	7	3	0	4	4	1	0	0	0	7	1	0	0	7
0	0	3	1	0	0	0	0	0	1	7	2	0	1	0	3	4	3	0	3	0	0	0	7	0	4	5	0	0	7	0	7	1	1	0	7
0	0	3	4	0	0	0	0	0	3	7	2	2	1	4	4	4	2	2	3	0	1	0	4	1	4	7	7	0	3	1	7	1	0	0	7
0	0	3	7	0	0	0	0	1	0	7	3	2	1	5	1	4	2	2	3	0	7	0	4	0	4	0	2	1	4	7	7	1	4	0	6
4	0	3	0	0	1	1	0	1	0	7	3	7	1	5	3	4	2	3	3	0	6	0	4	0	4	0	7	1	4	0	7	4	1	0	1
0	0	6	1	0	0	0	0	7	1	7	0	0	1	7	1	4	2	4	3	0	1	0	4	0	4	4	2	1	2	4	7	7	1	0	1
0	0	7	1	0	0	2	0	7	1	7	0	3	1	7	4	4	2	4	3	0	7	0	4	1	4	4	0	1	1	7	7	7	1	0	6
1	0	7	0	0	1	2	1	0	1	0	4	0	1	0	0	5	1	7	3	0	6	0	5	1	4	4	0	3	1	0	7	1	2	1	7
1	0	7	1	0	3	3	1	0	7	0	4	1	1	0	4	5	4	0	3	4	1	0	0	4	4	0	1	7	4	0	7	1	0	4	4
2	0	7	1	0	3	5	1	0	1	0	5	3	1	3	2	5	2	7	3	4	0	0	1	1	4	1	0	7	4	2	7	2	1	4	4
2	0	7	7	0	3	5	1	0	7	0	5	0	1	0	7	7	3	0	3	7	1	0	0	2	4	1	3	7	4	0	7	7	4	6	2
3	0	7	1	0	3	0	1	1	0	0	7	4	1	0	1	7	4	0	3	7	4	0	0	1	4	3	0	7	4						

Fig. 10. A transition function designed by evolution for the parallel replication of the small loop from Fig. 8a.

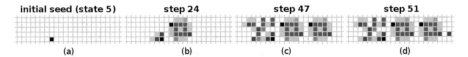

initial seed (state 5)	step 24	step 47	step 51
(a)	(b)	(c)	(d)

Fig. 11. A sample of the CA development from the seed according to the transition function from Fig. 10: (a) the initial seed, (b) the small loop is developed from the seed after step 24, leaving an undesirable structure on its left side, (c) the loop creates its first replica after step 47, the undesirable structure prevents from the replication on the left side, (d) the replication to the right in progress after step 51, the structure on the left no longer changes.

range of states used as the seed allows emerging the loop using various processes (i.e. sequences of CA states), which are totally different from the processes of replication from the complete loop. Although the state-5 seed does not enable to replicate the loop to both sides, the solution can be considered as robust because the undesirable structure does not cause the destruction of the loop that can subsequently replicate to the right side.

As an example of our research regarding the optimisation of replication techniques in cellular automata, a symmetric loop is considered as shown in Fig. 8c. Although the evaluation method applied to design the CA for this loop is out of the scope of this study, a result of a successful parallel replication will be presented, which demonstrates the potential of the GA in combination with the CMR encoding to discover novel techniques in cellular automata. As in the previous example, the goal of the experiment was to design transition rules for the parallel replication of the loop to the given directions. Since the loop is symmetric with respect to the arrangement of the cell states, it would be possible to adapt a single replication algorithm to perform the replication process simultaneously to various directions. Such adaptation is based on "rotating" the transition rules according to the ordering of cells in the cellular neighbourhoods with respect to the given directions as known from Byl's loop [4]. However, if the evaluation of the candidate solutions during evolution is performed with respect to the number and arrangement of the replicas only, then the GA can discover various independent replication algorithms as shown in Fig. 12. The corresponding transition function contains 137 table-based rules and is shown in Fig. 13. In this solution, not only the algorithms for the replication to the left and right side differ significantly, the number of steps needed to create the replica on the left side is nearly the double of the number of steps required for the replication to the right side. As evident from Fig. 12, the first replica of the initial loop is created on the right side after the 15th step, the first replica on the left side needs 26 steps to be completed. After the 27th step, the second replica on the right side is completed whilst the second copy on the left side has just started to develop. Such a process has never been observed before as regards the known replicating loops and hence it can be considered as an unconventional feature of the solution obtained in this experiment.

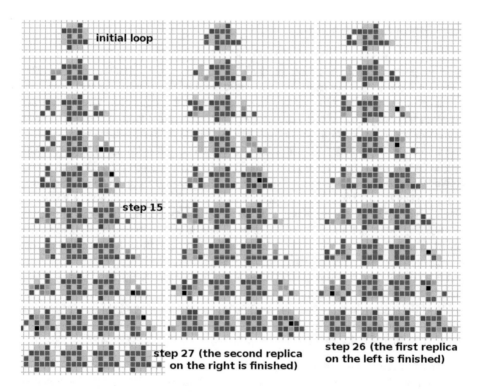

Fig. 12. A sample of the parallel replication of the symmetric loop from Fig. 8c according to the transition function shown in Fig. 13. Note that the number of steps needed to develop a replica on the right side is half the number of steps required to finish a replica on the left side.

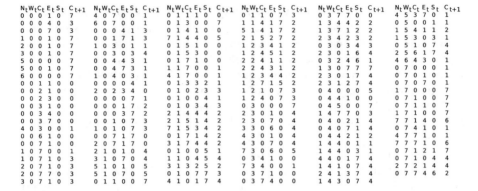

Fig. 13. The transition function designed by evolution for the parallel replication of the symmetric loop from Fig. 8c.

5.3 Summary and Discussion

Both the proposed loops proved the ability to replicate according to the given specification. It is worth to note that although the development of the loop from a seed was not explicitly required, the evaluation of the results obtained for both the loops showed that this ability is not rare. This means that the seed-based development may be evolved directly (without any initial loop available) in order the given loop can emerge autonomously. Some experiments were performed in order to validate this hypothesis, with the following observations. The GA is able to discover transition rules for the development of the given loop from the seed. However, no solution has yet been achieved that would be able to subsequently replicate the loop. One of the reasons for this issue may be the fact that the exact place in the cellular space, where the loop is developed from the seed, is hard to predict (it depends on the state of the seed, shape of the loop and the transition rules). Therefore, it is not evident how the replicas ought to be specified within the target CA state for the continuous replication. More research is needed in order to determine the necessary information provided to the GA, which would enable to solve this problem.

In order to perform a general evaluation of the results obtained within the context of computational features of cellular automata and with respect to the existing replicating loops, the following issues need to be clarified:

1. The objective was not to design self-replication. The loops with the ability to self-replicate contain the information of how to create a copy encoded in their "body" as a suitable arrangement of cell states. The transition rules interpret this information and calculate the appropriate state transitions of the CA in order to perform the replication process. In this work, however, the initial loop is considered as an object of a given shape that ought to be transformed onto a CA state that contains the copy of the loop. The goal was to find both the transition rules and the sequence of the CA states that lead to the emergence of the replica.

2. The resulting CA do not represent universal computing models (it was not a goal of the experiments). It means that a specific transition function, that was obtained as a result of a successful evolution, is dedicated to replicate the given loop only that was a subject of evaluation in the fitness function. Nevertheless, as the results showed, some transition functions are able to create the loops from a seed which was not explicitly required within the fitness evaluation.

Although the shape of the proposed loops was inspired by the existing (self-replicating) loops and the GA provided some successful results to replicate the loops with respect to the given specifications, no working solution has yet been achieved by the GA to replicate the existing loops (e.g. Byl's loop) with the exact shape and arrangement of the replicas. This issue can be caused by the fact that some of the self-replicating loops are dynamical structures even after the replica is finished (e.g. Byl's loop exhibits such feature). However, only static replicas were considered in our experiments. Another aspect may be the size of the loop. Large loops require a considerable number of steps to finish the replica (e.g. Langton's loop needs 151 steps), which makes the evaluation of such solutions very time-consuming. Finally, the information encoded in the loop body, that specifies the self-replication features, actually determines the replication algorithm (i.e. the CA development) which is specific for the given loop. If no more valid replication algorithms exist in the solution space for a given loop, then the GA may not be able to find the solution in a reasonable time.

6   Conclusions

In summary, the results presented in this work shows several facts related to the problem of replication in cellular automata. First, there are plenty of transition functions that are able to replicate a given loop. The experiments showed that it is possible to discover such functions routinely by means of the genetic algorithm even for complex multi-state cellular automata (herein demonstrated for CA working with 8 and 10 cell states). This was enabled by the utilisation of conditionally matching rules as a technique for the representation of the transition functions. Second, some unconventional features of the solutions were identified that cannot be observed in the known replicating loops and have never been published before. Specifically, in case of some solutions obtained, the CA can be initialised by a single-cell seed in a non-zero state, which allows developing the given loop that is subsequently able to replicate. Note that this ability was identified as an extra feature of the resulting cellular automata, which was not explicitly required by the specification for the evolutionary algorithm. This shows that some cellular automata are able, using a minimum information encoded in the initial state, to autonomously develop a complex emergent behaviour that is fully determined by the transition function and the state of a single cell only. Another feature, that was achieved by the evolution, is a parallel replication of the given loop into more directions, using different algorithms to create the replicas. The results showed that this behaviour is needed if the arrangement of the cell states in the loop is not fully symmetrical. However, an unconventional parallel replication can be observed even in case of a symmetric loop, where the difference is both in the way of the replication and the number of steps needed to create the replicas. Again, the evolution itself discovered such the behaviour just on the basis of the given target pattern containing the replicas of the initial loop.

The results obtained bring some open questions, the answers of which could be beneficial for the research of cellular automata in general. For example, can the seed-based development create a configuration in the CA that supports self-replication (or other useful features)? Are there other (simple) structures that support development of more complex (self-)replicating objects? Can evolutionary techniques be applied to the design of computationally universal CA-based models? Not only these questions represent ideas for our future work.

Acknowledgements. This work was supported by The Ministry of Education, Youth and Sports of the Czech Republic from the National Programme of Sustainability (NPU II); project IT4Innovations excellence in science - LQ1602.

References

1. Berlekamp, E.R., Conway, J.H., Guy, R.K.: Winning Ways for Your Mathematical Plays, vol. 4, 2nd edn. A K Peters/CRC Press, Boca Raton (2004)
2. Bidlo, M.: Evolving multiplication as emergent behavior in cellular automata using conditionally matching rules. In: 2014 IEEE Congress on Evolutionary Computation, pp. 2001–2008. IEEE Computational Intelligence Society (2014)
3. Bidlo, M., Vasicek, Z.: Evolution of cellular automata with conditionally matching rules. In: 2013 IEEE Congress on Evolutionary Computation (CEC 2013), pp. 1178–1185. IEEE Computer Society (2013)
4. Byl, J.: Self-reproduction in small cellular automata. Phys. D Nonlinear Phenom. **34**(1–2), 295–299 (1989)
5. Elmenreich, W., Fehérvári, I.: Evolving self-organizing cellular automata based on neural network genotypes. In: Bettstetter, C., Gershenson, C. (eds.) IWSOS 2011. LNCS, vol. 6557, pp. 16–25. Springer, Heidelberg (2011). doi:10.1007/978-3-642-19167-1_2
6. Fogel, L.J., Owens, A.J., Walsh, M.J.: Artificial Intelligence Through Simulated Evolution. Wiley, New York (1966)
7. Holland, J.H.: Adaptation in Natural and Artificial Systems. University of Michigan Press, Ann Arbor (1975)
8. Langton, C.G.: Self-reproduction in cellular automata. Phys. D Nonlinear Phenom. **10**(1–2), 135–144 (1984)
9. Mitchell, M., Hraber, P.T., Crutchfield, J.P.: Revisiting the edge of chaos: evolving cellular automata to perform computations. Complex Syst. **7**(2), 89–130 (1993)
10. Packard, N.H.: Adaptation toward the edge of chaos. In: Kelso, J.A.S., Mandell, A.J., Shlesinger, M.F. (eds.) Dynamic Patterns in Complex Systems, pp. 293–301. World Scientific, Singapore (1988)
11. Perrier, J.-Y., Sipper, M., Zahnd, J.: Toward a viable, self-reproducing universal computer. Phys. D **97**, 335–352 (1996)

12. Reggia, J.A., Armentrout, S.L., Chou, H.-H., Peng, Y.: Simple systems that exhibit self-directed replication. Science **259**(5099), 1282–1287 (1993)
13. Sapin, E., Adamatzky, A., Collet, P., Bull, L.: Stochastic automated search methods in cellular automata: the discovery of tens of thousands of glider guns. Natural Comput. **9**(3), 513–543 (2010)
14. Sapin, E., Bull, L.: Searching for glider guns in cellular automata: exploring evolutionary and other techniques. In: Monmarché, N., Talbi, E.-G., Collet, P., Schoenauer, M., Lutton, E. (eds.) EA 2007. LNCS, vol. 4926, pp. 255–265. Springer, Heidelberg (2008). doi:10.1007/978-3-540-79305-2_22
15. Sipper, M.: Quasi-uniform computation-universal cellular automata. In: Morán, F., Moreno, A., Merelo, J.J., Chacón, P. (eds.) ECAL 1995. LNCS, vol. 929, pp. 544–554. Springer, Heidelberg (1995). doi:10.1007/3-540-59496-5_324
16. Sipper, M. (ed.): Evolution of Parallel Cellular Machines. LNCS, vol. 1194. Springer, Heidelberg (1997)
17. Sipper, M., Goeke, M., Mange, D., Stauffer, A., Sanchez, E., Tomassini, M.: Online evolware. In: IEEE International Conference on Evolutionary Computation, pp. 181–186 (1997)
18. Stefano, G., Navarra, A.: Scintillae: how to approach computing systems by means of cellular automata. In: Sirakoulis, G.C., Bandini, S. (eds.) ACRI 2012. LNCS, vol. 7495, pp. 534–543. Springer, Heidelberg (2012). doi:10.1007/978-3-642-33350-7_55
19. Tempesti, G.: A new self-reproducing cellular automaton capable of construction and computation. In: Morán, F., Moreno, A., Merelo, J.J., Chacón, P. (eds.) ECAL 1995. LNCS, vol. 929, pp. 555–563. Springer, Heidelberg (1995). doi:10.1007/3-540-59496-5_325
20. von Neumann, J.: Theory of self-reproducing automata. In: Burks, A.W. (ed.) Essays on Cellular Automata. University of Illinois Press, Urbana and London (1966)
21. Yunès, J.-B.: Achieving universal computations on one-dimensional cellular automata. In: Bandini, S., Manzoni, S., Umeo, H., Vizzari, G. (eds.) ACRI 2010. LNCS, vol. 6350, pp. 660–669. Springer, Heidelberg (2010). doi:10.1007/978-3-642-15979-4_74

Adaptive Differential Evolution Supports Automatic Model Calibration in Furnace Optimized Control System

Miguel Leon[1(✉)], Magnus Evestedt[2], and Ning Xiong[1]

[1] School of Innovation, Design and Engineering,
Mälardalen University, Västerås, Sweden
`miguel.leonortiz@mdh.se`
[2] Industrial Systems, Prevas, Västerås, Sweden

Abstract. Model calibration represents the task of estimating the parameters of a process model to obtain a good match between observed and simulated behaviours. This can be considered as an optimization problem to search for model parameters that minimize the discrepancy between the model outputs and the corresponding features from the historical empirical data. This chapter investigates the use of Differential Evolution (DE), a competitive class of evolutionary algorithms, to solve calibration problems for nonlinear process models. The merits of DE include simple and compact structure, easy implementation, as well as high convergence speed. However, the good performance of DE relies on proper setting of its running parameters such as scaling factor and crossover probability, which are problem dependent and which can even vary in the different stages of the search. To mitigate this issue, we propose a new adaptive DE algorithm that dynamically adjusts its running parameters during its execution. The core of this new algorithm is the incorporated greedy local search, which is conducted in successive learning periods to continuously locate better parameter assignments in the optimization process. In case studies, we have applied our proposed adaptive DE algorithm for model calibration in a Furnace Optimized Control System. The statistical analysis of experimental results demonstrate that the proposed DE algorithm can support the creation of process models that are more accurate than those produced by standard DE.

Keywords: Differential evolution · Optimization · Model identification · Temperature estimation

1 Introduction

System modelling and identification provide an important basis for optimized process control in modern industrial scenarios. Its main goal is to identify a process model that is able to accurately predict the output of a system in response to a set of inputs. Generally, a process model can be constructed in two steps. In the first step, the structure of the model is determined in terms

© Springer International Publishing AG 2017
J.J. Merelo et al. (eds.), *Computational Intelligence*, Studies in Computational Intelligence 669,
DOI 10.1007/978-3-319-48506-5_3

of expert knowledge and insight into the process. The second step is model calibration that aims to estimate the parameters of the model to obtain a good match between observed and simulated behaviours. This can be considered as an optimization problem with the objective function being represented as the deviation with respect to the historical empirical data. In other words, automatic calibration is the process of searching for model parameters to minimize the discrepancy between the model outputs and the corresponding features from the empirical data.

Traditionally, the methods such as Least Mean Square (LMS) algorithm or Recursive Least Square Estimation (RLS) algorithms have been used to solve the model calibration problems. However, they are subject to two limitations. First, they were developed for linear system identification, i.e. when the process model is assumed to be linear. Second, they are essentially derivative-based optimization techniques and may fail to find the optimal solution when locating many (model) parameters in high dimensional spaces.

This paper advocates the application of Differential Evolution (DE) [1,2] algorithms to solve the calibration problems for nonlinear process models. DE presents a class of evolutionary computing techniques that perform population-based and beam search, thereby exhibiting strong global search ability in complex, non-linear and high dimensional spaces [3]. DE differs from many other evolutionary algorithms [4,5] in that mutation in DE is based on differences of individuals randomly selected from the population. Thus, the direction and magnitude of the search is decided by the distribution of solutions instead of a pre-specified probability density function. The merits of DE include simple and compact structure, easy implementation, as well as high convergence speed, which make it quite competitive in comparison with other evolutionary algorithms.

However, the performance of DE is largely dependent on its two running parameters: scaling factor and crossover rate, in real applications. Improper setting of such parameters will lead to low quality of solutions found by DE. Yet finding suitable values for them is by no means a trivial task, it involves a trial-and-error procedure that is time consuming.

In this paper we present a new adaptive algorithm for DE, which does not require good parameter values (scaling factor and crossover rate) to be specified by users in advance. Our new algorithm is established by integration of greedy local search into the standard DE algorithm. Greedy search is conducted repeatedly during the running of DE to reach better parameter assignments in the neighborhood. So far we have applied our adaptive DE algorithm for process model calibration in a Furnace Optimized Control System (FOCS). The experiment results revealed that our algorithm yielded process models that estimated temperatures inside a furnace more precisely than those produced by using the standard DE algorithm.

The remaining of the paper is organized as follows. Section 2 briefly describes the application scenario. The standard DE algorithm is outlined in Sect. 3, which is followed by the new adaptive DE algorithm in Sect. 4. Section 5 presents the results of experiments for model calibration in a Furnace Optimized Control System. Section 6 discusses some relevant works. Finally, concluding remarks are given in Sect. 7.

2   Problem Formulation

Energy consumption and environmental consideration are important issues to be handled within the steel industry today. In this respect, inventing and developing new ways to decrease fuel and emission levels in steel production and treating are crucial to production economy. The FOCS system was developed for reheating furnaces in the early 1980s, and has since grown to be the most commonly used system in the Scandinavian steel industry, [6].

Due to the harsh environment inside the reheating furnace, the temperature of the heated steel cannot be measured continuously. Therefore, the FOCS system core is a temperature calculation model that utilizes temperature sensor measurements in the walls inside the furnace as well as current fuel flow, to estimate the temperature in the heated material. In order to gain optimal control performance, it is crucial to estimate the temperatures accurately.

The temperature inside the material can be measured through the furnace by a test measurement setup. This is normally done 2–3 times every year to certify the furnace operation. The measurements are also used to calibrate the FOCS temperature calculation model, by changing the model parameters to fit the model output to the measurements. The calibration is performed manually and can be a tedious task due to many parameters and evaluation of several test measurements simultaneously.

We attempted to apply DE algorithms to facilitate automatic determination of the parameters of this process model. The goal is to find such a set of parameters to minimize the error between the estimated temperatures from the model and the actual temperatures obtained from the measurements. After the execution of DE, we acquire the optimal parameters of the model, as shown in Fig. 1. Subsequently this optimal model can be employed in future occasions to produce reliable estimates of temperatures based on input conditions.

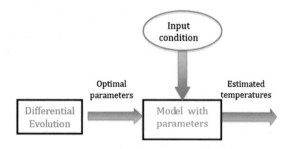

Fig. 1. DE, parameterized model, and temperature estimation.

3   Differential Evolution

DE is a stochastic and meta-heuristic technique that has been developed for solving optimization problems with real parameters [1]. It provides a powerful tool for searching for optimal solutions in high-dimensional spaces that are non-linear, non-differentiable, non-continuous, and containing multiple local optima. DE has become a highly competitive class of evolutionary algorithms in many practical applications.

A basic DE algorithm works with a population of NP vectors: $X_{i,G}$, $i = 1, \ldots NP$, where i is the index of solutions in the population, G stands for the generation and NP is the population size. A new generation of vectors is created in DE by applying three operators: mutation, crossover, and selection, which will be briefly introduced in the sequel.

Mutation is tasked to create a mutant vector for each target solution in the population by using the vector of difference between the current population members. Actually there are a number of variations to implement the mutation operation. Here we only introduce the most commonly used mutation strategy, which is notated as DE/rand/1. Other mutation strategies and their performance are described in [7]. According to the strategy DE/rand/1, the mutant vector $V_{i,G}$ for target vector $X_{i,G}$ is generated as follows

$$V_{i,G} = X_{r_1,G} + F \times (X_{r_2,G} - X_{r_3,G}) \tag{1}$$

where r_1, r_2; r_3 are random integers from 1 to N_P, and F is the scaling factor inside the interval $[0, 2]$. Figure 2 shows how this mutation strategy works, where d is the difference vector between $X_{r_2,G}$ and $X_{r_3,G}$.

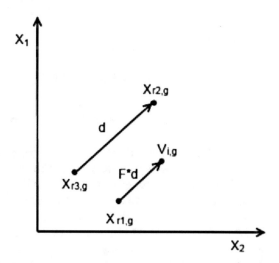

Fig. 2. Random mutation with one difference vector.

As can be seen from Eq. 1, it is possible for the mutant vector $V_{i,G}$ to have components violating pre-defined boundary constraints for the decision variables. To repair such illegal solution (if it emerges), we modify $V_{i,G}$ according to Eq. 2.

$$V_{i,G}[j] = \begin{cases} (Low[j]) & \text{if } V_{i,G}[j] < Low[j], \\ (Upper[j]) & \text{if } V_{i,G}[j] > Upper[j]. \end{cases} \tag{2}$$

where $V_{i,G}[j]$ denotes the jth component of vector $V_{i,G}$, and Low[j] and Upper[j] stand for the low and upper bounds of the jth decision variable respectively.

Crossover is used to combine a mutant vector with the corresponding target vector to create a new trial solution. When crossover is applied to the target vector $X_{i,G} = (X_{i,G}[1], X_{i,G}[2], \ldots, X_{i,G}[n])$ and its mutant vector $V_{i,G} = (V_{i,G}[1], V_{i,G}[2], \ldots, V_{i,G}[n])$, we obtain the trial solution $U_{i,G}$ as expressed below:

$$U_{i,G}[j] = \begin{cases} V_{i,G}[j] & \text{if } rand[0, 1] <= CR \text{ or } j = j_{rand} \\ X_{i,G}[j] & \text{otherwise} \end{cases} \tag{3}$$

where $U_{i,G}[j]$ is the jth component of the trial vector $U_{i,G}$, CR denotes the probability of crossover to be specified by a user, and jrand is a randomly selected index to ensure that $U_{i,G}$ contains at least one component from $V_{i,G}$.

Selection is performed to allow for competition between a trial vector $U_{i,G}$ and its associated target vector $X_{i,G}$. If the trial vector is better than the target solution as assessed by the objective function f, it replaces the target solution in the next generation. Otherwise, the trial solution is discarded and the target solution survives in the next generation. Therefore, for a minimization problem as example, the individuals in the new generation $G + 1$ are given as follows:

$$X_{i,G+1} = \begin{cases} U_{i,G} & \text{if } f(U_{i,G}) < f(X_{i,G}) \\ X_{i,G} & \text{otherwise} \end{cases} \tag{4}$$

The pseudocode of basic DE is given in Algorithm 1.

4   A New DE Algorithm with Parameter Adaptation

As is stated above, the scaling factor (F) and crossover rate (CR) are two important running parameters for DE that significantly affect the optimization performance. It is also recognized that the proper value of F may change with time in the evolutionary process, while the crossover rate CR is more dependent on the characteristics of the underlying problem. Hence it is important to automatically determine and adjust such parameters for DE when solving a practical problem. To this end we propose a new adaptive DE algorithm that dynamically adjusts its running parameters using greedy (local) search. In this section we shall first present the greedy scheme for parameter adjustment in Subsect. 4.1 and then we will discuss the integration of this scheme within a DE cycle in Subsect. 4.2.

Algorithm 1. Differential Evolution.

1: Initialize the population with randomly created individuals.
2: Calculate the objectives values of all vectors in the population.
3: **while** The termination condition is not satisfied **do**
4:     Create mutant vectors according to:
       $V_{i,G} = X_{r_1,G} + F \times (X_{r_2,G} - X_{r_3,G})$
5:     Create trial vectors by recombining target vectors with mutant vector:
$$U_{i,G}[j] = \begin{cases} V_{i,G}[j] & \text{if } rand[0,1] <= CR \quad \text{or} \quad j = j_{rand} \\ X_{i,G}[j] & \text{otherwise} \end{cases}$$
6:     Evaluate trial vectors with the objective function.
7:     Select vectors $X_{i,G+1}$ of the next generation by
$$X_{i,G+1} = \begin{cases} U_{i,G} & \text{if } f(U_{i,G}) < f(X_{i,G}) \\ X_{i,G} & \text{otherwise} \end{cases}$$
8: **end while**

4.1 Greedy Adjustment Scheme

Our basic idea is to perform local greedy search to adjust the values of control parameters (scaling factor and crossover probability) of DE to improve its performance. This means that at every step the current parameter assignment is compared with its neighbours and then moves to the best candidate in the neighbourhood. Nevertheless, the comparison of different DE parameters is not a trivial task. It is complicated by the stochastic characteristics of the mutation and crossover operators such that a good parameter assignment may also lead to undesired trial solutions created in the course of search.

It is advocated in the paper that a candidate for parameter assignment undergoing sufficient tests for reliable evaluation of its quality. The tests are made in a learning period comprising a specified number of generations to see how the candidate was useful to contribute to the creation of good trial solutions. We desire those parameter assignments that not only offer a high chance of survival for trial solutions but also enable substantial improvement of fitness in the next generation. In view of this, the relative improvement (RI) brought by a candidate assignment C (for either scaling factor or crossover probability) in test k is defined as:

$$RI(C,k) = \begin{cases} f(X^k) * 10^t - f(U^k) * 10^t, & \text{if } f(X^k) \geq f(U^k), \\ 0, & \text{otherwise} \end{cases} \quad (5)$$

where X^k and U^k represent respectively the parent and trial solutions in test k, and t is an integer such that $f(X^k) * 10^t$ lies in the interval [1,10]. Further, the progress rate (PR) for C is the average of the relative improvements from all the m tests of using C for producing trial solutions. Thus we can write

$$PR(C) = \frac{1}{m} \sum_{k=1}^{m} RI(C,k) \quad (6)$$

progress rate is used in this paper as the criterion to evaluate and compare candidates for DE parameter assignments.

In the greedy search procedure, the current parameter assignment and its two generated neighbours are randomly selected for being used in producing new trial solutions during the learning period. The best of them is then identified using the metric of progress rate as defined in Eq. 6. As proper values of control parameters can change over time, we perform life-long search from one learning period to the next to achieve continuous adjustment of parameters in the course of optimization. An algorithmic description of the greedy scheme for parameter adjustment is given in the following:

The greedy search for parameter adjustment:

1. $F_0 \leftarrow$ initial assignment for scaling factor.
2. $P_0 \leftarrow$ initial assignment for crossover probability.
3. Expand F_0: Creating its two neighbours F_1 and F_2.
4. Expand P_0: Creating its two neighbours P_1 and P_2.
5. count(x)=0, count(y)=0, sum(x)=0 and sum(y)=0 for all $x \in \{F_0, F_1, F_2\}, y \in \{P_0, P_1, P_2\}$
6. $i = 0$
7. **while** $(i \leq LP \times NP)$ % LP is the number of generations in the learning period
8.   Randomly select $x^* from \{F_0, F_1, F_2\}$
9.   Randomly select $y^* from \{P_0, P_1, P_2\}$
10.   Perform mutation and crossover using x^* and y^*.
11.   Derive RI(x^*) and RI(y^*) upon the trial and parent solutions
12.   count(x^*) = count(x^*)+1 and count(y^*) = count(y^*)+1
13.   sum(x^*) = sum(x^*) + RI(x^*) and sum(y^*) = sum(y^*) + RI(y^*)
14. $i = i + 1$;
15. **end while**
16. PR(x)=sum(x)/count(x) for all $x \in \{F_0, F_1, F_2\}$
17. PR(y)=sum(y)/count(y) for all $y \in \{P_0, P_1, P_2\}$
18. $F_0 = arg \max_{x \in \{F_0, F_1, F_2\}} PR(x)$;
19. $P_0 = arg \max_{y \in \{P_0, P_1, P_2\}} PR(y)$;
20. Go to step 3

4.2   Adaptive DE Algorithm with Greedy Search

The adaptive differential evolution algorithm is developed by incorporation of the greedy adjustment scheme into the basic DE algorithm. The whole evolutionary process is divided into a sequence of learning periods. In every learning period, three candidates (the current parameter assignment and its two neighbours) are tested in such a way that each of them gets an equal chance to be used in producing trial solutions. The evaluation of the candidates will be done after the learning period is ended, and the best candidate is treated as the current assignment and the search moves on to the next learning period. By this manner, greedy parameter adjustments are realized in successive learning periods to

Algorithm 2. GADE.

1: Set $CR_m = 0.5, F = 0.5, LP = 3, c_1 = c_2 = 0.01$;
2: $Z_F = \{F - c_1, F, F + c_1\}$;
3: $Z_{CR} = \{CR_m - c_2, CR_m, CR_m + c_2\}$;
4: $G = 1$;
5: Initialize the population $(X_{1,1}, X_{2,1}, \ldots, X_{NP,1})$
6: **while** The termination condition is not satisfied **do**
7:    **for** $i = 1$ to N_P **do**
8:       Set F_i by randomly selecting one element from Z_F.
9:       Set μ_{CR} by randomly selecting one element from Z_{CR}.
10:      $CR_i = Cauchy(\mu_{CR}, 0.2)$.
11:      Create the mutant vector using the random mutation strategy by:
        $V_{i,G} = X_{r_1,G} + F \times (X_{r_2,G} - X_{r_3,G})$
12:      Repair the mutant vector if it has values outside the boundaries:
$$V_{i,G}[j] = \begin{cases} (Low[j]) & \text{if } V_{i,G}[j] < Low[j], \\ (Upper[j]) & \text{if } V_{i,G}[j] > Upper[j]. \end{cases}$$
13:      Create the trial vector:
$$U_{i,G}[j] = \begin{cases} V_{i,G}[j] & \text{if } rand[0,1] \leq CR \text{ or } j = j_{rand} \\ X_{i,G}[j] & \text{otherwise} \end{cases}$$
14:      **if** $f(U_{i,G}) < f(X_{i,G})$ **then**
15:         $X_{i,G+1} = U_{i,G}$
16:      **else**
17:         $X_{i,G+1} = X_{i,G}$
18:      **end if**
19:    **end for**
20:    /* Update F*/
21:    **if** $G\%LP == 0$ **then**
22:      $F = arg \max_{x \in Z_F} PR(x)$;
23:      $Z_F = \{F - c_1, F, F + c_1\}$;
24:      $CR_m = arg \max_{y \in Z_{CR}} PR(y)$;
25:      $Z_{CR} = \{CR_m - c_2, CR_m, CR_m + c_2\}$;
26:    **end if**
27:    $G = G + 1$;
28: **end while**

facilitate continuous and dynamic adjustment of F and CR values during the execution of the DE algorithm.

The initial assignment for scaling factor is set as $F = 0.5$, and its two neighbours are $F + c_1$ and $F - c_1$ respectively, where c_1 is a user specified small positive number. The initial assignment for crossover rate is a Cauchy distribution with its centre $CR_m = 0.5$ and its scale parameter equal to 0.2. The two neighbours of this initial distribution are the shifted Cauchy distributions with their centres being located at $CR_m + c_2$ and $CR_m - c_2$ respectively, where c_2 is a small positive number specified by a user. Every current and neighbouring assignment (for both scaling factor and crossover rate) receives a probability of 1/3 to be selected for use in order to get a sufficient number of tests in the learning period. At the end of the learning period, a neighbouring assignment may replace the current one according to the assessed progress rates.

A more detailed description of our adaptive DE algorithm is given in the pseudocode below. Although the simple DE/rand/1 strategy is used in the present version of the algorithm, the principle and mechanism described here is generic and can be easily applied with other mutation strategies as well.

5   Experiments and Results

This section aims to examine the capability of our adaptive DE algorithm in a real industrial scenario. We applied our algorithm on the problems of model calibration in FOCS and then compared its results with those obtained by using the basic DE algorithm on the same problem.

5.1   Experimental Settings

Our adaptive DE algorithm and basic DE were tested in the experiments for comparison. Both algorithms use the binomial crossover operator, and both have three important running parameters: population size (NP), crossover rate (CR) and scaling factor (F). The parameters adopted for the basic DE are: $NP = 60$, $CR = 0.5$ and $F = 0.5$. The parameters used in our adaptive algorithm are: $NP = 60$, $CR_m = 0.5$, $F = 0.5$, $c_1 = c_2 = 0.01$, and the number of generations in a learning period $LP = 3$.

Both algorithms were executed 10 times in solving the model calibration problems associated with two data sets respectively. The maximal number of evaluations was set to a relatively low amount 2000, due to the high computational cost in fitness evaluations.

5.2   Results and Comparison in Problem 1

First, we applied the two DE algorithms to the problem of finding model parameters to mimic the empirical data as given in Data Set 1 (which consists of xx samples).

The errors of the process models found by the two algorithms in the 10 executions are listed in Table 1 for comparison. In the table, we also see the best, mean and worst errors from the 10 executions for each of the algorithms.

Table 2 shows the reduction of the best error (in absolute value and percentage), achieved by our adaptive DE algorithm in comparison to basic DE. This indicates a significant improvement in the quality of the acquired solutions.

5.3   Results and Comparison in Problem 2

Second, we applied the two DE algorithms (adaptive and basic) to the problem of model calibration to mimic the behaviour as specified in Data Set 2 (which

Table 1. Errors of the models from the two algorithms from problem 1.

Execution	DE	GADE
Exec 1	277.6	224.8
Exec 2	242.8	259.7
Exec 3	250.8	243.9
Exec 4	274	247
Exec 5	266.1	243.5
Exec 6	281.8	237.9
Exec 7	269.2	247.7
Exec 8	275.6	237.8
Exec 9	267.8	235.2
Exec 10	278.2	237.8
MEAN:	268.39	**241.53**
WORST:	281.8	**259.7**
BEST:	242.8	**224.8**

Table 2. The reduction of error in problem 1.

	Value	Percentage
Improvement	18	7.41 %

Table 3. Errors of the models from the two algorithms from problem 2.

Execution	DE	GADE
Exec 1	577	550.8
Exec 2	564.8	559.9
Exec 3	570.1	568.7
Exec 4	558.9	570.2
Exec 5	566.5	555.5
Exec 6	573.2	551.9
Exec 7	559.6	572.7
Exec 8	567.5	541.7
Exec 9	567.8	563.5
Exec 10	585.5	571.9
MEAN:	569.09	**560.68**
WORST:	585.5	**572.7**
BEST:	558.9	**541.7**

Table 4. The reduction of error in problem 2.

	Value	Percentage
Improvement	17.2	3.08 %

contains xx samples). Table 3 gives the results of both algorithms in the 10 executions together with the best, mean and worst values.

Table 4 shows the reduction of the best error achieved by our adaptive DE algorithm against the basic DE. It can be seen that the error of the model is further reduced by more than 17 using the adaptive DE algorithm, which means a reasonable improvement of model accuracy that would bring practical benefit to enhance the control system performance in steel production.

5.4   Evolution of the DE Parameters

In Fig. 3, we can see the evolution of the scaling factor (F) during the optimization process, in solving Problem 1. Since the individuals of the population were not sufficiently distinct from each other at the beginning, the value of F started to increase to enable big movement in the mutation.

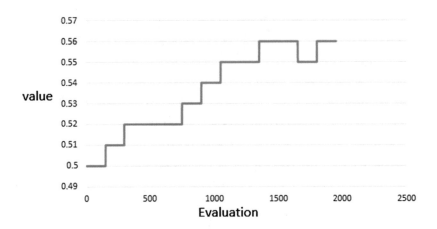

Fig. 3. The evolution of the scaling factor F.

In Fig. 4, we can observe the change of the distribution center (CR_m) for the crossover rate during the process. The value of CR_m increased for the same reason as stated for the scaling factor F, i.e., the population members did not have enough difference with each other. Hence we gave more chance to mutant vectors in crossover to increase the diversity of the population.

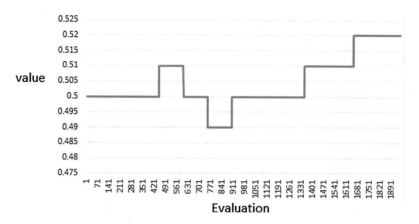

Fig. 4. The evolution of the distribution center for the crossover probability.

6   Relevant Work

DE has offered a powerful framework with tool and technique for applications in real-parameter optimization and system design. In [8] DE was employed for generating optimal set points and gain tuning in a power plant control system. The authors also showed that an additional evolutionary term could be added to the DE technique to increase the convergence speed of the algorithm in these applications. The investigations in [9] demonstrated the feasibility and effectiveness of two meta-heuristic techniques: DE and particle swarm optimization (PSO) to solve the reactive power and voltage control problems. The comparative study also reveled that PSO yielded in some cases slightly more reduction of power loss while DE was more economical in requiring a lower number of function evaluations. A multi-strategy DE algorithm [10], in which multiple mutation strategies were selected adaptively, was used to estimate the unknown parameters of the proton exchange membrane fuel cell (PEMFC) model. Mohanty et al. [11] applied DE to Load Frequency Control in a multi-source power system, where DE was used to find the optimal gains of Integral (I), Proportional Integral (PI) and Proportional Integral Derivative (PID) controllers. Therein the parameters of DE were tuned manually by executing multiple runs of the algorithm for each candidate set of parameters.

It is widely recognized that DE performance is largely affected by its control parameters (such as scaling factor and crossover rate), which are problem dependent and which can vary during the course of search. Self-adaptation of parameters has become a new trend in the research and development of competent DE algorithms in many industrial applications. Zou et al. [12] proposed dynamic modification of the scaling factor and crossover rate of DE to increase its exploration capability when solving the task assignment problem. More specifically, the scaling factor was adapted according to the objective values of candidate solutions, and crossover rate was adjusted with the increment of iterations. In

[13] the Self-Adaptive Differential Evolution (SADE) algorithm [14] was utilized to tune the parameters of a power system stabilizer (PSS), resulting in better damping under small and large disturbances than those tuned by classical DE.

7   Conclusion

Model calibration is an important step in the development of accurate and reliable process models that provide a fundamental basis for optimal decision and control in process automation. This chapter treats calibration as an optimization problem of searching for model parameters to minimize the errors of the model outputs with respect to empirical data. We present a new adaptive differential evolution algorithm that enables dynamic adaptation of its running parameters during the course of search. The case studies made in the scenario of the Furnace Optimized Control System indicate that the adaptive DE algorithm yields more precise results of model calibration than standard DE.

In spite of the successful results achieved, many works will be done in future to further enhance the proposed algorithm. One issue that is being considered is utilizing more advanced mutation strategy or allowing for adaptive selection of mutation strategies from a pool of candidates. The second possibility is to combine the adaptive DE approach with an effective local search mechanism such as greedy local search [15, 16], Quasi Newton method or variable neighborhood search [17].

Acknowledgements. The work is supported by the Swedish Knowledge Foundation (KKS) grant (project no. 16317).

References

1. Storn, R., Price, K.: Differential evolution - a simple and efficient heuristic for global optimization over continuous spaces. J. Global Optim. **11**(4), 341–359 (1997)
2. Das, S., Suganthan, N.: Differential evolution: a survey of the state-of-the-art. IEEE Trans. Evol. Comput. **15**, 4–31 (2011)
3. Xiong, N., Molina, D., Leon, M., Herrera, F.: A walk into metaheuristics for engineering optimization: principles, methods, and recent trends. Int. J. Comput. Intell. Syst. **8**(4), 606–636 (2015)
4. Kenedy, J., Eberhart, R.C.: Particle swarm optimization. In: Proceedings IEEE Conference on Neural Networks, pp. 1942–1948 (1995)
5. Goldberg, D.: Genetic Algorithm in Search, Optimization and Machine Learning. Addison-Wesley, New York (1989)
6. Norberg, P., Leden, B.: New developments of the computer control system FOCS-RF - application to the hot strip mill at SSAB, Domnarvet. Scanheating **II**, 31–60 (1988)
7. Leon, M., Xiong, N.: Investigation of mutation strategies in differential evolution for solving global optimization problems. In: Rutkowski, L., Korytkowski, M., Scherer, R., Tadeusiewicz, R., Zadeh, L.A., Zurada, J.M. (eds.) ICAISC 2014. LNCS (LNAI), vol. 8467, pp. 372–383. Springer, Heidelberg (2014). doi:10.1007/978-3-319-07173-2_32

8. Sickel, J.V., Lee, K., Heo, J.: Differential evolution and its applications to power plant control. In: International Conference on Intelligent Systems Applications to Power Systems, ISAP 2007, Toki Messe, Niigata, pp. 1–6. IEEE (2007)
9. Bakare, G., Krost, G., Venayagamoorthy, G., Aliyu, U.: Comparative application of differential evolution and particle swarm techniques to reactive power and voltage control. In: International Conference on Intelligent Systems Applications to Power Systems, ISAP 2007, Toki Messe, Niigata, pp. 1–6 (2007)
10. Gong, W., Cai, Z.: Parameter optimization of PEMFC model with improved multi-strategy adaptive differential evolution. Eng. Appl. Artif. Intell. **27**, 28–40 (2014)
11. Mohanty, B., Panda, S., Hota, P.K.: Controller parameters tuning of differential evolution algorithm and its application to load frequency control of multi-source power system. Int. J. Electr. Power Energy Syst. **54**, 77–85 (2014)
12. Zou, D., Liu, H., Gao, L., Li, S.: An improved differential evolution algorithm for the task assignment problem. Eng. Appl. Artif. Intell. **24**, 616–624 (2011)
13. Mulumba, T., Folly, K.: Application of self-adaptive differential evolution to tuning PSS parameters. In: 2012 IEEE Power Engineering Society Conference and Exposition in Africa (PowerAfrica), Johannesburg, pp. 1–5 (2012)
14. Qin, A., Suganthan, P.: Self-adaptive differential evolution algorithm for numerical optimization. In: 2005 IEEE Congress on Evolutionary Computation, vol. 2, pp. 1785–1791 (2005)
15. Leon, M., Xiong, N.: Differential evolution enhanced with eager random search for solving real-parameter optimization problems. Int. J. Adv. Res. Artif. Intell. **4**(12), 49–57 (2015)
16. Leon, M., Xiong, N.: Using random local search helps in avoiding local optimum in diefferential evolution. In: Proceedings of Artificial Intelligence and Applications, AIA2014, Innsbruck, Austria, pp. 413–420 (2014)
17. Mladenovic, N., Hansen, P.: Variable neighborhood search. Comput. Oper. Res. **24**(11), 1097–1100 (1997)

Self-configuring Ensemble of Multimodal Genetic Algorithms

Evgenii Sopov[⊠]

Department of Systems Analysis and Operations Research,
Siberian State Aerospace University, Krasnoyarsk, Russia
evgenysopov@gmail.com

Abstract. Ensemble methods are widely used to improve decision making in the field of statistics and machine learning. On average, the collective solution of multiple algorithms provides better performance than could be obtained from any of the constituent algorithms. The ensemble concept can be also used in the field of evolutionary algorithms. The main idea is to include many search algorithms in the ensemble and to design effective control of interaction of algorithms. Such interaction is implemented in different forms of island models, coevolutionary schemes, population-based algorithm portfolios and others. In this paper, a metaheuristic for designing multi-strategy genetic algorithm for multimodal optimization is proposed. Multimodal optimization is the problem of finding many or all global and local optima. In recent years many efficient multimodal techniques have been proposed in the field of population-based nature-inspired search algorithms. The majority of techniques are designed for real-valued problems. At the same time many real-world problems contain variables of many different types, including integer, rank, binary and others. In this case, a binary representation is used. There is a lack of efficient approaches for problems with binary representation. Moreover, binary and binarized problems are usually "black-box" optimization problems, thus there exists the problem of choosing a suitable algorithm and fine tuning it for a certain problem. The proposed approach contains many different multimodal genetic algorithms, which implement different search strategies. The metaheuristic adaptively controls the interactions of many search techniques and leads to the self-configuring solving of problems with a priori unknown structure. We present the results of numerical experiments for classical binary benchmark problems and benchmark problems from the CEC 2013 competition on multimodal optimization. We also present the results for some real-world problems.

Keywords: Metaheuristic · Multimodal optimization · Genetic algorithm · Niching · Self-configuration

1 Introduction

Many real-world problems have more than one optimal solution, or there exists only one global optimum and several local optima in the feasible solution space. Such problems are called multimodal. The goal of multimodal optimization (MMO) is to find all optima (global and local) or a representative subset of all optima.

© Springer International Publishing AG 2017
J.J. Merelo et al. (eds.), *Computational Intelligence*, Studies in Computational Intelligence 669,
DOI 10.1007/978-3-319-48506-5_4

Evolutionary and genetic algorithms (EAs and GAs) demonstrate good performance for many complex optimization problems. EAs and GAs are also efficient in the multimodal environment as they use a stochastic population-based search instead of the individual search in conventional algorithms. At the same time, traditional EAs and GAs have a tendency to converge to the best-found optimum losing population diversity.

In recent years MMO have become more popular, and many efficient nature-inspired MMO techniques were proposed. Almost all search algorithms are based on maintaining the population diversity, but differ in how the search space is explored and how optima basins are located and identified over a landscape. The majority of algorithms and the best results are obtained for real-valued MMO problems [3]. The main reason is the better understanding of landscape features in the continuous search space. Thus many well-founded heuristics can be developed.

Unfortunately, many real-world MMO problems are usually considered as black-box optimization problems and are still a challenge for MMO techniques. Moreover, many real-world problems contain variables of many different types, including integer, rank, binary and others. In this case, usually binary representation is used. Unfortunately, there is a lack of efficient approaches for problems with binary representation. Existing techniques are usually based on general ideas of niching and fitness sharing. Heuristics from efficient real-valued MMO techniques cannot be directly applied to binary MMO algorithms because of dissimilar landscape features in the binary search space.

In this study, a novel approach based on a metaheuristic for designing multi-strategy MMO GA is proposed. Its main idea is to create an ensemble of many MMO techniques and adaptively control their interactions. Such an approach would lead to the self-configuring solving of problems with a priori unknown structure.

The rest of the paper is organized as follows. Section 2 describes related work. Section 3 describes the proposed approach. In Sect. 4 the results of numerical experiments and real-world MMO problems solving are discussed. In the Conclusion the results and further research are discussed.

2 Related Work

The problem of MMO has exists since the first EAs. The first MMO techniques were applied in EAs and GAs for improvement in finding the global optimum in the multimodal environment.

The MMO, in general, can have at least 3 goals [14]:

- to find a single global optimum over the multimodal landscape only;
- to find all global optima;
- to find all optima (global and local) or a representative subset of all optima.

It is obvious that the second and the third goals are more interesting from both a theoretical and a practical point of view.

Over the past decade interest for this field has increased. The recent approaches are focused on the goal of exploring the search space and finding many optima to the problem. Many efficient algorithms have been proposed. In 2013, the global completion on MMO was held within IEEE CEC 2013 [8].

The list of widespread MMO techniques includes [3, 4, 10]:

- General Techniques:
  - Niching (parallel or sequential)
  - Fitness sharing, Clearing and Cluster-based niching
  - Crowding and Deterministic crowding
  - Restricted tournament selection (RTS)
  - Mating restriction
  - Species conservation
- Special Techniques:
  - Niching memetic algorithm
  - Multinational EA
  - Bi-objective MMO EA
  - Clustering-based MMO EA
  - Population-based niching
  - Topological algorithms
- Other Nature-inspired Techniques:
  - PSO, ES, DE, Ant Colony Optimization and others

The main advantage of the general techniques is that they do not use any specific information about the objective landscape and features of the search space. Thus they can be applied for a wide range of MMO problems with different representations.

Binary and binarized MMO problems are usually solved using the GA based on general techniques. Also special techniques are applied, but some of their features can be lost in the binary space. Unfortunately, many efficient nature-inspired MMO algorithms have no binary version and cannot be easily converted to binary representation.

As we can see from many studies, there is no universal approach that is efficient for all MMO problems. Many researches design hybrid algorithms, which are generally based on a combination of search algorithms and some heuristic for niching improvement. For example, here are four top-ranked algorithms from the CEC 2013 competition on MMO: Niching the CMA-ES via Nearest-Better Clustering (NEA2), A Dynamic Archive Niching Differential Evolution algorithm (dADE/nrand/1), CMA-ES with simple archive (CMA-ES) and Niching Variable Mesh Optimization algorithm (N-VMO) [8].

Another way is combining many basic MMO algorithms to run in parallel, migrate individuals and combine the results. In [2] an island model is applied, where islands are iteratively revised according to the genetic likeness of individuals. In [21] four MMO niching algorithms run in parallel to produce offspring, which are collected in a pool to produce a replacement step. In [15] the same scheme is realized using the clearing procedure.

The conception of designing MMO algorithms in the form of an ensemble seems to be perspective. A metaheuristic that includes many different MMO approaches (different search strategies) can deal with many different MMO problems. And such a metaheuristic can be self-configuring due to the adaptive control of the interaction of single algorithms during the problem solving.

In [19] a self-configuring multi-strategy genetic algorithm in the form of a hybrid of the island model, competitive and cooperative coevolution was proposed. The approach is based on a parallel and independent run of many versions of the GA with many search strategies, which can deal with many different features of optimization problems inside the certain optimization class. The approach has demonstrated good results with respect to multi-objective and non-stationary optimization.

3 Metaheuristic for MMO GA Ensemble Control

In the field of statistics and machine learning, ensemble methods are used to improve decision making. On average, the collective solution of multiple algorithms provides better performance than could be obtained from any of the constituent algorithms. This concept can be also used in the field of EA. The main idea is to include different search strategies in the ensemble and to design effective control of algorithm interaction. Our hypothesis is that different EAs are able to deal with different features of the optimization problem, and the probability of all algorithms failing with the same challenge in the optimization process is low. Moreover, the interaction of algorithms can provide the ensemble with new options for optimization, which are absent in stand-alone algorithms.

The general structure of the self-configuring multi-strategy genetic algorithm proposed in [19] is called Self*GA (the star sign corresponds to the certain optimization problem) and it is presented in Fig. 1.

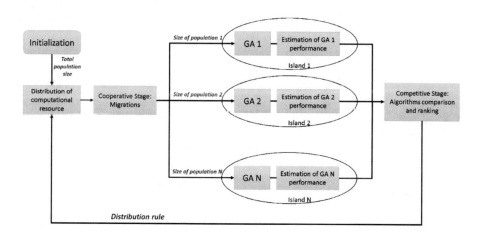

Fig. 1. The Self*GA structure.

The total population size (or the sum of populations of all stand-alone algorithms) is called the computational resource. The resource is distributed between algorithms, which run in parallel and independent over the predefined number of iterations (called the adaptation period). All algorithms have the same objective and use the same encoding (solution representation). All populations are initialized at random. After the distribution, each GA included in Self*GA has its own population which does not overlap with populations of other GAs. At the first iteration, all algorithms get an equal portion of the resource. This concept corresponds to the island model, where each island realizes its own search strategy.

After the adaptation period, the performance of individual algorithms is estimated with respect to the objective of the optimization problem. After that algorithms are compared and ranked. Search strategies with better performance increase their computational resource (the size of their populations). At the same time, all algorithms have a predefined amount of resource that is not distributed to give a chance for algorithms with low performance. This concept corresponds to the competitive coevolution scheme.

Finally, migrations of the best solutions are set to equate the start positions of algorithms for the run with the next adaptation period. According to the optimization problem, such a migration can be deterministic, selection-based or random. This concept corresponds to cooperative coevolution.

Such a technique eliminates the necessity to define an appropriate search strategy for the problem as the choice of the best algorithm is performed automatically and adaptively during the run.

Now we will discuss the design of a Self*GA for MMO problems that can be named SelfMMOGA.

At the first step, we need to define the set of individual algorithms included in the SelfMMOGA. In this study we use six basic techniques, which are well-studied and discussed [3, 17], and they can be used with binary representation with no modification. Algorithms and their specific parameters are presented in Table 1. All values for radiuses and distances in Table 1 are in the Hamming metric for binary problems and in the Euclidean metric for continuous problems.

Table 1. The SelfMMOGA component algorithms.

	Algorithm	Parameters
Alg1	Clearing	Clearing radius, Capacity of a niche
Alg2	Sharing	Niche radius, α
Alg3	Clustering	Number of clusters, min distance to centroid, max distance to centroid
Alg4	Restricted Tournament Selection (RTS)	Window size
Alg5	Deterministic Crowding	–
Alg6	Probabilistic Crowding	–

The motivation of choosing certain algorithms is that if the SelfMMOGA performs well with basic techniques, we can develop the approach with more complex algorithms in further works.

The adaptation period is a parameter of the SelfMMOGA. Moreover, the value depends on the limitation of the computational resource (total number of fitness evaluations).

The key point of any coevolutionary scheme is the performance evaluation of a single algorithm. For MMO problems performance metrics should estimate how many optima were found and how the population is distributed over the search space. Unfortunately, good performance measures exist only for benchmark MMO problems, which contain knowledge of the optima. Performance measures for black-box MMO problems are still being discussed. Some good recommendations can be found in [13]. In this study, the following criteria are used.

The first measure is called Basin Ratio (*BR*). The *BR* calculates the number of covered basins, which have been discovered by the population. It does not require knowledge of optima, but an approximation of basins is used. The *BR* can be calculated as

$$BR(pop) = \frac{l}{k} \tag{1}$$

$$l = \sum_{i=1}^{k} min\left\{1, \sum_{\substack{x \in pop \\ x \neq z_i}} b(x, z_i)\right\}$$

$$b(x, z) = \begin{cases} 1, & \text{if } x \in basin(z) \\ 0, & otherwise \end{cases}$$

where *pop* is the population, k is the number of identified basins by the total population, l is the indicator of basin coverage by a single algorithm, b is a function that indicates if an individual is in basin z.

To use the metric (1), we need to define how to identify basins in the search space and how to construct the function $b(x, z)$.

For continuous MMO problems, basins can be identified using different clustering procedures like Jarvis-Patrick, the nearest-best and others [12]. In this study, for MMO problems with binary representation we use the following approach. We use the total population (the union of populations of all individual algorithms in the SelfMMOGA). For each solution, we consider a predefined number of its nearest neighbours (with respect to the Hamming distance). If the fitness of the solution is better, it is denoted as a local optima and the centre of the basin. The number of neighbours is a tunable parameter. For a real-world problem, it can be set from some practical point of view. The simplified basin identification procedure is described using a pseudo-code as follows:

```
Z=Ø;
for all (x ∈ total population)
{
    for i=1,..,S
        yᵢ=define nearest neighbour(x);

    for all yᵢ
        if (fitness(x) > fitness(yᵢ))
        {
            Z=Z+x;
        };

};
```

The function $b(x, z)$ can be easily evaluated by defining if individual x is in a predefined radius of basin centre z. The radius is a tunable parameter. In this study, we define it as

$$radius = \frac{total\ population\ size}{k} \tag{2}$$

where k is the number of identified basins ($k = |Z|$).

The second measure is called Sum of Distances to Nearest Neighbour (*SDNN*). The *SDNN* penalizes the clustering of solutions. This indicator does not require knowledge of optima and basins. The *SDNN* can be calculated as

$$SDNN(pop) = \sum_{i=1}^{popsize} d_{nn}(x_i, pop) \tag{3}$$

$$d_{nn}(x_i, pop) = \min_{y \in pop \setminus \{x_i\}} \{dist(x_i, y)\}$$

where d_{nn} is the distance to the nearest neighbour, *dist* is the Hamming distance.

Finally, we combine the *BR* and the *SDNN* in an integrated criterion K:

$$K = \alpha \cdot BR(pop) + (1 - \alpha) \cdot \overline{SDNN}(pop) \tag{4}$$

where $\overline{SDNN}$ is a normalized value of *SDNN*, α defines weights of the BR and the *SDNN* in the sum ($\alpha \in [0, 1]$).

Next, we need to design a scheme for the redistribution of computational resources. New population sizes are defined for each algorithm. In this study, all algorithms give to the "winner" algorithm a certain percentage of their population size, but each algorithm has a minimum guaranteed resource that is not distributed. The guaranteed resource can be defined by the population size or by problem features.

At the cooperative stage, in many coevolutionary schemes, all individual algorithms begin each new adaptation period with the same starting points (such a migration scheme is called "the best displaces the worst"). For MMO problems, the best solutions are defined by discovered basins in the search space. As we already have evaluated the approximation of basins (Z), the solutions from Z are introduced in all populations replacing the most similar individuals.

Stop criteria in the SelfMMOGA are similar to those in the standard GA: maximum number of objective evaluations, the number of generations with no improvement (stagnation), etc.

4 Experimental Results

To estimate the approach performance, we have used the following list of benchmark and real-world problems

- Six binary MMO problems are from [21]. These test functions are based on the unitation functions, and they are massively multimodal and deceptive.

- Eight real-valued MMO problems are from CEC 2013 Special Session and Competition on Niching Methods for Multimodal Function Optimization [9].
- Fuzzy rule base classification system design using MMO GA.
- Designing loan portfolios for the Bank of Moscow.

We have denoted the functions as in the source papers. Some details of the benchmark problems are presented in Table 2.

Table 2. Test suite.

Problem	Number of desirable optima	Problem dimensionality[a]
binaryF11	32 global	30
binaryF12	32 global	30
binaryF13	27 global	24
binaryF14	32 global	30
binaryF15	32 global	30
binaryF16	32 global	30
cecF1	2 global + 3 local	9, 12, 15, 19, 22
cecF2	5 global	4, 7, 10, 14,17
cecF3	1 global + 4 local	4, 7, 10, 14,17
cecF4	4 global	14, 22, 28, 34, 42
cecF5	2 global + 2 local	11, 17, 24, 31, 37
cecF6	18 global + 742 local	16, 22, 30, 36, 42
cecF7	36 global	14, 20, 28, 34, 40
cecF8	12 global	8, 14, 20, 28, 34

[a]Real-valued problems have been binarized using the standard binary encoding with 5 accuracy levels proposed in the CEC 2013 competition rules.

In all comparisons, all algorithms have equal maximum number of the objective evaluations, but may differ in population sizes.

The following criteria for estimating the performance of the SelfMMOGA over the benchmark problems are used for continuous problems:

- Peak Ratio (*PR*) measures the percentage of all optima found by the algorithm (5).
- Success Rate (*SR*) measures the percentage of successful runs (a successful run is defined as a run where all optima were found) out of all runs.

$$PR = \frac{|\{q \in Q | d_{nn}(q, pop) \le \varepsilon\}|}{k} \tag{5}$$

where $Q = \{q_1, q_2, \ldots, q_k\}$ is a set of known optima, ε is accuracy level.

The maximum number of function evaluation and the accuracy level for the *PR* evaluation are the same as in CEC completion rules [9]. The number of independent runs of the algorithm is 50.

In the case of binary problems, we cannot define the accuracy level in the *PR*, thus the exact points in the search space have to be found. This is a great challenge for search algorithms, thus we have substituted the *SR* measure with Peak Distance (*PD*). The *PD* indicator (6) calculates the average distance of known optima to the nearest individuals in the population [13].

$$PD = \frac{1}{k}\sum_{i=1}^{k} d_{nn}(q_i, pop) \tag{6}$$

To demonstrate the control of algorithm interaction in the SelfMMOGA, we have chosen an arbitrary run of the algorithm on the cecF1 problem and have visualized the distribution of the computational resource (see Fig. 2).

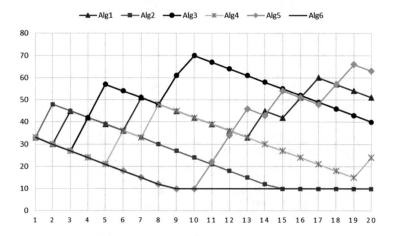

Fig. 2. Example of the SelfMMOGA run.

The total population size is 200 and the minimal guaranteed amount of the computational recourse is 10. All algorithms are initialized with 33 individuals. The maximum number of generations is 200 and the size of the adaptation period is 10, thus the horizontal axis contains numeration of 20 periods.

As we can see, there is no algorithm that wins all the time. At the first two periods, Sharing (Alg2) and Clearing (Alg1) had better performance. The highest amount of the resource was won by Clustering (Alg3) at the 10th period. At the final stages, Deterministic Crowding (Alg5) showed better performance.

4.1   Experimental Results for Binary Benchmark Problems

The results of estimating the performance of the SelfMMOGA with the pack of binary problems are presented in Table 3. The table contains the values of the *PR*, the *SR* and the *PD* averaged over 50 independent runs. We also have compared the results with Ensemble of niching algorithms (ENA) proposed in [21]. There is only the *SR* value for the ENA.

Table 3. Detailed results for binary problems.

	ENA	Alg1	Alg2	Alg3	Alg4	Alg5	Alg6	Mean	SelfMMOGA
Problem: binaryF11									
PR	–	0.94	0.84	0.91	1.00	0.97	0.78	0.91	1.00
SR	1.00	0.90	0.84	0.88	1.00	0.94	0.80	0.89	1.00
PD	–	2.40	3.37	2.40	0.00	2.33	3.30	2.30	0.00
Problem: binaryF12									
PR	–	0.97	0.97	1.00	1.00	0.97	0.84	0.96	1.00
SR	1.00	0.96	0.98	1.00	1.00	0.94	0.84	0.95	1.00
PD	–	2.00	1.00	0.00	0.00	1.67	3.62	1.38	0.00
Problem: binaryF13									
PR	–	1.00	0.96	0.96	0.93	0.96	0.89	0.95	1.00
SR	1.00	1.00	0.96	0.94	0.90	0.94	0.84	0.93	1.00
PD	–	0.00	2.50	2.67	2.80	2.67	3.37	2.34	0.00
Problem: binaryF14									
PR	–	0.91	0.81	0.91	1.00	0.94	0.75	0.89	1.00
SR	1.00	0.92	0.92	0.90	1.00	0.94	0.80	0.91	1.00
PD	–	3.25	2.50	2.60	0.00	2.67	3.20	2.37	0.00
Problem: binaryF15									
PR	–	0.88	0.88	0.84	0.88	0.88	0.72	0.84	1.00
SR	1.00	0.88	0.86	0.84	0.86	0.84	0.64	0.82	1.00
PD	–	2.33	2.57	2.62	2.71	2.37	3.06	2.61	0.00
Problem: binaryF16									
PR	–	0.84	0.75	0.84	0.88	0.78	0.56	0.78	1.00
SR	0.99	0.84	0.80	0.86	0.84	0.76	0.66	0.79	1.00
PD	–	3.25	2.80	3.00	2.87	3.08	3.47	3.08	0.00

The setting for the SelfMMOGA are:

- Maximum number of function evaluation is 50000 (as for the ENA);
- Total population size is 200 (the ENA uses 500);
- Adaptation period is 10 generations (25 times);
- All specific parameters of individual algorithms are self-tunable using the concept from [16].

As we can see, binary problems are not too complex for the SelfMMOGA and the ENA. Therefore, we will analyze the results in details. In Table 3, the results for the ENA, stand-alone algorithms, the average of 6 stand-alone algorithms and the Self-MMOGA (6 algorithms ensemble) are presented. The average value ("Mean" column) can be viewed as the average performance of a randomly chosen algorithm. Such an estimate is very useful for black-box optimization problems, because we have no information about problem features and, consequently, about what algorithms to use. If the performance of the SelfMMOGA is better that the average of its component, we can conclude that on average the choice of the SelfMMOGA will be better.

As we can see from Table 3, the SelfMMOGA always outperforms the average of its stand-alone component algorithms for binary problems. Moreover, the ensemble can provide results that are better than obtained from component algorithms. For example, for problems F15 and F16 none of component algorithms has a *SR* value equal to 1, but the SelfMMOGA does.

4.2 Experimental Results for Continuous Benchmark Problems

The results of estimating the performance of the SelfMMOGA with the pack of continuous problems are presented in Tables 4 and 5. Table 4 shows a comparison of results averaged over all problems with other techniques. Table 5 contains ranks of algorithms by separate criteria.

All problems and settings are as in the rules of the CEC 2013 competition on MMO. For each problem there are 5 levels of accuracy of finding optima ($\varepsilon = \{1e-01, 1e-02, 1e-03, 1e-04, 1e-05\}$). Thus, each problem has been binarized 5 times. The dimensionalities of binarized problems are presented in Table 2.

We have compared the results of the SelfMMOGA runs with some efficient techniques from the competition. The techniques are DE/nrand/1/bin and Crowding DE/rand/1/bin [9], N-VMO [11], dADE/nrand/1 [5], and PNA-NSGAII [1].

The settings for the SelfMMOGA are:

- Maximum number of function evaluation is 50000 (for cecF1-cecF5) and 200000 (for cecF6-cecF8);
- Total population size is 200;
- Adaptation period is 10 generations 25 times (for cecF1-cecF5) and 25 generations 40 times (cecF6-cecF8);
- All specific parameters of individual algorithms are self-tunable.

As we can see from Tables 4 and 5, the SelfMMOGA shows results comparable with popular and well-studied techniques. It yields to dADE/nrand/1 and N-VMO, but we should note that these algorithms are specially designed for continuous MMO problems, and have taken 2nd and 4th places [8], respectively, in the CEC competition. At the same time, the SelfMMOGA has very close average values to the best two algorithms, and outperforms PNA-NSGAII, CrowdingDE and DE, which have taken 7th, 8th and 9th places in the competition respectively [8].

In this study, we have included only basic MMO search techniques in the Self-MMOGA. Nevertheless, it performs well due to the effect of collective decision making in the ensemble. The key feature of the approach is that it operates in an automated, self-configuring way. Thus, the SelfMMOGA can be a good alternative for complex black-box MMO problems.

Table 4. Average PR and SR for each algorithm.

ε	Self MMOGA		DE/nrand/ 1/bin		cDE/rand/ 1/bin		N-VMO		dADE/ nrand/1		PNA- NSGAII	
	PR	SR	PR	SR	PR	SR	PR	SR	PR	SR	PR	SR
1e–01	0.962	0.885	0.850	0.750	0.963	0.875	1.000	1.000	0.998	0.938	0.945	0.875
1e–02	0.953	0.845	0.848	0.750	0.929	0.810	1.000	1.000	0.993	0.828	0.910	0.750
1e–03	0.943	0.773	0.848	0.748	0.847	0.718	0.986	0.813	0.984	0.788	0.906	0.748
1e–04	0.907	0.737	0.846	0.750	0.729	0.623	0.946	0.750	0.972	0.740	0.896	0.745
1e–05	0.816	0.662	0.792	0.750	0.642	0.505	0.847	0.708	0.835	0.628	0.811	0.678
Average	0.916	0.780	0.837	0.750	0.822	0.706	0.956	0.854	0.956	0.784	0.893	0.759

Table 5. Algorithms ranking over cecF1-cecF8 problems.

Rank by PR criterion	Algorithm	Rank by SR criterion	Algorithm
1	N-VMO and dADE/nrand/1	1	N-VMO
2	SelfMMOGA	2	dADE/nrand/1
3	PNA-NSGAII	3	SelfMMOGA
4	DE/nrand/1/bin	4	PNA-NSGAII
5	cDE/rand/1/bin	5	DE/nrand/1/bin
–	–	6	cDE/rand/1/bin

4.3 Experimental Results for Real-World Problem of Fuzzy Rule Base Classification System Design

Modern machine learning methods often use evolutionary computation techniques as a design tool, which is universal and can be applied for various structures. These evolutionary algorithms applied for machine learning problems are often called genetics-based machine learning algorithms. The fuzzy rule-based classification systems (FRBCSs) are effective approaches in machine learning, as they can provide easy-to-understand models for the end users [6].

Traditional GAs applied to the FRBCSs design have a tendency to converge to the best-found optimum losing population diversity. Such single best-found solution usually has very good accuracy, but may have a structure that is not convenient for human understanding and analysis. Thus there is a good idea to find many (or all) global and acceptable local optima which represent different solutions to the problem. In a case of the FRBCS, such optima, while saving comparable accuracy, may contain different rules in the rule base and/or different fuzzy term structures.

The number of rules in computational experiments was fixed and equal to 12. The FRBCS method, which have been implemented, is based on a simple rule base encoding into the GA chromosome. The chromosome contains fuzzy sets assigned to input variables in the premise part and class labels assigned to output variables in the conclusion part of each rule in the rule base. The number of fuzzy sets for granulation was fixed and equal to 5 + 1. Additional fuzzy term is the "Don't care" condition (corresponding input variable is ignored). Including this term allows decreasing the size of the rule base and increasing the rules' generalization ability. If all terms in a certain rule are set to "Don't care" (DC), the rule is considered as empty and not used in classification, so the algorithm is capable of decreasing the number of rules.

The fitness function includes two values: error on the training set with weight 1 and the complexity of the rule base with weight 0.1. The complexity of the rule base was calculated as the ratio of number of non-empty fuzzy sets to the total number of possible fuzzy sets in the rule base. Including complexity of the rule base into the fitness function allows creating of simpler rule bases. The distance between two rule bases for the MMO GA was calculated as the number of different fuzzy sets for these rule bases. More detailed information can be found in [18].

The computational experiments for the fuzzy classification were performed on 7 datasets from UCI and KEEL repositories [7, 20]. Table 6 contains the information about the datasets.

Table 6. Datasets description.

Dataset	Number of instances	Number of features	Number of classes
Australian credit	690	14	2
Banknote	1372	4	2
Column 2c	310	6	2
Column 3c	310	6	3
Ionosphere	351	34	2
Liver	345	6	2
Seeds	210	7	3

The Table 7 contains the classification results for the test sample obtained with the standard GA and three best solutions obtained with the SelfMMOGA.

Table 7. Classification results for test sample.

Dataset	Standard GA	SelfMMOGA Solution 1	SelfMMOGA Solution 2	SelfMMOGA Solution 3
Australian	0.839	0.862	0.867	0.816
Banknote	0.947	0.892	0.867	0.862
Column 2c	0.773	0.789	0.768	0.751
Column 3c	0.668	0.741	0.674	0.619
Ionosphere	0.747	0.680	0.656	0.665
Liver	0.567	0.586	0.597	0.598
Seeds	0.874	0.793	0.691	0.621

As we can see, for three datasets the standard GA allows finding most accurate solutions. However, the SelfMMOGA outperforms the standard GA on 4 datasets out of 7. Moreover, the best solution is not always the first one – for example, for datasets Australian and Liver, the best solution was second or even third. Thus, using this method, several local optima have been found, and the researcher is able to select one of them.

As an example, we provide three rule bases for Liver dataset with the best accuracy, obtained on the last iteration. The rule bases are presented in Figs. 3, 4 and 5. Each row is a single rule, where every position contains the fuzzy term for corresponding variable, and the last position is the assigned class label. The DC ("Don't Care") term means that corresponding variable is ignored in a rule.

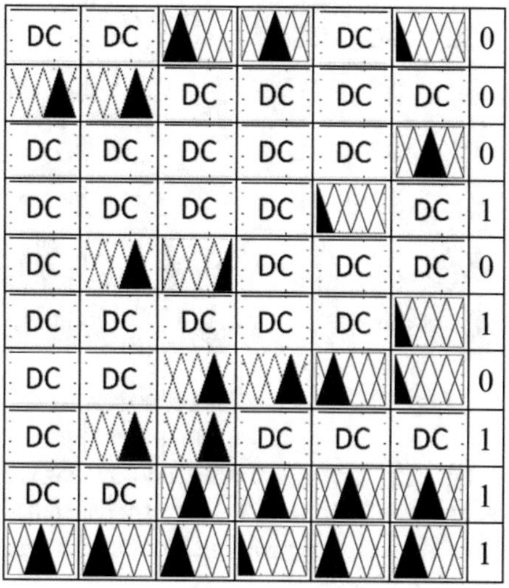

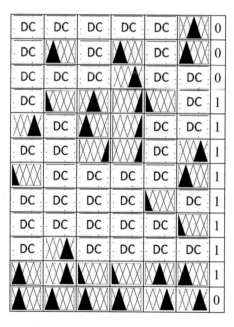

Fig. 3. Solution 1 for Liver dataset.

Fig. 4. Solution 2 for Liver dataset.

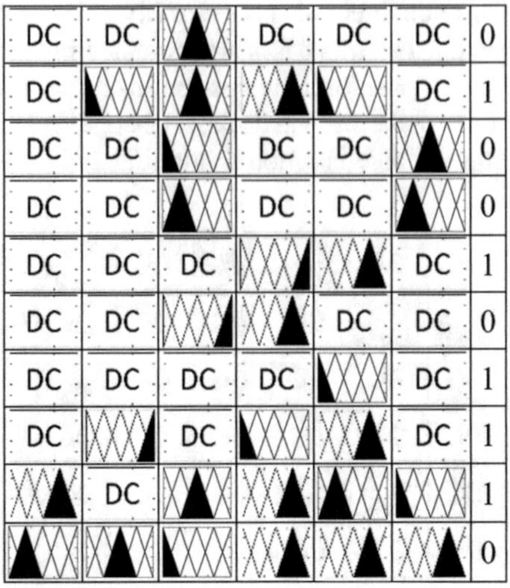

Fig. 5. Solution 3 for Liver dataset.

These rule bases contain 10, 12 and 10 rules and are very different, although they have almost the same accuracy about 0.544. We suggest that the results can help the human experts in a field of the solving problem to obtain better (or may be very new) information about the problem features.

4.4 Experimental Results for Real-World Problem of Designing Loan Portfolios for the Bank of Moscow

The problem of bank loan portfolio design is an optimization problem of maximizing the profit of the bank with some constraints on the amount of free liabilities, the amount of credit requested, periods of credits, credit interests and so on. Input data to the problem is a set of credit requests from loan borrowers. The bank portfolio is a subset of requests that are approved by the bank.

In this paper, the loan portfolio based on data presented by Krasnoyarsk department of the Bank of Moscow is discussed. The following profit model (optimization objective) is used (7):

$$Profit(X) = \sum_{j=1}^{N} k_j \cdot (1 + d_j \cdot t_j) \cdot x_j \rightarrow max \tag{7}$$

$$Risk(X) = \frac{1}{\sum_{j=1}^{N} x_j} \cdot \sum_{j=1}^{N} P_j \cdot x_j \leq \rho$$

$$\sum_{j=1}^{N} k_j \cdot x_j \leq F$$

$$X = (x_1, x_2, \ldots, x_N), x_i \in \{0, 1\}$$

where F – the amount of free liabilities held by the Bank at a given time; N – the number of borrowers; k_j – the amount of credit requested by the j-th borrower $j = 1, N$; t_j – the period for which the j-th borrower takes a loan; x_j – Boolean variable taking the value 1, if the k_j loan is issued, and 0 otherwise; d_j – interest (%) on j-th credit; P_j – probability of non-payment of loan and interest on the loan; ρ – limitation on the total riskiness of the loan portfolio.

As a candidate solution is binary vector, there is no need to encode it to chromosome. The fitness function is defined as sum of the Profit and penalty functions for given constraints.

The initial information about credit requests and their characteristics is presented in Table 8.

Table 8. Initial data for the loan portfolio design problem.

Request no.	Request amount	Loan rate (%)	Period	Risk
1	10 000 000	25	75	0.042
2	5 300 000	28	80	0.039
3	2400000	25	91	0.029
4	50 000 000	23	84	0.033
5	1 000 000	28	64	0.026
6	500 000	30	76	0.046
7	250 000	37	91	0.044
...				
48	9 000 000	27	86	0.024
49	22 000 000	29	91	0.016
50	350 000	27	69	0.026

Total sum of requests = 256 695 000

The amount of free liabilities = 188 500 000

The length of the chromosome is 50. The search space contains 2^{50} ($\approx 10^{15}$) different portfolios. The maximum number of the fitness evaluation is set to 10^6 that is 8 * 10^{-10} % of the cardinality of the search space.

The results of the bank portfolio design (global and three local solutions) are presented in Table 9.

Table 9. Results for the loan portfolio problem.

Solutions (the structure of the loan portfolio)	Profit of portfolio	Rest of free liabilities	Total riskiness
01111011111111101111010001111011101000010101010111	199734518.9	30000	0.0292
11011100111110110110011011101011101010100101011110	199691164	15000	0.0286
01011110101111000111011101111110110011110110011110	199668728.9	15000	0.028
00110011010001010110111000110110101011111101011011111	195593407.3	10000	0.0276

As we can see from Table 9, solutions obtained with the SelfMMOGA have very close values of the profit, but have very different structures. Thus these portfolios can be used as alternative solutions or as additional information for the portfolio analysis.

The problem has been also solved using the brute-force search. The first best solution founded by the SelfMMOGA is the exact global solution to the problem.

5 Conclusions

In this study, a metaheuristic for control of MMO GA ensemble (called SelfMMOGA) is proposed. It involves many different search strategies in the process of MMO problem solving and adaptively control their interactions.

The SelfMMOGA allows complex MMO problems to be dealt with, which are the black-box optimization problems (a priori information about the objective and its features are absents or cannot be introduces in the search process). The algorithm uses binary representation for solutions, thus it can be implemented for many real-world problems with variables of arbitrary (and mixed) types.

We have included 6 basic MMO techniques in the SelfMMOGA realization to demonstrate that it performs well even with simple core algorithms. We have estimated the SelfMMOGA performance with a set of binary benchmark MMO problems and continuous benchmark MMO problems from CEC 2013 Special Session and Competition on Niching Methods for Multimodal Function Optimization. The proposed approach has demonstrated a performance comparable with other well-studied techniques.

Experimental results show that the SelfMMOGA outperforms the average performance of its stand-alone algorithms. It means that it performs better on average than a randomly chosen technique. This feature is very important for complex black-box optimization, where the researcher has no possibility of defining a suitable search algorithm and of tuning its parameters.

We have also applied the SelfMMOGA for solving some real-world problems to demonstrate the effect of identifying many optima to the problem. The key feature of the approach is that it operates in an automated, self-configuring way. Thus, the SelfMMOGA can be a good alternative for solving complex black-box MMO problems.

In further works, we will investigate the SelfMMOGA using more advanced component techniques.

Acknowledgements.   The research was supported by President of the Russian Federation grant (MK-3285.2015.9).

References

1. Bandaru, S., Deb, K.: A parameterless-niching-assisted bi-objective approach to multimodal optimization. In: Proceedings of 2013 IEEE Congress on Evolutionary Computation (CEC 2013), pp. 95–102 (2013)
2. Bessaou, M., Pétrowski, A., Siarry, P.: Island model cooperating with speciation for multimodal optimization. In: Schoenauer, M., Deb, K., Rudolph, G., Yao, X., Lutton, E., Merelo, J.J., Schwefel, H.-P. (eds.) PPSN 2000. LNCS, vol. 1917, pp. 437–446. Springer, Heidelberg (2000). doi:10.1007/3-540-45356-3_43
3. Das, S., Maity, S., Qub, B.-Y., Suganthan, P.N.: Real-parameter evolutionary multimodal optimization: a survey of the state-of-the art. Swarm Evol. Comput. **1**, 71–88 (2011)
4. Deb, K., Saha, A.: Finding multiple solutions for multimodal optimization problems using a multi-objective evolutionary approach. In: Proceedings of the 12th Annual Conference on Genetic and Evolutionary Computation, GECCO 2010, pp. 447–454. ACM, New York (2010)
5. Epitropakis, M.G., Li, X., Burke, E.K.: A dynamic archive niching differential evolution algorithm for multimodal optimization. In: Proceedings of 2013 IEEE Congress on Evolutionary Computation (CEC 2013), pp. 79–86 (2013)
6. Ishibuchi H.: Hybridization of fuzzy GBML approaches for pattern classification problems. IEEE Trans. Syst. Man Cybern. B Cybern. **35**(2), 359–365 (2005)
7. KEEL (Knowledge Extraction based on Evolutionary Learning). http://www.keel.es
8. Li, X., Engelbrecht, A., Epitropakis, M.: Results of the 2013 IEEE CEC competition on niching methods for multimodal optimization. Report presented at 2013 IEEE Congress on Evolutionary Computation Competition on: Niching Methods for Multimodal Optimization (2013)
9. Li, X., Engelbrecht, A., Epitropakis, M.G.: Benchmark functions for CEC 2013 special session and competition on niching methods for multimodal function optimization. Evolutionary Computation and Machine Learning Group, RMIT University, Melbourne, VIC, Australia. Technical report (2013)
10. Liu, Y., Ling, X., Shi, Z., Lv, M., Fang, J., Zhang, L.: A survey on particle swarm optimization algorithms for multimodal function optimization. J. Softw. **6**(12), 2449–2455 (2011)
11. Molina, D., Puris, A., Bello, R., Herrera, F.: Variable mesh optimization for the 2013 CEC special session niching methods for multimodal optimization. In: Proceedings of 2013 IEEE Congress on Evolutionary Computation (CEC 2013), pp. 87–94 (2013)
12. Preuss, M., Stoean, C., Stoean, R.: Niching foundations: basin identification on fixed-property generated landscapes. In: Proceedings of the 13th Annual Conference on Genetic and Evolutionary Computation, GECCO 2011, pp. 837–844 (2011)
13. Preuss, M., Wessing, S.: Measuring multimodal optimization solution sets with a view to multiobjective techniques. In: Emmerich, M., et al. (eds.) EVOLVE – A Bridge Between Probability, Set Oriented Numerics, and Evolutionary Computation IV. AISC, vol. 227, pp. 123–137. Springer, Heidelberg (2013)
14. Preuss, M.: Tutorial on multimodal optimization. In: Proceedings of the 13th International Conference on Parallel Problem Solving from Nature, PPSN 2014, Ljubljana, Slovenia (2014)
15. Qu, B., Liang, J., Suganthan, P.N., Chen, T.: Ensemble of clearing differential evolution for multi-modal optimization. In: Tan, Y., Shi, Y., Ji, Z. (eds.) ICSI 2012. LNCS, vol. 7331, pp. 350–357. Springer, Heidelberg (2012). doi:10.1007/978-3-642-30976-2_42

16. Semenkin, E., Semenkina, M.: Self-configuring genetic algorithm with modified uniform crossover operator. In: Tan, Y., Shi, Y., Ji, Z. (eds.) ICSI 2012. LNCS, vol. 7331, pp. 414–421. Springer, Heidelberg (2012). doi:10.1007/978-3-642-30976-2_50
17. Singh, G., Deb, K.: Comparison of multi-modal optimization algorithms based on evolutionary algorithms. In: Proceedings of the Genetic and Evolutionary Computation Conference, Seattle, pp. 1305–1312 (2006)
18. Sopov E., Stanovov V., Semenkin E.: Multi-strategy multimodal genetic algorithm for designing fuzzy rule based classifiers. In: Proceedings of 2015 IEEE Symposium Series on Computational Intelligence (IEEE SSCI 2015), Cape Town, South Africa, pp. 167–173 (2015)
19. Sopov, E.: A self-configuring metaheuristic for control of multi-strategy evolutionary search. In: Tan, Y., Shi, Y., Buarque, F., Gelbukh, A., Das, S., Engelbrecht, A. (eds.) ICSI 2015. LNCS, vol. 9142, pp. 29–37. Springer, Heidelberg (2015). doi:10.1007/978-3-319-20469-7_4
20. UC Irvine Machine Learning Repository. http://archive.ics.uci.edu/ml/
21. Yu, E.L., Suganthan, P.N.: Ensemble of niching algorithms. Inf. Sci. **180**(15), 2815–2833 (2010)

SPSO Parallelization Strategies
for Electromagnetic Applications

Anton Duca[1(✉)], Laurentiu Duca[2(✉)], Gabriela Ciuprina[1(✉)],
and Daniel Ioan[1(✉)]

[1] Faculty of Electrical Engineering, Politehnica University of Bucharest,
Bucharest, Romania
{anton.duca,gabriela.ciuprina,daniel.ioan}@upb.ro
[2] Faculty of Computer Science, Politehnica University of Bucharest,
Bucharest, Romania
laurentiu.duca@cs.pub.ro

Abstract. Two parallelization techniques, GPGPU and Pthreads for multiprocessor architectures, are used to implement a SPSO algorithm in order to solve electromagnetic optimization problems. Several configurations for the GPGPU implementation are tested and a new full parallel minimum branching implementation is proposed. The best GPGPU approaches are then compared with a Pthreads implementation in terms of speed up and solution quality. To test the efficiency of the parallelization techniques two electromagnetic optimization problems were chosen, namely the TEAM22 benchmark and Loney's solenoid. In the end the paper provides suggestions regarding what parallelization technique should be used considering the implementation features of the optimization function.

Keywords: SPSO · GPGPU · Pthreads · Electromagnetic field · Optimization

1 Introduction

Electromagnetic optimizations problems are well known for their complex objective functions which for evaluation involve solving electromagnetic field equations. The objective function is most of the times multidimensional, with several local minimum and a wide search area, while the optimization variables often have to meet difficult constraints. For this reasons the evaluation of the objective function is usually computational intensive, requires a large number of subroutine calls (sometimes recursive), having a high level of branching, and many instructions [1, 2].

Since deterministic approaches like the gradient descent or conjugate gradient can not be used because of the multiple local minimum, in the past years stochastic methods based on, simulated annealing, tabu search, genetic algorithms, or swarm optimization, were widely adopted as standard methods for solving electromagnetic problems [3, 4]. The advantages of the heuristics based methods are their ability to find the global optimum, usually without knowing the objective function derivatives, and their robustness. The disadvantage of the stochastic methods is the large number of

© Springer International Publishing AG 2017
J.J. Merelo et al. (eds.), *Computational Intelligence*, Studies in Computational Intelligence 669,
DOI 10.1007/978-3-319-48506-5_5

evaluations for the objective function, essential for real world optimization problems which for the evaluation is often time consuming.

To decrease the solving time there are the following options: to reduce the number of objective function evaluations by using more efficient stochastic methods [5, 6], to implement parallel and/or distributed optimization algorithm architectures [7], or to decrease the objective function evaluation time using methods specific to electromagnetic problems [8].

In this paper two different parallelization techniques, Pthreads (POSIX threads) for multiprocessor architectures, and GPGPU (General Purpose Computation on Graphics Processing Units), are investigated for accelerating an optimization algorithm, namely SPSO (Standard Particle Swarm Optimization).

At first, a new GPGPU parallel implementation is proposed and several GPGPU configurations are compared. The proposed implementation is designed in order to deal with the specific implementation aspects of the electromagnetic objective functions. Afterwards, the most efficient GPGPU approach is compared with a multiprocessor implementation, in terms of speed up and solution fitness, for different SPSO swarm sizes. The parallel implementation for the GPU (Graphics Processing Unit) will use the CUDA language while the implementation for the multiprocessor architecture will use Pthreads. To test and compare the SPSO parallelization techniques two electromagnetic optimization benchmark problems, with different implementation features, have been chosen, TEAM22 and Loney's solenoid [9, 10].

2 SPSO Algorithm

Initially proposed by Kennedy and Eberhart [11], PSO (Particle Swarm Optimization) is an iterative optimization algorithm which has the roots in biology and is inspired from the social behavior inside a bird flock or a fish school. Each particle in the swarm is described by position and velocity. The position encapsulates the potential solution of the optimization problem (its coordinates in the searching space) while the velocity describes the way the position is modified.

At iteration (time) $t + 1$ the position x_i and the velocity v_i of each particle i in the swarm are computed as follows:

$$x_i(t+1) = x_i(t) + v_i(t+1), \tag{1}$$

$$v_i(t+1) = w_v \cdot v_i(t) + w_{PB,i} \cdot r_1 \cdot \Delta x_{PB\,i}(t) + w_{GB} \cdot r_2 \cdot \Delta x_{GB}(t), \tag{2}$$

$$\Delta x_{PB\,i}(t) = x_{PB\,i}(t) - x_i(t), \; \Delta x_{GB}(t) = x_{GB}(t) - x_i(t), \tag{3}$$

where x_{PB}, x_{GB} are the best personal position and the best position in the group (swarm), w_v, w_{PB}, w_{GB} are the weights for velocity, "cognitive" term and "social" term, and r_1, r_2 two random numbers distributed uniformly in the interval [0, 1). So the time step is considered 1 and the velocity vector is computed as a weighted average, assuring a random but enough smooth movement of particles, attracted to the best known position.

The main issues of the original PSO are the high probability of being trapped in local minima and the large number of objective function evaluations needed to find the global solution. During time, for improving the performance of the PSO different approaches were proposed. Some of the most efficient PSO based algorithms available today are IPSO (Intelligent PSO) [5], SPSO (Standard PSO) [12], QPSO (Quantum-behaved PSO) [13] and DPSO (Discrete PSO) [14].

Currently at its third version [15], SPSO modifies the classical algorithm in terms of initialization, velocity/position update equations, neighborhood and confinement. In the case of SPSO, the particles of the swarm are connected, each connection representing a link between two different particles. A connection has an informed and an informing particle, the first particle knowing the personal best and the position of the second particle. Thus, each informed particle has a set of informing particles called neighborhood. SPSO uses a random topology which changes the connections graph at each unsuccessful iteration (when the global best solution is not improved).

The initializations for position and velocity are made to avoid leaving the search area, especially when the optimization variables number is high. The position coordinates are generated randomly for each direction (d) using a uniform distribution, while the velocity coordinates are generated taken into consideration the generated position coordinates:

$$x_i(0) = U(\min_d, \ \max_d), \ v_i(0) = U(\min_d - x_{i,d}(0), \ \max_d - x_{i,d}(0)). \tag{4}$$

The velocity formula introduces a new term, the center of gravity, for obtaining "exploration" and "exploitation". The center of gravity depends on three terms: the current position, a term relative to the previous best $x_{PB,i}$, and a term relative to the previous best in the neighborhood $x_{LB,i}$. Thus, the update equations for velocity and positions are changed comparing with the original PSO algorithm, as follows:

$$v_i(t+1) = w \cdot v_i(t) + x_i'(t) - x_i(t), \tag{5}$$

$$x_i(t+1) - x_i(t) + v_i(t+1) = w \cdot v_i(t) + x_i'(t), \tag{6}$$

where x_i' is a random point inside a hypersphere of radius $\|G_i - x_i\|$ and center G_i, with G_i being the center of gravity for the particle i:

$$G_i = \frac{x_i + (x_i + c \cdot (x_{PB\,i} - x_i)) + (x_i + c \cdot (x_{LB\,i} - x_i))}{3}, \tag{7}$$

or if the particle i is the best particle in its neighborhood (has the best fitness value):

$$G_i = \frac{x_i + (x_i + c \cdot (x_{PB\,i} - x_i))}{2}. \tag{8}$$

Another feature of the SPSO algorithm is the confinement. If during the iterative process a particle moves outside the search space on some coordinate d, its velocity and position are modified as follows:

$$\text{if } (x_{i,d}(t) < \min_d) \quad \text{then } \{ x_{i,d}(t) = \min_d; \ v_{i,d}(t) = -0.5 \cdot v_{i,d}(t) \}, \qquad (9)$$

$$\text{if } (x_{i,d}(t) > \max_d) \quad \text{then } \{ x_{i,d}(t) = \max_d; \ v_{i,d}(t) = -0.5 \cdot v_{i,d}(t) \}. \qquad (10)$$

The main disadvantage of stochastic methods is the large number of objective function evaluations, especially in real problems when the objective function evaluation cost is significant. In this case, the solving time for the sequential implementations is significant, the need for a parallel optimization algorithm being obvious.

3   SPSO Parallelization – GPGPU Approach

Due to market demand for high-definition 3D graphics, and realtime processing, the GPU evolved into a parallel, multithreaded, and manycore processor with high computational power and memory bandwidth [16]. If a CPU focuses on flow control and data caching, a GPU is designed for parallel computational applications, like graphics rendering, and is suitable for problems where the same program is run in parallel on many and different sets of data. In order to use the GPU for general purpose computation and to solve complex computational problems from different and various domains (not only graphics rendering) several programming models such as CUDA and OpenCL have been created.

3.1   Existing Approaches

The idea of using implementations based on GPGPU for PSO is not new. [17] GPGPU implementation of SPSO 2007 showed acceleration up to 11 times compared with traditional CPU implementations.

In [18] the authors focus on the data representation in memory (especially on the best global position/local) such that reading/writing operations to be carried out effectively. The obtained acceleration was up to 100 times comparing to the sequential CPU implementation, for a problem with 100 variables and PSO algorithms with 3 sub-swarms.

In [19, 20] the authors study the results quality of the GPGPU implementations depending on where the random numbers are generated (CPU or GPU). Both studies suggest ways of generating random numbers on the GPU, the results having a good quality.

In [21] the authors study the parallelization of a multi-swarm PSO algorithm to solve combinatorial problems such as the allocation of tasks. Again, it was observed that the GPGPU implementation led to an acceleration of 37 times compared with the sequential version of the algorithm, especially for large problems.

In [22] a multi-objective PSO version which uses one subswarm for each objective is parallelized. The GPGPU implementation performed 3 to 7 times faster than CPU implementation.

Other GPGPU implementations of the PSO based algorithms are proposed in [23, 24].

Most of the proposed solutions are tested (like many others) on functions with simple analytical expressions (with many local minimum but not computational intensive), and focus on the influence over the performance of: the data transfer between the host and the device (GPU), the manner and the place of generating random values, the type of implementation synchronous/asynchronous, etc. Unfortunately the solutions do not address specific aspects of the objective function implementation such as the level of branching or the code complexity.

3.2   The Proposed Full Parallel Minimum Branching (FPMB) Implementation

To implement the parallel version of the SPSO algorithm the CUDA-C language was chosen. Introduced by Nvidia, CUDA [16] is a programming model a parallel computing platform. The CUDA developer kit allows software developers to create general purpose parallel applications with languages such as Java, C++, C Fortran and others.

Because of the hardware variety of the GPUs, which can have a different stream multiprocessors number, CUDA was built as a scalable software programming model. Thus, a CUDA software program can be executed (compiled) on any GPU device independent of the multiprocessors number.

The CUDA programming model has as its core the following three key concepts: a memory model, synchronization mechanisms and a hierarchy of thread blocks. These concepts help the developer to split the task into smaller tasks which can be solved separately by different blocks of threads. For solving a task, the threads inside a CUDA program can work independently or can cooperate.

In order to solve a problem the threads can use barrier mechanisms to synchronize their execution. These barrier mechanisms can only be used to synchronize threads from the same block, and can not synchronize blocks. To synchronize blocks the software developer must split the program into smaller sections and implement those with different functions (kernels).

For implementing the SPSO parallel algorithm with CUDA there are the following two possible options: an implementation for configurations with all threads in a single block (one thread block), or an implementation for cases with multiple thread blocks. In both implementations each thread simulates a particle's behavior and calls functions as evaluation, movement, personal/local best calculation, etc.

In first case, the SPSO parallel implementation is done using one kernel. The synchronization between particles (threads), necessary at certain steps, is obtained using `__syncthreads` function (a barrier mechanism for synchronization). This strategy has as main advantage the avoidance of kernels relaunch. The disadvantages of this implementation are: the maximum number of particles is 1024 (a block may have at most 1024 threads), multiple warps of threads (if the particles number exceeds the warp size – 32), and threads branching possibility (Fig. 1), depending on the objective function.

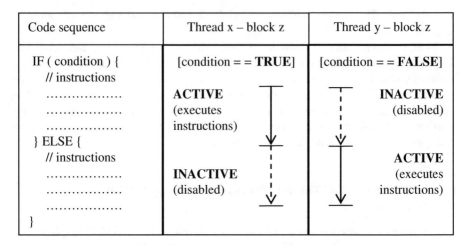

Fig. 1. Branching effect inside the same warp.

The branching behavior can occur for threads executing in the same warp, but not for threads belonging to different blocks. Threads inside a warp execute one instruction at a time, so the divergence of the threads can appear if a conditional (data dependent) instruction has to be executed. The warp serially executes the instructions on each path: threads on the current execution path are active, while threads which are not on the current execution path are disabled. The threads within the warp join the same execution path only after all possible branching paths are finished. Because the branching has as the main effect the serialization, the outcome is that not all the threads in the same warp are executed in parallel at the same time and this leads to a penalty in the performance (execution time). If the objective function has a high level of branching the execution time can be severely influenced, even more for single kernel implementations [2].

For the second implementation strategy, because the barrier synchronization mechanism provided by CUDA can only be applied for threads in the same block, the synchronization of particles is done by implementing each particle function as a kernel. The main disadvantage is the delay introduced by the relaunch of kernels at each SPSO iteration. The advantages comparing with the first strategy are: the option to run all threads in parallel (for one warp configurations, with no more than 32 threads per block), and the swarm size is no longer limited to 1024 particles.

Another advantage of the multiple kernel implementation is that for objective functions with a high level of branching, the branching effect in the parallel code can be avoided if a configuration with one thread per block is used. Even in some cases this configuration proved its efficiency [2], the main disadvantage is that the maximum number of blocks which can run in parallel at a time is limited by the number of stream multiprocessors of the GPU hardware.

In order to run all the threads in parallel and minimize the branching effect, the threads can be distributed on a configuration with the maximum number of blocks which can run in parallel. If the number of particles does not divide exactly to the number of blocks some threads will have to be invalidated. The invalidation is done based on thread ID which has to be smaller than the maximum number of particles in the swarm. If the IDs for the threads are calculated using the classical approach there might be cases when the hardware potential might not be used efficiently in order to minimize the branching, and entire blocks of threads might be invalidated (Fig. 2).

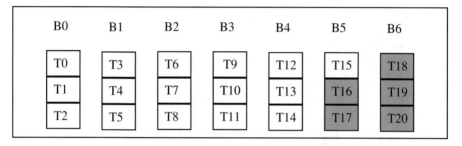

Fig. 2. Thread IDs in the classical approach (16 particles, 7 blocks, 3 threads per block).

In order to take advantage of the whole hardware potential and minimize the branching effect, while maintaining a full parallelism, this paper proposes a new Full Parallel Minimum Branching (FPMB) implementation. For the FPMB approach the threads are distributed as shown in Fig. 3. Comparing with the classical approach, in the FPMB implementation the invalidated threads are distributed between blocks, thus the number of active threads per block is smaller (for the given example only 2 blocks have 3 threads comparing with 5 in the classic approach) and as a result the branching probability is smaller.

B0	B1	B2	B3	B4	B5	B6
T0	T3	T6	T8	T10	T12	T14
T1	T4	T7	T9	T11	T13	T15
T2	T5	T16	T17	T18	T19	T20

Fig. 3. Thread IDs for the FPMB approach (16 particles, 7 blocks, 3 threads per block).

To implement such an allocation for the threads IDs a kernel function (`calcu-lateThreadsIDs`) was implemented. The kernel receives as a parameter and sets the threads IDs in a vector stored in the global memory of the GPU device. At first the function calculates the threads IDs for the blocks which will have all the threads active, using the classic approach. For the remaining blocks the number of active threads will be smaller (block dimension minus one), each of this blocks containing one thread which will be disabled and will be allocated an ID greater than or equal to the number of particles.

```
__global__ void calculateThreadsIDs (int *threadsIDs) {
  int tid; //thread ID

  if (blockIdx.x < PARTICLES_NUMBER % B) {
    tid = blockIdx.x * blockDim.x + threadIdx.x;
  } else {
    if(threadIdx.x < PARTICLES_NUMBER / B) {
      tid = blockIdx.x * (blockDim.x-1) + threadIdx.x;
    } else {
      tid = PARTICLES_NUMBER +
            (blockIdx.x - PARTICLES_NUMBER % B);
    }
  }
  threadsIDs[blockIdx.x * blockDim.x + threadIdx.x] = tid;
}
```

As opposed to the classical approach, in the FPMB implementation the threads IDs are calculated only once (not at each iteration) before the SPSO multi kernel main loop. The `calculateThreadsIDs` function is called once, thus the branching introduced by it is insignificant:

```
calculateThreadsIDs<<B,TpB>>(threadIDs);
... //initialization, evaluation, global best calculation
... //topology generation, local best calculation
for(int i=0; i < SPSO_ITERATIONS_NUMBER; i++) {
  moveParticles<<B,TpB>>(
        threadsIDs, particles, particleGB, varsMin, varsMax);
  //evaluateParticles also updates PB
  evaluateParticles<<B,TpB>>(threadsIDs, particles);
  findGlobalBest<<B,TpB>>(
        threadsIDs, particles, particleGB, improvedGB);
  generateTopology<<B,TpB>>(threadsIDs, particles, improvedGB);
  findLocalBest<<B,TpB>>(threadsIDs, particles);
}
```

where B is the bocks number and TpB is the threads number per block.

The kernels variables are global variables and they are stored on the device (GPU). The varMin, varMax arrays contain the domain limits (minimum and maximum values) for each search space coordinate. The variable improvedGB has a boolean type and is used to decide if the swarm topology will be changed (if the global best value is not improved at a certain iteration the generateTopology kernel is called). The swarm particles are stored in the particles variable, which is an array of type Particle:

```
typedef struct {
  double coords[PROBLEM_SIZE];
  double fitnessValue;
  double velocity[PROBLEM_SIZE];
  double gravityCenter[PROBLEM_SIZE];
  int indexLB;
  int neighbours[PARTICLES_NUMBER];
} Particle;
```

The moveParticles function computes the new particles positions, while the evaluate function computes the fitness value, updates the personal best (position and fitness value) for each particle. The functions called inside evaluate (paramsCorrection, objectiveFunction, findPersonalBest) are device functions which have the __device specifier. Each of these device functions is executed in parallel (just like evaluate) for all the swarm particles. The first function checks the coordinates restrictions (imposed by the problem) and, if is needed, changes the particle's coordinates to meet the constraints. The second function, the optimization problem (TEAM22 or Loney's solenoid), has a sequential implementation and computes the fitness value for a particle. Like all the other SPSO kernel functions, in the beginning, it computes the index for accessing the threadsIDs global variable, where will find its thread ID.

```
__global__ void evaluateParticles(
                          int *threadsIDs, Particle *particles){
  //calculate thread ID -- classic approach
  //int tid = blockIdx.x * blockDim.x + threadIdx.x;
  //obtain thread ID - proposed FPMB approach
  int tidIndex = blockIdx.x * blockDim.x + threadIdx.x
  int tid = threadsIDs[tidIndex];

  if(tid < PARTICLES_NUMBER) {
    parametersCorrection(&particles[tid]);
    particles[tid].fitnessValue -
                          objectiveFunction(particles[tid]);
    findPersonalBest(particles);
  }
}
```

The findGlobalBest updates the best particle of the swarm, and the improvedGB variable (to true or false if the fitness value for the best particle was or was not improved at the current step). The generateTopology creates a new topology (new connections between the swarm particles) if the global best value was not improved at the current step. Based on the new topology, the findLocalBest calculates the index of the local best for the neighborhood of each particle. The indexLB data field is then used to establish whether the particle is the best particle in its neighborhood, in order to choose the formula for determining the new particle's coordinates.

```
__global__ void findLocalBest(
                          int *threadIDs, Particle *particles){
  //calculate thread ID -- classic approach
  //int tid = blockIdx.x * blockDim.x + threadIdx.x;

  //obtain thread ID - proposed FPMB approach
  int tidIndex = blockIdx.x * blockDim.x + threadIdx.x
  int tid = threadsIDs[tidIndex];

  if(tid < PARTICLES_NUMBER) {
    particles[tid].indexLB = tid;
    for(int i = 0; i < PARTICLES_NUMBER; i++) {
      if(particles[tid].neighbours[i] == 1) {
        if(particles[i].fitnessValuePB
          < particles[particles[tid].indexLB].fitnessValuePB) {
          particles[tid].indexLB = i;
        }
      }
    }
  }
}
```

4   SPSO Parallelization – Pthreads Approach

In shared memory multiprocessor architectures, threads can be used to implement parallelism. POSIX Threads [25], usually referred as Pthreads, is a POSIX (Portable Operating System Interface) standard for threads [26] which defines an API implemented on many Unix like operating systems as Linux, Solaris, FreeBSD and MacOS.

In such operating systems, a process requires a significant amount of overhead, containing information about program resources and program execution state: process ID, user ID, environment, program instructions, registers, stack, heap, file descriptors, signal actions, shared libraries, inter-process communication tools (message queues, pipes, semaphores and shared memory), etc.

Unlike a process, a thread is an independent stream of instructions that can be scheduled to run by the operating system. In a Unix environment, a thread exists within a process, uses the process resources, and has its own independent flow of control. A thread duplicates only the essential resources needed to be independently schedulable: stack pointer, registers, scheduling properties (policy and priority), and set of pending and blocked signals. Because most of the overhead has already been accomplished through the creation of its process, a thread is lightweight when compared to the cost of creating and managing a process, and can be created with much less operating system overhead. Therefore managing threads requires fewer system resources than managing processes.

When running in shared-memory model, each thread has access to its on private data but also has access to the global (shared) memory. Because the threads belonging to a process share their resources, changes of global resources made by one thread will be seen by all threads. This is why the read /write operations to the same memory location require explicit synchronization, synchronization which can be implemented using mechanisms as barriers and mutexes.

Comparing to other parallelization options for multi-processor architecture with shared memory, like MPI or OpenMP, Pthreads was created to achieve optimum performance [27]. While MPI [28] and OpenMP [29] are simpler parallelization options (easier to use) requiring a smaller amount of work, Pthreads provides more flexibility and it offers more control over the parallelization.

4.1 Existing Approaches

Just as in the CUDA case, there is a significant number of PSO parallel implementations based on the shared memory multiprocessor architectures. While the optimization algorithms are used to solve a variety of applications most of the programs are based on MPI and OpenMP because of the implementation simplicity [30–32].

In [33] the authors use a PSO OpenMP implementation to design a class E amplifier. The speed up obtained by parallelization is up to 5 times. In [34] a MPI implementation is used for solving the optimum capacity allocation of distributed generation units and an 3 times acceleration is obtained comparing to the serial implementation.

In [35] the authors make a comparison between a PSO CUDA implementation and a PSO MPI implementation used to solve an optimization problem from the area of power electronics. Both implementations are faster than the sequential PSO, the GPGPU CUDA implementation being 32 times faster than the multiprocessor MPI implementation.

In our opinion there is no single best parallel implementation strategy for the PSO based algorithms. As we will see from our simulations results, the performances depend on many factors as PSO parameters and especially the objective function to be optimized and its implementation features (like the code complexity and the level of branching).

4.2 Proposed POSIX Threads Implementation

Just as in the CUDA implementation, in the Pthreads case we implemented the behavior of each particle in the swarm using a dedicated thread. The threads management is done explicitly. The threads are created and launched using pthread_create library function. The function receives as parameters a reference to the thread, thread attributes (NULL means defaults are applied), the function to be executed by the thread, and the thread ID:

```
pthread_t threads[PARTICLES_NUMBER];
int i, tid[PARTICLES_NUMBER];
...
for(i = 0; i < PARTICLES_NUMBER; i++) {
  tid[i] = i;
  pthread_create(&threads[i],NULL,&jobForOneThread,&tid[i]));
}
```

After creation, each thread executes the code corresponding to the function jobForOneThread. The function contains the SPSO main loop and performs the basic operations: particle movement and evaluation, personal/global best calculation, reset/generate new topology, and local best calculation:

```
void* jobForOneThread (void *params) {
  int tid = *((int*)params);
  ...
  for(i = 0; i < SPSO_ITERATIONS_NUMBER; i++)   {
    moveEvaluateUpdatePersonalBest(tid); barrier();
    findGlobalBest(tid); barrier();
    generateTopology(tid); barrier();
    findLocalBest(tid); barrier();
  }
  ...
}
```

The variable passed to the SPSO basic functions is only the thread ID. The code of these functions is the same as in the CUDA implementation. The variables varMin, varMax, improvedGB, particles (which were passed as function parameters in the CUDA implementation and were stored in the GPU device memory) are now global variables stored in the host computer memory, all threads having access to them.

The particles synchronization (necessary after each operation) is achieved using a barrier mechanism based on the pthread_barrier_wait library function:

```
void barrier() {
  int rc = pthread_barrier_wait(&barr);
  if(rc != 0 && rc != PTHREAD_BARRIER_SERIAL_THREAD) {
    printf("Could not wait on barrier\n"); exit(-1);
  }
}
```

The barr parameter is a variable of type pthread_barrier_t which contains several data members as the current number of threads reaching the barrier, the size of the barrier (the necessary number of threads to unlock the barrier), a mutex (for exclusive access to data members), etc. The variable is defined and initialized before the thread creation and execution using the pthread_barrier_init function:

```
// Barrier initialization - before the thread creation loop
pthread_barrier_t barr;
if(pthread_barrier_init(&barr, NULL, PARTICLES_NUMBER)) {
    printf("Could not init barrier\n"); exit(1);
}
```

5   Electromagnetic Problems

The parallel implementations were tested on two benchmark problems defined by the computational electromagnetics community.

5.1   The TEAM22 Benchmark Problem

Two coaxial coils carry current with opposite directions (Fig. 4), operate under superconducting conditions and offer the opportunity to store a significant amount of energy in their magnetic fields, while keeping within certain limits the stray field [6].

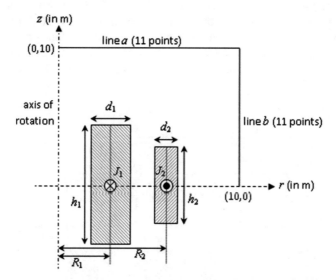

Fig. 4. TEAM22 problem configuration.

An optimal design of the device should therefore couple the value of the energy E to be stored by the system with a minimum stray field B_{stray}. The two objectives are combined into one objective function:

$$OF = \frac{B_{stray}^2}{B_{norm}^2} + \frac{|E - E_{ref}|}{E_{ref}}, \quad B_{stray}^2 = \frac{\sum_{i=1}^{22} |B_{stray,i}|^2}{22} \tag{11}$$

where $E_{ref} = 180$ MJ, and $B_{norm} = 3$ μT.

The objective function has as parameters, the radii (R_1, R_2), the heights (h_1, h_2), the thicknesses (d_1, d_2) and the current densities (J_1, J_2). Besides domain restrictions, the problem must take into account the following conditions: the solenoids do not overlap each other $(R_1 + d_1/2 < R_2 - d_2/2)$, and the superconducting material should not violate the quench condition that links together the value of the current density and the maximum value of magnetic flux density $(|J| \leq (-6.4 \cdot |B| + 54.0) \text{ A/mm}^2)$. It is a constrain imposed to the current densities.

The evaluation method of the objective function is based on the Biot-Savart-Laplace formula in which the elliptic integrals are computed by using the King algorithm and numerical integration. Moreover, the optimization problem is reformulated as a one with six parameters, since for a given geometry and a stored energy, the values of the current densities can be computed by deterministic quadratic optimization as in [9].

5.2   The Loney's Solenoid Problem

The Loney's solenoid benchmark problem, formulated in [10] consists of a main coil (Fig. 5), with given dimensions $(r_1 = 11 \text{ mm}, r_2 = 29 \text{ mm}, h = 120 \text{ mm})$ and two identical correction coils, having fixed radii $(r_3 = 30 \text{ mm}, r_4 = 36 \text{ mm})$. A constant current flows through the coils such that they current density is the same. The aim is to produce a constant magnetic flux density in the middle of the main coil. The parameters to be optimized are the length of the correction coils (s) and the axial distance between them (l).

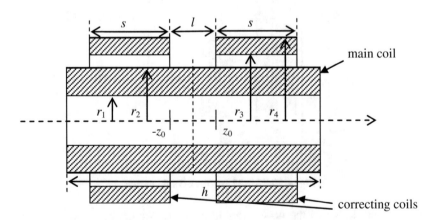

Fig. 5. Loney's solenoid problem configuration.

The objective function is of minmax type, i.e. minimize the maximum difference between the values of the magnetic flux density along a straight segment in the middle of the main solenoid, i.e. minimize $(B_{max} - B_{min})/B_0$, where B_0 is the magnetic field density in the middle of the main coil $(r = 0, z = 0)$. The maximum and minimum values are sought along the segment $[-z_0, z_0]$, where $z_0 = 2.5$ mm. Tests done by the authors of this benchmark revealed that the problem is non convex and ill conditioned [36]. The electromagnetic field problem is easily solved, in a magnetostatic regime, by discretizing the coils in elementary coils without thickness and by applying well known analytical formulas for the field along the solenoid axis.

6 Results

To solve the electromagnetic optimization problems two parallel SPSO implementations have been used, a multiple kernel CUDA implementation and a Pthreads implementation. In both implementations one thread is mapped to one particle of the swarm.

The objective functions for the TEAM 22 and Loney's solenoid have sequential implementations and they were written in C. For a given set of parameters, the evaluation of one objective function in case of TEAM22 problem consists in executing hundreds of thousands of lines of code (thousands of subroutines calls) with a very high level of branching, while in the case of Loney's solenoid one evaluation consists of hundreds lines of code with a lower level of branching.

The CUDA SPSO code was tested on a NVIDIA M2070 GPU with 448 cores, compute capability 2.0 and 1.13 GHz core processors. The Pthreads SPSO code was tested on a multiprocessor hardware architecture with two Intel Xeon X5650 CPUs (2.67 GHz), each processor with 6 cores and each core being able to run in parallel 2 independent threads. In total only 12 threads can run in parallel at a time on the multiprocessor architecture, significantly smaller than in the GPU case.

6.1 GPGPU Configurations

For the parallel GPGPU implementation three different configurations have been tested: a full warp configuration (FW), a single thread per block configuration (ST), and a new full parallel minimum branching configuration (FPMB).

The full warp configuration is a classic approach and has the main advantage the optimal use of the GPU hardware resources per multiprocessor (no core is idle). For this configuration the number of blocks equals the swarm size divided by 32. The main problem of this approach is the highest level of branching from all configurations.

The second configuration has one thread in each block, with the number of blocks being equal with the swarm size. The main advantage of the configuration is the avoidance of branching, while the main problem is that not all the threads can run in parallel in the same time, the limit being imposed by the number of the multiprocessors.

The third configuration, is the new FPMB configuration especially proposed for problems affected by branching. The configuration distributes the threads between blocks such as all the threads run in parallel in the same time while maintaining the lowest possible branching level (from the configurations which run all threads in parallel in the same time). The number of blocks equals the number of multiprocessors, while the maximum number of threads per block depends on the swarm size, being equal to the number of particles divided to the number of blocks. Because in some cases the number of particles can not be precisely divided to the number of blocks, as the number of multiprocessors is hardware specific, some threads are invalidated, as described in Sect. 3.

Tables 1 and 2 present the average execution time of the GPGPU – SPSO implementation for 30 independent runs (tests), for each configuration, and for different swarm sizes of the SPSO algorithm. For each run (test) the stop criteria was the maximum iterations number corresponding to 2560 evaluations of the objective function.

Table 1. Average execution times for TEAM 22 problem.

Swarm size (S)		32	64	128
GPGPU configuration	FW (32 TpB)	491 s	312 s	179 s
	FPMB (3/5/10 TpB)	**278** s	233 s	163 s
	ST (1TpB)	327 s	**198** s	**144** s

Table 2. Average execution times for Loney's solenoid problem.

Swarm size		32	64	128
GPGPU configuration	FW (32 TpB)	25 ms	16 ms	11 ms
	FPMB (3/5/10 TpB)	**16** ms	**10** ms	**6.5** ms
	ST (1TpB)	21 ms	15 ms	10 ms

For the TEAM22 electromagnetic optimization problem the proposed FPMB out-performs the FW configuration for each swarm size. When compared with the ST configuration the FPMB approach is better only for small swarm sizes. The explanation is that the TEAM22 objective function implementation is complex and when the swarm size increases, the number of the threads per block in the FPMB configuration increases and the branching significantly influences the performance.

For the Loney's solenoid problem, the proposed FPMB configuration is the most suitable approach outperforming all the other configurations for each swarm size. The implementation of the objective function for Loney's solenoid is less complex than in the TEAM22 problem case, and as a result the increase of the swarm size (which means the increase of the number of threads per block in the FPMB configuration) does not lead to a high level of branching. Although the branching is present (the ST config-uration performs similar with the FW configuration), it does not have a dramatic impact over the performances of the FPMB configuration when the swarm size increases (as in the case of TEAM22).

6.2   GPGPU vs. Pthreads

Tables 3 and 4 present the average execution time of the Pthreads – SPSO imple-mentation for 30 independent runs (tests) for different swarm sizes. In the same time the tables present the best values obtained with the GPGPU – SPSO implementation, from all three tested configurations. Just as in the GPGPU implementation case, for each run (test) the stop criteria was the maximum iterations number corresponding to 2560 evaluations of the objective function.

Table 3. Average execution times for TEAM 22 problem.

Swarm size		32	64	128
Algorithm	GPGPU – SPSO	278 s	198 s	144 s
	Pthreads – SPSO	**19** s	**17** s	**15** s

Table 4. Average execution times for Loney's solenoid problem.

Swarm size		32	64	128
Algorithm	GPGPU – SPSO	**16** ms	**10** ms	**6.5** ms
	Pthreads – SPSO	71 ms	79 ms	82 ms

For the TEAM 22 optimization problem the Pthreads implementation is faster than the CUDA implementation for each swarm size. The speed up obtained for Pthreads with respect to GPGPU implementation is from 9 times, in the case of 128 particles, to 17 times, in the case of 32 particles.

Even if in the CUDA case the number of threads running in parallel in the same time is higher than in the Pthreads case, the Pthreads implementation is faster because of the complexity of the TEAM22 objective function implementation (high level of branching and large number of instructions). The main explanation is that the GPU cores have lower clock rates, no branch prediction and no speculative execution comparing with the multiprocessor cores.

For the Loney's solenoid problem the situation is reversed, the CUDA implementation being the fastest. The speedup for GPGPU with respect to Pthreads implementation is from 4 times, when the swarm has 32 particles, to 10 times, when the number of particles is 128. The explanation once again is related to the objective function implementation, which in this case has a much lower number of instructions and a lower branching level comparing with the TEAM22 case. The advantages of the multiprocessor architecture (the higher clock rates, the bigger cache level, the branch prediction, the speculative execution, etc.) can not compensate the disadvantage of the larger number of threads running in parallel on the GPU architecture.

In terms of solution fitness (Tables 5 and 6) the results obtained with the parallel Pthreads implementation are slightly better than those obtained with the CUDA code, for both electromagnetic optimization problems. For both implementations the random numbers necessary for the SPSO algorithm are generated at each step, on host in the case of Pthreads and on device/GPU in the case of CUDA.

For the Loney's solenoid problem the best performances are offered when the size of the swarm is small (32 particles), for both implementations. For TEAM 22 benchmark problem the optimum swarm size is between 32 and 64 when Pthreads

Table 5. Objective function values and standard deviation for TEAM 22.

Algorithm	Swarm size	Min - best of value	Max - best of value	Mean - best of value	Standard deviation
GPGPU – SPSO	32	3.15 E–3	17.40 E–3	6.49 E–3	3.89 E–3
	64	3.53 E–3	11.30 E–3	5.83 E–3	2.16 E–3
	128	3.37 E–3	9.11 E–3	6.37 E–3	1.74 E–3
Pthreads – SPSO	32	**3.06** E–3	8.46 E–3	**5.21** E–3	1.47 E–3
	64	3.34 E–3	**8.09** E–3	5.22 E–3	**1.23** E–3
	128	3.75 E–3	12.24 E–3	6.89 E–3	1.93 E–3

Table 6 Objective function values and standard deviation for Loney's solenoid.

Algorithm	Swarm size	Min - best of value	Max - best of value	Mean - best of value	Standard deviation
GPGPU – SPSO	32	1.31 E–8	1.61 E–8	1.52 E–8	**0.06** E–8
	64	1.51 E–8	2.07 E–8	1.66 E–8	0.15 E–8
	128	1.49 E–8	6.61 E–8	3.32 E-8	1.56 E–8
Pthreads – SPSO	32	**1.25** E–8	**1.59** E–8	**1.51** E–8	0.07 E–8
	64	1.31 E–8	2.44 E–8	1.67 E–8	0.19 E–8
	128	1.34 E–8	18.63 E–8	3.84 E–8	3.29 E–8

implementation is used, while for the CUDA implementation it does not seem to be an optimal size (32 offers best solution, 64 best mean, 128 best standard deviation).

7   Conclusions

The present paper have studied two of the most popular parallelization techniques, GPGPU and POSIX threads for multiprocessor architectures, for accelerating a SPSO algorithm to solve optimization problems from electromagnetism. The aim of the paper was to provide insights on what parallelization strategy should be adopted taking into consideration the implementation features of the objective function. In order to achieve this and find the most suitable approach, the SPSO parallel implementations have been tested on two different electromagnetic problems: TEAM22 and Loney's solenoid. While TEAM22 has a complex implementation, with a high level of branching, and needs a large number of subroutines calls for one evaluation of the objective function, the Loney's solenoid has a simpler implementation, with a low/medium level of branching.

Firstly, several GPGPU configurations have been tested and compared, and a new full parallel minimum branching (FPMB) implementation was proposed. The proposed implementation proved to be the most efficient for problems with a less complex implementation (such as Loney's solenoid). In the same time the configuration is suitable for problems with a more complex implementation (such as TEAM22) but only when the number of particles is low.

Secondly, the best GPGPU results were compared with results provided by a Pthreads implementation. For the TEAM22 benchmark the fastest approach has been the implementation for multiprocessor architecture. The Pthreads implementation outperformed the GPGPU CUDA based implementation up to 17 times. In the case of Loney's solenoid the fastest solution was the GPGPU implementation. The CUDA based implementation running in a full parallel minimum branching configuration was up to 10 times faster.

Regarding solution fitness, the best approach was the implementation based on Pthreads, but the solutions supplied by the CUDA implementation are very close, the difference being insignificant. A priori generation of the random numbers on host machine, followed by a transfer to the GPU device, could further improve the solution quality for CUDA implementation. In most cases, the best solutions have been achieved for a small size of the particles swarm.

As we have shown, there is no single best parallelization method and the performances dependent of the objective problem to be solved, and especially its implementation features such as level of branching, number of instructions, the necessary number of subroutines calls, recursivity, etc. While for problems with complex implementations (such as TEAM 22), the most efficient approach is based on Pthreads, in the case of problems with a less complex implementation, like Loney's solenoid, the best approach is the GPGPU.

Acknowledgements. This work has been supported by the Politehnica University of Bucharest in the frame of the UPB Grant of Excellence, no. 254/2016, the Romanian Government in the frame of the PN-IIPT-PCCA-2011-3 program, no. 5/2012 (managed by CNDI–UEFISCDI, ANCS), and in the frame of the RO-BE bilateral project, no. xx/2016.

References

1. Takagi, T., Fukutomi, H.: Benchmark activities of eddy current testing for steam generator tubes. In: Pavo, J., Albanese, R., Takagi, T., Udpa, S.S. (eds.) Electromagnetic Nondestructive Evaluation (IV), vol. 17, pp. 235–252. IOS Press, Amsterdam (2001)
2. Duca, A., Duca, L., Ciuprina, G., Yilmaz, A.E., Altinoz, O.T.: PSO Algorithms and GPGPU Technique for Electromagnetic Problems. In: International Workshops on Optimization and Inverse Problems in Electromagnetism (OIPE), Delft, The Netherlands (2014)
3. Duca, A., Rebican, M., Janousek, L., Smetana, M., Strapacova, T.: PSO based techniques for NDT-ECT inverse problems. In: Capova, K., Udpa, L., Janousek, L., Rao, B.P.C. (eds.) Electromagnetic Nondestructive Evaluation (XVII), vol. 39, pp. 323–330. IOS Press, Amsterdam (2014)
4. Li, Y., Udpa, L., Udpa, S.S.: Three-dimensional defect reconstruction from eddy-current NDE signals using a genetic local search algorithm. IEEE Trans. Magn. **40**(2), 410–417 (2004)
5. Ciuprina, G., Ioan, D., Munteanu, I.: Use of intelligent-particle swarm optimization in electromagnetics. IEEE Trans. Magn. **38**(2), 1037–1040 (2002)
6. Ioan, D., Ciuprina, G., Szigeti, A.: Embedded stochastic-deterministic optimization method with accuracy control. IEEE Trans. Magn. **35**, 1702–1705 (1999)

7. Duca, A., Tomescu, F.M.G.: A distributed hybrid optimization system for NDET inverse problems. In: Proceedings of the International Symposium of Nonlinear Theory and Its Applications (NOLTA), Bologna, Italy, pp. 1059–1062 (2006)

8. Chen, Z., Rebican, M., Yusa, N., Miya, K.: Fast simulation of ECT signal due to a conductive crack of arbitrary width. IEEE Trans. Magn. **42**, 683–686 (2006)

9. TEAM22 benchmark problem (2015). http://www.compumag.org/jsite/team.html

10. DiBarba, P., Gottvald, A., Savini, A.: Global optimization of Loney's solenoid: a benchmark problem. Int. J. Appl. Electromagn. Mech. **6**(4), 273–276 (1995)

11. Kennedy, J., Eberhart, R.C.: Particle swarm optimization. In: Proceedings of IEEE International Conference on Neural Networks, pp. 1942–1948 (1995)

12. Bratton, D., Kennedy, J.: Defining a standard for particle swarm optimization. In: Proceedings of the IEEE Swarm Intelligence Symposium (2007)

13. Sun, J., Fang, W., Palade, V., Wua, X., Xu, W.: Quantum-behaved particle swarm optimization with Gaussian distributed local attractor point. Appl. Math. Comput. **218**, 3763–3775 (2011)

14. Pan, Q.K., Tasgetiren, M.F., Liang, Y.C.: A discrete particle swarm optimization algorithm for the no-wait flowshop scheduling problem with makespan and total flowtime criteria. J. Comput. Oper. Res. **35**, 2807–2839 (2008)

15. Clerc: Standard particle swarm optimization. open access archive HAL (2012). http://clerc.maurice.free.fr/pso/SPSO_descriptions.pdf

16. Nvidia CUDA C programming guide (2015) http://docs.nvidia.com/cuda/cuda-c-programming-guide

17. Zhou, Y., Tan, Y.: GPU-based parallel particle swarm optimization. In: Proceedings of the IEEE Congress on Evolutionary Computation (CEC 2009), pp. 1493–1500 (2009)

18. Mussi, L., Cagnoni, S.: Particle swarm optimization within the CUDA architecture. In: Proceedings of the 11th Annual Conference on Genetic and Evolutionary Computation (GECCO 2009) (2009)

19. Mussi, L., Cagnoni, S., Daolio, F.: GPU-Based Road Sign Detection using Particle Swarm Optimization. In: Proceedings of the Ninth International Conference on Intelligent Systems Design and Applications (ISDA 2009), pp. 152–157 (2009)

20. Bastos-Filho, C.J., Oliveira, M.A., Nascimento, D.N., Ramos, A.D.: Impact of the random number generator quality on particle swarm optimization algorithm running on graphic processor units. In: Proceedings of the 10th International Conference on Hybrid Intelligent Systems, pp. 85–90 (2010)

21. Solomon, S., Thulasiraman, P., Thulasiram, R.: Collaborative multi-swarm PSO for task matching using graphics processing units. In: Proceedings of the 13th Annual Conference on Genetic and Evolutionary Computation (GECCO 2011) (2011)

22. Hung, Y., Wang, W.: Accelerating parallel particle swarm optimization via GPU. Optim. Methods Softw. **27**(1), 33–51 (2012)

23. Castro-Liera, I., Castro-Liera, M., Antonio-Castro, M.: Parallel particle swarm optimization using GPGPU. In: Proceedings of the 7th Conference on Computability in Europe (CIE-2011) (2011)

24. Mussi, L., Daolio, F., Cagnoni, S.: Evaluation of parallel particle swarm optimization algorithms within the CUDA architecture. Inf. Sci. **181**(20), 4642–4657 (2011)

25. POSIX Threads (2015). http://en.wikipedia.org/wiki/POSIX_Threads

26. POSIX Threads standard (2008). http://standards.ieee.org/findstds/standard/1003.1-2008.html

27. Pthreads tutorial (2015). https://computing.llnl.gov/tutorials/pthreads

28. MPI (2015). http://en.wikipedia.org/wiki/Message_Passing_Interface

29. OpenMP (2015). http://www.openmp.org

30. Wang, D., Wang, D., Yan, Y., Wang, H.: An adaptive version of parallel MPSO with OpenMP for Uncapacitated Facility Location problem. Control and Decision Conference (CCDC), pp. 2387–2391 (2008)
31. Liu, Z.-H., Zhao, J.-X., Tan,W.: Multi-core based parallelized cooperative PSO with immunity for large scale optimization problem. In: Conference on Cloud Computing and Internet of Things, pp. 96–100 (2014)
32. Han, X.G., Wang, F., Fan, J.W.: The research of PID controller tuning based on parallel particle swarm optimization. Appl. Mech. Mater. Artif. Intell. Comput. Algorithms **433**, 583–586 (2013)
33. Tanji, Y., Matsushita, H., Sekiya, H.: Acceleration of PSO for designing class E amplifier. In: International Symposium on Nonlinear Theory and Its Applications (NOLTA), pp. 491–494 (2011)
34. Thomas, R., Pattery, J.M., Hassaina, R.: Optimum capacity allocation of distributed generation units using parallel PSO using message passing interface. Int. J. Res. Eng. Technol. **2**, 216–219 (2013)
35. Roberge, V., Tarbouchi, M.: Comparison of parallel particle swarm optimizers for graphical processing units and multicore processors. Int. J. Comput. Intell. Appl. **12**, 1350006 (2013)
36. DiBarba, P., Savini, A.: Global optimization of Loney's solenoid by means of a deterministic approach. Int. J. Appl. Electromagn. Mech. **6**(4), 247–254 (1995)

Bio-inspired Strategies for the Coordination of a Swarm of Robots in an Unknown Area

Nunzia Palmieri[1,2(✉)], Floriano de Rango[1], Xin She Yang[2], and Salvatore Marano[1]

[1] Department of Computer Engineering Modeling, Electronics and System Science, University of Calabria, Rende (CS), Italy
{n.palmieri,derango,marano}@dimes.unical.it
[2] School of Science and Technology, Middlesex University, The Burroughs, London, UK
{n.palmieri,x.yang}@mdx.ac.uk

Abstract. This paper addresses the problem of searching mines in an unknown area and disarming them in a cooperative manner. We describe two bio-inspired mechanisms that allow the robots to initiate the coordination with other robots when a mine is discovered. We model this problem as a multi-objective exploration and disarming problem. Specifically we propose a modified version of the Ant Colony Optimization (ATS-RR) and the Firefly Algorithm (FTS-RR). The proposed approaches have been implemented and evaluated in several simulated environments varying the parameter of the problems in term of team sizes, the number of mines disseminated in the area, the dimension of area. Our approaches have been implemented in simulation environments and have been compared with Particle Swarm Optimization (PSO). The results demonstrate the efficiency of the FTS-RR over others.

Keywords: Swarm intelligence · Bio-inspired algorithm · Multi-robot system · Coordination

1 Introduction

In the past new years, the attention of researchers fovuses on the idea of creating groups of mobile robots that are able to collaborate in order to accomplish one or more predefined tasks such as aerial surveillance and reconnaissance, unmanned search and rescue, exploration and so on. Multi-robot systems can provide improved performance, fault-tolerance and robustness in those tasks through parallelism and redundancy. The key issue concerning collective robotics is how to specify the rules of behavior and interaction at the level of individual robots such that coordination can be achieved automatically at the global level. This is called the coordination problem [1, 2].

A central aspect of the coordination of the swarm is to ensure that the robots are distributed in efficient manner into the area in order to ensure rapid and efficient completion of the tasks.

Swarm robotics is a new approach to the coordination of large numbers of relatively simple robots. The approach takes its inspiration from the system-level functioning of

© Springer International Publishing AG 2017
J.J. Merelo et al. (eds.), *Computational Intelligence*, Studies in Computational Intelligence 669,
DOI 10.1007/978-3-319-48506-5_6

social insects which demonstrate three desired characteristics for multi-robot systems: robustness, flexibility and scalability [3].

Algorithms in swarm robotics mostly rely on cooperation and simple interactions between robots, rather than on complex individual behaviours that require powerful sensory capabilities. Several researchers have developed simple information sharing techniques for multi-robot systems using simple, nature inspired models such as stigmergy to enable coordination among the robots in dynamic environments [24].

In this paper, we address a problem of the coordination of a swarm of robots. We consider a situation where a swarm of mobile robots is deployed in an unknown, mined area where some mines are disseminated at an unknown locations. The objective is to explore overall area trying to distribute the robots in the different locations in order to minimize the exploration time and the time to disarm all mines and the number of accesses in the cells.

The exploration strategy has been investigated in previous work [4] here we are interested to investigate the coordination strategy for disarming the mines in better way. Since a mine needs to be disarmed by a certain number of robots, there is an issue on how the robots inform the swarm about the mines and try to recruit the necessary number of robots so as to disarm the mine safely. The robots that receive more recruitment requests need to decide which location that robot will go to. The question we address is how this can be done efficiently in a distributed way.

For this problem we propose two bio- inspired strategies. The first is based on indirect communication through pheromone based on inspiration from the Ant Colony Optimization; when a robot detect a mine it sprays some pheromone, like the ant in nature, at certain distance, the other robots that sense the substance are attracted and try to reach the mine locations following this scent.

The second mechanism is based on the explicit communication of structured information. When a mine is detected, the robot, becomes like a firefly, sends the information about its location to the others in the wireless range; the robots try to reach the location according to the Firefly algorithm. If a robot receives more than one request it chooses the best firefly that is the firefly at minimum distance.

We compare both mechanisms in terms of their efficiency, evaluating the performance in comparison with the well-known PSO.

In our collective construction task, there are some mines randomly distributed in an unknown area. The robots should first search for these mines individually, but for the disarming task, multiple robots needed to work together. The problem is not a pure exploration: on the one hand, it is required for robots to cover as much area as possible in the minimum amount of time, avoiding any overlapping area. On the other hand, the problem needs to allocate more robots in the same area to disarm a mine. The problem is a bi-objective optimization problem where robots have to make decisions whether to explore the area or to help other robots to disarm the detected mines.

Because the problem of the unknown lands with the constraint to disarm mine is a NP hard problem, we proposed a combined approach using two bio-inspired meta-heuristic approaches such as Ant Colony Optimization (ACO) and Firefly algorithm (FA) to perform the coordination task among robots.

Basically, each robot consists of two phases during the task: searching and disarming. When there is no detected mine, the robot status should be in the searching phase, where robots are exploring the area and searching for mines, taking into account the quantity of pheromone perceived in the cells. Once mine is detected either by the robot itself or by its neighbours, the robot status should be switched to the disarming phase, under specific conditions. The strategy for the exploration task is designed according to the main ideas of the ant system [5]. While the robots navigate, they deposit a specific substance, the pheromone (the analogue of the pheromone in biological ant systems), into the environment. At each time/iteration, each robot receives information from the pheromone and makes a navigation decision: it chooses the area in which it perceives a less quantity of pheromone because this area has a greater probability to be unexplored [4, 6]. The algorithm for exploration has been previously validated and this paper presents the analysis of the recruiting strategies in order to disarm the mines. The first is based on the exploration strategy and uses the pheromone to attract the robots in the area where the mine is placed. The second strategy is based on the new recent bio-inspired technique called Firefly Algorithm (FA) where the robots that detect the mines become the fireflies and try to attract the other robots according to a certain formula [7–9] These strategies were compared to the well known Particle Swarm Optimization in order to evaluate the better coordination mechanism for this problem. This contribution can be effective because the recruiting strategy can affect the exploration task and the overall bi-objective exploring and recruiting tasks.

The paper is organized as follows. Section 2 introduces the related work. Section 3 present the problem statement. In Sect. 4 we present the distributed cooperative algorithms for a multi-robot disarming task. Section 5 presents the simulation results using a java-based platform. To conclude the paper, Sect. 6 outlines the main research conclusions and discusses topics for future work.

2   Related Work

Multi-robot exploration has received much attention in the research community. Swarm robotic searching algorithm is one of the most concerns of the researchers for solving those basic tasks. The swarm intelligence shows a great ability in scalability, flexibility and robustness and it is suitable for real life applications with the aid of various existing strategies. Within the context of swarm robotics, most work on cooperative exploration is based on biologically behaviour and indirect stigmergic communication (rather than on local information, which can be applied to systems related to GPS, maps, wireless communications). This approach is typically inspired by the behaviour of certain types of animals, like the ants, that use chemical substances known as pheromone to induce behavioural changes in other members of the same species [10–14].

Other authors experimented with chemical pheromone traces, e.g. using alcohol [15] or using a special phosphorescent glowing paint [16]. Another approach is the pheromone robotics where robots spread out over an area and indicate the direction to a goal robot using infrared communication [17]. In our approach, during the exploration the robots sign/mark the crossed cell through the scent that can be detected by the other robots; the robots choose the cell that has the lowest quantity of substance to allow the exploration of the unvisited cells in order to cover the overall area in less time [4, 6].

The self-organizing properties of animal swarms such as insects have been studied for better understanding of the underlying concept of decentralized decision-making in nature, but it also gave a new approach in applications to multi-agent systems engineering and robotics. Bio-inspired approaches have been proposed for multi-robot division of labour in applications such as exploration and path formation, or cooperative transport and prey retrieval. Within the context of swarm robotics, most work on cooperative tasks is based on social behaviour like Ant Colony Optimization [5, 18] Particle Swarm Optimization [19–21] Bee Algorithm [22].

For sharing information and accomplishing the tasks there are, basically, three ways of information sharing in the swarm: direct communication (wireless, GPS), communication through environment (stigmergy) and sensing. More than one type of interaction can be used in one swarm, for instance, each robot senses the environment and communicates with their neighbour. Balch [23] discussed the influences of three types of communications on the swarm performance and Tan [3] presents an accurate analysis of the different type of communication and the impact in the behaviour of swarm.

In this paper, we have considered the spatial and temporal dispersion of the pheromone to make the scenario more realistic [4]. While walking, the robots leave pheromone, which marks the cells they took. This chemical substance can be detected by other robots. After a while, the concentration of pheromone decreases due to the evaporation and diffusion associated with the distance and with the time; in this way we can allow continuous coverage of an area via implicit coordination. The other robots, through proper sensors, smell the scent in the environment and move in the direction with a minimum amount of pheromone that corresponds to an area less occupied and probably an unexplored area. On the other hand, in order to deactivate the mines, the first robot that detects a mine (recruiter) in a cell, sprays another scent, different from the pheromone used for the exploration, perceived by the robots; in this case the robots move into the cells with a higher concentration of pheromone and reach the area where to deactivate the mines. In this attraction strategy of the recruiter, another recent and novel bio-inspired approach inspired by other insects such as fireflies has been investigated in this work so as to see the effectiveness of the algorithm and potential use of different insect behaviour on the robot coordination task and their performance. The algorithm inspired by fireflies is called Firefly algorithm (FA).

3   Problem Definition

We consider an environment assuming that it is discretized into equally spaced cells that contains a certain number of mines. Each cell has the potential to consider three states: free, occupied by mine, occupied by robot. Robots can move among cells and they can have just local information about robots (neighbors) or regions to explore (neighbor cells) in order to provide a scalable strategy. The considered scenario is presented under this assumption:

1. The search space is unknown a priori by the agents.
2. A task corresponds to a mine detection and disarm a mine. The distribution of the mines that need to be disarmed by agents is not known.

3. A single robot is only capable of discovering and partially executing tasks; the disarming task can be completed only if multiple agents work together.
4. To recruit the other agents required to complete a task, an agent that discovers a mine communicates to the others, this communication has to be done in a distributed manner without using a central location or shared memory to facilitate information exchange among agents.
5. The robots are equipped with proper sensors that are able to deposit and smell the chemical substances (pheromones) leaved by the other robots; for exploration task they make probabilistic decision based on amount of pheromone in the cells. The exploration strategy is the same for the recruiting strategies.
6. The robots are equipped with proper sensors to detect the mines.
7. The robots can move on a cell-by-cell basis to explore new cells or to go towards the mine.

The robots during the exploration can spray a scent (pheromone) into the cells to support the navigation of the others. In the algorithm, the robots decide the direction of the movement relying on a probabilistic law inherited by swarm intelligence and swarm robotics techniques. The scent evaporates not only due to diffusion effects in the time, but also in the space according to the distance; this allows a higher concentration of scent in the cell where the robot is moving and a lower concentration depending on the distance [4].

Let M be the matrix of size $m \times n$ representing the unknown area of size $m \times n$. Let $M(i,j)$ be a cell in the matrix with row i and column j. Let z be the number of mines on a set MS to distribute on the grid in a random fashion (e.g., it is applied a uniform distribution on X and Y axes). The MS set is characterized by the coordinates of the mines. For example, $MS = \{(3,4),(5,10),(7,12)\}$ indicates that there are 3 mines in the area with the coordinates (3, 4), (5, 10) and (7, 12). The robots can be placed on the same initial cell or can be randomly distributed on the grid area. It is assumed that each robot in a cell $M(i,j)$ can move just in the neighbor cells through discrete movements. Let t_e be the time necessary for a robot to consider a cell, and let t_d be the time necessary to disarm a mine once it has been detected. It is assumed that a fixed number of robots (rd_{min}) are necessary to disarm a mine; this means that for the exploration task robots can be distributed among the area because each robot can independently explore the cells, whereas for the mine detection, more robots need to be recruited in order to perform the task. x_{ij} is a variable representing the number of robots (accesses) that passed through the cell (i,j).

For the problem, we define an bi-objective function as both the time to detect and the disarming the mine through the exploration on the overall grid.

$$\min\left(\sum t_e + \sum_{k=0}^{z} t_{d,k}\right) \text{ and } \min \sum_{i=1}^{m}\sum_{j=1}^{n} x_{ij} \tag{1}$$

subject to

$$\sum_{i=1}^{m}\sum_{j=1}^{n} x_{ij} \geq 1 i = 1\ldots m; j = 1\ldots n \ \ with (i,j) \in M$$

$$\sum_{i=1}^{m}\sum_{j=1}^{n} x_{ij} \geq rd_{min} \ with \ (i,j) \in MS$$

The law used by the robots to choose the cells during the movements is presented below [4].

We consider a robot in a cell s and it will attribute to the set of next cells v_i following a probability as:

$$p(v_i|s) = \frac{\left[\tau_{v_i,t}\right]^{\varphi} \cdot \left[\eta_{v_i,t}\right]^{\vartheta}}{\sum_{i \in N(s)} \left[\tau_{v_i,t}\right]^{\varphi} \cdot \left[\eta_{v_i,t}\right]^{\vartheta}}, \forall \ v_i \in N(s) \tag{2}$$

where $(p(v_i|s)$ represents the probability that the robot, that is in the cell s, chooses the cell v_i; $N(s)$ is the set of neighbors to the cells, $\tau_{vi,t}$ is the amount of pheromone in the cell v_i; $\eta_{yi,t}$ is the heuristic parameter introduced to make the model more realistic. In addition, φ and θ are two parameters which affect respectively the pheromone and heuristic values.

Taking into account the spatial dispersion of the scent and the temporal dispersion in the amount of pheromone in the cell v where the robot will move during the exploration is:

$$\tau_{v,t+1}(d) = \tau_{v,t} + \tau_v(d) \tag{3}$$

In order to explore different areas of the environments, the robots choose the cell with a minimum amount of pheromone, corresponding to cells that probably are less frequented and therefore not explored cells. The chosen cell will be selected according with Eq. (2):

$$v_{next} = min[p(v_i|s)] \forall v_i \in N(s) \tag{4}$$

4   Coordination Methods to Disarming Task

4.1   Ant-based Based Team Strategy for Robots Recruitment (ATS-RR)

To realize the coordination mechanism in our system, we use the stigmergetic activity of social insects such as ants [25]. Stigmergy is a reinforcement learning mechanism that reinforces solutions in a solution space (for e.g. food for ants searching for food) with a chemical substance called pheromone. Pheromone provides positive reinforcement to future ants, and, ants searching for the food later on get attracted to the pheromone to locate and possibly consume the food. In our system, when a robot detects a mine, it deposits a certain amount of synthetic pheromone to mark the mine's location, this scent spreads into the environment and it is perceived, at certain distances, by other robots of the swarm.

In our problem we assume that the robots are equipped with sensor that perceived a pheromone, different by the pheromone used for the exploration. The robots communicate with others through the environment (indirect communication). We considered

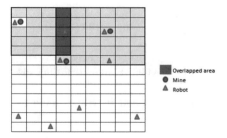

Fig. 1. ATS-RR Strategy. The robots that detect a mine during their movements spray some pheromone perceived at a certain distance by others.

that the mine disarming time is equal to the total evaporation time of substance (scent); in this way when the mine is disarmed, the robots involved in this operation will not be affected by scent trails (Fig. 1).

We assume t is the time in which the robot r detected a mine and it deposits the substance. The robot r continues to spray until all necessary robots reach its position.

If m is the time needed to disarm the mine, the law for the evaporation of the scent is the following:

$$\xi_{t+1} = \xi_t - \frac{1}{m}\xi_{t_0} \tag{5}$$

where ζ_{t_0} is the substance sprayed when the robot detects a mine. At the beginning $\zeta_t = \zeta_{t_0}$.

In this way after m steps ζ should be zero so the scent will not affect any more the movement of the other robots. This assures that all robots will cover other new space and disarm other mines completing the task in an efficient and distributed manner.

4.2   Firefly Based Team Strategy for Robots Recruitment (FTS-RR)

Firefly Algorithm (FA) is a nature-inspired stochastic global optimization method that was developed by Yang [7, 8]. The FA tries to mimic the flashing behavior of swarms of fireflies. In the FA algorithm, the two important issues are the variation of light intensity and the formulation of attractiveness. The brightness of a firefly is determined by the landscape of the object function. Attractiveness is proportional to brightness and, thus, for any two flashing fireflies, the less bright one move towards the brighter one. The light intensity decays with the square of the distance, the fireflies have limited visibility to other fireflies. This plays an important role in the communication of the fireflies and the attractiveness, which may be impaired by the distance. Some simplifications are assumed such as:

(a) it is assumed that all fireflies are unisex so they will be attracted to each other regardless of their sex;
(b) the attractiveness is proportional to their brightness and they both decrease as the distance increases;

(c) in the case of no existence of no brighter firefly on then, the fireflies will move randomly;

(d) the brightness of firefly is affected by its fitness.

The distance between any two fireflies i and j, at positions X_i and X_j, respectively, can be defined as the Cartesian or Euclidean distance as follows:

$$r_{ij} = \|X_i - X_j\| = \sqrt{\sum_{k=1}^{D} (X_{i,k} - X_{j,k})^2} \tag{6}$$

where $X_{i,k}$ is the k-th component of the spatial coordinate X_i of the i-th firefly and D is the number of dimensions.

In the firefly algorithm, as the attractiveness function of a firefly j, one should select any monotonically decreasing function of the distance to the chosen firefly, e.g., the exponential function:

$$\beta = \beta_0 e^{-\gamma r_{ij}^2} \tag{7}$$

where r_{ij} is the distance defined as in Eq. (6), β_0 is the initial attractiveness at r_0, and γ is an absorption coefficient at the source which controls the decrease of the light intensity.

The movement of a firefly i which is attracted by a more attractive (i.e., brighter) firefly j is governed by the following evolution equation:

$$x_i = x_i + \beta_0 e^{-\gamma r_{ij}^2}(x_i - x_j) + \alpha\left(\sigma - \frac{1}{2}\right) \tag{8}$$

where the first term on the right-hand side is the current position of the firefly, in our case a mine, the second term is used for considering the attractiveness of the firefly to light intensity seen by adjacent fireflies, and the third term is used for the random movement of a firefly in case there are not any brighter ones. The coefficient α is a randomization parameter determined by the problem of interest, while σ is a random number generator uniformly distributed in the space $[0, 1]$.

Furthermore, we look at Eq. (8), thus this non linear equation provides much richer characteristics. Firstly, if γ is very large, then attractiveness decreases too quickly, this means that the second term in (8) becomes negligible, leading to the standard simulated annealing (SA). Secondly, if γ is very small (i.e. $\gamma \to 0$), then the exponential factor $e^{-\gamma r_{ij}^2} \to 1$ and FA reduces to a variant of particle swarm optimization (PSO). In addition, the randomization term can be extended to other distributions such as Lévy flight. Furthermore, FA uses a non linear updating equation, which can produce rich behavior and higher convergence than linear updating equation used for example in standard PSO. Regarding the parameters setting, parametric studies suggest that $\beta_0 = 1$ can be used for most application; γ should related to the scaling L. In general, we can set $\gamma = \frac{1}{L}$ [9].

The Firefly Algorithm has been proved very efficient and it has three key advantages [9]:

- Automatic subdivision of the whole population into subgroups so that each subgroup can swarm around a local mode. Among all the local modes, there exists the global optimality. Therefore, FA can deal with multimodal optimization naturally.
- FA has the novel attraction mechanism among its multiple agents and this attraction can speed up the convergence. The attractiveness term is nonlinear, and thus may be richer in terms of dynamical characteristics.
- FA can include PSO, DE, and SA as its special cases. Therefore, FA can efficiently deal with a diverse range of optimization problems.

For our disarming task when a robot finds a mine, during the exploration task, it becomes a recruiter (firefly) of the other robots in order to disarm the mine and it tries to attract the other robots on the basis of the mine position.

In this case, the robots are assumed to have transmitters and receivers, using which they can communicate messages to each other. The messages are mostly coordinate positions of the detected mines. However, the robots are assumed to be able to broadcast messages in their wireless range; in this way, a robot can transmits its position only to its neighbours directly and there is not propagation of the messages (one hop communication).

The original version of FA is applied in the continuous space [7], and cannot be applied directly to tackle discrete problem, so we modified the algorithm in order to fit with our problem. In our case, a robot can move in a discrete space because it can go just in the contiguous cells step-by-step. This means that when a robot perceives, in its wireless range, the presence of a firefly (the recruiter robot) and it is in a cell with coordinates x_i and y_i, it can move according with the FA attraction rules such as expressed below:

$$\begin{cases} x_i^{t+1} = x_i^t + \beta_o e^{\gamma r_{ij}^2}(x_j - x_i) + \alpha(\sigma - \frac{1}{2}) \\ y_i^{t+1} = y_i^t + \beta_o e^{\gamma r_{ij}^2}(y_j - y_i) + \alpha(\sigma - \frac{1}{2}) \end{cases} \tag{9}$$

where x_j and y_j represent the coordinates of detected mine translated in terms of row and column of the matrix area, r_{ij} is the Euclidean distance between mine (or firefly) according to the Eq. (6) and the robot that moves towards the mine. The robots movements are conditioned by mine (firefly) position, in the second term of the formula, and it depends on attractiveness of the firefly such as expressed in Eq. (7) and by a random component in the third term of Eq. (8). The coefficient α is a randomization parameter determined by a problem of interest. The σ coefficient is a random number generator uniformly distributed in the space $[0, 1]$ and it is useful to avoid that more robots go towards the same mine if more robots are recruited by the same firefly and enabling to the algorithm to jump out of any local optimum (Fig. 2). In order to modify the FA to a discrete version, the robot movements have been considered through three possible value updates for each coordinates: $\{-1, 0, 1\}$ according to the following condition:

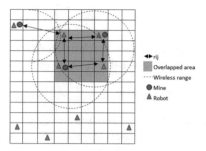

Fig. 2. FTS-RR Strategy: two robots during the exploration receive more recruiting requests because they are in an overlapped area. The FA tries to coordinate the robots in the disarming task avoiding that more robots go towards the same cells.

$$\begin{cases} x_i^{t+1} = x_i^t + 1 & if\left[\beta_o e^{\gamma r_{ij}^2}(x_j - x_i) + \alpha\left(\sigma - \tfrac{1}{2}\right) > 0\right] \\ x_i^{t+1} = x_i^t - 1 & if\left[\beta_o e^{\gamma r_{ij}^2}(x_j - x_i) + \alpha\left(\sigma - \tfrac{1}{2}\right) < 0\right] \\ x_i^{t+1} = x_i^t + 0 & if\left[\beta_o e^{\gamma r_{ij}^2}(x_j - x_i) + \alpha\left(\sigma - \tfrac{1}{2}\right) = 0\right] \end{cases} \tag{10}$$

$$\begin{cases} y_i^{t+1} = y_i^t + 1 & if\left[\beta_o e^{\gamma r_{ij}^2}(y_j - y_i) + \alpha\left(\sigma - \tfrac{1}{2}\right) > 0\right] \\ y_i^{t+1} = y_i^t - 1 & if\left[\beta_o e^{\gamma r_{ij}^2}(y_j - y_i) + \alpha\left(\sigma - \tfrac{1}{2}\right) < 0\right] \\ y_i^{t+1} = y_i^t + 0 & if\left[\beta_o e^{\gamma r_{ij}^2}(y_j - y_i) + \alpha\left(\sigma - \tfrac{1}{2}\right) = 0\right] \end{cases} \tag{11}$$

A robot r, that stores in the cell (x_i, y_i) as depicted in Fig. 3, can move in eight possible cells according with the three possible values attributed to x_i and y_i. For example if the result of the Eqs. (10–11) is $(-1, 1)$ the robot will move in the cell $(x_i - 1, y_i + 1)$.

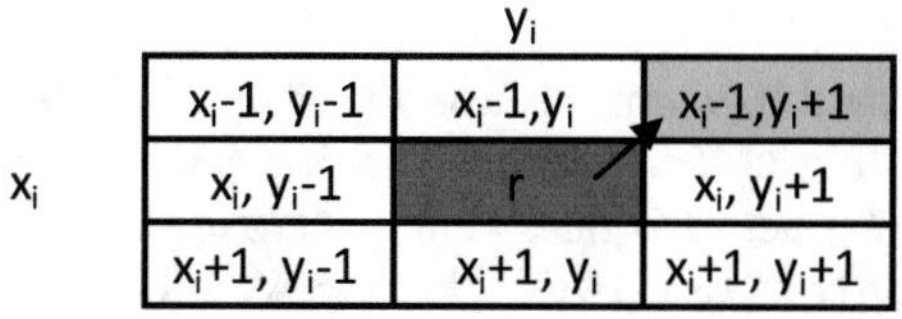

	y_i		
	x_i-1, y_i-1	x_i-1,y_i	x_i-1,y_i+1
x_i	x_i, y_i-1	r	x_i, y_i+1
	x_i+1, y_i-1	x_i+1, y_i	x_i+1, y_i+1

Fig. 3. Possible movement for a robot r stores in a cell (x_i, y_i).

In the described problem, the firefly algorithm is executed as follows:

- Step 1: Get the list of the detected mines (fireflies) and initialize algorithm's parameters: z number of fireflies, the attractiveness coefficient β_0, the light absorption coefficient γ, randomization parameter α.
- Step 2: Get the list of the robots in the wireless range of the fireflies.

- Step 3: For each robot calculate the distance r_{ij} from the fireflies in its range using the Euclidian distance.
- Step 4: For each robot find the firefly at minimum distance (the best firefly) and try to move the robot from its location to the location of the best firefly according to the Eqs. (10)–(11).
- Step 5: Terminate if the all detected mines are disarmed.

This steps are executed when the robots is recruited by others, indeed when no fireflies are detected or if the new location of the robots is outside of the wireless range of the fireflies, the robots explore in independent manner the area according the Eq. (2). This happens because the nature of the problem is bi-objective and the robots have to balance the two tasks.

5 Performance Evaluation and Comparison

In this section, we evaluate the performance of the two proposed algorithms in comparison with the well known PSO.

To highlight the performance benefits, we use random positions of the mines and the robots in the area, varying the number of robots operating in the 2-D, the size of grid map and the number of robots needed to disarm a mine, in order to study performance and scalability of the proposed algorithms (Table 1).

Table 1. Parameters for each experimental setup.

Grid size		Swarm size	Number of mines	Number of robots needed to disarm a mine
Scenario 1	30 × 30	20 40 60 80 100	5	4 5 6
Scenario 2	40 × 40	20 40 60 80 100	8	4 5 6
Scenario 3	50 × 50	20 40 60 80 100	12	4 5 6

Each of the numerical experiments was repeated 50 times in order to perform a statistical analysis of the results. Specific FTS-RR parameters were set as follows: $\beta = 1$; $\alpha = 0.2$; $\Upsilon = 1/L$ where L is max$\{m, n\}$ where m and n are the numbers of rows and columns of the matrix M, respectively. For the ATS-RR, the parameter values were set as follow: $\varphi = 1$, $\theta = 1$; $\eta = 0.9$.

Experiment setup was created to check the influence of the disarming strategies in solutions evaluation the relative error measured defined as the follow. We define $\sum_{i=1}^{m} \sum_{j=1}^{n} x_{ij}$ as the total number of accesses in the cell in one experiment.

x_{ij}^{*} is the optimum value.

This means that if we have a grid area of $m \times n$ with z mines and k is the minimum number of robots that needs to disarm a mine, we have:

$$x_{ij}^* = (m * n) - z + k * z \qquad (12)$$

In the best final configuration of the swarm in the area we want to have one access in all cell of the grid area and k accesses in the cells where the mines are located.

We define the relative error as follows:

$$E_{rel} = \frac{\left| \left(\sum_{i=1}^{m} \sum_{j=1}^{n} x_{ij} \right) - x_{ij}^* \right|}{x_{ij}^*} \qquad (13)$$

The best strategy has the smallest E_{rel}.

The optimum value depends on the dimension of grid, the number of mines disseminated in the area and the minimum number of robot needed to disarm a mine.

The effectiveness of the algorithms in terms of relative error is shown in Figs. 4, 5 and 6. The results show that the relative error decreases for a larger swarm. It should be noticed that the coordination of the robots and the distribution of the robots in the cells of the area depend on the number of robots needed to disarm a mine considering the same area at the same number of disseminated mines. Increasing the number of robots needed to disarm a mine the recruiting task is more complex and the robots pass more time in the same cell in order to reach the mines location.

The difference of the three strategies in terms of relative error is high, especially, when the size of swarm in the operative area is low and tends to be comparable when the number of robots increases.

When the complexity of the task increases Figs. 3(c), 4(c) and 5(c) it is possible that more robots in an overlapped region that is more robots receive the same request, go towards the same mine, passing more time in the same cells in the area creating, a not necessary redundancy.

However, in all cases the FTS-RR exhibits superior performance because this strategy is able to distribute better the robots in the area especially in comparison with the ATS-RR. Regarding the difference between the FTS-RR and PSO; in the Firefly based strategy the measure of the relative error is comparable and not significant difference when the number of robots to coordinate is low. Instead, increasing the number of robots and the mines the FTS-RR has the smallest relative error. This demonstrates that applying the same exploration strategy the disarming technique affects significantly the final distribution of the robots in the mission area.

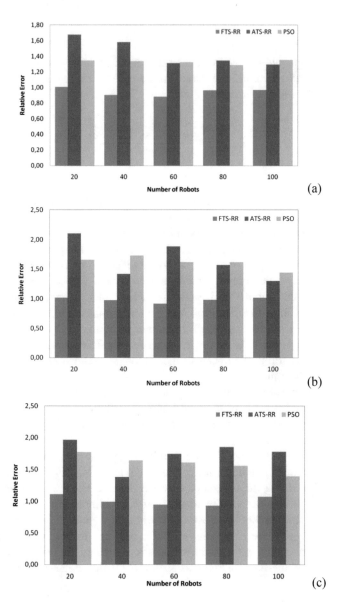

Fig. 4. Relative error comparison in a grid area 30×30 and 5 mines to disarm. (a) 4 robots needed to disarm a min. (b) 5 robots needed to disarm a mine. (c) 6 robots needed to disarm a mine.

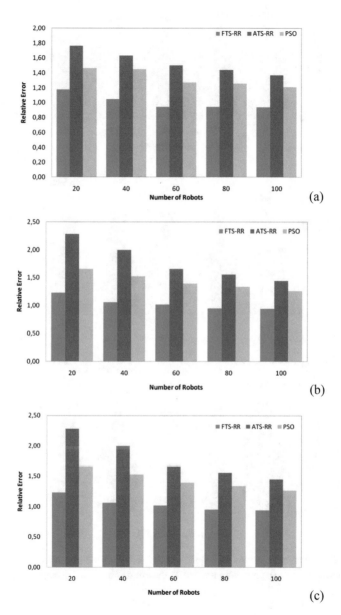

Fig. 5. Relative error comparison in a grid area 40 × 40 and 8 mines to disarm. (a) 4 robots needed to disarm a mine. (b) 5 robots needed to disarm a mine. (c) 6 robots needed to disarm a mine.

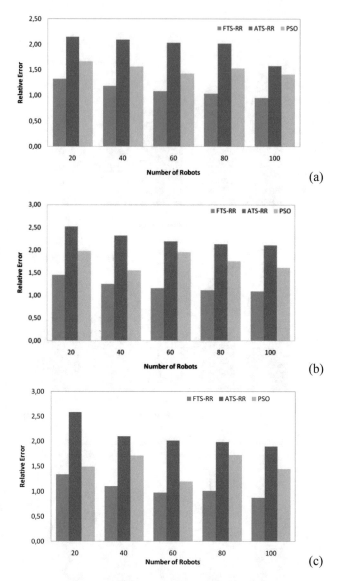

Fig. 6. Relative error comparison in a grid area 50×50 and 12 mines to disarm. (a) 4 robots needed to disarm a min. (b) 5 robots needed to disarm a mine. (c) 6 robots needed to disarm a mine.

6 Conclusions

Swarm intelligence based algorithms are very efficient in solving a wide range of optimization problems in diverse applications in science and engineering.

In this article, its application for a recruiting task in a swarm of mobile robots is investigated using Firefly Algorithm and Ant Colony Optimization.

The quality of solution is analyzed using a defined performance metric, which in our case was a relative error referred to the number of accesses in the cells that gives a measure about how the distribution of the robots in the area is efficient.

Our experiments through simulation showed that applying the FTS-RR the relative error is lower than the other two strategies. In particular the results demonstrate that the relative error applied to the total number of accesses is higher for the stigmergic approach, especially, when the number of robots needed to disarm a mine increases. The PSO and the FTS-RR methods are comparable when the task is not complex, but the difference is evident when the coordination task requires more requests of recruitment.

Future work will consider the continuous movement of the robots in the area of interest. In addition, we will introduce obstacles in the area and dropping wireless connection.

References

1. Bellingham, J.D., Godin, M.: Robotics in remote and hostile environments. Science **318**, 1098–1102 (2007)
2. Ducatelle, F., Di Caro, G.A., Förster, A., Bonani, M., Dorigo, M., Magnenat, S., Mondada, F., O'Grady, R., Pinciroli, C., Rétornaz, P., et al.: Cooperative navigation in robotic swarms, Swarm Intell., 1–33 (2014)
3. Tan, Y., Zheng, Z.-Y.: Research advance in swarm robotics. Defence Technol. **9**(1), 18–39 (2013)
4. De Rango, F., Palmieri, N.: A swarm-based robot team coordination protocol for mine detection and unknown space discovery. In: 8th International Conference on Wireless Communications and Mobile Computing (IWCMC), pp. 703–709 (2012)
5. Dorigo, M., Birattari, M., Stutzle, T.: Ant colony optimization. IEEE Comput. Intell. Mag. **1** (4), 28–39 (2006)
6. De Rango, F., Palmieri N., Yang, X.S., Marano, S: Bio-inspired exploring and recruiting tasks in a team of distributed robots over mined regions. In: International Symposium on Performance Evaluation of Computer and Telecommunication System, Chicago (2015)
7. Yang, X.S.: Firefly algorithms for multimodal optimization. In: Watanabe, O., Zeugmann, T. (eds.) SAGA 2009. LNCS, vol. 5792, pp. 169–178. Springer, Heidelberg (2009). doi:10. 1007/978-3-642-04944-6_14
8. Yang, X.S.: Firefly algorithm, stochastic test functions and design optimisation. Int. J. Bio-Inspired Comput. **2**(2), 78–84 (2010)
9. Yang, X.S.: Cuckoo Search and Firefly Algorithm Theory and Applications. Book on Studies in Computational Intelligence, ISBN: 978-3-319-02140-9 (2014)
10. Russell, R.: Ant trails - an example for robots to follow. In: Proceedings of IEEE International Conference on Robotics and Automation (ICRA), pp. 2698–2703 (1999)
11. Sugawara, K., Kazama, T., Watanabe, T.: Foraging behavior of interacting robots with virtual pheromone. In: Proceedings of the IEEE/RSJ International Conference on Intelligent Robots and Systems (IROS), pp. 3074–3079 (2004)

12. Garnier, S., Tache, F., Combe, M., Grimal, A., Theraulaz, G.: Alice in pheromone land: an experimental setup for the study of ant-like robots. In: Proceedings of the IEEE Swarm Intelligence Symposium (SIS), pp. 37–44, Washington, DC, USA (2007)
13. Ducatelle, F., Di Caro, G.A., Pinciroli, C., Gambardella, L.M.: Self organized cooperation between robotic swarms. Swarm Intell. **5**(2), 73–96 (2011)
14. Masár, M.: A biologically inspired swarm robot coordination algorithm for exploration and surveillance. In: Proceedings of 17th IEEE International Conference on Intelligent Engineering Systems INES, Budapest, pp. 271–275, ISBN 978-1-4799-0830-1 (2013)
15. Fujisawa, R., Dobata, S., Kubota, D., Imamura, H., Matsuno, F.: Dependency by concentration of pheromone trail for multiple robots. In: Dorigo, M., Birattari, M., Blum, C., Clerc, M., Stützle, T., Winfield, A.F.T. (eds.) ANTS 2008. LNCS, vol. 5217, pp. 283–290. Springer, Heidelberg (2008). doi:10.1007/978-3-540-87527-7_28
16. Mayet, R., Roberz, J., Schmickl, T., Crailsheim, K.: Antbots: a feasible visual emulation of pheromone trails for swarm robots. In: Dorigo, M., et al. (eds.) ANTS 2010. LNCS, vol. 6234, pp. 84–94. Springer, Heidelberg (2010). doi:10.1007/978-3-642-15461-4_8
17. Payton, D., Daily, M., Estowski, R., Howard, M., Lee, C.: Pheromone robotics. Auton. Robots **11**(3), 319–324 (2001)
18. Labella, T.H., Dorigo, M., Deneubourg, J.L.: Division of labour in a group inspired by ants' foraging behaviour. ACM Trans. Auton. Adapt. Syst. **1**(1), 4–25 (2006)
19. Meng, Y., Gan, J.: A distributed swarm intelligence based algorithm for a cooperative multi-robot construction task. In: IEEE Swarm Intelligence Symposium (2008)
20. Pugh, J., Martinoli, A.: Multi-robot learning with particle swarm optimization. In: Proceedings of Fifth International Joint Conference on Autonomous Robots and Multirobot Systems, pp. 441–448, Japan (2006)
21. Hereford, J.M., Siebold, M., Nichols, S.: Using the particle swarm optimization algorithm for robotic search applications. In: IEEE Swarm Intelligence Symposium (2007)
22. Jevtic, A., Gutiérrez, A., Andina, D., Jamshidi, M.: Distributed Bees algorithm for task allocation in swarm of robots. IEEE Syst. J. **6**(2), 296–304 (2012)
23. Balch, T.: Communication, diversity and learning: cornerstones of swarm behavior. In: Şahin, E., Spears, W.M. (eds.) SR 2004. LNCS, vol. 3342, pp. 21–30. Springer, Heidelberg (2005). doi:10.1007/978-3-540-30552-1_3
24. Mohan, Y., Ponnambalam, S.: An extensive review of research in swarm robotics. In: Nature and Biologically Inspired Computing, NaBIC, Word Congress (2009)
25. Bonabeau, E., Dorigo, M., Theraulaz, G.: Swarm Intelligence: From Natural to Artificial Systems. Oxford University Press, New York (1999)

GEML: A Grammatical Evolution, Machine Learning Approach to Multi-class Classification

Jeannie M. Fitzgerald[✉], R. Muhammad Atif Azad, and Conor Ryan

Biocomputing and Developmental Systems Group,
University of Limerick, Limerick, Ireland
{jeannie.fitzgerald,atif.azad,conor.ryan}@ul.ie

Abstract. In this paper, we propose a hybrid approach to solving multi-class problems which combines evolutionary computation with elements of traditional machine learning. The method, *Grammatical Evolution Machine Learning* (GEML) adapts machine learning concepts from decision tree learning and clustering methods and integrates these into a Grammatical Evolution framework. We investigate the effectiveness of GEML on several supervised, semi-supervised and unsupervised multi-class problems and demonstrate its competitive performance when compared with several well known machine learning algorithms. The GEML framework evolves human readable solutions which provide an explanation of the logic behind its classification decisions, offering a significant advantage over existing paradigms for unsupervised and semi-supervised learning. In addition we also examine the possibility of improving the performance of the algorithm through the application of several ensemble techniques.

Keywords: Multi-class classification · Grammatical evolution · Evolutionary computation · Machine learning

1 Introduction

Evolutionary algorithms (EAs) are algorithms which are inspired by biological evolution and which are constructed to emulate aspects of evolution, such as genetic mutation and recombination and the notion of natural selection. Genetic Programming (GP) [29] is an evolutionary algorithm which has been successful on a wide range of problems from various diverse domains [19], achieving many *human competitive* results [4]. However, a significant proportion of previous work has concentrated on supervised learning tasks and, aside from some notable exceptions, studies on unsupervised and semi-supervised learning have been left to the wider machine learning (ML) community.

Two of the most important problems types which benefit from the application of ML techniques are regression and classification, and GP has proven itself as an effective learner on each of these: achieving particularly competitive results on symbolic regression and binary classification tasks. Although many studies have

© Springer International Publishing AG 2017
J.J. Merelo et al. (eds.), *Computational Intelligence*, Studies in Computational Intelligence 669,
DOI 10.1007/978-3-319-48506-5_7

been undertaken, multi-class classification (MCC) remains a problem which is considered challenging for traditional tree based GP [11].

While we are concerned with multi-class classification generally, an important motivation for the current investigation is the requirement for an algorithm which can be applied to multi-class grouping/categorisation tasks involving both labelled and unlabelled inputs from the *medical domain*, where the unsupervised algorithm must be able to supply *human interpretable* justification for categorisation decisions.

Clustering is a natural choice for this type of task, but standard clustering algorithms generally fail to satisfy the requirement of providing the reasoning behind cluster allocations in a human readable form. In the medical domain, it is usually important that the learner has the capability to provide human understandable explanations of its decisions so that human experts can have confidence in the system. In this respect, decision trees (DTs) have the attractive property that the induced DT itself provides an easily comprehensible explanation of all decisions. Unfortunately, traditional DTs rely on ground truth information to make decisions and use of this information is not permissible in an unsupervised context. For these reasons, although each of these methods have attractive properties, we conclude that neither DTs nor clustering approaches are, *in their normal mode of use,* appropriate for unsupervised categorisation tasks which require an explanation from the learner.

Although there is some important existing work in the area of unsupervised classification in the medical domain, including for example [6,22,32], the subject remains relatively unexplored. This paper takes up a triple challenge: it investigates MCC in a supervised, semi-supervised as well as in an unsupervised context.

We hypothesise that it may be possible to combine the desirable qualities of both algorithms by taking the underlying concepts and wrapping them in an evolutionary framework – specifically a grammatical evolution (GE) [40] framework. This approach is appealing due to its symbiotic nature: a GE grammar is used to generate human readable decision tree like solutions and the evolutionary process is applied to the task of optimising the resulting cluster assignments – thus emulating both the decision making behaviour of DTs and the iterative operation of traditional clustering approaches. Not only does GE produce human readable solutions, but it has been shown [3] that the paradigm seems to be able to avoid bloated, over-complex ones.

While we can hypothesise that this hybrid approach might be a good idea, objective evaluation of algorithm performance is required before concluding that the resulting models are likely to be of any practical use. One approach to accomplishing this is to compare results of the hybrid method with other unsupervised algorithms using some common metric of cluster performance. However, it could be argued that without ground truth information any method of comparison is flawed. Another possible approach for evaluating the effectiveness of the proposed method would be to apply it to data about which something is already known, where that knowledge is not used in the learning process for the purpose of evaluating and comparing performance afterwards.

Considering our original objective, we were also interested in learning about potential performance differences may expected between our unsupervised system and, supervised and semi-supervised approaches using the same data. Thus, we choose to construct this study such that it would be possible to compare the performance and behaviour of the hybrid unsupervised learner with another state of the art unsupervised algorithm as well as with supervised and semi-supervised learners on the same data. Rather than using our original medical dataset at this point, we chose to carry out this first study using controllable synthetic data as outlined in Sect. 4.2, with the intention of optimising the GEML system based on lessons learned, if results of these preliminary experiments prove encouraging. Once optimised, the system can be applied to the more challenging medical datasets in the future.

In this work, we investigate the hypothesis that combining ML concepts with GE can facilitate the development of a new hybrid algorithm with three important properties: the ability to learn multi-class problems in both supervised and unsupervised environments, and the capability of producing human readable results. However, due to the way in which we have designed the experiments – so that meaningful evaluation of the proposed algorithm would be possible, the resulting system delivers much more than initially planned – functioning in supervised, unsupervised and semi-supervised domains.

In summary, we investigate a hybrid GE system which incorporates ideas from two well known ML techniques: decision tree learning which is often applied to supervised tasks, and clustering methods which are commonly used for unsupervised learning tasks. The proposed system which we call GEML is applied to supervised, unsupervised and semi-supervised MCC problems. Its performance is compared with several state of the art algorithms and is shown to outperform its ML counterparts and to be competitive with the best performing ML algorithm, on the datasets studied.

In this work we extend and describe in greater detail, the GEML framework as previously proposed in [17]. In the remainder of this section we will briefly explain some of the concepts employed.

1.1   Clustering

Clustering involves the categorisation of a set of samples into groups or subsets called clusters, such that samples allocated to the same cluster are similar in some way. There are various types of clustering algorithms capable of generating different types of cluster arrangements, such as flat or hierarchical clustering. One of the best known clustering algorithms is K-means clustering which works in an iterative fashion by creating a number of centroids (aspirationally cluster centres). The algorithm groups samples depending on their proximity to these centres and then measuring the distance between the data and the nearest centroids – K-means iteratively minimises the sum of squared distances by changing the centroid in each iteration and reassigning samples to possibly different groups. Of the EC work in the existing literature which combines GP or GE with unsupervised methods, K-means is the most popular of those used, as is outlined in Sect. 2.

1.2   Decision Tree Learning

A decision tree is a hierarchical model that can be used for decision-making. The tree is composed of internal decision nodes and terminal leaf nodes. In the case of classification for example, internal decision nodes represent attributes, whereas the leaf nodes represent an assigned class label. Directed edges connect the various nodes forming a hierarchical, tree-like structure. Each outgoing edge from an internal node corresponds to a value or a range of values of the attribute represented by that particular node. Tree construction is a filtering and refining process which aims to gradually separate samples into the various classes with possibly multiple routes through the decision process for a particular class assignment.

1.3   Unsupervised, Semi-supervised and Supervised Learning

In simple terms, supervised, semi-supervised and unsupervised learning methods are differentiated by the amount of ground truth information that is available to the learning system: in supervised learning systems the 'answer' which may, for example, be a target variable or a class label is known to the system; semi-supervised systems may have access to such information for a limited number of samples or may involve revalidation of the automated prediction with expert knowledge; and unsupervised learners do not have any ground truth information with which to guide the learning process.

Although classification and clustering are conceptually similar, in practice the techniques are usually used in fundamentally different ways: clustering methods are generally applied to unsupervised tasks and do not require either training data or ground truth label information, whereas classification is usually a supervised task which requires both. At a basic level the goal of clustering is to group similar things together without reference to the name of the group or what membership of a group represents, other than the fact that the members are similar in some way, whereas the objective of classification is to learn, from examples, relationships in the data which facilitate the mapping of training instances to class labels, such that when presented with a new unseen instance the classification system may assign a class label to that instance based on rules/relationships learned in the training phase.

2   Previous Work

An exhaustive review of the application of EAs to clustering methods and decision tree induction is beyond the scope of this paper. Here we have chosen to focus on the most recent work and that which we determined to be most relevant to the current study. For a comprehensive survey of EAs applied to clustering, the interested reader is directed to [23], whereas a detailed review of EAs applied to decision tree induction can be found in [5].

Relative to the volume of existing research on supervised learning in the field of Evolutionary Computation (EC), unsupervised and semi-supervised learning have received little attention. Of the existing work, a significant proportion in the area of unsupervised learning recommends the use of clustering methods for feature selection [28,31,33], and the majority of this work recommends a traditional K-means approach.

[36] used clustering was used in an interesting way whereby a Differential Evolution (DE) algorithm with built-in clustering functionality was proposed. They studied its effectiveness on an image classification task, and compared their results with several well known algorithms such as K-means but reported statistically indistinct results.

A different unsupervised GP approach was proposed in [35] where a novel fitness function was used in feature selection for the purpose of identifying redundant features. The authors reported superior results when performance was compared with several state of the art algorithms. GP was again employed in [21] where it was used to develop low level thin edge detectors. In that work the authors demonstrated that edge detectors trained on a single image (without ground truth) could outperform a popular edge detector on the task of detecting thin lines in unseen images.

Another novel application of K-means was proposed by [25] who integrated it into GP and used this hybrid approach for problem decomposition – grouping fitness cases into subsets. They applied their strategy to several symbolic regression problems and reported superior results to those achieved using standard GP. They later developed a similar approach [26] for time series prediction.

On the subject of multi-class classification problems, there have been several interesting approaches using tree based GP including strategies for decomposing the task into multiple binary problems [48], treating MCC problems as regression tasks [11] and experimenting with various thresholding schemes such as [51]. Other methods have been proposed which utilise GP variants including multi level GP (MLGP) [50], Parallel linear GP [16] and probability based GP [47] to name a few.

There have been several other evolutionary approaches to MCC including self-organising swarm (SOSwarm) which was described in [38]. In that work, particle swarm optimisation (PSO) was used to generate a mapping which had some similarities to a type of artificial neural network known as a self organising map. SOSwarm was studied on several well known classification problems and while the average performance seemed to degrade as the number of classes increased – the best performing solutions were competitive with the state of the art.

DTs have previously been combined with GE in [13]. The algorithm was applied to the binary classification task of detecting gene-gene interactions in genetic association studies. The researchers reported good results when their GE with DT (GEDT) system was compared with the C.45 DT algorithm. Our suggested approach has some similarities to this work. However, that research focused on a supervised binary task where attribute values were restricted to a common set of 3 items.

Clustering methods have also been applied to MCC problems. A hybrid method which combined a GA with local search and clustering was suggested in [41], where it was applied to a multi-class problem on gene expression data. The results of that investigation showed that their method (HGACLUS) delivered a competitive performance when compared with K-means and several earlier GA approaches described in [12,30]. GP was again combined with K-means clustering for MCC in [1] where the researchers used the K-means algorithm to cluster the GP program semantics in order to determine the predicted class labels.

Competitive results were also reported in [34] in which K-Means clustering was again used with GP for MCC. There, clustering was combined with a multigenic GP approach in which each individual was composed of several solution parse trees having a common root node.

Concerning DTs, [5] concluded that good performance of EAs for decision tree induction in terms of predictive accuracy had been empirically established. They recommended that investigation of these algorithms on *synthetic data* should be pursued and also the possibility of using evolutionary computation for the evolution of decision tree induction *algorithms*. In this paper we address the first of these recommendations. The candidate solutions evolved by GE are computer programs which emulate decision trees, and these computer programs are produced using a grammar template capable of generating a multitude of different solutions. Thus, it could be argued that the proposed approach does, at least in some sense, also meet the second objective – the evolution of DT induction algorithms.

The novel contributions of this study are the proposal of a technique for unsupervised learning using an EA where the evolved learning hypotheses are in human readable form, and the extension of this to the development of a hybrid GE framework which can also be used for supervised and semi-supervised learning. The new system which we call GEML is successfully applied to the problem of multi-class classification. We also investigate the effectiveness of extending the GEML method through the construction of ensemble, majority vote learners.

3   Proposed Method

In ML DTs generally employ the concept of *information gain* to inform branching decisions during the construction of a decision tree, where the measure of information gain used relies on knowledge of the ground truth labels. While it is normal practice to make use of the ground truth information in the training data for a supervised learner, this is not possible for unsupervised methods and to a limited extent for semi-supervised methods as there is no such data available in the unsupervised case, and only limited aces to ground truth labels in the semi-supervised domain. Instead, we construct of an *if then else* structure where the if component may be used to test various conditions pertaining to the data, whereby the learning system has access to both the attribute values and also to the variance of each attribute on the training data. Thus, by design our system does not currently implement DTs according to a strict definition of the algorithm, as using label information precludes unsupervised learning.

3.1 Grammatical Evolution

Grammatical Evolution (GE) [40] is a flexible EC paradigm which has several advantages over other evolutionary methods including standard GP. In common with its traditional GP relative, GE involves the generation of candidate solutions in the form of executable computer programs. The difference is that GE does this using powerful grammars whereas GP operates with a much more limited tool-set.

A key aspect of the GE approach is *genotype phenotype separation* whereby the genotype is usually (but not necessarily) encoded as a vector of integer *codons*, some or all of which are *mapped* to production rules defined in a user specified grammar (usually in Backus-Naur-Form). This mapping results in a phenotype executable program (candidate solution). GE facilitates focused search through the encoding of domain knowledge into the grammar and the separation of search and solution space such that the search component is independent of the representation and may, in principle, be carried out using any suitable algorithm – a genetic algorithm is often used but other search algorithms such as PSO [39] and DE [37] have also been used to good effect.

The role of the user-defined grammar is key to guiding the evolutionary search towards desirable solutions. The grammar is essentially a specification of what *can be evolved* and it is up to the evolutionary system to determine which of the many possible solutions which can be generated using the specification *should be evolved* [45]. A small change in a grammar may induce drastically different behaviour. In this work we have designed a grammar, shown in Fig. 1, which facilitates the assignment of data instances to clusters based on the results of applying simple 'if then else' decision rules. While the individual rules are quite simple, the grammar allows for the construction of powerful expressions capable of representing both simple and complex relationships between attributes as demonstrated in Fig. 2.

3.2 Objective Functions

For each of the three learning problems: supervised, unsupervised and semi-supervised, we employ a different objective function to drive evolutionary progress. In the supervised case we use classification accuracy which is simply the proportion of instances correctly classified by the system. Since this study uses balanced data sets, we simply use the number of correct predictions to measure system performance. Accuracy values range between 0 and 1 where

```
<expr> ::=<label> if <cond> else <label>
    | <label>if<cond>else<expr>
    | <label>if<cond>and<cond>else <expr>
    | <label>if<cond>or<cond>else<expr>
    | <expr> if <cond> else <expr>
    | <expr> | <label>

<label> ::= 0 | 1 | 2 | 3 | 4

<cond> ::= <attr><relop><const>
        | <subExpr><relop><const>
        | <subExpr><relop><subExpr>
        | <var><relop><var>
        | <attr><relop><attr>

<subExpr> ::= <attr><op><const>
        | <attr><op><var>
        | <attr><op><attr>

<op>   ::= + | - | * | /
<relop> ::= <=
<const> ::= <digit>.<digit>
        | -<digit>.<digit>

<digit> ::= 0 | 1 | 2 | 3 | 4
        | 5 | 6 | 7 | 8 | 9

<attr> ::= x[0] | x[1] | x[2]
<var>  ::= v[0] | v[1] | v[2]
```

Fig. 1. Example grammar for five class problem with three attributes ($<attr>$). The $<var>$ entries represent the variance in the training data across each attribute. The 'then if else' format is designed to simplify the syntax required in a python environment – as the system evolves python expressions. The division operation is protected in the implementation.

```
2 if x[0]*v[1]<=-0.2 and v[2]<=v[1]
else 1 if (x[1]*v[0])<=(x[0]*-0.8)
else 0 if x[2]/v[2]<= x[0]*-2.0 or x[2]<=x[1]
else 1 if x[1]<=x[0]
else 1 if x[0]*x[2]<=x[1]/x[0] and x[1]/v[2]<=-0.3)
else 2 if x[2]/v[2]<=x[0]*2.9 or x[2]<=x[1]
else 1 if x[0]<=-0.1 else 0
```

Fig. 2. Example expression generated for a three class, three attribute classification task.

1 represents perfect classification. The system is configured with an objective function designed to maximise fitness.

For the unsupervised task we have chosen to use a metric of clustering performance known as a *silhouette co-efficient* or *silhouette score* (SC) as the objective function. The SC is a metric which does not require knowledge of the ground truth which makes it suitable for use in an unsupervised context. For each data point two measures are calculated: a. the average distance (according to some distance metric) between it and every other point in the same cluster and b. the mean distance between it an all of the points in the nearest cluster that is not it's own cluster. The silhouette score over all points is calculated according to the formula shown in Fig. 3. Our system tries to maximise this value during the evolutionary process. Note that this approach, which aims at optimising the SC rather than cluster centroids, is quite different from the other EC methods outlined in Sect. 2 where clustering tasks have generally been tackled using the K-means algorithm.

The silhouette score ranges between −1 and 1, where a negative value implies that samples are not assigned to the correct clusters, a value close to 0 indicates that there are overlapping clusters and a score close to 1 means that clusters are cohesive and well separated.

$$(b-a)/max(a,b) \qquad\qquad (1)$$

Fig. 3. Silhouette co-efficient.

To calculate the silhouette co-efficient it is first necessary to choose an appropriate distance metric from the many and varied options available in the literature. In this work we have used *cosine distance* also known as *cosine similarity* as it is suitable for determining the similarity between vectors of features and obviates the need for data normalisation. Also, we experimented with several metrics including euclidean and mahalanobis distance before choosing *cosine distance* – as its use resulted in the best results over a range of synthetic classification problems. The results for cosine distance were better in terms of classification accuracy when cluster assignments were converted to class labels using a standard approach.

Semi-supervised learning is suitable for classification situations where some but not all ground truth labels are available. It may be the case, for example, that scarce or expensive human expertise is required to determine the labels. In these cases, it is usually possible to improve unsupervised performance by adding a small number of labelled examples to the system. Although, our synthetic data is fully labelled, we simulate partial labelling by only considering a random subset of training data (20 % of the full data set) to be labelled; the rest of the data set is treated as unlabelled. We compute prediction accuracy on the labelled data and the silhouette score on the unlabelled set. We then add the two measures to get a final score and strive to maximise this score during evolution.

To summarise, candidate solutions are generated using a specification described in a grammar such as the one shown in Fig. 1, and the same grammar is used for all of the GEML problem configurations. Then, applying the decision rules defined in the grammar problem instances are assigned to clusters as shown in Fig. 4 and then depending on whether the task is supervised, unsupervised, or semi-supervised the system tries to optimise the classification accuracy, the silhouette score or a combination of the two. The key point here, is that the same grammar is used for each type of learning – only the objective function is different.

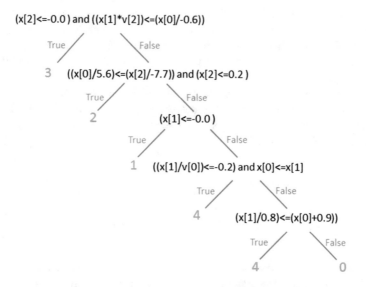

Fig. 4. Example expression tree.

Figure 4 illustrates the expression tree of an example solution for a five-class task. Similar to a DT, the internal nodes of the tree represent branching decision points and the terminal nodes represent cluster assignments.

4   Experiments

In this section we outline the construction of our experiments including the parameters, datasets and benchmarks used. We also detail the results of these experiments together with the results achieved on the same problems with our chosen benchmark algorithms. Details of the naming convention for the various experimental configurations are shown in Table 1.

4.1   Benchmark Algorithms

As we incorporate ideas from DT learning and clustering methods into our hybrid GEML approach, it is appropriate that we benchmark the proposed approach against Decision Trees [10] as a supervised method and against the K-means clustering algorithm [49] as an unsupervised paradigm. For semi-supervised learning we compare with a label propagation (LP) [43] algorithm. The idea behind LP is similar to k-Nearest Neighbours (k-NN) [2] and was originally proposed for detecting community structures in networks.

We also compare with support vector machines (SVMs) [8] for supervised learning as the method may provide a useful benchmark as it is known to achieve good results with balanced datasets, which is the case here, and while SVMs are inherently binary classifiers they can perform multi-class classification in various ways, most commonly using a "one versus all" strategy.

For comparison purposes we choose simple classification accuracy as a performance metric. It has been empirically established in the GP literature that simple classification accuracy is not a reliable measure of classification on *unbalanced* datasets [7], and that other measures such as average accuracy or Matthews Correlation Co-efficient might be more appropriate especially if combined with a sampling approach [18]. However, in this preliminary investigation, the classes *are balanced* which allows us to consider simple classification accuracy as a reasonable measure, particularly as we want to be able to observe differences in performance across the various levels of learning.

Table 1. Experimental configurations.

Configuration	Explanation
GEML-SUP	Supervised GEML
GEML-SEMI	Semi-supervised
GEML-UN	Unsupervised GEML
DT	Decision Tree Learning
LP	Label Propagation
KM	K-means Clustering
SVM	Support Vector Machine

We adopt a popular mechanism to determine which predicted label represents which a priori class label: the predicted class is mapped to the a priori class which has the majority of instances assigned to it, e.g. for a binary task with 1000 training instances, if predicted class 1 has 333 members of ground truth class 1 assigned to it and 667 instances of class 0, then predicted class 1 is determined to represent the a priori class 0. It is important to note that this method is used to calculate the accuracy metric and used only for reporting and comparison purposes across all tasks and methodologies. The measure (classification accuracy) is the main driver of the evolutionary process in the supervised tasks, is used on only a percentage of the training instances in the semi-supervised case. In all cases, the same mapping from cluster assignment to class label determined during the training phase also applies when evaluating performance on test data.

For each of the GEML methods the evolved solutions have similar form to the example shown in Fig. 2. This is essentially a python expression that can be evaluated for each training and each test instance. The result of evaluating the expression on a given instance is an integer which is converted into first a cluster assignment and then a class label, using the method previously described. Although the objective functions used to determine fitness and drive evolution differ according to the type of learning model as detailed in Sect. 3.2, we calculate the classification accuracy for each unevaluated individual at each generation on training and test data. At no time is the test data used in the learning process.

For each problem, for each dataset and each learner, the algorithm was run fifty times using the same synthetic datasets and train/test splits. A different random seed was used for each run of the same algorithm and these same random seeds were used for the corresponding run of each algorithm. The popular scikit-learn [42] python library for machine learning was used for all of these ML experiments.

4.2 Datasets

The various algorithms were tested on several synthetic multi-class datasets which were produced using the scikit-learn library [42] which provides functionality for the generation of datasets with the aid of various configurable parameters. For this study we investigate balanced multi-class problems of two, three, four and five classes each. The library facilitates user control of the number, type and nature of features selected for experiments. For example, features can be informative, duplicate or redundant. We have chosen to use informative features only for the current work.

Given a problem configuration (number of classes), for each run of each algorithm a dataset of 1000 instances was generated and then split into training and test sets of 700 and 300 instances respectively. Identical random seeds were used for the corresponding run for each configuration, such that the same dataset was generated for each setup for a particular run number.

We have chosen to use synthetic datasets: 1000 instances were generated, without added noise, and with few features, each of which is informative. Employing synthetic datasets allows us to configure the data to have *informative* features such that it is, as far as possible, amenable to being clustered or classified. We have made these choices in an effort to ensure that the data is not biased to favour any particular algorithm or learning paradigm. For example, DTs are known to over-fit and not generalise well where there are a large number of features and few instances.

The decision to use synthetic datasets also delivers on the recommendation of [5] to use synthetic data for decision tree induction, as described in Sect. 2.

Table 2. Evolutionary parameters.

Parameter	Value
Population size	500
Replacement strategy	Generational
Number of generations	100
Crossover probability	0.9
Mutation probability	0.01

4.3 Evolutionary Parameters

Important parameters used in these experiments are outlined in Table 2. Evolutionary search operators in GE are applied at the genotypic level, and in this work each individual's genotype is a linear genome represented by a vector of integers. The mutation operator operates by replacing a single integer with a new one randomly generated within a predefined range. One point crossover is used, whereby a single crossover point is randomly and independently selected from each of the two parents (that is, the two points are likely to correspond to different locations) and two new offspring are created by splicing parental segments together. In both cases, these operations take place in the effective portion of the individual, i.e. the segment of the integer vector that was used in the genotype to phenotype mapping process – sometimes a complete phenotype is generated before requiring the full integer vector.

4.4 Experimental Results

Results for average and best training and test accuracy can be seen in Table 3, where for convenience the best result in each category is in bold text. For comparison purposes we are interested in comparing the supervised methods with each other and the unsupervised approaches with the other unsupervised methods etc. Thus we compare GEML-SUP with both DT and SVM, GEML-UNS with KM and GEML-SEMI with LP. However, we are also interested in observing the relative performances of the three different levels of learning.

Looking first at the supervised approaches, we can see that the SVM approach performs well across all of the problems studied with regard to average classification accuracy on both training and test data. Encouragingly, the GEML-SUP configuration is very competitive with SVM on the first three problems and outperforms DT on each of those tasks.

On the semi-supervised experiments GEML-SEMI outperforms LP on all problems for both training and test data in terms of average classification accuracy.

Finally, with regard to the unsupervised set-ups GEML-UN outperforms KM for average accuracy on training and test data on all problems.

For each algorithm, the performance of the various configurations degrades as the number of classes increases which is not surprising as adding more classes increases the difficulty of the problem to be solved. Overall, the SVM algorithm suffered least from this issue, which is again not surprising given that the implementation used here [42] solves MCC problems using a binary decomposition strategy.

Reviewing the results in Table 3, we see that values for *best overall* training and test accuracy on the binary and three class tasks for each of the GEML methods are not competitive with the other approaches. For example, on the three class task, the average test accuracy for K-means is 0.74 whereas the best result is 0.99 compared with GEML-UN which has an average test accuracy of 0.75 and a best result of 0.86 and the GEML-SUP which has an average test accuracy of 0.92 and a best result of 0.95. The results for each of the GEML setups show that the reported standard deviation is lower than for the other algorithms.

It is unclear to us whether this phenomenon is associated with the GE paradigm itself, the nature of the MCC problem or some other factors. However, it could be argued that the behaviour is not necessarily a negative result, as having a larger standard deviation with a higher extreme value can also mean that the algorithm is unreliable. After all, a good test set performance is only valuable if it is consistently achieved, not as an exceptional case.

Due to the stochastic nature of GE one might hypothesise that there is a higher probability of many individuals achieving good results across many runs on the easier one and two class problems than on the more difficult problems where individuals have to learn to incorporate a larger number of class labels: due to the added complexity there are likely to be fewer fit solutions early in the evolutionary process and thus fewer opportunities to improve through the application of genetic operators. One can easily imagine that there could be significant variability across runs depending on the quality of the initial population, and the existence of fewer highly fit solutions reduces the probability of truly excellent ones emerging.

Looking at the generalisation performance of each method in terms of the variance component, we adopt a simple measure whereby the variance error is simply the difference in performance of the various learning hypothesis between training and test data. In this respect, of the algorithms studied only the LP approach exhibits high variance. The various GEML methods all produce good generalisation performance. This is quite interesting as its close relation GP is known to exhibit a low bias high variance behaviour [27]. We can hypothesise that possible contributing factor to this contrast in behaviour is due to the grammar used, even if it contains recursive rules it is likely to constrain the size of the evolved programs. [3] demonstrated empirically that while program size tends to increase steadily during GP runs, the average size of GE genomes remains roughly static after an initial period of growth or shrinkage. In those experiments, GP genomes were consistently larger than GE genomes after only fifty generations. It may be the case that these effects are preventing the evolved models from becoming over complex. Recent results presented in [3] would suggest that GE does not over-fit on regression problems, where the required grammar would be fundamentally different, either.

If we analyse the difference in performance between the various supervised, semi-supervised and unsupervised algorithms, it is not surprising that in all cases the supervised algorithms produced the best results and that the semi-supervised algorithms performed better than the unsupervised ones. Of course, the unsupervised and semi-supervised methods are not usually evaluated in the same way as supervised classification approaches: using accuracy as a performance metric. We have chosen to do so here as a convenient and practical way to gain some insight into the likely relative performance of our hybrid technique when it is applied to the three learning approaches.

The results suggest that while the performance of all of the algorithms deteriorates as the number of classes increases, this effect is even more evident for the unsupervised and semi-supervised methods where the performance of GEML-UN drops from 90 % on the binary task to 66 % for the five class problem, although this is still better than the corresponding LP algorithm. Again, we can hypothesis that while adopting a binary decomposition approach may seem attractive, this would be very challenging in an unsupervised context. However, there may be some scope for the strategy in the semi-supervised paradigm.

Statistical Analysis. We carried out tests for statistical significance on the test results using the non-parametric Mann-Whitney U-Test. This revealed that statistical significance of results sometimes varied depending on the problem. Any differences between SUP and SVM were not significant for the two and three class problems but for the four and five 5 ones SVM is significantly better with 99 % confidence. Comparing SUP against DT, the differences are significant at the 95 %, 99 % and 99 % confidence levels for the two, three and four class problems respectively (SUP is better), but not significant for the five class task. For the semi-supervised tasks, any differences are not significant for the binary task but the GEML-SEMI results are significantly better at the 99 % confidence level for the other three problems. Finally, the analysis comparing GEML-UNS with K-Means behaves similarly, where GEML-UNS is significantly better on the two, four and five class tasks having confidence levels of 99 %, 95 % and 95 % respectively, and with a p-value of 0.58 there was no significant difference on the three class task.

5   Ensemble Approaches

The results demonstrate that while the GEML approach is competitive with the best ML algorithms on the two and three class supervised tasks, SVMs outperform GEML on the four and five class problems. As GE is a non-deterministic algorithm we hypothesised that it may be possible to improve its relative performance on the more difficult four and five class tasks, by generating GE *majority voting classifiers* which have previously [9,15,20] been shown to be effective in

Table 3. Average and best classification accuracy on training and test data.

Task	Method	Avg. training accuracy	StdDev	Best training accuracy	Avg. test accuracy	StdDev	Best test accuracy
C2	GEML-SUP	**0.96**	0.01	0.98	**0.96**	0.01	0.97
	DT	0.93	0.04	0.99	0.93	0.04	0.99
	SVM	0.95	0.03	0.99	0.95	0.03	0.99
	GEML-SEMI	0.90	0.01	0.93	0.91	0.01	0.92
	LP	0.90	0.06	0.99	0.88	0.07	0.99
	GEML-UN	0.90	0.01	0.92	0.91	0.01	0.93
	KM	0.84	0.08	0.99	0.84	0.08	0.99
C3	GEML-SUP	**0.93**	0.01	0.94	**0.92**	0.02	0.95
	DT	0.87	0.04	0.97	0.88	0.05	0.99
	SVM	0.92	0.03	0.98	**0.92**	0.04	0.99
	GEML-SEMI	0.88	0.04	0.94	0.87	0.04	0.92
	LP	0.83	0.05	0.96	0.79	0.08	0.93
	GEML-UN	0.76	0.04	0.87	0.75	0.04	0.86
	KM	0.75	0.07	0.91	0.74	0.08	0.99
C4	GEML-SUP	0.86	0.01	0.88	0.86	0.02	0.89
	DT	0.82	0.04	0.94	0.83	0.04	0.93
	SVM	**0.88**	0.03	0.94	**0.88**	0.03	0.96
	GEML-SEMI	0.78	0.05	0.84	0.78	0.05	0.85
	LP	0.77	0.05	0.88	0.71	0.07	0.85
	GEML-UN	0.71	0.04	0.79	0.71	0.04	0.81
	KM	0.65	0.06	0.83	0.67	0.07	0.84
C5	GEML-SUP	0.77	0.06	0.85	0.75	0.04	0.83
	DT	0.77	0.04	0.88	0.79	0.05	0.89
	SVM	**0.85**	0.03	0.93	**0.86**	0.03	0.94
	GEML-SEMI	0.72	0.04	0.76	0.75	0.06	0.82
	LP	0.71	0.05	0.84	0.65	0.07	0.83
	GEML-UN	0.66	0.06	0.77	0.69	0.07	0.82
	KM	0.63	0.05	0.78	0.63	0.06	0.79

improving classifier generalization. In general, these approaches operate by combining a large number of classifiers and then classifying each instance with the class label of the class which receives the greatest number of votes.

A *weak learner* is one whose accuracy in labelling examples may be only slightly better than random guessing whereas a strong learner is one whose predictions are strongly correlated with the true labels. It has been well established in the literature [24, 46] that rather surprisingly, *weak learners* can be combined to produce much stronger models. Indeed the strategy is so successful that it has been widely adopted in evolutionary computation and other ML algorithms. See [44] for a comprehensive review.

We have chosen to investigate two different strategies for combining our GEML models. In the first instance we combine the predictions of the best of run models of each run giving a total of fifty models to form an ensemble and we call this configuration ensBest. Secondly, we explored an approach whereby we combined every model from each generation whose accuracy exceeded a predefined threshold and we refer to this approach as ensPop. Using this second approach, which may generate thousands of models to add to the ensemble we are interested to discover if the combined approach may achieve better accuracy that our single best model.

For these initial experiments within the GE runs, we initially chose to apply a weak threshold of 0.60 accuracy whereby any individual whose fitness was greater would have its predictions added to the ensemble. However, the accuracy scores produced by these generated ensembles were not at all encouraging – often several percent worse than the best individual score. In the final experiments we set the threshold to be a value which was 10 % lower than the average training accuracy of the population as per Table 3.

As the evolutionary system converges the population becomes dominated over time by very similar or identical individuals. We chose not to allow duplicates to join the ensemble. Once an ensemble is constructed, we examine the prediction correlation between the candidate predictions using the Pearson correlation coefficient and then eliminate potential solutions which were greater than 90 % correlated with more than two thirds of their ensemble mates. These choices were designed to promote diversity in the ensemble which has been determined to be a necessary condition for constructing an effective ensemble [14]. However, the values chosen are somewhat ad-hoc, and it is likely that they could be improved upon. The construction and modification of the ensemble membership is carried out on the training data and then the test predictions of the final ensemble members are compared with the ground truth.

The results obtained shown in Table 4 indicate that the ensembles constructed from the fifty best-of-run individuals, for the four class problem, achieved the same average test accuracy but did not improve on the result for the single best individual previously reported. For the five class task the ensBest ensemble was significantly better that the average test result and produced a slight improvement on the overall best individual score.

Table 4. Ensemble results.

Task	Average test	Best test	ensBest	Average ensPop
4Class	0.86	0.89	0.86	0.87
5Class	0.75	0.83	0.84	0.80

Similarly, the larger ensembles, which may be constructed from predictions of thousands of members, even after the duplicates and highly correlated ones have been removed, produce test results which are better than the end-of-run population average. There may be potential to further improve the performance of the ensPop ensemble construction by, for example, determining which candidates are well correlated with the true training labels and assigning a higher weight to the predictions of those individuals on the test data. It is interesting to note that both ensemble approaches did comparatively better on the more difficult five class task than on the four class problem.

5.1   Discussion

This is a simple study into the potential of the GEML system to tackle multi-class classification tasks which may be supervised, unsupervised or semi-supervised in nature. Although the results are quite encouraging we feel that there is potential for improvement in the existing system. The obvious place to look for improvement is the all-important grammar. Our next steps will be to examine this to see how we can make it more effective. As a first move in that direction we will analyse the best individuals from our existing runs to determine which rules are contributing most and which are not performing. We will then modify the system applying this new information and use it to tackle a large, potentially noisy real-world medical dataset.

In the results section of this paper we have compared with several multi-class classification algorithms, and the reported results demonstrate that the most successful supervised technique is SVM. However, it is perhaps fair to point out that SVMs are not inherently multi-class, rather the algorithm usually (but not always) implements multi-class problems in either a "one versus one" or a "one versus all" approach, which in fact was how SVMs were implemented in this study. Thus, the performance of GEML and SVMs are, in one sense, not directly comparable. Given that the average performance of GEML on the two and three class problems is very competitive with that of the SVM, it is reasonable to hypothesise that equivalent performance to SVMs which use binary decomposition, might be expected on problems with greater numbers of classes if the GEML method were adapted to also perform multi-classification by way of binary decomposition. It may also be worth re-iterating that unlike SVMs the GEML setups all provide human readable solutions, which is an important consideration in many problem domains.

We have seen in this investigation that the GEML system which incorporates ideas from decision tree and cluster based learning has produced some statistically significant results. As GE is such a flexible paradigm there is no reason why alternative ML algorithms could not be incorporated instead. Once the candidate ML algorithm has some aspect which requires optimisation it should be suitable for an evolutionary approach.

6 Conclusions

In this paper we described a novel hybrid approach for solving multi-class problems in supervised, semi-supervised and unsupervised domains. The system which we call GEML, combines elements of decision tree logic and clustering techniques and incorporates these into a flexible grammatical evolution framework.

We have described the GEML framework in detail together with a set of experiments comparing GEML with several other state of the art ML algorithms. Results of these experiments were presented and discussed and we noted that the proposed system delivered competitive and generalizable accuracy which was shown to be statistically significant on all of the problems studied.

Our initial ensemble approaches have delivered encouraging results and in future work we may investigate other strategies which may be used to optimise the various parameters including the threshold values used both to determine which models to include in the original ensemble and which correlated models to exclude. Finally, it may be interesting to determine if improved classification accuracy may be achieved through creating a master ensemble from the separate ensembles generated from each GP run.

Acknowledgement. We gratefully acknowledge the support of Science Foundation Ireland. Grant number 10/IN.1/I3031.

References

1. Al-Madi, N., Ludwig, S.A.: Improving genetic programming classification for binary and multiclass datasets. In: Hammer, B., Zhou, Z.H., Wang, L., Chawla, N. (eds.) IEEE Symposium on Computational Intelligence and Data Mining, CIDM 2013. pp. 166–173. Singapore, 16–19 April 2013
2. Altman, N.S.: An introduction to kernel and nearest-neighbor nonparametric regression. Am. Stat. **46**(3), 175–185 (1992)
3. Azad, R.M.A., Ryan, C.: The best things don't always come in small packages: constant creation in grammatical evolution. In: Nicolau, M., Krawiec, K., Heywood, M.I., Castelli, M., García-Sánchez, P., Merelo, J.J., Rivas Santos, V.M., Sim, K. (eds.) EuroGP 2014. LNCS, vol. 8599, pp. 186–197. Springer, Heidelberg (2014). doi:10.1007/978-3-662-44303-3_16
4. Banzhaf, W.: Evolutionary computation and genetic programming. In: Lakhtakia, A., Martin-Palma, R.J. (eds.) Engineered Biomimicry, chap. 17, pp. 429–447. Elsevier, Boston (2013). http://www.sciencedirect.com/science/article/pii/B9780124159952000179
5. Barros, R.C., Basgalupp, M.P., De Carvalho, A.C., Freitas, A., et al.: A survey of evolutionary algorithms for decision-tree induction. IEEE Trans. Syst. Man Cybern. Part C Appl. Rev. **42**(3), 291–312 (2012)

6. Belhassen, S., Zaidi, H.: A novel fuzzy c-means algorithm for unsupervised heterogeneous tumor quantification in pet. Med. Phy. **37**(3), 1309–1324 (2010)
7. Bhowan, U., Johnston, M., Zhang, M.: Developing new fitness functions in genetic programming for classification with unbalanced data. IEEE Trans. Syst. Man Cybern. Part B Cybern. **42**(2), 406–421 (2012)
8. Boser, B.E., Guyon, I.M., Vapnik, V.N.: A training algorithm for optimal margin classifiers. In: Proceedings of the Fifth Annual Workshop on Computational Learning Theory, pp. 144–152. ACM (1992)
9. Breiman, L.: Bagging predictors. In: Machine Learning, pp. 123–140 (1996)
10. Breiman, L., Friedman, J., Stone, C.J., Olshen, R.A.: Classification and Regression Trees. CRC Press, New York (1984)
11. Castelli, M., Silva, S., Vanneschi, L., Cabral, A., Vasconcelos, M.J., Catarino, L., Carreiras, J.M.B.: Land cover/land use multiclass classification using GP with geometric semantic operators. In: Esparcia-Alcázar, A.I. (ed.) EvoApplications 2013. LNCS, vol. 7835, pp. 334–343. Springer, Heidelberg (2013). doi:10.1007/978-3-642-37192-9_34
12. Cowgill, M.C., Harvey, R.J., Watson, L.T.: A genetic algorithm approach to cluster analysis. Comput. Math. Appl. **37**(7), 99–108 (1999)
13. Deodhar, S., Motsinger-Reif, A.: Grammatical evolution decision trees for detecting gene-gene interactions. In: Pizzuti, C., Ritchie, M.D., Giacobini, M. (eds.) EvoBIO 2010. LNCS, vol. 6023, pp. 98–109. Springer, Heidelberg (2010). doi:10.1007/978-3-642-12211-8_9
14. Dietterich, T.: Ensemble methods in machine learning. In: Maimon, O., Rokach, L. (eds.) MCS 2000. LNCS, vol. 1857, pp. 1–15. Springer, Heidelberg (2000)
15. Dietterich, T.G., Bakiri, G.: Solving multiclass learning problems via error-correcting output codes. J. Artif. Intell. Res. **2**(1), 263–286 (1995)
16. Downey, C., Zhang, M., Liu, J.: Parallel linear genetic programming for multiclass classification. Genet. Programm. Evolvable Mach. **13**(3), 275–304 (2013). Special issue on selected papers from the 2011 European conference on genetic programming
17. Fitzgerald, J., Azad, R.M.A., Ryan, C.: GEML: Evolutionary unsupervised and semi-supervised learning of multi-class classification with grammatical evolution. In: Rosa, A., Merelo, J.J., Dourado, A., Cadenas, J.M., Madani, K., Ruano, A., Filipe, J. (eds.) ECTA. 7th International Conference on Evolutionary Computation Theory and Practice, paper 31. SCITEPRESS - Science and Technology Publications, Lisbon, Portugal, 12–14 November 2015
18. Fitzgerald, J., Ryan, C.: A hybrid approach to the problem of class imbalance. In: Matousek, R. (ed.) 19th International Conference on Soft Computing, MENDEL 2013, pp. 129–137, Brno, Czech Republic, 26–28 June 2013
19. Fogel, D.B.: What is evolutionary computation? IEEE Spectr. **37**(2), 26–28 (2000)
20. Freund, Y., Schapire, R.E., et al.: Experiments with a new boosting algorithm. In: ICML, vol. 96, pp. 148–156 (1996)
21. Fu, W., Johnston, M., Zhang, M.: Unsupervised learning for edge detection using genetic programming. In: Coello, C.A.C. (ed.) Proceedings of the 2014 IEEE Congress on Evolutionary Computation, pp. 117–124, Beijing, China, 6–11 July 2014
22. Greene, D., Tsymbal, A., Bolshakova, N., Cunningham, P.: Ensemble clustering in medical diagnostics. In: 17th IEEE Symposium on Computer-Based Medical Systems, CBMS 2004, Proceedings, pp. 576–581. IEEE (2004)

23. Hruschka, E.R., Campello, R.J., Freitas, A., De Carvalho, A.C., et al.: A survey of evolutionary algorithms for clustering. IEEE Trans. Syst. Man Cybern. Part C: appl. Rev. **39**(2), 133–155 (2009)

24. Ji, C., Ma, S.: Combinations of weak classifiers. IEEE Trans. Neural Netw. **8**(1), 32–42 (1997)

25. Kattan, A., Agapitos, A., Poli, R.: Unsupervised problem decomposition using genetic programming. In: Esparcia-Alcázar, A.I., Ekárt, A., Silva, S., Dignum, S., Uyar, A.Ş. (eds.) EuroGP 2010. LNCS, vol. 6021, pp. 122–133. Springer, Heidelberg (2010). doi:10.1007/978-3-642-12148-7_11

26. Kattan, A., Fatima, S., Arif, M.: Time-series event-based prediction: an unsupervised learning framework based on genetic programming. Inf. Sci. **301**, 99–123 (2015). http://www.sciencedirect.com/science/article/pii/S0020025515000067

27. Keijzer, M., Babovic, V.: Genetic programming, ensemble methods and the bias/variance tradeoff – introductory investigations. In: Poli, R., Banzhaf, W., Langdon, W.B., Miller, J., Nordin, P., Fogarty, T.C. (eds.) EuroGP 2000. LNCS, vol. 1802, pp. 76–90. Springer, Heidelberg (2000). doi:10.1007/978-3-540-46239-2_6

28. Kim, Y., Street, W.N., Menczer, F.: Feature selection in unsupervised learning via evolutionary search. In: Proceedings of the Sixth ACM SIGKDD International Conference on Knowledge Discovery and Data Mining, pp. 365–369. ACM (2000)

29. Koza, J.R.: Genetic programming: a paradigm for genetically breeding populations of computer programs to solve problems. Technical report (1990)

30. Maulik, U., Bandyopadhyay, S.: Genetic algorithm-based clustering technique. Pattern Recogn. **33**(9), 1455–1465 (2000)

31. Mierswa, I., Wurst, M.: Information preserving multi-objective feature selection for unsupervised learning. In: Proceedings of the 8th Annual Conference on Genetic and Evolutionary Computation, pp. 1545–1552. ACM (2006)

32. Mojsilović, A., Popović, M.V., Nešković, A.N., Popović, A.D.: Wavelet image extension for analysis and classification of infarcted myocardial tissue. IEEE Trans. Biomed. Eng. **44**(9), 856–866 (1997)

33. Morita, M., Sabourin, R., Bortolozzi, F., Suen, C.Y.: Unsupervised feature selection using multi-objective genetic algorithms for handwritten word recognition. In: 2013 12th International Conference on Document Analysis and Recognition, vol. 2, pp. 666–666. IEEE Computer Society (2003)

34. Muñoz, L., Silva, S., Trujillo, L.: M3GP – multiclass classification with GP. In: Machado, P., Heywood, M.I., McDermott, J., Castelli, M., García-Sánchez, P., Burelli, P., Risi, S., Sim, K. (eds.) EuroGP 2015. LNCS, vol. 9025, pp. 78–91. Springer, Heidelberg (2015). doi:10.1007/978-3-319-16501-1_7

35. Neshatian, K., Zhang, M.: Unsupervised elimination of redundant features using genetic programming. In: Nicholson, A., Li, X. (eds.) AI 2009. LNCS (LNAI), vol. 5866, pp. 432–442. Springer, Heidelberg (2009). doi:10.1007/978-3-642-10439-8_44

36. Omran, M.G., Engelbrecht, A.P., Salman, A.: Differential evolution methods for unsupervised image classification. In: The 2005 IEEE Congress on Evolutionary Computation, vol. 2, pp. 966–973. IEEE (2005)

37. O'Neill, M., Brabazon, A.: Grammatical differential evolution. In: Arabnia, H.R. (ed.) Proceedings of the 2006 International Conference on Artificial Intelligence, ICAI 2006, vol. 1, pp. 231–236, CSREA Press, Las Vegas, Nevada, USA, 26-29 June 2006. http://citeseerx.ist.psu.edu/viewdoc/summary?doi=10.1.1.91.3012

38. O'Neill, M., Brabazon, A.: Self-organizing swarm (SOSwarm): a particle swarm algorithm for unsupervised learning. In: IEEE Congress on Evolutionary Computation, CEC 2006, pp. 634–639. IEEE (2006)

39. O'Neill, M., Leahy, F., Brabazon, A.: Grammatical swarm: a variable-length particle swarm algorithm. In: Nedjah, N., de Macedo Mourelle, L. (eds.) Swarm Intelligent Systems, Studies in Computational Intelligence, vol. 28, pp. 59–74. Springer,
Heidelberg (2006) Chap. 5
40. O'Neill, M., Ryan, C.: Automatic generation of programs with grammatical evolution. In: Bridge, D., Byrne, R., O'Sullivan, B., Prestwich, S., Sorensen, H. (eds.)
Artificial Intelligence and Cognitive Science AICS 1999, No. 10, University College
Cork, Ireland, 1–3 September 1999. http://ncra.ucd.ie/papers/aics99.ps.gz
41. Pan, H., Zhu, J., Han, D.: Genetic algorithms applied to multi-class clustering for
gene expression data. Bioinf. Genom. Proteomics **1**(4), 279–287 (2003)
42. Pedregosa, F., Varoquaux, G., Gramfort, A., Michel, V., Thirion, B., Grisel, O.,
Blondel, M., Prettenhofer, P., Weiss, R., Dubourg, V., Vanderplas, J., Passos, A.,
Cournapeau, D., Brucher, M., Perrot, M., Duchesnay, E.: Scikit-learn: machine
learning in Python. J. Mach. Learn. Res. **12**, 2825–2830 (2011)
43. Raghavan, U.N., Albert, R., Kumara, S.: Near linear time algorithm to detect
community structures in large-scale networks. Phys. Rev. E **76**(3), 036106 (2007)
44. Ren, Y., Zhang, L., Suganthan, P.: Ensemble classification and regression-recent
developments, applications and future directions. IEEE Comput. Intell. Mag.
11(1), 41–53 (2016)
45. Ryan, C., O'Neill, M.: How to do anything with grammars. In: Barry, A.M. (ed.)
GECCO 2002: Proceedings of the Bird of a Feather Workshops, Genetic and Evolutionary Computation Conference. pp. 116–119. AAAI, New York, 8 Jul 2002.
http://www.grammatical-evolution.org/gews2002/howto.ps
46. Schapire, R.E.: The strength of weak learnability. Mach. Learn. **5**(2), 197–227
(1990)
47. Smart, W., Zhang, M.: Probability based genetic programming for multiclass
object classification. Technical report, CS-TR-04-7, Computer Science, Victoria
University of Wellington, New Zealand (2004). http://www.mcs.vuw.ac.nz/comp/
Publications/archive/CS-TR-04/CS-TR-04-7.pdf
48. Smart, W., Zhang, M.: Using genetic programming for multiclass classification by simultaneously solving component binary classification problems. In:
Keijzer, M., Tettamanzi, A., Collet, P., Hemert, J., Tomassini, M. (eds.) EuroGP
2005. LNCS, vol. 3447, pp. 227–239. Springer, Heidelberg (2005). doi:10.1007/
978-3-540-31989-4_20
49. Steinhaus, H.: Sur la division des corps matériels en parties. Bull. Acad. Pol. Sci.
Cl. III **4**, 801–804 (1957)
50. Wu, S.X., Banzhaf, W.: Rethinking multilevel selection in genetic programming. In: Krasnogor, N., Lanzi, P.L., Engelbrecht, A., Pelta, D., Gershenson, C.,
Squillero, G., Freitas, A., Ritchie, M., Preuss, M., Gagne, C., Ong, Y.S., Raidl, G.,
Gallager, M., Lozano, J., Coello-Coello, C., Silva, D.L., Hansen, N., Meyer-Nieberg,
S., Smith, J., Eiben, G., Bernado-Mansilla, E., Browne, W., Spector, L., Yu, T.,
Clune, J., Hornby, G., Wong, M.L., Collet, P., Gustafson, S., Watson, J.P., Sipper,
M., Poulding, S., Ochoa, G., Schoenauer, M., Witt, C., Auger, A. (eds.) GECCO
2011: Proceedings of the 13th Annual Conference on Genetic and Evolutionary
Computation, pp. 1403–1410. ACM, Dublin, Ireland, 12–16 July 2011, best paper
51. Zhang, M., Smart, W.: Multiclass object classification using genetic programming.
In: Raidl, G.R., Cagnoni, S., Branke, J., Corne, D.W., Drechsler, R., Jin, Y.,
Johnson, C.G., Machado, P., Marchiori, E., Rothlauf, F., Smith, G.D., Squillero,
G. (eds.) EvoWorkshops 2004. LNCS, vol. 3005, pp. 369–378. Springer, Heidelberg
(2004). doi:10.1007/978-3-540-24653-4_38

Adjudicated GP: A Behavioural Approach to Selective Breeding

Jeannie M. Fitzgerald$^{(\boxtimes)}$ and Conor Ryan

Biocomputing and Developmental Systems Group,
University of Limerick, Limerick, Ireland
{jeannie.fitzgerald,conor.ryan}@ul.ie

Abstract. For some time, there has been a realisation among Genetic Programming researchers that relying on a single scalar fitness value to drive evolutionary search is no longer a satisfactory approach. Instead, efforts are being made to gain richer insights into the complexity of program behaviour. To this end, particular attention has been focused on the notion of *semantic* space. In this paper we propose and unified hierarchical approach which decomposes program behaviour into *semantic*, *result* and *adjudicated* spaces, where adjudicated space sits at the top of the behavioural hierarchy and represents an abstraction of program behaviour that focuses on the success or failure of candidate solutions in solving problem sub-components. We show that better, smaller solutions are discovered when crossover is directed in adjudicated space. We investigate the effectiveness of several possible adjudicated strategies on a variety of classification and symbolic regression problems, and show that both of our novel *pillage* and *barter* tactics significantly outperform both a standard genetic programming and an enhanced genetic programming configuration on the fourteen problems studied. The proposed method is extremely effective when incorporated into a standard Genetic Programming structure but should also complement several other semantic approaches proposed in the literature.

Keywords: Program semantics · Selective breeding · Genetic programming

1 Background

Previously, research effort concerned with directed crossover has focused primarily on structural considerations when determining suitable crossover points in genetic programming trees (GP) [9]. See, for example [14,15].

One example of this effort can be seen with Context Aware Crossover (CAC) which was proposed in [18]. In this method, after two parents have been selected for crossover, one sub-tree is randomly chosen in the first parent and this sub-tree is then crossed over into *all* possible locations in the second parent and all generated children are evaluated. The best child (based on fitness) is selected and copied to the next generation. An advantage of such context-based crossovers is

© Springer International Publishing AG 2017
J.J. Merelo et al. (eds.), *Computational Intelligence*, Studies in Computational Intelligence 669,
DOI 10.1007/978-3-319-48506-5_8

increased probability of producing children which are better than their parents. On the other hand, it can be time consuming to evaluate the context of each sub-tree.

The notion of a potentially unifying, representation independent *geometric* crossover operator was initially explored in [20–22] in which the authors proposed viewing solution space as a geometric discrete space rather than a graph structure as was previously the norm. This new view of solution space supports the concept of *distance* by which we can imagine measuring somehow the distance between candidate solutions in the solution space or the distance between a solution and the global maximum/minimum.

These ideas provided a platform for looking at genetic operators such as crossover in a profoundly different way, where the emphasis is shifted away from the structure of solutions and focuses instead on their meaning as expressed by their semantics. Taking this approach facilitates the measurement and utilisation of distances in *semantic space* both between candidate solutions and between those solutions and the desired target.

There is currently no definitive agreement on the exact meaning of the term *semantics* in GP. However, a fairly widely adopted one [5,11,19], which we also adopt here, is that the semantics of a GP program is the vector of outputs that GP program produces on training data: i.e. each value in the output vector represents the result of evaluating the GP program on a single training instance.

Semantically Driven Crossover (SDC) was suggested in [3] in which they applied a technique which removed redundant and unreachable arguments from boolean GP programs and produced Reduced Order Binary Decision Diagrams (ROBBDs) which could then be used to compare program semantics. In that work, crossovers were discarded unless the offspring were *semantically different* from the parents. They reported superior performance and less code bloat using SDC and observed that bloat may be partially a result of intron creation during crossover.

This ideas of SDC was extended for real valued symbolic regression (SR) problems in [24,29] which proposed Semantic Aware Crossover (SAC), and investigated several possible scenarios in which they compared the semantics of offspring with their parents, and depending on the outcome accepted either or both offspring and/or parents into the new population. They also examined the effectiveness of a method which compared the semantics of sub-trees at proposed crossover points, only accepting offspring into the new population if the sub-trees were *not semantically equivalent*. They investigated SAC on several real-world SR problems and concluded that the sub-tree approach was the most effective of those trials, and that SAC was a useful technique for maintaining diversity a GP population.

[11] developed an approach to semantic crossover that utilised a type of brood recombination. In this method, called approximately geometric semantic crossover (SX), a pool of offspring is produced using sub-tree crossover for each mating pair, and the offspring whose semantics are closest to its parents is

selected unless there is a child with higher fitness than both parents, in which case it is selected regardless of semantics.

The alluring appeal of geometric semantic crossover is that effective operators of this type can provide a guarantee that the fitness of the offspring produced will be no worse than the fitness of its parent with the worst fitness, providing the semantics of the offspring lie between the semantics of its parents in solution space. The challenge is to design operators that have this property but which are also usable in practice. Against this background, Krawiec [10] investigated two approaches for generating offspring GP individuals whose semantics are *medial* (intermediate) with respect to the semantics of their parents. Both methods concentrated on approximating mediality by determining semantic similarity of sub-programs and basing crossovers on that - an approach that was much more computationally realistic than trying to deal with whole programs.

A novel approach influenced by Quantitative Genetics which the researchers called *phenotypic crossover* was suggested in [2]. This method aimed at maximising heritability by forcing offspring to have similar traits to their ancestors. The method delivered improved results over standard GP on several problems.

[23,28] adopted a strategy which abstracted one or two levels beyond semantic space into what they referred to as *behaviour space*. They explored the idea of *behaviour* based search using several binary classification problems, where rather than using an explicit fitness function they used open ended evolution guided by a type of novelty search (NS) [16,17]. In this approach, selection was based on the relative novelty of individual behaviour, where behaviour was represented by a binary descriptor. They experimented with two different binary descriptors, each of which was a vector of zeros and ones: one which captured whether the individual correctly predicted each class label or not (*accuracy descriptor*), and the other which captured the predicted class labels (*class descriptor*). They reported that NS outperformed standard GP on difficult problems but did slightly worse on trivial ones. Interestingly, they also observed that their application of NS seemed to eliminate or at least control bloat.

ESAGP (Error Space Alignment GP) was presented by [26] who explored mechanisms for finding compatible individuals based on their alignment in *error space*. In other work, [13] have recently proposed behavioural programming GP (BPGP), an approach which involves decomposing and archiving for later use, *sub-programs* which have good *utility*, where utility captures both the error of the sub-program and its perceived usefulness according to a decision tree methodology. They reported excellent results on a wide variety of problems.

A closely related but quite different idea was explored by [12] who applied a clustering technique to test based problems. Their Discovery of Objectives by Clustering (DOC) system clustered GP programs together if they had similar behaviour on the same test cases. They constructed *interaction matrices*, obtaining derived objectives to approximately represent this common behaviour and produce more effective search drivers. The method was compared with several other optimised GP algorithms and was shown to deliver statistically better results on a range of problems.

The notion of behaviour space has its origins in the area of robotics research [4] for which the terminology would seem to be eminently suitable. We propose to further refine and unify the terminology for GP such that behaviour space as defined in [23, 28] is decomposed into *semantic space, result space* and *adjudicated space*. In this view, taking classification as an example, result space maps to the class descriptor described in [23] and adjudicated space to the accuracy descriptor. With regard to symbolic regression, result space is equivalent to error space as described in [26].

An exhaustive description of the relevant literature is beyond the scope of this paper. For in depth reviews of semantic approaches the reader is directed to [25, 31].

In this work we extend and describe in greater detail the concept of *adjudicted* GP as previously proposed in [8].

2   Adjudicated GP (AGP)

Our method is analagous to the process of *selective breeding* (sometimes called artificial or unnatural selection), whereby humans breed animals or plants for certain traits – typically in order to domesticate them. Typically this approach involves several steps:

1. Decide which characteristics are important
2. Choose parents that show these characteristics
3. Select the best offspring from parents to breed the next generation
4. Repeat the process continuously

We have studied the effectiveness of our proposed approach on both classification and SR problems. The strategy is essentially the same for both tasks, but is, of necessity, slightly more complex in the case of symbolic regression. For the moment, we will explain the basic method as it applies to classification and defer description of symbolic details for later.

If we take a hypothetical example of a binary classification problem using GP, where each candidate solution is evaluated on the same ten problem instances. Supposing we have an individual which produces the semantics shown in Fig. 1 and we apply a threshold whereby if the semantic is $<= 50$ the instance is classified as belonging to class 1 and otherwise to class 2.

| 10 | 23 | 126 | 4 | 78 | 33 | 279 | 8 | 67 | 22 |

Fig. 1. Semantic descriptor.

This thresholding gives rise to the result descriptor shown in Fig. 2, where 0 represents instances of class 1 and 1 represents instances of class 2.

| 0 | 0 | 1 | 0 | 1 | 0 | 1 | 0 | 1 | 0 |

Fig. 2. Result descriptor.

Now, if we consider the ground truth for the 10 instances as shown in Fig. 3, we can *adjudicate*, i.e. make a judgement on the success or failure of our hypothetical individual on *each problem instance*, resulting in the adjudicated descriptor shown in Fig. 4. The adjudicated representation provides a fine grained view of individual performance compared to a scalar fitness value such as classification accuracy. We can easily imagine that even for a ten instance problem there may be many individuals with exactly the same fitness score, each of whom are correctly classifying a *different* set of instances.

| 1 | 1 | 1 | 1 | 1 | 0 | 0 | 0 | 0 | 1 |

Fig. 3. Ground truth.

As Krawiec et al. [13] pointed out, the reliance on a scalar fitness value to drive evolution "may be crippling because one cannot expect difficult learning and optimization problems to be efficiently solved by heuristic algorithms that are driven by a scalar objective function which provides low-information feedback".

| 0 | 0 | 1 | 0 | 1 | 1 | 0 | 1 | 0 | 0 |

Fig. 4. Adjudicated descriptor.

Figure 5 illustrates our proposed unified view of behaviour space in GP, encapsulating some of the important semantic concepts of behaviour space [13, 28], accuracy descriptor [28], class descriptor [23] and error space [26], previously described in Sect. 1.

As the behavioural hierarchy is ascended from the bottom up, the depth of knowledge about a candidate solution increases.

Semantic space is at the bottom of the hierarchy, representing the simple semantics of a genetic program as previously outlined in [11]. While the information available here may be helpful in understanding program behaviour, it does not tell us anything very useful about the candidate solution until a quantitative problem dependent step has been completed, such as thresholding in the case of classification or determining an error vector for symbolic regression tasks.

Once that algorithmic step has been completed, the resulting vector in *result space* provides more useful information into program behaviour than is available

Behaviour Space

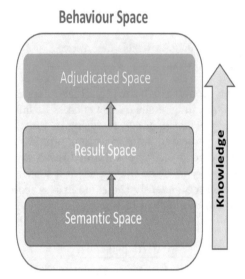

Fig. 5. Unified hierarchical view.

in semantic space. However, result space does not yet provide insight into the quality of the solution: if the program has an absolute error of 0.3 for a given training example, is that good or bad?, if the program predicts class 1 for training example 4 what does that mean for the quality of the solution?

At the highest level of the behavioural hierarchy is *adjudicated space* which realizes a fine-grained, insightful view of program behaviour.

Thus, we choose to pursue the goal of *effectively* navigating the solution space by focusing on program behaviour in *adjudicated space*. We are not concerned with program syntax or representation - simply on identifying which GP programs can solve which problem instances and using this information to determine a mating strategy inspired by selective breeding.

Thus, for each individual we decompose its adjudicated descriptor into a *for sale* list which is a list identifying the problem instances that it is able to correctly predict and a *wanted* list which details those instances which it has failed to correctly predict. See Figs. 6 and 7.

In traditional GP approaches, individuals are selected for mating based on fitness, where very unfit individuals usually have very limited opportunities to participate in crossover. In contrast, much of the research effort outlined in Sect. 1 explores various strategies for finding pairs or groups of individuals which are well-matched. according to some measure of semantic compatibility before combining them to produce new candidate solutions. Similar to other recent work on semantic aware crossover, we choose to explore the idea that it is more important that individuals are *compatible* in other, potentially more important ways than fitness.

The system that we propose simplifies the search for compatible mates by focusing on individual behaviour in *adjudicated space*. Once an adjudication has been made based on an individual's results, and the for sale and wanted lists have been populated, we can select a mate for that individual by choosing a prospective partner whose for sale list advertises the ability to solve instances that are on its wanted list.

As long as all individuals are adjudicated in the same way, if the for sale list of an individual contains a reference to an instance which is on the wanted list of another, then that pair of individuals are defined to be compatible to some degree and suitable candidates for selective breeding.

Table 1. Symbolic Regression Benchmarks. Where X is one of 20 values uniformly distributed between -1 *and* $+1$.

Name	Description
Nyg2 [30]	$X^4 + X^3 + X^2 + X$
Nyg3 [30]	$X^5 + X^4 + X^3 + X^2 + X$
Nyg4 [30]	$X^6 + X^5 + X^4 + X^3 + X^2 + X$

As we have already described, the adjudication process for classification tasks is quite straightforward regardless of the number of classes: each semantic is converted into a result (predicted class label) and this is judged to be correct or incorrect – for sale or wanted. The process is slightly more complicated for symbolic regression problems as the notion of correctness is not as clear cut. There are various ways that this issue could be approached including, for example, using the idea of "hits" where some defined minimum level of error on a training instance constitutes a hit. Preliminary experiments confirmed the intuition that setting the threshold value too low was unhelpful, particularly early in the evolutionary process. Thus we choose to use the population median mean absolute error (MAE) as the threshold for determining whether an instance is put on the for sale or wanted list. That is, for a given individual, if its error for a given training instance is less than the population median error for that instance it is adjudicated as being a success and the fitness case is put on the individual's for sale list, whereas if the error is greater than or equal to the population median error, the individual is adjudicated to have failed on that fitness case and the instance is put on the wanted list. This is an aspect that requires further experimentation and analysis.

Fig. 6. For sale list.

0	1	3	6	8	9

Fig. 7. Wanted list.

Once the for sale and wanted lists have been created for each individual in the population, there are probably many different strategies which could be adopted in order to maximise compatibility. For this preliminary study we have chosen to explore two different strategies which we call *pillage* and *barter*. Each of these strategies aims to find a mating pair which are approximately optimally compatible according to slightly differing objectives.

2.1   Pillage

The pillage method is a selfish strategy whereby for each individual the system seeks out and chooses that mate which offers the best return in terms of satisfying the wanted list of the first individual. For both SR and classification tasks, the wanted list is compared with the for sale list of every other individual and the individual which has greatest number of elements in the intersection of the two lists is selected.

2.2   Barter

As the name suggests, the barter approach is a more congenial strategy whereby each participating individual has the opportunity to gain from the transaction. When the barter tactic is employed, directed crossover only happens if each prospective parent lists instance/s on their for sale list which the other has on their wanted list.

At each generation the compatibility of each individual with every other individual is determined by calculating a *barter rate* which is analogous to the balanced accuracy measure used in classification. Similar to the pillage approach, the mate with highest compatibility is selected.

2.3   MuLambda GP (mlGP)

For the mlGP configuration crossover and mutation operate as for stdGP, however the selection process is slightly different: similar to the selection method explained in [6] where μ individuals from the initial population are used to generate λ offspring, and the best μ individuals from the entire $\mu + \lambda$ pool are selected to form the new population. In this instance $\lambda = 2 * \mu$; each crossover operation produces two offspring.

2.4   AGP Selection

In traditional GP a mating pool is often created by pre-selecting individuals according to some selection algorithm. Tournament selection is a popular approach, whereby the larger the tournament the more elitist the selection process.

We do not consider this method appropriate for Adjudicated GP (AGP) as the overall fitness score of any individual is largely irrelevant for the purpose of directing crossover. For example, we can easily imagine that an otherwise unfit individual may have the capability to correctly solve some small set of fitness cases. Thus, each individual in the population has the opportunity to participate in crossover events and we perform *post selection* at each generation, once mating is completed.

This is achieved by adopting a $\mu + \lambda$ approach similar to the mlGP method outlined above: a population of μ candidate solutions is used to produce a pool of λ new individuals consisting of parents and offspring, from which μ individuals are chosen by tournament selection to form the next generation. In the current implementation $\lambda = 2 * \mu$; each individual program participates in crossover with its compatible mate and each crossover, which occurs at a predetermined probability, produces two offspring.

Table 2. GP parameters. For classification problems a tournament size of 3 applies to standard and Mu Lambda (ML) experiments whereas tournaments of 7 candidates were used for the AGP setups.

Parameter	Value	Value
Problem type	**Classification**	**SR**
Population size	200	200
Max. generations	30	250
Max init depth	6	6
Max depth	16	16
Tournament size	3/9	7
Crossover prob.	0.9	0.9
Mutation prob.	0.1	0.1
Evolutionary model	Generational	Generational

3   Experimental Analysis

We choose to compare our proposed AGP variants (pillage and barter) with a standard GP (stdGP) set-up. In addition, and in order to isolate any potential effects we also compare with a basic $\mu + \lambda$ approach (mlGP) to determine if the selection strategy confers any benefits in and of itself.

3.1   Problems

We have selected several well known classification and symbolic regression benchmark problems on which to evaluate our proposed method. Classification problems consist of eight binary and three multi-class problems with varying numbers

Table 3. Classification benchmarks [1].

Dataset	Acronym	Instances	Attributes	Classes
Blood transfusion	BT	684	3	2
Liver disorders	BUPA	256	6	2
Caravan insurance	CAR	5946	85	2
German credit	GC	750	25	2
Haberman's survival	HS	255	4	2
Ionosphere	ION	348	35	2
Parkinsons disease	PK	195	22	2
Wisconsin breast cancer	WBC	452	9	2
Iris	IR	150	4	3
Vertebral column	VC	310	6	3
Wine	WN	178	9	3

of instances and attributes as outlined in Table 3, whereas the three symbolic regression tasks chosen are described in Table 1.

3.2 Parameters

Details of the function sets used are shown in Table 4. Note that constants are not used for any of the problems studied. Details of other relevant parameter settings are detailed in Table 2.

Table 4. Function sets used. Division, log, exp are protected.

Type	Function set
Classification	$+, -, *, /$
Symbolic regression	$+, -, *, /, sin, cos, log, exp, neg$

For the regression tasks the objective function aims to minimise MAE, whereas for all of the classification problems balanced accuracy is the objective function which the system strives to maximise. *Balanced accuracy* also known as *Average accuracy* which is a well know performance measure used in classification. This method modifies the calculation for overall accuracy to better emphasise the performance of each individual on *each* class as shown in Fig 8. The true positive (TP) rate is the proportion of positive instances which the individual classifies as positive, whereas the true negative (TN) rate is the proportion of negative instances which are classified as negative. The false positive (FP) and false negative (FN) rates are the proportions of negatives which are wrongly classified as positive and the proportion of positive instances which are incorrectly

$$BalAcc = 0.5 * \left(\frac{TP}{(TP+FN)} + \frac{TN}{(TN+FP)} \right)$$

Fig. 8. Balanced accuracy.

classified as negative. Generally, positive and negative instances correspond to instances of the minority and majority classes respectively.

3.3 Results

For classification benchmarks we report the average training and test balanced accuracy and the program size. Looking at the plots in Figs. 9, 10, 11 and 12 we can see that a consistent pattern emerges: the Barter approach produces the best performance on all of the benchmarks studied and stdGP delivers the weakest results overall. While the success of the barter approach compared to pillage is philosophically satisfying it is nevertheless somewhat surprising given

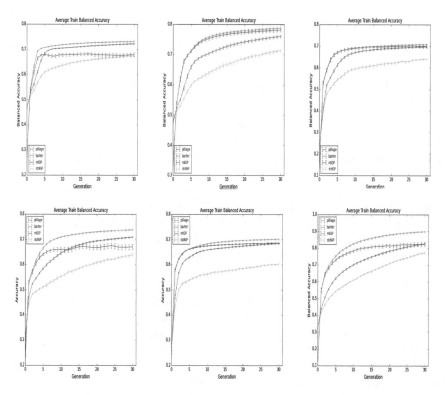

Fig. 9. Balanced training accuracy for (from top to bottom and left to right) BT, BUPA, CAR, GC, HS and ION data.

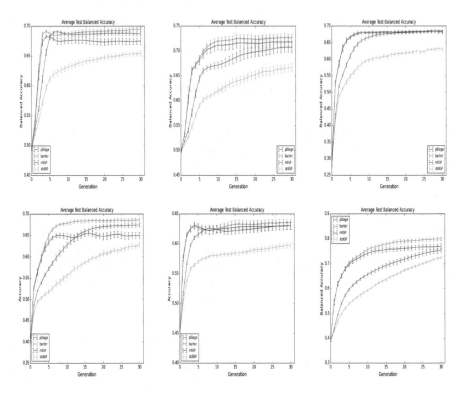

Fig. 10. Balanced test accuracy for (from left to right) BT, BUPA, CAR, GC. HS and ION data.

that there is almost inevitably a compromise associated with using the barter method. Interestingly, the mlGP set-up produces results which are much better than stdGP.

Turning our attention to the symbolic regression tasks we report both the number of successful runs together with the median MAE of the best of run individuals in Table 5. We use the same criteria for a successful run as in [30] which defines a successful run as one where any individual scores hits on all fitness cases – where a hit occurs when the absolute error is less than 0.01 for a single fitness case.

Looking at the results in Table 5 we can see that, similar to the classification performances, of the two AGP configurations, the Barter configuration delivers superior results in terms of the number of successful runs on all three problems, also outperforming both stdGP and mlGP, having almost twice as many successful runs as stdGP on all problems. When it comes to average MAE the situation is reversed, with both stdGP and mlGP producing the lowest median error, although the difference is not significant.

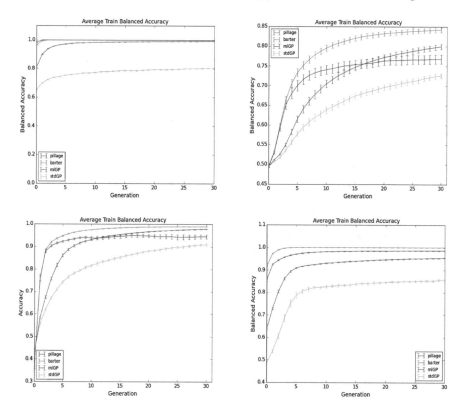

Fig. 11. Training accuracy for (from left to right) IRIS, PARK, WBC and WINE data.

Program Size. For *all* of the classification problems studied program growth during evolution was much more modest when either of the AGP variants were employed as can be seen in Fig. 13. Of course, there is some computational cost to the proposed AGP method as compatibility has to be determined for each prospective mate. However, this is strongly mitigated by the fact that solutions evolved using AGP are significantly smaller than those produced by stdGP or mlGP.

Smaller solutions are also produced by the AGP methods for the SR problems. This may partly be explained by the fact that evolution terminates if a perfect solution is found, and there are more of these discovered during AGP runs. Thus, one possible reason for smaller solutions is that the average size may be smaller when there are more early terminations.

Aside from the empirical evidence we do not currently have any solid explanation as to why solutions evolved using AGP are so much smaller than those produced using the canonical GP on the classification problems. However, we can hypothesise that the targeted nature of the method may reduce the possibility of intron development. In this regard, we note the similarity with the behaviour reported in [27] and also in [3].

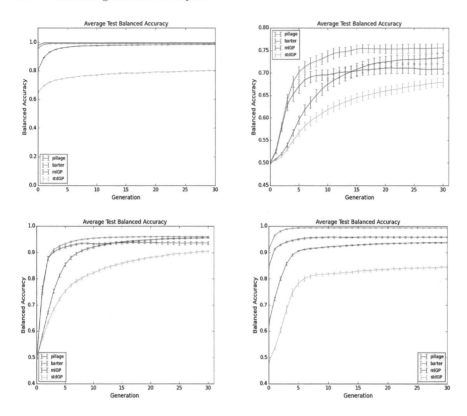

Fig. 12. Test accuracy for (from left to right) IRIS, PARK, WBC and WINE data.

AGP Growth Pattern. Looking at the plots for program size in Fig. 13 we can observe an interesting phenomenon: whereas the plots for stdGP exhibit a very similar gradient for each of the problems, those of the aGP approaches vary according to the problem. For example, if we compare program growth of barter on the CAR and IRIS problems we can see that on the CAR problem the rate of growth is almost as high as stdGP, whereas for the IRIS task, barter barely grows at all. As the CAR problem is a difficult one for GP (due to the larger number of attributes) and IRIS is known to be an easy problem, it seems possible that for aGP the rate of program growth may be somewhat correlated with problem difficulty. If this were the case, it would be a very useful algorithmic attribute.

Accordingly, we carried out an investigation on the IRIS dataset whereby we attempted to increase the problem difficulty by adding noise to a percentage of the attribute values. This dataset has 4 attributes, and we first added noise to the first attribute for a random 20 % of training instances, then increased the difficulty by *also* adding noise to 20 % of the attributes of the second instance, and so on.

Table 5. Correct solutions, median error and nodes used for, best-of-run individuals over 100 evolutionary runs.

	Method	Correct	Median MAE	Nodes
Nyg2	Barter	33	0.02	70.9
	Pillage	20	0.02	63.3
	mlGP	18	0.02	72.4
	stdGP	16	0.02	95.6
Nyg3	Barter	20	0.03	84.0
	Pillage	9	0.03	81.1
	mlGP	8	0.02	97,2
	stdGP	6	0.02	88.1
Nyg4	Barter	13	0.03	67.5
	Pillage	13	0.03	64.1
	mlGP	1	0.02	103.2
	stdGP	4	0.02	108.7

The results of this experiment which can be seen in Fig. 14 are a little surprising. We can see that the two configurations with the least and most amount of added noise achieve the best average accuracy on the training data and the one with the greatest added noise also does best on test data. Although it should be pointed out that the difference in the plots is over a very small scale. The result on the test data is not unexpected as an established technique to prevent over-fitting is to ad noise to the training data. However the the difference in training performances somewhat surprising.

When it comes to program growth, the plots suggest that added noise does contribute to an increase in program size, although the configuration with the most noise added does not on average produce the largest programs.

As we have carried out this preliminary experiment on a single dataset and, as previously mentioned, the IRIS classification task is known to be a relatively easy one – we should not draw any firm conclusions about a potential relationship between problem difficulty and the rate of program growth in aGP without extensive further investigation. The result is interesting nonetheless.

3.4   Statistical Analysis

To determine statistical significance, we carried out the non-parametric Friedman test which is regarded as a suitable test for the empirical comparison of the performance of different algorithms [7] as shown in Fig. 15. Using this approach, which does not simply count wins, but rather takes into account the relative performance of each algorithm compared with every other algorithm on all of the problems tackled, makes it easier to gain a clear insight into which are most effective. Results demonstrated that the AGP barter approach performed

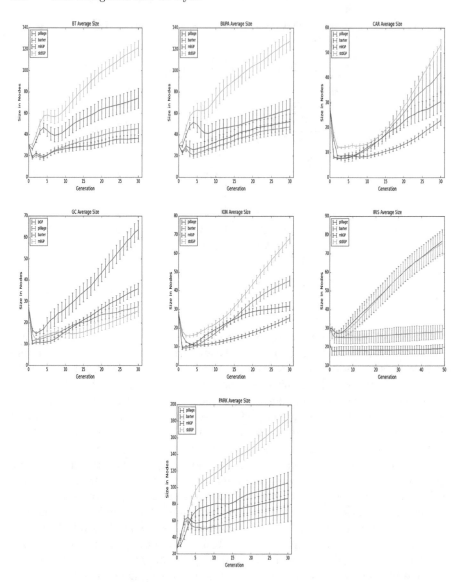

Fig. 13. Average program size for (from left to right, top to bottom) BT, BUPA, CAR, GC, ION, IRIS and PARK data.

significantly better than the other methods investigated on the selected benchmarks as post-hoc tests produced very small p-values (0.002 and 0.00006) for the differences between it and mlGP and stdGP respectively. A p-value of 0.003 was reported for the difference between pillage and stdGP.

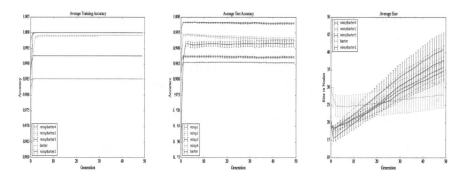

Fig. 14. Comparison of average training and test accuracy and average program size of barter strategy on IRIS data with added noise.

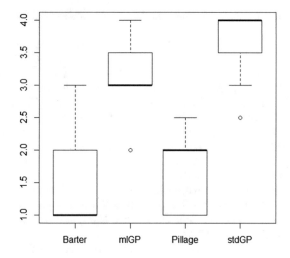

Fig. 15. Friedman plot of test accuracy on classification. Methods ranked from 1 to 4 where 1 is better.

3.5   Conclusions

In this paper, we have described a novel approach for selecting compatible candidates recombination using crossover in GP. Our approach which we call aGP operates in adjudicated space which is a high level of abstraction of program semantics. The results of the experiments that we have described illustrate that aGP is a promising methodology for evolutionary computation. Both aGP strategies performed consistently well across the range of benchmarks studied, with the barter method performing best overall.

The selective breeding process of aGP has several important advantages: it is relatively simple to implement; produces small programs showing no evidence of bloat and, most importantly, is independent of the chosen representation and could theoretically be applied to any evolutionary algorithm which uses a crossover operator and is applied to classification or regression problems.

The method works well when added to a standard GP structure but we see no reason why it could not also enhance several other semantic approaches in the literature, as its focused approach to recombination may ameliorate some of the known issues relating to size. Thus, as a next step we will investigate possible effects of incorporating aGP into some other semantic paradigm.

Acknowledgement. We gratefully acknowledge the support of Science Foundation Ireland. Grant No. 10/IN.1/I3031.

References

1. Bache, K., Lichman, M.: UCI machine learning repository (2013). http://archive.ics.uci.edu/ml
2. Bassett, J., Kamath, U., De Jong, K.: A new methodology for the GP theory toolbox. In: Soule, T., Auger, A., Moore, J., Pelta, D., Solnon, C., Preuss, M., Dorin, A., Ong, Y.S., Blum, C., Silva, D.L., Neumann, F., Yu, T., Ekart, A., Browne, W., Kovacs, T., Wong, M.L., Pizzuti, C., Rowe, J., Friedrich, T., Squillero, G., Bredeche, N., Smith, S., Motsinger-Reif, A., Lozano, J., Pelikan, M., Meyer-Nienberg, S., Igel, C., Hornby, G., Doursat, R., Gustafson, S., Olague, G., Yoo, S., Clark, J., Ochoa, G., Pappa, G., Lobo, F., Tauritz, D., Branke, J., Deb, K. (eds.) Proceedings of the Fourteenth International Conference on Genetic and Evolutionary Computation Conference, GECCO 2012, Philadelphia, Pennsylvania, USA, 7–11 July 2012, pp. 719–726. ACM (2012)
3. Beadle, L., Johnson, C.: Semantically driven crossover in genetic programming. In: Wang, J. (ed.) Proceedings of the IEEE World Congress on Computational Intelligence, Hong Kong, 1–6 June 2008. IEEE Computational Intelligence Society, pp. 111–116. IEEE Press (2008)
4. Brooks, R.A.: Cambrian Intelligence: The Early History of the New AI. MIT Press, Cambridge (1999)
5. Castelli, M., Vanneschi, L., Silva, S.: Prediction of the unified parkinsons disease rating scale assessment using a genetic programming system with geometric semantic genetic operators. Expert Syst. Appl. **41**(10), 4608–4616 (2014)
6. Deb, K., Pratap, A., Agarwal, S., Meyarivan, T.: A fast and elitist multiobjective genetic algorithm: NSGA-II. IEEE Trans. Evol. Comput. **6**(2), 182–197 (2002)
7. Demšar, J.: Statistical comparisons of classifiers over multiple data sets. J. Mach. Learn. Res. **7**, 1–30 (2006)
8. Fitzgerald, J.M., Ryan, C.: For sale or wanted: directed crossover in adjudicated space. In: Rosa, A., Merelo, J.J., Dourado, A., Cadenas, J.M., Madani, K., Ruano, A., Filipe, J. (eds.) Proceedings of the 7th International Joint Conference on Computational Intelligence, ECTA 2015, Paper No. 32, Lisbon, Portugal, 12–14 November 2015. SCITEPRESS - Science and Technology Publications (2015)
9. Koza, J.R.: Genetic programming: a paradigm for genetically breeding populations of computer programs to solve problems. Technical report (1990)
10. Krawiec, K.: Medial Crossovers for Genetic Programming. In: Moraglio, A., Silva, S., Krawiec, K., Machado, P., Cotta, C. (eds.) EuroGP 2012. LNCS, vol. 7244, pp. 61–72. Springer, Heidelberg (2012). doi:10.1007/978-3-642-29139-5_6

11. Krawiec, K., Lichocki, P.: Approximating geometric crossover in semantic space. In: Raidl, G., Rothlauf, F., Squillero, G., Drechsler, R., Stuetzle, T., Birattari, M., Congdon, C.B., Middendorf, M., Blum, C., Cotta, C., Bosman, P., Grahl, J., Knowles, J., Corne, D., Beyer, H.G., Stanley, K., Miller, J.F., van Hemert, J., Lenaerts, T., Ebner, M., Bacardit, J., O'Neill, M., Di Penta, M., Doerr, B., Jansen, T., Poli, R., Alba, E. (eds.) Proceedings of the 11th Annual Conference on Genetic and Evolutionary Computation, GECCO 2009, Montreal, 8–12 July 2009, pp. 987–994. ACM (2009)

12. Krawiec, K., Liskowski, P.: Automatic derivation of search objectives for test-based genetic programming. In: Machado, P., Heywood, M.I., McDermott, J., Castelli, M., García-Sánchez, P., Burelli, P., Risi, S., Sim, K. (eds.) EuroGP 2015. LNCS, vol. 9025, pp. 53–65. Springer, Heidelberg (2015). doi:10.1007/978-3-319-16501-1_5

13. Krawiec, K., O'Reilly, U.M.: Behavioral programming: a broader and more detailed take on semantic GP. In: Proceedings of the 2014 Conference on Genetic and Evolutionary Computation, pp. 935–942. ACM (2014)

14. Langdon, W.B.: Directed crossover within genetic programming. Research Note RN/95/71, University College London, UK. http://www.cs.ucl.ac.uk/staff/W.Langdon/ftp/papers/directed_crossover.pdf

15. Langdon, W.B.: Size fair and homologous tree genetic programming crossovers. In: Banzhaf, W., Daida, J., Eiben, A.E., Garzon, M.H., Honavar, V., Jakiela, M., Smith, R.E. (eds.) Proceedings of the Genetic and Evolutionary Computation Conference, Orlando, Florida, USA, 13–17 July 1999, vol. 2, pp. 1092–1097. Morgan Kaufmann (1999). http://www.cs.ucl.ac.uk/staff/W.Langdon/ftp/papers/WBL.gecco99.fairxo.ps.gz

16. Lehman, J., Stanley, K.O.: Exploiting open-endedness to solve problems through the search for novelty. In: ALIFE, pp. 329–336 (2008)

17. Lehman, J., Stanley, K.O.: Efficiently evolving programs through the search for novelty. In: Branke, J., Pelikan, M., Alba, E., Arnold, D.V., Bongard, J., Brabazon, A., Branke, J., Butz, M.V., Clune, J., Cohen, M., Deb, K., Engelbrecht, A.P., Krasnogor, N., Miller, J.F., O'Neill, M., Sastry, K., Thierens, D., van Hemert, J., Vanneschi, L., Witt, C. (eds.) Proceedings of the 12th Annual Conference on Genetic and Evolutionary Computation, GECCO 2010, Portland, Oregon, USA, 7–11 July 2010, pp. 837–844. ACM (2010)

18. Majeed, H., Ryan, C.: Using context-aware crossover to improve the performance of GP. In: Keijzer, M., Cattolico, M., Arnold, D., Babovic, V., Blum, C., Bosman, P., Butz, M.V., Coello Coello, C., Dasgupta, D., Ficici, S.G., Foster, J., Hernandez-Aguirre, A., Hornby, G., Lipson, H., McMinn, P., Moore, J., Raidl, G., Rothlauf, F., Ryan, C., Thierens, D. (eds.) Proceedings of the 8th Annual Conference on Genetic and Evolutionary Computation, GECCO 2006, Seattle, Washington, USA, 8–12 Jul 2006, vol. 1, pp. 847–854. ACM Press. http://www.cs.bham.ac.uk/~wbl/biblio/gecco2006/docs/p847.pdf

19. Moraglio, A., Krawiec, K., Johnson, C.G.: Geometric semantic genetic programming. In: Coello, C.A.C., Cutello, V., Deb, K., Forrest, S., Nicosia, G., Pavone, M. (eds.) PPSN 2012. LNCS, vol. 7491, pp. 21–31. Springer, Heidelberg (2012). doi:10.1007/978-3-642-32937-1_3

20. Moraglio, A., Poli, R.: Topological interpretation of crossover. In: Deb, K. (ed.) GECCO 2004. LNCS, vol. 3102, pp. 1377–1388. Springer, Heidelberg (2004). doi:10.1007/978-3-540-24854-5_131

21. Moraglio, A., Poli, R.: Geometric landscape of homologous crossover for syntactic trees. In: Proceedings of the 2005 IEEE Congress on Evolutionary Computation (CEC 2005), Edinburgh, 2–4 September 2005, vol. 1, pp. 427–434. IEEE (2005) http://privatewww.essex.ac.uk/~amoragn/cec2005fin.PDF

22. Moraglio, A., Poli, R., Seehuus, R.: Geometric crossover for biological sequences. In: Collet, P., Tomassini, M., Ebner, M., Gustafson, S., Ekárt, A. (eds.) EuroGP 2006. LNCS, vol. 3905, pp. 121–132. Springer, Heidelberg (2006). doi:10.1007/11729976_11

23. Naredo, E., Trujillo, L., Martínez, Y.: Searching for novel classifiers. In: Krawiec, K., Moraglio, A., Hu, T., Etaner-Uyar, A.Ş., Hu, B. (eds.) EuroGP 2013. LNCS, vol. 7831, pp. 145–156. Springer, Heidelberg (2013). doi:10.1007/978-3-642-37207-0_13

24. Nguyen, Q.U., Nguyen, X.H., O'Neill, M.: Semantic aware crossover for genetic programming: the case for real-valued function regression. In: Vanneschi, L., Gustafson, S., Moraglio, A., Falco, I., Ebner, M. (eds.) EuroGP 2009. LNCS, vol. 5481, pp. 292–302. Springer, Heidelberg (2009). doi:10.1007/978-3-642-01181-8_25

25. Pawlak, T.P., Wieloch, B., Krawiec, K.: Review and comparative analysis of geometric semantic crossovers. Genet. Program. Evolvable Mach. 16, 351–386 (2014)

26. Ruberto, S., Vanneschi, L., Castelli, M., Silva, S.: ESAGP – a semantic GP framework based on alignment in the error space. In: Nicolau, M., Krawiec, K., Heywood, M.I., Castelli, M., García-Sánchez, P., Merelo, J.J., Rivas Santos, V.M., Sim, K. (eds.) EuroGP 2014. LNCS, vol. 8599, pp. 150–161. Springer, Heidelberg (2014). doi:10.1007/978-3-662-44303-3_13

27. Trujillo, L., Muñoz, L., Naredo, E., Martínez, Y.: NEAT, there's no bloat. In: Nicolau, M., Krawiec, K., Heywood, M.I., Castelli, M., García-Sánchez, P., Merelo, J.J., Rivas Santos, V.M., Sim, K. (eds.) EuroGP 2014. LNCS, vol. 8599, pp. 174–185. Springer, Heidelberg (2014). doi:10.1007/978-3-662-44303-3_15

28. Trujillo, L., Naredo, E., Martinez, Y.: Preliminary study of bloat in genetic programming with behavior-based search. In: Emmerich, M., Deutz, A., Schuetze, O., Bäck, T., Tantar, E., Tantar, A.-A., Del Moral, P., Legrand, P., Bouvry, P., Coello, C.A. (eds.) EVOLVE - A Bridge between Probability, Set Oriented Numerics, and Evolutionary Computation IV. Advances in Intelligent Systems and Computing, vol. 227, pp. 293–305. Springer, Cham (2013)

29. Uy, N.Q., Hoai, N.X., O'Neill, M., McKay, B., Galván-López, E.: An analysis of semantic aware crossover. In: Cai, Z., Li, Z., Kang, Z., Liu, Y. (eds.) ISICA 2009. CCIS, vol. 51, pp. 56–65. Springer, Heidelberg (2009). doi:10.1007/978-3-642-04962-0_7

30. Uy, N.Q., Hoai, N.X., O'Neill, M., McKay, R.I., Galván-López, E.: Semantically-based crossover in genetic programming: application to real-valued symbolic regression. Genet. Program. Evolvable Mach. 12(2), 91–119 (2011)

31. Vanneschi, L., Castelli, M., Silva, S.: A survey of semantic methods in genetic programming. Genet. Program. Evolvable Mach. 15(2), 195–214 (2014). http://link.springer.com/article/10.1007/s10710-013-9210-0

Demand-Side Management: Optimising Through Differential Evolution Plug-in Electric Vehicles to Partially Fulfil Load Demand

Edgar Galván-López[1]([✉]), Marc Schoenauer[2], Constantinos Patsakis[3], and Leonardo Trujillo[4]

[1] School of Computer Science and Statistics, Trinity College Dublin, Dublin, Ireland
edgar.galvan@scss.tcd.ie
[2] TAO Project, INRIA Saclay & LRI - Univ. Paris-Sud and CNRS, Orsay, France
marc.schoenauer@inria.fr
[3] Department of Informatics, University of Piraeus, Piraeus, Greece
kpatsak@unipi.gr
[4] Doctorado en Ciencias de la Ingeniería,
Instituto Tecnológico de Tijuana, Tijuana, Mexico
leonardo.trujillo@tectijuana.edu.mx

Abstract. In this paper, we investigate the use of an stochastic optimisation bio-inspired algorithm, differential evolution, and proposed two fitness (cost) functions that can automatically create an intelligent scheduling for a demand-side management system so that it can use plug-in electric vehicles's (PEVs) batteries to partially and temporarily fulfil electricity requirements from a set of household units. To do so, we proposed two fitness functions that aim: (a) to use the most amount of energy from the batteries of PEVs while still guaranteeing that they can complete a journey, and (b) to enrich the previous function to reduce peak loads.

Keywords: Differential evolution · Demand-side management systems · Plug-in electric vehicles

1 Introduction

Evolutionary Algorithms (EAs) [1,2], also known as Evolutionary Computation systems, are influenced by the theory of evolution by natural selection. These algorithms have been with us for some decades and are very popular due to robust theoretical works [3–6] developed around them that have helped us to understand why they work (e.g., representations' properties) and due to their successful application in a variety of different and challenging problems, ranging from the automated design of an antenna carried out by NASA [7], the automated optimisation of game controllers [8], the automated evolution of Java code [9], up to the automated design of combinational logic circuits [10,11]. EAs can be considered a "black-box", as they do not require any specific knowledge

© Springer International Publishing AG 2017
J.J. Merelo et al. (eds.), *Computational Intelligence*, Studies in Computational Intelligence 669,
DOI 10.1007/978-3-319-48506-5_9

of the fitness function. They work even when, for example, it is not possible to define a gradient on the fitness function or to decompose the fitness function into a sum of per-variable objective functions.

In this work, we are interested in investigating the applicability of EAs in a dynamic and challenging problem in Demand-Side Management (DSM) Systems taken from Smart Grids where, in summary, the goal is to automatically create fine-grained solutions that indicate the amount of energy that can be taken from electric vehicles' (PEVs) batteries to partially satisfy energy demand in residential areas and reducing electricity peaks, whenever possible. The proposed approach and fitness functions used in our work (described in Sect. 2) is not amenable to analytic solution or simple gradient-based optimisation, hence search algorithms such as EAs are required.

DSM is normally considered as a mechanism or program, implemented by utility companies to control the energy consumption at the customer side [12]. DSM is an important research area in the Smart Grid (SG) community as shown by the increasing number of publications over the years (e.g., more than 2,000 papers have been published in this area where more than two thirds have been published since 2010 [13]).

DSM programs include different approaches (e.g., manual conservation and energy efficiency programs [14], Residential Load Management (RLM) [15,16]), where RLM programs based on smart pricing are amongst the most popular methods. The idea behind smart pricing is to encourage users to manage their loads, so that they can reduce electricity prices while, at the same time, the utility companies achieve a reduction in the peak-to-average ratio (PAR)[1] in load demand by shifting consumption whenever possible [13,15,17].

One of the major limitations of smart pricing is the fact that the electricity price is proportional to the electricity demand (i.e., a high number of appliances/devices connected to the grid results in having high electricity costs). To alleviate this problem, we propose the development of a demand-side *autonomous intelligent* management system that exploit plug-in electric vehicles' (PEVs) batteries. More precisely, our system uses the PEV's batteries to partially and temporarily fulfil the demand of end-use consumers instead of using only the electricity available from a substation transformer. This is possible thanks to the vehicle to grid technology (V2G), which is described as a system in which electric-drive vehicles can feed power to the grid with the appropriate communication/connection technologies acting as mobile generators of limited output [18,19].

The deployment of such a system implies several significant challenges, e.g. different driving patterns resulting in the amount of energy needed at the time of departure, amount of energy taken from the PEVs' batteries. To tackle this problem, we use an optimisation EA.

---

[1] Peak-to-average ratio is calculated by the maximum load demand for a period of time over the average load demand, so a lower PAR is normally preferred due to e.g. maintenance costs [16].

Thus, the main contribution of this research is a novel approach to balance the load demand from dozens of household units using both a substation transformer and PEVs' batteries as mobile energy storage units[2] by considering the *automatic* generation of solutions via the use of EAs. To this end, we are interested in maximising, in general, the use of available energy from the PEVs' batteries while ensuring that each of the PEVs can complete a journey to work, where the PEVs can be charged, and in particular, helping in the reduction of peak loads at the transformer level by using the most quantity of energy from the PEVs' batteries during these peak periods. This problem would be simple enough if it was not for the dynamicity associated to the problem and if we would not care about keeping the PAR relatively low.

To achieve this, we allow the DSM system to make fine-grained decisions (i.e., variable amount of energy requested) by using a continuous representation instead of using a discrete representation (i.e., turning a device/appliance on or off resulting in feeding/getting a constant amount of energy) as normally adopted in DSM [20].

To this end, we use a form of EAs, called Differential Evolution (DE) [21], that allows us to achieve this. More specifically, DE uses a vector of real-valued functions and we use them to represent an individual (potential solution) that specifies an energy consumption scheduling vector, which in turn indicates the amount of energy that should be taken from the PEVs' batteries aiming at fulfilling the goals previously described (e.g., maximising the energy consumption available from the batteries while at the same time reducing peak loads at the transformer level with associated constraints such as guaranteeing that each PEV would complete a journey to work). Details on how this algorithm works and its adoption in this research are described in Sect. 2.

1.1 Significance of This Research

From the 1980s, DSM has been studied extensively by the research community. Analysing the research carried on DSM is difficult if we consider that there are more than 2,000 scientific papers published only in the IEEE Xplore database. Inspired by the work conducted by Poli [22], where the author analysed titles, keywords and abstract of hundreds of papers, we also carried a similarity analysis relationship between hundred of papers[3] that discussed DSM and key terms of these papers for a quick and useful interpretation of the research carried out in this area.

As we will see, the research conducted in DSM over the last decades has evolved significantly, and due to space constraints, we only show the visual representation[4] of the research conducted from 1985 until 2009 (572 papers were analysed) and from 2010 until 2015 (1,841 were analysed), shown in Fig. 1.

---

[2] In this work, we use the terms "substation transformer" and "PEV's batteries" to differentiate between the two sources of energy.

[3] Source: IEEE Xplore database searching for "Demand-side Management". Last accessed date: 22/01/2015.

[4] Details on how these figures were produced can be found in [22].

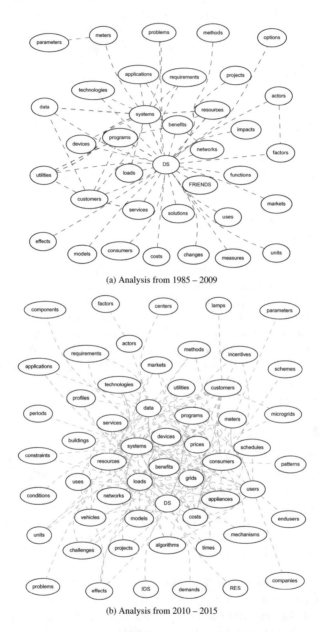

(a) Analysis from 1985 – 2009

(b) Analysis from 2010 – 2015

Fig. 1. Analysis of the publications on 'Demand Side Management' Systems from 1985 until 2009 (top) and from 2010 until 2015 (bottom). Only links (similarities) with strength greater than 40 in (a) and 60 (b) were passed to neato. The rest-length for repulsive forces between nodes was set to 9.

It is clear to see that some areas remain of vital importance in DSM, such as the benefits that DSM can offer to both customers and utility companies. There are, however, other areas of research emerging in DSM as shown at the bottom of Fig. 1 (research conducted over the last five years). Note, for example, the interest of investigating the impact/integration of electric vehicles in DSM. This is shown in the very core of Fig. 1 regarding the analysis from 2010 to 2015 (bottom of the figure). Other elements worth observing are data, users, devices.

The research presented in this work deals with these elements and shows its importance to DSM. Specifically, as mentioned previously, we are interested in using PEVs' batteries as mobile energy storage units to help the SG by designing an intelligent autonomous DSM.

Energy storage units, such as pumped hydroelectric energy storage units and compressed air energy storage units, have been with us around for several decades [23] via [24] and they have been used to provide both energy and ancillary services. Their use, however, have not been massively popular mainly because there is a cost associated with their acquisition and their corresponding installation. However, with the emergence of relatively new technologies (e.g., PEVs) and their relatively "easy" integration into the grid, it is necessary to account for autonomous and intelligent algorithms to exploit their capabilities. This in consequence can bring substantial benefits to both end-use consumers and to the grid (e.g., reduction of peak loads, savings in electricity costs; see [25–27] for a more detailed discussion of energy storage units' benefits).

The rest of this paper is organised as follows. In the following section we briefly introduce differential evolution and present our proposed approach. In Sect. 3, we present the experimental setup used in this work and Sect. 4 discusses the findings of our approach. Finally, in Sect. 5 we draw some conclusions.

2   Proposed Approach

2.1   Background

There are multiple EAs methods, such as Genetic Algorithms (GAs) [28], Genetic Programming (GP) [29], Differential Evolution (DE) [21]. All these methods use evolution as an inspiration to automatically generate potential solutions for a given problem. They differ, mainly, in the representation used (i.e., encoding of a solution). For example, the typical representation used in GAs is fixed bitstrings, GP's typical representation is tree-like structures, DE uses a vector of real-valued functions.

In this work, we use a DE algorithm given its natural representation (i.e., real-valued functions). Other bio-inspired algorithms can also use this type of representation, however, in this work we decided to use a DE given its efficiency for global optimisation over continuous search spaces [21]. By using this type of representation, we can have a more fine-grained action granularity (e.g., in this work, each element in the vector represents how much energy will be taken from the PEVs' batteries to feed electricity to household units), instead of using a more limited representation such as a bitstring representation that could indicate to take a pre-defined amount of energy (i.e., on or off) from PEVs' batteries to partially fulfil energy consumption from household units. We further discuss this later in this section.

The goal of DE is to evolve NP D-dimensional parameter vectors $x_{i,G} = 1, 2, \cdots, NP$, so-called population, which encode the potential solutions (individuals), i.e., $x_{i,G} = \{x^1_{i,G} \cdots, x^D_{i,G}\}, i = 1, \cdots, NP$ towards the global optimum solution (e.g., highest values when maximising a cost function). The initial population is randomly generated and this should be done by spreading the points across the entire search space (e.g., this could be achieved by distributing each parameter on an individual vector with uniform distribution between lower and upper bounds x^l_j and x^u_j). To automatically evolve these potential solutions over generations via the definition of a fitness function, DE uses the most common bio-inspired operators as commonly carried out in EAs: mutation and crossover to find the global optimum solution. Each of these operators is briefly explained in the following lines (refer to [21, 30] for a detailed description on how they work).

The mutation operator generates a mutant vector following one of the following strategies:

DE/rand/1

$$v_{i,G} = x_{r^i_1,G} + F \cdot (x_{r^i_2,G} - x_{r^i_3,G})$$

DE/best/1

$$v_{i,G} = x_{best,G} + F \cdot (x_{r^i_1,G} - x_{r^i_2,G})$$

DE/rand-to-best/1

$$v_{i,G} = x_{i,G} + F \cdot (x_{best,G} - x_{i,G}) + F \cdot (x_{r^i_1,G} - x_{r^i_2,G})$$

DE/best/2

$$v_{i,G} = x_{best,G} + F \cdot (x_{r^i_1,G} - x_{r^i_2,G}) + F \cdot (x_{r^i_3,G} - x_{r^i_4,G})$$

DE/rand/2

$$v_{i,G} = x_{r^i_1,G} + F \cdot (x_{r^i_2,G} - x_{r^i_3,G}) + F \cdot (x_{r^i_4,G} - x_{r^i_5,G})$$

where indexes $r_1, r_2, r_3, r_4 \in \{1, 2, \cdots, NP\}$ are random and mutually different. F is a real and constant factor $\in [0, 2]$ for scaling differential vectors and $x_{best,G}$ is the individual with best fitness value (e.g., highest value for a maximisation function) in the population at generation G.

The crossover operator increases the diversity of the mutated parameter vectors and is defined by:

$$v_{i,G+1} = (v_{1i,G+1}, v_{2i,G+1}, \cdots, v_{Di,G+1})$$

where:

$$v_{ji,G+1} = \begin{cases} v_{ji,G+1} & \text{if } randb(j) \leq CR \text{ or } j = rnbr(i), \\ x_{ji,G} & \text{otherwise} \end{cases}$$

where $j = 1, \cdots, D$, $randb(j)$ is the jth evaluation of a uniform random number generator with outcome $\in [0, 1]$. CR is the constant crossover rate $\in [0, 1]$. $rnbr(i)$ is a randomly chosen index $\in 1, 2, \cdots, D$ which ensures that $u_{i,G+1}$ receives at least one parameter value from $u_{i,G+1}$.

The performance of the DE algorithm depends on different factors, such as the values associated to the parameters (e.g., population size) as well as the variant of the operator used (e.g., variant of the mutation operator). This, intuitively means, that some preliminary runs would be normally required to determine which variant of an operator performs better on a given problem. We further discuss this in the following section.

2.2   Proposed Representation and Fitness Function

We now extend the natural DE representation to tackle the problem described throughout the paper and proceed to define the fitness (cost) function that allows the algorithm to automatically guide the evolutionary search.

Let N denote the number of household units (users), where the number of household units is $N \triangleq |\ N\ |$. For each household $n \in N$, let l_n^t denote the total load at time $t \in T \triangleq \{t_i, \cdots, t_f\}$. Without loss of generality, we assume that time granularity is 15 min. The load for household n, from t_i to t_f, is denoted by:

$$l_n \triangleq [l_n^{t_i}, \cdots, l_n^{t_f}] \tag{1}$$

From this, we can calculate the load across all household units N at each time $t \in [t_i, t_f]$ as follows:

$$L_t \triangleq \sum_{n \in N} l_n^t \tag{2}$$

Similarly, let M denote the number of plug-in electric vehicles available in N. For each electric vehicle $m \in M$, let E_m^t denote the energy that can be taken from the PEV at time $t \in T \triangleq \{t_i, \cdots, t_f\}$. Without loss of generality, we assume that time granularity is again 15 min. The total energy taken from an PEV from t_i until t_f is denoted by:

$$E_m \triangleq [E_m^{t_i}, \cdots, E_m^{t_f}] \tag{3}$$

We use this as a foundation to represent an individual that specifies an energy consumption scheduling vector. More specifically, an individual is represented by:

$$
E_M \triangleq
\begin{bmatrix}
E_{m_1}^{t_i}, \cdots, E_{m_1}^{t_f} \\
E_{m_2}^{t_i}, \cdots, E_{m_2}^{t_f} \\
\vdots \\
E_{m_M}^{t_i}, \cdots, E_{m_M}^{t_f}
\end{bmatrix}
\tag{4}
$$

where each E_m^t is a real value representing the quantity of energy taken from an PEV's battery. Each row represents the behaviour of a single PEV over the full period; each column represents the behaviour of all PEVs at a single time-slot. An individual in the EA is just a matrix E_M, unrolled to give a vector of real-valued functions, that is:

$$
E_1^{t_i}, \cdots, E_1^{t_f}, E_2^{t_i}, \cdots, E_2^{t_f}, \cdots, E_M^{t_i}, \cdots, E_M^{t_f}
\tag{5}
$$

Based on these definitions, the total energy taken across all M PEVs at each $t \in [t_i, t_f]$ can be calculated as:

$$
E_t \triangleq \sum_{m \in M} E_m^t
\tag{6}
$$

To automatically find good energy consumption scheduling solutions, defined in Eq. 4, we need to define a fitness (cost) function that indicates the quality of our evolved solution. First, we focus our attention in designing a cost function that tries to create valid solutions in terms of using the maximum allowed energy from each PEV (i.e., guaranteeing that a minimum state of charge (SoC) is left at the time of departure t_f).

From Eq. 3, we know the amount of energy available from $m \in M$ at any given period of time t denoted by E_m^t. Because each PEV can be charged at work and the distance from home to work remains constant, it is fair to assume the knowledge of a minimum SoC expressed in kW, denoted as m_{SoC}, at the time of departure t_f for each $m \in M$, so that it can reach work and be recharged at a lower rate. From this, we let the DE to assess a potential solution, denoted in Eq. 4, measuring the amount of energy taken from the PEVs. This is defined as:

$$
f_l(E_M) \triangleq \textbf{maximise} \ \frac{1}{\#\{m \in M\}} \sum_{m \in M} \frac{E_m + (E_m + 1)(m_{SoC} - E_m^{t_i})}{m_{SoC} \left(E_m^{t_i} - m_{SoC} \right)}
\tag{7}
$$

Equation 7 guides evolutionary search towards a local optimum solution since it only encourages the finding of solutions that maximise the use of allowable energy taken from PEVs' batteries. Thus, there is a necessity to further enrich this equation, so that a higher quantity of energy is taken from the PEVs' batteries whenever deemed necessary (e.g., higher consumption during high peak periods). We achieve this by using Eqs. 2 and 6 that indicate the load across all household units L_t at time t and the total energy taken across all PEVs E_t at time t, respectively; and we define a degree of importance for each time slot as t_r. Putting everything together we have:

$$f_g(E_M) \triangleq f_l(E_M) + \textbf{maximise} \ \frac{1}{\#\{m \in M\}} t_r \sum_{t=t_i}^{t_f} \frac{E_t}{L_t t_r} \forall t_r < T_r - \frac{1}{\#\{m \in M\}} t_r \sum_{t=t_i}^{t_f} \frac{E_t}{L_t t_r} \forall r \geq T_r$$
(8)

where T_r is a threshold that denotes the number of time slots that are considered critical (i.e., high peak period). In this work, as defined in this section and we discuss further afterwards, a number of time slots is defined by t_i and t_f, where a third is considered critical ($T_r = 20$).

3 Experimental Setup

3.1 Household Units

To test the scalability of our proposed approach, we simulated the consumption of 40 and 80 household units, where each of them uses between 10 and 20 appliances. As indicated throughout the paper, the goal is to use PEVs' batteries in an intelligent way to partially satisfy energy demand from the end-use consumers (recall that we work under the assumption that the PEVs can be charged at work).

To this end, we simulated that around 20 % of household units account for an PEV. To make this problem dynamic, we allowed the patterns of arrival (t_i), departure (t_f) and initial State of Charge (SoC) for each of the PEVs to vary for each of the 30 simulated working days. More specifically, the arrival and departure time for each of the PEVs have a 90-minute time frame starting at $t_i = 17{:}00$ and $t_f = 6{:}30$, respectively (i.e., arrival time could be between 17:00 and 18:30, whereas departure time could be between 6:30 and 8:00). The initial SoC_{t_i} for each of the PEVs for each of the simulated days is set between 48 % and 60 % and the final SoC_{t_f} is set between 30 % and 35 % to allow each PEV to reach work. Table 1 summarises the parameters used to simulate our scenario. We ran our simulations for a period of 30 days of simulated time.

3.2 Scenarios

As indicated in Sect. 2, we defined a bottom-up approach, where we defined, first, a fitness function that tries to maximise the energy that can be taken from the PEVs' batteries while ensuring that each of them reaches work, described in Eq. 7. We then enriched the fitness function by trying to also reduce the highest load demands at the substation transformer, described in Eq. 8 (i.e., use the most amount of energy from the batteries at high-peak time while at the same time ensuring the PAR remains low). We tested both fitness functions for 40 and 80 household units, resulting in four different scenarios.

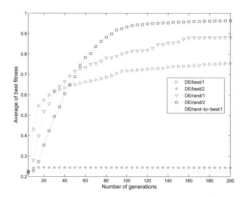

Fig. 2. Average of best fitness values of 30 independent runs for each of the five types of mutation operators tested in this work (see Sect. 2), using 500 individuals and 200 generations, to maximise energy consumption from electric vehicles' batteries (Eq. 7). Higher values are preferred.

3.3   Differential Evolution

As mentioned in Sect. 2, differential evolution's performance, as any other evolution-based algorithm, depends, among other things, on the values associated to the parameters that need to be specified for the algorithm (e.g., population size, number of generations), in general, and in the type of operator used, in particular.

No *a priori* knowledge is available to presume which mutation operator will perform better in the previously defined problem. To this end, we executed 30 independent runs of our proposed approach for each of the mutation variants, e.g., DE/rand/1, DE/best/1 (150[5] independent runs in total to find only the best mutation strategy) using the first proposed fitness function (Eq. 7) which maximises the energy taken from 11 PEVs' batteries to complement the energy consumption of 40 household units averaged over 30 days. Figure 2 shows the performance by measuring the average of best fitness per generation for each of the five mutation variants, using a population size of 500 individuals and 200 generations.

Clearly, the mutation strategy DE/rand/2 achieved the best performance and we used it to run our experiments to automatically find a (nearly) optimal solution. To obtain meaningful results, we performed 30 independent runs for each of the scenarios explained in the previous paragraphs (we executed 30 * 4 runs in total[6]). number of generations was reached.

---

[5] 30 independent runs * 5 variants of the mutation operator.

[6] 30 independent runs, 4 different scenarios (i.e., 40 and 80 household units, trying to maximise: (a) energy consumption from PEVs, and (b) energy consumption from PEVs while also considering reducing highest load peaks; for each of the set of household units used in this work).

Table 1. Summary of parameters used for our smart grid system.

Parameter	Value
Number of household units	40, 80
Number of appliances	Uniform in [10, 20]
Number of PEVs	≈20 % of houses have one PEV
Arrival and departure time	$t_i = [17{:}00,18{:}30]$ $t_f = [6{:}30,8{:}00]$
Frequency of making a decision	15 min
Number of times slots T	60
State of Charge at t_i	Uniform in [48, 60]
State of Charge at t_f	Uniform in [30, 35]

Table 2. Summary of parameters used for our evolutionary algorithm.

Parameter	Value
Population size	500
Length of the individual	T (see Table 1)
Height of the individual	Number of PEVs (see Table 1)
Generations	200
Crossover rate	0.5
Mutation strategy	DE/rand/2
Elitism	1 individual
Termination criterion	Maximum number of generations
Independent runs	30

As mentioned in Sect. 2, every element of the DE vector represents how much energy can be taken from the batteries of the PEVs. We make a decision every 15 min. Thus, the length of the individual that represent the solution is the number of time slots defined between 17:00 and 8:00am, whereas the height is defined by the number of electric vehicles used, as defined in Eq. 4. The parameters used in our experiments are summarised in Table 2.

4   Results

In the following paragraphs, we will analyse: (a) how the PEVs' batteries were used to partially satisfy the demand of a set of household units, (b) when the highest consumption from PEVs' batteries occurred, and finally, (c) the implications of the new consumption model via the analysis of the peak-to-average-ratio.

4.1   Maximising Energy Consumption from PEVs' Batteries

Let us start analysing our approach on how the batteries of the PEVs helped to partially satisfy the consumption demand from a set of household units. The averaged consumption over a period of 30 days of these household can be seen in Fig. 3(a, b) and (c, d) for 40 and 80 houses, respectively.

In the left-hand side of this figure, we show the distribution of consumption of both transformer and PEVs' batteries proposed by the differential evolution algorithm, when trying to maximise the consumption of energy from the PEVs' batteries via Eq. 7. More specifically, it aims at using all the possible energy available from the batteries while guaranteeing that each PEV has a minimal SoC at the time of departure (see Table 1) that guarantees that each PEV will reach work. The white-filled bars represent the electricity taken from the substation transformer whereas the remaining consumption to fulfil the load demand is taken from the PEVs' batteries. The latter is shown by the black-filled bars.

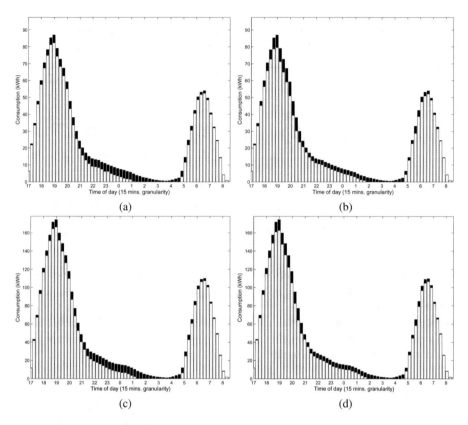

Fig. 3. Average of 30-day energy consumption for 40 (top) and 80 (bottom) household units, each using between 10–20 appliances. The consumption of energy from the transformer alone is shown by the white-filled bars whereas the black-filled bars represent the consumption taken from electric vehicles' batteries. Maximising energy consumption from electric vehicles only and maximising energy consumption from electric vehicles while considering also reducing highest load peaks are shown in the left-hand side and right-hand side of the figure, respectively.

Because we are interested in using the PEVs' batteries as mobile energy storage units, we are particularly interested in seeing how the energy consumption from these is managed by the differential evolution algorithm. In the first instance of our algorithm (i.e., maximising the energy consumption from the batteries of PEVs with associated constraints as formally described in Eq. 7, as mentioned previously), it is expected that the energy taken from the batteries would not follow a particular pattern (e.g., there is no correlation between the amount of energy consumption from PEVs and the energy needed by a number of household units). Indeed, this is the case as seen in the left-hand side of Fig. 3. For example, notice how the consumption from PEVs' is proportionally similar during both high-peak (e.g., 18:30–19:30) and low-peak periods (e.g., 22:00–23:00).

40 houses trying to maximise the use of PEVs' batteries

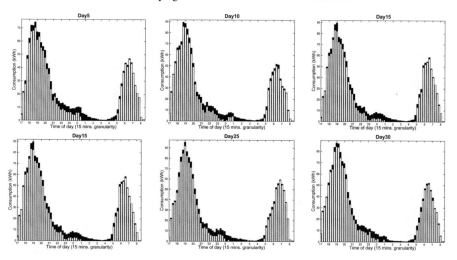

40 houses trying to maximise the use of PEVs' batteries while attempting to reduce high peak loads

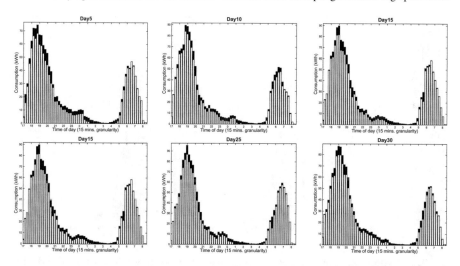

Fig. 4. Consumption per day (only 6 days chosen randomly) for 40 household units using each between 10 and 20 appliances. The consumption of energy from the transformer alone is shown by the white-filled bars whereas the black-filled bars represent the consumption taken from electric vehicles batteries. Maximising energy consumption from electric vehicles only is shown in the first two rows and maximising energy consumption from electric vehicles while considering also reducing highest load peaks is shown in the last two rows.

The situation is more encouraging when we consider the second instance of our algorithm (i.e., maximising energy consumption from PEVs' batteries while

80 houses trying to maximise the use of PEVs' batteries

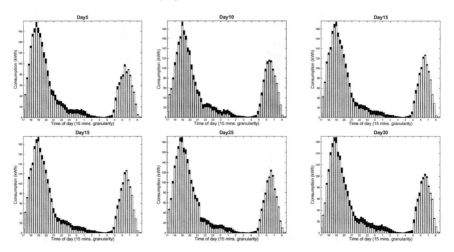

80 houses trying to maximise the use of PEVs' batteries while attempting to reduce high peak loads

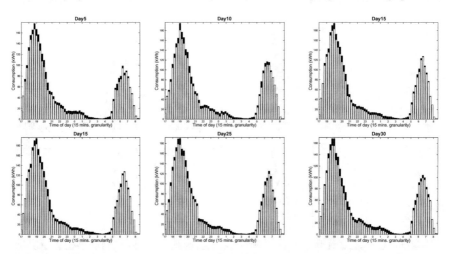

Fig. 5. Consumption per day (only 6 days chosen randomly) for 80 household units using each between 10 and 20 appliances. The consumption of energy from the transformer alone is shown by the white-filled bars whereas the black-filled bars represent the consumption taken from electric vehicles batteries. Maximising energy consumption from electric vehicles only is shown in the first two rows and maximising energy consumption from electric vehicles while considering also reducing highest load peaks is shown in the last two rows.

considering high-peak periods as formally described in Eq. 8), shown in the right-hand side of Fig. 3. As it can be observed, the proposed enriched fitness function is able to automatically produce results that can reduce the load peaks from the substation transformer by using more electricity from the PEVs' batteries.

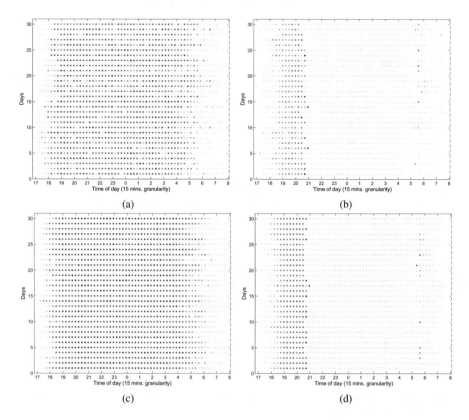

Fig. 6. Energy quantity taken from 11 (a, b) and 21 (c, d) electric vehicles over the range of time period studied in this work, from 17:00 until 8:00 (shown in the x-axis), for 30 days (shown in the y-axis) to help with the energy consumption of 40 (a, b) and 80 (c, d) household units. Darker-filled circles represent higher energy quantity taken from the PEVs' batteries. The enriched cost function, described in Eq. 8, follows a well-defined desired pattern (b, d), whereas the cost function that tends to find local optimum solutions, described in Eq. 7, tends to have a rather undesirable random pattern (a, c).

For example, notice how the consumption of energy from batteries is higher during high-peak periods (e.g., 18:30–19:30) and lower during low-peak periods (e.g., 22:00–23:00). Details on the consumption, per day for six days, can be seen in Figs. 4 and 5 when using 40 and 80 household units, respectively. The first two rows and the last two rows of these figures show the behaviour observed when using Eqs. 7 and 8, respectively.

4.2 Consumption from PEV's Batteries

In the previous paragraphs, we discussed and showed the results obtained by our approach using two fitness (cost) functions, formally described in Eqs. 7 and 8.

It is clear that the latter function is able to use a higher quantity of energy from the PEVs' batteries during high-peak periods compared to the effects observed when using the former function, as shown in the right-hand and left-hand side of Fig. 3, respectively, using 40 and 80 household units. This averaged result over a period of 30 simulated working days, however, does not inform us in detail when the highest consumption from batteries occurred (e.g., when and how much consumption from the batteries for every of the simulated days occurred). Some insight can be gained when analysing some days (see Figs. 4 and 5) but this still is limited since, due to page-limit constraints, not all days can be shown.

To this end, we kept track of the consumption from the PEVs' batteries during the simulated period of time (i.e., 17:00–8:00) for every day of the simulated days. The patterns of such consumption are shown in Fig. 6(a, b) and (c, d) for 40 and 80 household units, respectively.

Let us start our analysis when maximising the energy that can be taken from the batteries while ensuring that each PEV has the minimum SoC at the time of departure, defined in Eq. 7. The consumption pattern of this is shown in Fig. 6(a) and (c) for 40 and 80 household units, respectively. It should be noted that the higher the consumption from batteries is, the darker the dot. We can see that a random pattern is achieved by the cost function shown in Eq. 7. That is, for every recorded day, shown in the y-axis, the amount of energy taken from the batteries is rather random regardless of the period time, shown in the x-axis, except from 17:00–18:30 and 6:30–8:00, where the consumption from batteries is low. This can be explained due to the availability of PEVs during these periods. That is, as indicated in Sect. 3, each PEV has its own time of arrival and departure which varies during these periods of time.

We continue our analysis on the proposed enriched maximisation cost function, see Eq. 8, that aims at using the most amount of energy from the batteries of the PEVs while ensuring that each has a minimum SoC at the time of departure, and that tries to reduce the highest peak loads. The consumption pattern from the batteries is shown in Fig. 6(b) and (d) for 40 and 80 household units, respectively. This is a mirror image of what we discussed in the previous paragraph. That is, there is a well-defined pattern for each of the simulated days, shown in the y-axis, during the period of study, shown in the x-axis of the figure. We can observe that this cost function indeed achieves at using the most amount of energy when it is needed the most (high-peaks) as shown by the darker-filled squares while ensuring that the constraints are not violated (e.g., minimum SoC at the time of departure).

4.3 Peak-to-Average Ratio

As indicated previously, the peak-to-average ratio (PAR) is calculated by the maximum load demand for a period of time over the average load demand for the same period. It has been shown that a lower PAR is preferred [16].

We calculated the PAR considering the consumption from the substation transformer. Figure 7 shows the PAR for 40 (left-hand side) and 80 (right-hand side) household units for each of the 30 working simulated days

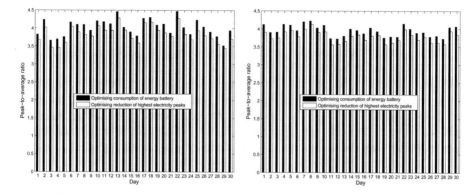

Fig. 7. Peak-to-average ratio (PAR) load demand achieved by our proposed approach when trying to maximise energy consumption from PEVs' batteries (black-filled bars) vs. when trying to maximise energy consumption from PEVs' batteries while aiming at reducing highest load peaks (white-filled bars), for 40 and 80 household units shown at the left-hand side and right-hand side of the figure, respectively. A lower PAR is preferred.

using our proposed approach. It is easy to observe that a higher PAR is achieved by the fitness (cost) function formally defined in Eq. 7, which goal is to use the most amount of energy from PEVs' batteries while at the same time aims at guaranteeing that each PEV has a minimum SoC at the time of departure compared to that PAR achieved by the enriched fitness function formally described in Eq. 8 that is built on the top of Eq. 7, which also tries to reduce the highest peak loads.

This, in fact, is to be expected given that the fitness function described in Eq. 8 does consider an associated ranking system (recall that a third of time slots are considered critical, i.e., high peak period) that is able to reflect smoothly the consumption from the substation transformer as shown by the low PAR achieved by this enriched fitness function for each day of the 30 simulated days, denoted by the white-filled bars in Fig. 7.

5   Conclusions

Demand-Side Management (DSM) refers to programs that aim to control the energy consumption at the customer side of the meter. Different techniques have been proposed to achieve this. The most popular techniques are those based on smart pricing (e.g., critical-peak pricing, real-time pricing). One major limitation of smart pricing is the fact that the electricity price is proportional to the electricity demand. This is particularly true for the time-of-use smart pricing adopted in some countries, where there is a financial incentive to use the electricity at night given its lower cost compared to its cost during day time. To alleviate this problem, we proposed the development of a demand-side *autonomous intelligent* management system that exploit plug-in electric vehicles' (PEV) batteries. More precisely, our system uses the PEV's batteries to partially and temporarily fulfil the demand of end-use consumers instead of using only the electricity available from a substation transformer.

To this end, we used an stochastic bio-inspired method, differential evolution, given its natural representation (encoding of a solution) that allows to make fine-grained decision in terms of the exact energy that can be taken from PEVs' batteries to partially and temporarily fulfil energy requirements from a set of household units. To effectively do so, we proposed two fitness (cost) functions that achieve: (a) to use the maximum allowed energy from PEVs while still guaranteeing they can complete a journey, and (b) to use the maximum energy consumption from PEVs batteries while considering reducing high-peak periods.

From experimental results, it is clear that the enriched fitness function is able to use the most amount of energy from PEVs, it is also able to reduce peak loads and it is also able to achieve a lower PAR compared to the other 'simple' fitness function proposed in this work.

Acknowledgements. Edgar Galván López's research is funded by an ELEVATE Fellowship, the Irish Research Council's Career Development Fellowship co-funded by Marie Curie Actions. The first author would also like to thank the TAO group at INRIA Saclay & LRI - Univ. Paris-Sud and CNRS, Orsay, France for hosting him during the outgoing phase of the ELEVATE Fellowship. The authors would like to thank all the reviewers for their useful comments that helped us to significantly improve our work.

References

1. Bäck, T., Fogel, D.B., Michalewicz, Z. (eds.): Evolutionary Computation 1: Basic Algorithms and Operators. IOP Publishing Ltd., Bristol (1999)
2. Eiben, A.E., Smith, J.E.: Introduction to Evolutionary Computing. Springer Verlag, Heidelberg (2003)
3. Galván-López, E., McDermott, J., O'Neill, M., Brabazon, A.: Defining locality in genetic programming to predict performance. In: IEEE Congress on Evolutionary Computation, pp. 1–8. IEEE (2010)
4. Fagan, D., O'Neill, M., Galván-López, E., Brabazon, A., McGarraghy, S.: An analysis of genotype-phenotype maps in grammatical evolution. In: Esparcia-Alcázar, A.I., Ekárt, A., Silva, S., Dignum, S., Uyar, A.Ş. (eds.) EuroGP 2010. LNCS, vol. 6021, pp. 62–73. Springer, Heidelberg (2010). doi:10.1007/978-3-642-12148-7_6
5. Galván-López, E., Dignum, S., Poli, R.: The effects of constant neutrality on performance and problem hardness in GP. In: O'Neill, M., Vanneschi, L., Gustafson, S., Esparcia Alcázar, A.I., Falco, I., Cioppa, A., Tarantino, E. (eds.) EuroGP 2008. LNCS, vol. 4971, pp. 312–324. Springer, Heidelberg (2008). doi:10.1007/978-3-540-78671-9_27. http://dl.acm.org/citation.cfm?id=1792694.1792723
6. McDermott, J., Galván-Lopéz, E., O'Neill, M.: A fine-grained view of GP locality with binary decision diagrams as ant phenotypes. In: Schaefer, R., Cotta, C., Kołodziej, J., Rudolph, G. (eds.) PPSN 2010. LNCS, vol. 6238, pp. 164–173. Springer, Heidelberg (2010). doi:10.1007/978-3-642-15844-5_17
7. Lohn, J., Hornby, G., Linden, D.: An evolved antenna for deployment on Nasa's space technology 5 mission. In: O'Reilly, U.-M., Yu, T., Riolo, R., Worzel, B. (eds.) Genetic Programming Theory and Practice II, vol. 8, pp. 301–315. Springer, New York (2005). doi:10.1007/0-387-23254-0_18

8. Galván-López, E., Swafford, J.M., O'Neill, M., Brabazon, A.: Evolving a Ms. Pac-Man controller using grammatical evolution. In: Chio, C., et al. (eds.) EvoApplications 2010. LNCS, vol. 6024, pp. 161–170. Springer, Heidelberg (2010). doi:10. 1007/978-3-642-12239-2_17

9. Cody-Kenny, B., Galván-López, E., Barrett, S.: locoGP: improving performance by genetic programming java source code. In: Proceedings of the Companion Publication of the 2015 on Genetic and Evolutionary Computation Conference, GECCO Companion 2015, pp. 811–818. ACM, New York (2015). doi:10.1145/ 2739482.2768419

10. Galván-López, E., Poli, R., Coello, C.A.C.: Reusing code in genetic programming. In: Keijzer, M., O'Reilly, U.-M., Lucas, S., Costa, E., Soule, T. (eds.) EuroGP 2004. LNCS, vol. 3003, pp. 359–368. Springer, Heidelberg (2004). doi:10.1007/ 978-3-540-24650-3_34

11. Galván-López, E.: Efficient graph-based genetic programming representation with multiple outputs. Int. J. Autom. Comput. **5**(1), 81–89 (2008). doi:10.1007/ s11633-008-0081-4

12. Masters, G.M.: Renewable and Efficient Electric Power Systems. Wiley-Interscience, Hoboken (2004)

13. Galván-López, E., Harris, C., Trujillo, L., Vázquez, K.R., Clarke, S., Cahill, V.: Autonomous demand-side management system based on Monte Carlo tree search. In: IEEE International Energy Conference (EnergyCon). IEEE Press, Dubrovnik, Croatia, pp. 1325–1332 (2014)

14. Pacific Northwest GridWise Testbed Demonstration Projects, Part I. Olympic Peninsula Project, October 2007

15. Galvan, E., Harris, C., Dusparic, I., Clarke, S., Cahill, V.: Reducing electricity costs in a dynamic pricing environment. In: Proceedings of Third IEEE International Conference on Smart Grid Communications (SmartGridComm). IEEE Press, Tainan, Taiwan, pp. 169–174 (2012)

16. Mohsenian-Rad, A., Wong, V., Jatskevich, J., Schober, R., Leon-Garcia, A.: Autonomous demand-side management based on game-theoretic energy consumption scheduling for the future smart grid. IEEE Trans. Smart Grid **1**(3), 320–331 (2010). doi:10.1109/TSG.2010.2089069

17. Galván-López, E., Curran, T., McDermott, J., Carroll, P.: Design of an autonomous intelligent demand-side management system using stochastic optimisation evolutionary algorithms. Neurocomputing **170**, 270–285 (2015). doi:10.1016/j.neucom.2015.03.093. http://www.sciencedirect.com/science/article/ pii/S0925231215009303

18. Kempton, W., Letendre, S.E.: Electric vehicles as a new power source for electric utilities. Transp. Res. Part D: Transp. Environ. **2**(3), 157–175 (1997). doi:10.1016/ S1361-9209(97)00001-1

19. Kempton, W., Tomic, J.: Vehicle-to-grid power fundamentals: calculating capacity and net revenue. J. Power Sources **144**(1), 268–279 (2005). doi:10.1016/j.jpowsour. 2004.12.025

20. Brooks, A., Lu, E., Reicher, D., Spirakis, C., Weihl, B.: Demand dispatch: using real-time control of demand to help balance generation and load. IEEE Power Energy Mag. **8**, 20–29 (2010)

21. Storn, R., Price, K.: Differential evolution a simple and efficient heuristic for global optimization over continuous spaces. J. Global Optim. **11**(4) 341–359 (1997). doi:10.1023/A:1008202821328

22. Poli, R.: Analysis of the publications on the applications of particle swarm optimisation. J. Artif. Evol. Appl. **2008**, 4:1–4:10 (2008). doi:10.1155/2008/685175

23. Cheung, K., Cheung, S., Silva, R., Juvonen, M., Singh, R., Woo, J.: Large-Scale Energy Storage Systems ISE2. Imperial College London, London (2003)
24. Wang, Z., Gu, C., Li, F., Bale, P., Sun, H.: Active demand response using shared energy storage for household energy management. IEEE Trans. Smart Grid **4**(4), 1888–1897 (2013). doi:10.1109/TSG.2013.2258046
25. Eyer, J.M., Corey, G.P.: Energy Storage for the Electricity Grid: Benefits and Market Potential Assessment Guide. A study for the DOE Energy Storage Systems Program. Prepared by Sandia National Laboratories
26. Eyer, J.M., Iannucci, J.J., Corey, G.P.: Energy storage benefits, market analysis handbook: a study for the DOE Energy Storage Systems Program. Prepared by Sandia National Laboratories
27. Mohd, A., Ortjohann, E., Schmelter, A., Hamsic, N., Morton, D.: Challenges in integrating distributed energy storage systems into future smart grid. In: IEEE International Symposium on Industrial Electronics, 2008, ISIE 2008, pp. 1627–1632 (2008). doi:10.1109/ISIE.2008.4676896
28. Goldberg, D.E.: Genetic Algorithms in Search, Optimization and Machine Learning, 1st edn. Addison-Wesley Longman Publishing Co., Inc., Boston (1989)
29. Koza, J.R.: Genetic Programming: On the Programming of Computers by Means of Natural Selection. MIT Press, Cambridge (1992)
30. Qin, A.K., Huang, V.L., Suganthan, P.N.: Differential evolution algorithm with strategy adaptation for global numerical optimization. Trans. Evol. Comp. **13**(2), 398–417 (2009). doi:10.1109/TEVC.2008.927706

Particle Convergence Time in the Deterministic Model of PSO

Krzysztof Trojanowski and Tomasz Kulpa[(⊠)]

Faculty of Mathematics and Natural Sciences, School of Exact Sciences,
Cardinal Stefan Wyszyński University in Warsaw,
Wóycickiego 1/3, 01-938 Warsaw, Poland
{k.trojanowski,tomasz.kulpa}@uksw.edu.pl

Abstract. A property of particles in Particle Swarm Optimization (PSO), namely, particle convergence time (*pct*) is a subject of theoretical and experimental analysis. For the model of PSO with inertia weight a new measure for evaluation of *pct* is proposed. The measure evaluates number of steps necessary for a particle to obtain a stable state defined with any precision. For this measure an upper bound formula of *pct* is derived and its properties are studied. Four main types of particle behaviour characteristics are selected and discussed. In the experimental part of the research effectiveness of swarms with different characteristics of their members are verified. A new type of swarm control improving efficiency of a swarm in escaping traps of local optima is proposed and experimentally verified.

1 Introduction

Particle swarm optimization (PSO) [1] belongs to a big family of modern heuristic optimization methods. A number of versions of PSO has already been proposed sharing the same paradigm of stochastic, population-based method of exploration in the given space of solutions in searching for the best one. In our research we selected one of the earlier versions of PSO proposed in [2]. Like in other methods, the population consists of members called here particles which represent solutions from the given space. Particles are also equipped with memories which store attractors, that is, solutions best found so far by the particles. A working group of particles controlled by the method is called a swarm. After the initialization of a swarm the cycle of iterations performs the search process. The distinctive features of PSO are: (1) application of particle memory as well as the mechanism of memory sharing by groups of neighbouring solutions, (2) the method of finding new solutions based on the idea of displacement originated from the real-world. Unlike other metaheuristics, every iteration consists of two main steps: particles memory update and the displacement of particles within the space of solutions. In PSO less-fit particles do not die, that is, there is no "survival of the fittest" mechanism typical for the evolutionary approach. The rules of displacement make use of the information from the memory and are expressed by equations which may differ to each other for different versions

© Springer International Publishing AG 2017
J.J. Merelo et al. (eds.), *Computational Intelligence*, Studies in Computational Intelligence 669,
DOI 10.1007/978-3-319-48506-5_10

of PSO. Particularly, in the version of PSO which we selected for analysis the rules of displacement use the inertia weight parameter.

Numerous applications of PSO confirmed its usefulness and potential but also motivate for studying their theoretical properties. Particularly, a particle stability analysis is a subject of great interest. One of the main aims is estimation of particle parameter ranges guaranteing the convergent movement within the given boundaries of the search space. For the purpose of theoretical analysis some assumptions concerning randomness have always to be made. The most restricted deterministic approach simply eliminates stochastic coefficients from the velocity equation [3]. Other approaches implement expected values of the particle locations [4,5] (which is called a first order stability analysis), or the variance of the locations (a second order stability analysis) [6–8].

In the presented research we study behaviour of a particle which parameters belong to the ranges guaranteing the convergent movement, particularly, we evaluate the time necessary for a particle to enter the convergent state. This kind of a swarm property was already investigated for swarms consisting of a number of particles [9]. In a series of experiments for different particle configurations authors evaluated number of iterations necessary to satisfy the assumed convergence condition. However, in our paper we propose a new method of evaluation of a particle convergence time based on the deterministic model of PSO with inertia weight [5] and a new convergence condition. This means that the analysis concerns a particle model based on the following assumptions:

1. the particle moves in one-dimensional search space — there is no need to consider n-dimensional velocity vectors due to the fact, that all the velocity parameters are evaluated individually for each of the search space coordinates and they do not influence to each other in any way,
2. random values in the velocity equation are replaced by constant values, thus the rules of the particle movement become deterministic,
3. both the local and the global attractor remain in the same place of the search space over the entire time of the modelled particle behaviour,
4. there is just one particle to observe — due to the previous assumption that global attractor remains unchanged, no communication between particles exists in fact,
5. values of parameters in the velocity equation belong to the ranges guaranteeing convergent movement of the modelled particle.

Thus stability is defined as:

$$\lim_{t \to \infty} \mathbf{x}(t) = \mathbf{y} \tag{1}$$

where $\mathbf{y}$ is a constant point in the search space.

The selected model based on the five assumptions allows to generate convergent trajectories of particle locations over space. However, it has to be stressed that the shape of the trajectory does not influence the proposed measure and the only important information is the number of steps necessary for the particle to get and stay in the sufficiently close neighborhood of $\mathbf{y}$.

The paper consists of six sections. In Sect. 2 the model of PSO with inertia weight is briefly described. Section 3 presents the proposed new measure of particle convergence time. Discussion of the new measure properties can be found in Sect. 4. Section 5 presents the results of experiments with swarms consisting of particles with different types of characteristic. Section 6 concludes the paper.

2   The PSO Model

The PSO model with inertia weight implements the following velocity and position equations:

$$\begin{cases} v_{t+1} = w \cdot v_t + \varphi_1(y_t - x_t) + \varphi_2(y_t^* - x_t), \\ x_{t+1} = x_t + v_{t+1} \end{cases} \tag{2}$$

where $\varphi_1 = r_1 c_1$, $\varphi_2 = r_2 c_2$, and c_1, c_2 represent acceleration coefficients, $r_1, r_2 \sim U(0, 1)$. In the further analysis the stochastic components φ_1 and φ_2 are substituted by constant values. We also assume that both attractors are constant over time.

From this pair of equations a recursive formula can be derived [5]:

$$x_{t+1} = (1 + w - \varphi_1 - \varphi_2)x_t - wx_{t-1} + \varphi_1 y + \varphi_2 y^* \tag{3}$$

which allows to evaluate the particle location, assuming that its two previous locations and its attractor are known. This way a basic simplified dynamic system can be defined:

$$\mathbf{P}_{t+1} = M \times \mathbf{P}_t, \tag{4}$$

where:

- $\mathbf{P}^t$ — the particle state made up of its current position x_t and the previous one x_{t-1}.
- M — the dynamic matrix whose properties determine the transformations of the particle state.

Results from dynamic system theory say that the transformations of the particle state depend on the eigenvalues of M. Further analysis of the dynamic matrix originated from Eq. (3) allowed to define the region in the parameters space where eigenvalues of M are smaller than 1. All the configuration parameters sets originated from this region guarantee that the particles do not diverge during the process of search.

In [5] authors show that the particle equilibrium point is a weighted average of its personal best y and global best y^* positions: $\frac{\varphi_1 y + \varphi_2 y^*}{\varphi_1 + \varphi_2}$. However, just for simplicity of calculations and without loss of generality we can assume, that $y^* = y$. In this case we can substitute ϕ for $\varphi_1 + \varphi_2$ and Eq. (3) is reformulated as follows:

$$x_{t+1} = (1 + w - \phi)x_t - wx_{t-1} + \phi y \tag{5}$$

Eventually, the following *stable region*, that is, a set of convergent configurations satisfies the following system of inequalities was derived:

$$\begin{cases} w > 0 \wedge w < 1, \\ \phi > 0, \\ w > 0.5\,\phi - 1 \end{cases} \tag{6}$$

Since the first presentation of the above-mentioned boundaries of the stable region a number of publications appeared discussing the problem of boundaries definition based on different assumptions concerning stochastic components in the velocity equations and stability of attractors. For more details the reader is referred to [6,7,10–12]. Particularly, in [12] a set of inequalities coinciding with Eq. (6) has been derived. In our research presented in the further text we implement the stable region as it is defined by Eq. (6) having in mind that constraint $w > 0$ represents just the intuitive assumption that inertia of a moving object should not be negative.

3   The Proposed Measure

3.1   Particle Convergence Time

Even if the stable region is given, it is also interesting to know the number of steps necessary for the particle to obtain its stable state for different configurations (ϕ, w). In this case "obtaining stable state" means that the distance between current and the next location of the particle is never greater than the given threshold value δ.

Lets define a set of natural numbers $S(\delta)$ for a given $\delta > 0$ such that:

$$s \in S(\delta) \iff |x_{t+1} - x_t| < \delta \text{ for all } t \geq s. \tag{7}$$

We define the *particle convergence time (pct)* for given $\delta > 0$ as follows:

$$pct(\delta) = \min\{s \in S(\delta)\}. \tag{8}$$

The particle convergence time *pct* is the minimal number of steps necessary for the particle to obtain its stable state as defined above. For estimation of the particle convergence time we use Eq. (3).

3.2   Upper Bound Formula for *pct*

Recurrent equations are difficult to analyse, however, an explicit closed form of the recurrence relation Eq. (5) is also known [5]:

$$x_t = k_1 + k_2\lambda_1^t + k_3\lambda_2^t, \tag{9}$$

where

$$k_1 = y, \tag{10}$$

$$k_2 = \frac{\lambda_2(x_0 - x_1) - x_1 + x_2}{\gamma(\lambda_1 - 1)}, \tag{11}$$

$$k_3 = \frac{\lambda_1(x_1 - x_0) + x_1 - x_2}{\gamma(\lambda_2 - 1)}, \tag{12}$$

$$x_2 = (1 + w - \phi)x_1 - wx_0 + \phi y, \tag{13}$$

$$\lambda_1 = \frac{1 + w - \phi + \gamma}{2}, \tag{14}$$

$$\lambda_2 = \frac{1 + w - \phi - \gamma}{2}, \tag{15}$$

$$\gamma = \sqrt{(1 + w - \phi)^2 - 4w}. \tag{16}$$

Thus, the distance between two subsequent values of the particle locations x_{t+1} and x_t equals:

$$|x_{t+1} - x_t| = |k_2\lambda_1^t(\lambda_1 - 1) + k_3\lambda_2^t(\lambda_2 - 1)|. \tag{17}$$

From the triangle inequality it follows that:

$$|x_{t+1} - x_t| \le |k_2||\lambda_1|^t|\lambda_1 - 1| + |k_3||\lambda_2|^t|\lambda_2 - 1|. \tag{18}$$

We are interested in the minimal number of steps s after which the condition

$$|x_{t+1} - x_t| < \delta \tag{19}$$

is satisfied for all $t \ge s$. To obtain this we employ the fact, that:

$$|a| < \delta/2 \land |b| < \delta/2 \Rightarrow |a + b| < \delta \tag{20}$$

whore $|\cdot|$ is the absolute value.

Thus, we look for such t_1 and t_2, that:

$$|k_2||\lambda_1|^{t_1}|(\lambda_1 - 1)| < \delta/2, \tag{21}$$

$$|k_3||\lambda_2|^{t_2}|(\lambda_2 - 1)| < \delta/2. \tag{22}$$

and we get:

$$t_1 > \frac{\ln \delta - \ln(2|k_2||\lambda_1 - 1|)}{\ln |\lambda_1|}, \tag{23}$$

$$t_2 > \frac{\ln \delta - \ln(2|k_3||\lambda_2 - 1|)}{\ln |\lambda_2|}. \tag{24}$$

Now, we define $s = \max(t_1, t_2)$, where t_1 and t_2 are minimal natural number satisfying Eqs. (23) and (24) respectively. From (20), (21) and (22) it follows that for all $t \ge s$ the condition (19) is satisfied.

In the case where γ is a complex number consisting of just an imaginary value, that is, when $(1 + w - \phi)^2 < 4w$, the reasoning presented above may be simplified. In this case the following is satisfied: $|\lambda_1| = |\lambda_2|$ and $|\lambda_1 - 1| = |\lambda_2 - 1|$. Let's denote: $|\lambda| = |\lambda_1| = |\lambda_2|$ and $|\lambda - 1| = |\lambda_1 - 1| = |\lambda_2 - 1|$. Then, Eq. (18) can be expressed as:

$$|x_{t+1} - x_t| \leq |\lambda|^t |\lambda - 1|(|k_2| + |k_3|). \tag{25}$$

In this case we look for such t that:

$$|\lambda|^t |\lambda - 1|(|k_2| + |k_3|) < \delta, \tag{26}$$

which is equivalent to

$$t > \frac{\ln \delta - \ln(|\lambda - 1|(|k_2| + |k_3|))}{\ln |\lambda|}, \tag{27}$$

Now, we define s as a minimal natural number t satisfying Eq. (27). From (25) and (27) it follows that for all $t \geq s$ the condition (19) is satisfied.

For both cases, that is, real and imaginary value of γ, the defined number of steps s satisfies condition (7). Due to the fact, that $pcs(\delta)$ is defined as a minimal number satisfying condition (7), we get $pcs(\delta) \leq s$.

Thus, Eqs. (23), (24) and (27) give us the analytic upper bounds for the particle convergence time, which is denoted as $pctb(\delta)$. The explicit formula for $pctb(\delta)$ is

$$pctb(\delta) = \max \left(\frac{\ln \delta - \ln(2|k_2||\lambda_1 - 1|)}{\ln |\lambda_1|}, \right.$$
$$\left. \frac{\ln \delta - \ln(2|k_3||\lambda_2 - 1|)}{\ln |\lambda_2|} \right) \tag{28}$$

for real value of γ and

$$pctb(\delta) = \frac{\ln \delta - \ln(|\lambda - 1|(|k_2| + |k_3|))}{\ln |\lambda|} \tag{29}$$

for imaginary value of γ.

4   Visualizations of *pctb* Characteristics

Particle convergence time depends on three groups of parameters: values of factors in a velocity update rule, initial localization and velocity and fitness landscape. Parameters from the first group, that is, ϕ and w define character (or temperament) of a particle. An example graph of $pctb(\phi, w)$ is presented in the subsection below. The next subsection presents example graphs of $pctb(x_0, x_1)$, that is, convergence times of particles with selected characters respectively to their starting conditions. Particle trajectories for respective types of character are also presented. The third subsection shows how $pctb(x_0, x_1)$ and $pctb(x_0, v)$ graphs vary respectively to the changes in a particle character.

4.1   Particle Convergence Time for Different Types of Particles

The characteristics of *pctb* as a function of particle configuration parameters ϕ and w share common shape presented in Fig. 1. The Figure depicts the $pctb(\phi, w)$ characteristic obtained from a grid of evaluation points starting from a configuration $[\phi = 0.025,\ w = 0.044]$ and changing with step 0.05 in both directions. This choice of method for the function graph generation is due to the fact, that γ appears in the denominator of Eqs. (11) and (12), so, it cannot equal zero. Unfortunately, this is the case, when $w = 1 + \phi - 2\sqrt{\phi}$, that is, there exist points in the stable region for which the upper bound for their convergence time can be evaluated neither with formula (28) nor (29). For better visibility the $pctb(\phi, w)$ axis has logarithmic scale and the evaluation points from outside the stable region have assigned the constant value 5000.

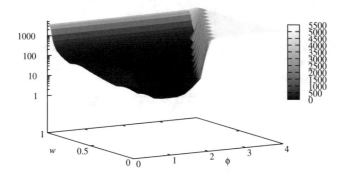

Fig. 1. Particle convergences $pctb(\phi, w)$ for example starting conditions: $x_0 = 1$ and $x_1 = -8.1$.

Figure 1 shows that when the inertia weight w is low the convergence times are also low and increase as the inertia grows. Additionally, *pctb* increases also for the cases when ϕ approaches boundary values, both left and right, however, for the right boundary the increase is much higher than for the left.

4.2   *pctb* as a Function of Initial Location and Velocity

For ϕ and w values satisfying Eq. (6) the shapes of $pctb(x_0, x_1)$ can be classified into four main types (samples for $\delta = 0.0001$ are depicted in Fig. 2).

A: convergence is fast when the velocity is low (x_1 close to x_0) and the initial location x_0 is irrelevant in every case;
B: a transitional state between states A and C;
C: convergence is fast when the velocity is adjusted to the location and directed toward the attractor;
D: the particle has almost no inertia, so, the less distance from x_1 to the attractor, the less value of *pctb*.

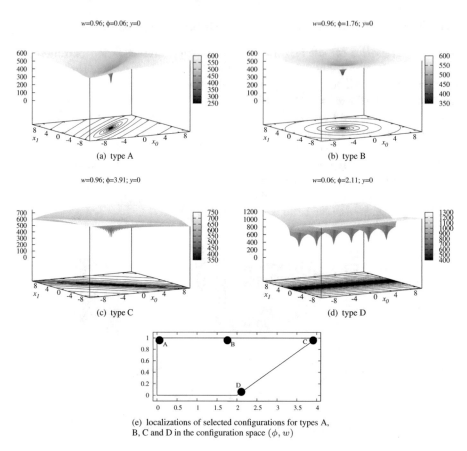

(a) type A   (b) type B

(c) type C   (d) type D

(e) localizations of selected configurations for types A, B, C and D in the configuration space (ϕ, w)

Fig. 2. Graphs of $pctb(x_0, x_1)$ for selected configurations (ϕ, w) which represent four types of characteristics: A, B, C and D.

Figure 3 shows subsequent locations over time for particle configurations selected for presentation in Fig. 2 and for three different starting locations each. Graphs of particle trajectories similar to the ones presented in Fig. 3 can be also found in [4], however, in that case they were obtained for different particle parameter space. Graphs with trajectories can be also found in other publications, particularly in [5], however, they are not classified respectively to the subarea in the stable region of the configuration space they appear.

In Fig. 3(a) "A" particles are represented by three cases: with low (starting points x_0 and x_1 at (8,8.1)) and high initial velocity: ((8,1.1) and (1,8.1)). High inertia and weak attraction toward y make the movement smooth and the subsequent steps short in every case. For the high initial velocity oscillations around the attractor are higher. In the case of "B" particles (Fig. 3(b)) oscillations appear in every graph, however, the length of subsequent steps is irregular: when the particle moves away from y with high velocity, sometimes

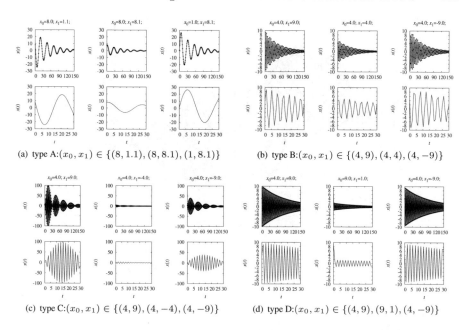

Fig. 3. Particle trajectories for the four types of characteristics: A, B, C and D, and for three example starting locations; a view of 150 locations (top figures) and a close-up of the first 30 locations (bottom figures).

the attracting force almost stops it, velocity decreases and the particle turns back slowly, whereupon runs toward the attractor with a high velocity again. Figure 3(c) presents a "zig-zag" trajectories of "C" particles which amplitude cyclically increases and decreases. The amplitude of oscillations is less when the initial velocity is adjusted to the initial location and directed toward the attractor. Clearly, the fastest convergence of $pctb$ is obtained when x_1 has the same absolute value as x_0 but the opposite sign. Figure 3(d) also presents a "zig-zag" trajectories of "D" particles but without cycles in the magnitude of amplitude. In this case particle also converges to the attractor faster when the initial velocity is adjusted to the initial location, however, in this case the velocity has to be adjusted so as to locate x_1 in the nearest neighborhood of the attractor.

Finally, it is worth noting that different types of trajectories appear for different types of particle characteristics, which confirms the proposed selection of types and allows one to assume that none of the selected types is a subtype of any other.

4.3 Properties of $pctb(\phi, w)$

One can observe that for all the selected configurations (ϕ, w) graphs of $pctb(x_0, x_1)$ printed in Fig. 2 have regular, symmetric shape. For every case the minimum value $pctb(x_0, x_1)$ is in $(0, 0)$ which is obvious when we remind, that this

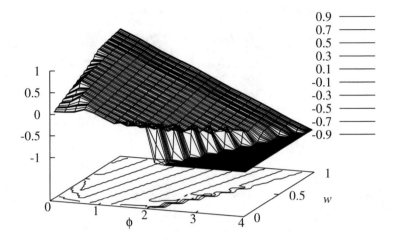

Fig. 4. Angle coefficient for major axis $a_{ma}(\phi, w)$.

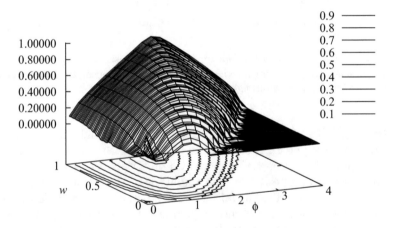

Fig. 5. Eccentricity $\varepsilon(\phi, w)$.

is the location of the attractor y^*. Clearly, when the first and the second location of a particle is the same as its attractor, the particle has already obtained its equilibrium state and no further move of the particle will be observed, that is, its convergence time equals zero.

The contours of $pctb(x_0, x_1)$ printed on the plane $X_0 X_1$ resemble ellipses. An ellipse can be characterised by two parameters: an angle coefficient for the major axis (a_{ma}) and an eccentricity (ε), that is, a ratio between the minor and the major radius. We assume, that for a given configuration (ϕ, w) all the contours are ellipses and have the center in the point $x_0 = 0; x_1 = 0$. One can observe, that the values of a_{ma} and ε for different configuration (ϕ, w) differ to each other. Pictures of the two characteristics: $a_{ma}(\phi, w)$ and $\varepsilon(\phi, w)$ are presented in Figs. 4 and 5 and described below.

The Angle Coefficient for the Major Axis $a_{\mathrm{ma}}(\phi, w)$. For estimation of $a_{\mathrm{ma}}(\phi, w)$ a linear regression method was used. For every configuration (ϕ, w) a set of 1000 random pairs (x_0, x_1) is generated such that $pctb(x_0, x_1, \phi, w)$ for each of them is not greater than a fixed limit L. This way one can obtain a cloud of points having the shape of a contour for $pctb(x_0, x_1, \phi, w) = L$ on the plane $X_0 X_1$. L is always selected so as to fit all the points in the cloud into the range $[-10, 10]$ for both x_0 and x_1. Then, the equation $y = ax + b$ for the linear regression line of this group of points is calculated. Obtained coefficient a in the equation represents just the requested $a_{\mathrm{ma}}(\phi, w)$ value. It is worth noting, that for all the configurations (ϕ, w) in the grid the obtained coefficient b was always equal zero, that is, the assumption concerning central localization of ellipses (contours) has never been denied.

Figure 4 shows the mean values of the angle coefficient for major axis a_{ma} of contours of $pctb(x_0, x_1)$ obtained from a series of 100 independent evaluations. The means are generated for a grid of configurations (ϕ, w) starting from $[\phi = 0.04, w = 0.03]$ and changing with step 0.05 in both directions (which gave 80×20 points). All means for the configurations (ϕ, w) from the stable region fit the range $[-1, 1]$. The configurations from outside the stable region have assigned a constant value of minus one.

The Eccentricity $\varepsilon(\phi, w)$. A graphical visualisation of the eccentricity of contours of $pctb(x_0, x_1)$ is presented in Fig. 5. This figure is also generated for a grid of configurations (ϕ, w) starting from $[\phi = 0.04, w = 0.03]$ and changing with step 0.05 in both directions and shows means obtained from a series of 100 independent evaluation. The contour of $pctb(x_0, x_1)$ for the selected L can be sketched quite precisely by a convex hull of the cloud of points already generated for evaluation of a_{ma} from the equation of the linear regression line. The convex hull is found with a method of Graham scan [13]. Among the points from the convex hull two of them located in the shortest and the longest distance from the centre of the cloud, that is, from the point $(0, 0)$ are selected. The two distances approximate lengths of minor (r_{mi}) and major (r_{ma}) radius of the ellipse respectively. Thus, the eccentricity equals: $\varepsilon = r_{\mathrm{mi}}/r_{\mathrm{ma}}$. As in the case of depicting $a_{\mathrm{ma}}(\phi, w)$, $\varepsilon(\phi, w)$ for the configurations from outside the stable region have assigned a constant value, however, in this case they are set to 0.00001.

Classification of Particles. The parameters $a_{\mathrm{ma}}(\phi, w)$ and $\varepsilon(\phi, w)$ allow to identify the four types of particles defined in Subsection 4.2. Type B particles are characterised by high values of eccentricity (close to one) whereas the remaining types have lower values. The remaining three types can be identified respectively to the value of $a_{\mathrm{ma}}(\phi, w)$. The highest values close to one of $a_{\mathrm{ma}}(\phi, w)$ have particles of type A, the lowest values close to minus one – particles of type C, and the moderate values, around zero, has particles of type D.

4.4   Transformations of *pctb* Characteristics

Figures 4 and 5 show the $a_{\mathrm{ma}}(\phi, w)$ and $\varepsilon(\phi, w)$ characteristics which allows to imagine the way of the ellipse shape transformation when the ϕ and w parameters vary. For better understanding of these transformations, below, three example series of *pctb* figures are depicted. The examples show, how the contours change from one to another. Example series: Q1, Q2 and Q3 of *pctb* graph pairs: $pctb(x_0, x_1)$ and $pctb(x_0, v)$ for $\delta = 0.0001$ are presented in Figs. 7, 8 and 9 respectively. Localizations of selected series of configurations: Q1, Q2 and Q3 in the configuration space (ϕ, w) can be found in Fig. 6.

In Fig. 7 the first series of figures called Q1 shows the transformations when the inertia weight w is high, that is, $w = 0.96$ and ϕ varies from minimal to maximal values within the stability region: $\phi \in \{0.06, 0.46, 2.46, 3.91\}$. For small values of ϕ the most important for *pctb* is the initial velocity: when it is small, the

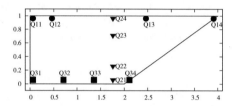

Fig. 6. Localizations of configuration series presented in top three rows of pictures: Q1 (marked as circles), Q2 (triangles), and Q3 (squares) in the configuration space (ϕ, w).

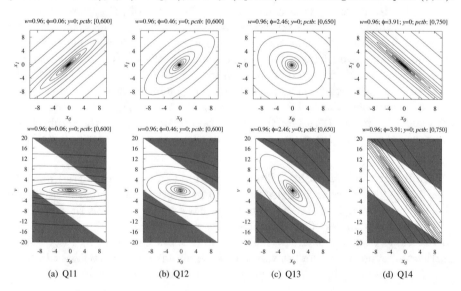

Fig. 7. Particle convergence times *pctb* for a series Q1: fixed $w = 0.96$ and $\phi \in \{0.06, 0.46, 2.46, 3.91\}$; the top figures: $pctb(x_0, x_1)$; the bottom figures: $pctb(x_0, v)$; the white area in figures for $pctb(x_0, v)$ maps to the domain defined for $pctb(x_0, x_1)$.

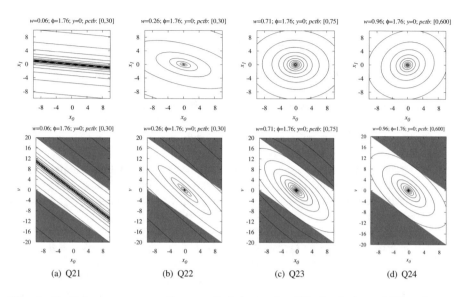

Fig. 8. Particle convergence times *pctb* for a series Q2: fixed $\phi = 1.76$ and $w \in \{0.06, 0.26, 0.71, 0.96\}$; the top figures: $pctb(x_0, x_1)$; the bottom figures: $pctb(x_0, v)$; the white area in figures for $pctb(x_0, v)$ maps to the domain defined for $pctb(x_0, x_1)$.

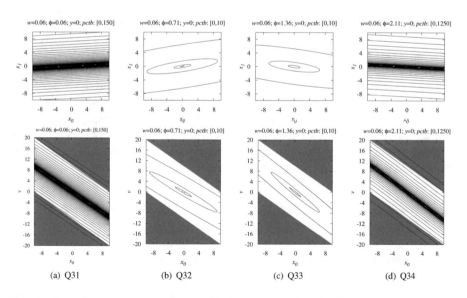

Fig. 9. Particle convergence times *pctb* for a series Q3: fixed $w = 0.06$ and $\phi \in \{0.06, 0.71, 1.36, 2.11\}$; the top figures: $pctb(x_0, x_1)$; the bottom figures: $pctb(x_0, v)$; the white area in figures for $pctb(x_0, v)$ maps to the domain defined for $pctb(x_0, x_1)$.

pctb is low, otherwise, the number of steps necessary to reach the attractor grows rapidly. On the opposite end of series Q1 one can observe the case when for small values of *pctb* the velocity should be adjusted to the distance to the attractor. The further is the particle from the attractor, the higher initial velocity is needed to reach the attractor in small number of steps. In every case the velocity must be directed toward the attractor.

In Fig. 8 the series Q2 is presented. The attractor coefficient is fixed, that is, $\phi = 1.76$ and the inertia weight varies: $w \in \{0.06, 0.26, 0.71, 0.96\}$. In every case for the sake of *pctb* minimization the initial velocity should be adjusted to the initial location of the particle. However, for small values of the inertia weight a small error in adjustment causes large increase of *pctb* value, whereas, large values of inertia make this change less abrupt, that is, the system is more stable.

The series of characteristics Q3 is depicted in Fig. 9. In this case the inertia weight w is low, that is, $w = 0.06$ and $\phi \in \{0.06, 0.71, 1.36, 2.11\}$. As it is in the series Q1, when ϕ is small the initial location is almost negligible and the most influential parameter is velocity: when v is close to zero, the *pctb* is the smallest. In the Q3 series the boundary cases represent configurations sensitive to the error of velocity vs. location adjustment, that is, the stability of these configurations is low. The most stable configurations are the ones in the middle of the range.

Finally, note, that the three series have two shared configurations. Q21 may belong also to Q3: this configuration can be located between Q33 and Q34. Q24 may belong to Q1 and located between Q12 and Q13.

When we take a look at all the series, one can also observe that in most cases the *pctb* is sensitive to an error in the adjustment particularly for the largest values of ϕ both for small and high values of w (particularly, the examples Q14, Q21, and Q34). The most stable configurations, that is, resistant to lack of appropriate adjustment of parameters can be found in the middle of the series Q2, particularly Q24. It is worth noting here, that one of the popular choices of particle parameters: $c_1 = c_2 = 1.49445$ and $w = 0.72984$ (in [14] authors showed that the two values lead to satisfying results for a series of benchmark functions) belongs to the area of such a stable configurations. On the other side, for the smallest values of ϕ the initial location of a particle has no significant influence and *pctb* depends on just the velocity: the smaller v the less *pctb*.

5   Swarms with Different Profiles of Particle Type Membership – Experimental Research

In the experimental part of the research we verified a thesis that types of particles defined in Subsect. 4.2 can be useful in construction of effective heterogenous swarms. The PSO model with inertia weight (Eq. (2)) was implemented with eleven particle type profiles of a swarm: swarms with particles which represent just a single type A, or B, or C or D, swarms where a half of particles is of one type and the remaining part — of another type: A and B, A and C, A and D, B and C, B and D, C and D, and the last profile where all the types A, B, C and D are represented with equal number of particles.

The experimental part consists of two groups of experiments. For both groups tested swarm sizes are: 4, 8, 12, 16, 20, 24, 28, and 56 particles. The outcome of a single run of a swarm is the number of **fitness function calls** (f.f.c.) necessary to find a solution located in a distance from the global optimum not greater than 0.1. A single configuration of an experiment is identified by a couple: the swarm profile and the swarm size. For every configuration 1000 runs were executed and respectively 1000 values obtained. It is assumed, that when a swarm is not able to reach the requested quality of the result the run is finished after 100000 f.f.c.

5.1   The First Group of Experiments: Unimodal Optimization Problem

In the first group the swarms are tested on a unimodal optimization problem represented by a one-dimensional Gaussian function: $f(x) = 1/(\sqrt{2\pi\sigma^2})\exp(-\frac{(x-\mu)^2}{2\sigma^2})$ where $\mu = 0$, $\sigma^2 = 1$, that is, optimum $x^* = 0$. All the particles in a swarm start from the same point $x_0 = -5$ but are initialized with different velocities $v_0 = U(-0.1, 0.1)$.

Distributions of the numbers of f.f.c. in the form of a five-number summary (min, first quartile, median, third quartile and max) are depicted as box-and-whisker plots in Figs. 10 and 11. Figure 10 shows statistics for seven profiles where location and spread of observation are quite reasonable, that is, there

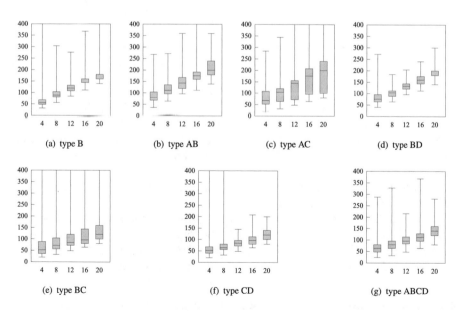

Fig. 10. Distributions of the numbers of f.f.c. for a series of 1000 runs of the swarms which cope with the unimodal optimization problem.

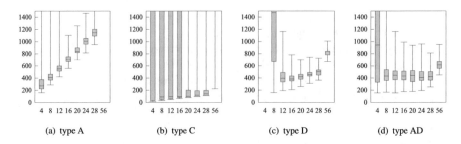

(a) type A   (b) type C   (c) type D   (d) type AD

Fig. 11. Distributions of the numbers of f.f.c. for a series of 1000 runs of the swarms which needed more time to reach the target.

Table 1. Unimodal problem: numbers of runs when swarm failed to reach the target within series of 1000 runs.

Profile vs. size	4	8	12	16	20	24	28	56
A	64	0	0	0	0	0	0	0
B	0	0	0	0	0	0	0	0
C	547	451	356	261	196	155	133	6
D	1000	133	0	0	0	0	0	0
AB	0	0	0	0	0	0	0	0
AC	0	0	0	0	0	0	0	0
AD	495	2	0	0	0	0	0	0
BC	59	19	3	1	0	0	0	0
BD	0	0	0	0	0	0	0	0
CD	75	2	0	0	0	0	0	0
ABCD	0	0	0	0	0	0	0	0

exist a swarm profile for which the majority of runs needed less than a hundred f.f.c. to reach the target. In this Figure due to effective behaviour of the swarms and for better visibility of box-plots the range of tested swarm sizes is decreased, thus the maximum size equals 20. Worse results for the full range of swarm sizes, from 4 to 56, are presented in Fig. 11.

Numbers of runs when swarm failed to reach the target within the time of 100000 f.f.c. are presented in Table 1.

In this group of experiments the four configurations of swarms which returned best results are: B, BD, CD, and ABCD. In Fig. 10 one can see that the least median is obtained for CD-swarms, however, cost of a solution, that is, f.f.c. is more stable for BD-swarms, particularly, for small-sizes of a swarm.

5.2   The Second Group of Experiments: Saddle Crossing Problem

In this group the swarms are tested on an one-dimensional function being a sum of two Gaussian functions, the first one with $\mu = 0.038$ and $\sigma^2 = 0.5$ (the first

Table 2. Saddle crossing problem: numbers of runs when swarm failed to reach the target within series of 1000 runs

Profile vs. size	4	8	12	16	20	24	28	56
A	933	880	805	758	710	673	639	400
B	24	0	0	0	0	0	0	0
C	705	532	396	316	234	165	121	6
D	857	735	646	556	481	422	325	122
AB	0	0	0	0	0	0	0	0
AC	257	67	17	5	1	1	0	0
AD	901	824	732	667	558	554	474	242
BC	109	15	3	0	0	0	0	0
BD	9	0	0	0	0	0	0	0
CD	328	66	18	5	1	0	0	0
ABCD	5	0	0	0	0	0	0	0

local optimum which is also a global optimum $x^* = 0.038$) and the second one with $\mu = -2.51$ $\sigma^2 = 1$ (the second local optimum $x^* = -2.51$). All the particles in a swarm start from the same point $x_0 = -2.5$, that is, from the second local optimum, and are initialized with different velocities $v_0 = U(-0.5, 0.5)$.

In the case of saddle-crossing it is assumed that swarm is in the local optimum trap and the main problem is to encourage particles to start searching in the wider surrounding than usual. This was obtained by moderate increase of velocity: in the previous case of a unimodal problem optimization the initial velocity was set to $v_0 = U(-0.1, 0.1)$ whereas in the case of saddle-crossing – $v_0 = U(-0.5, 0.5)$. In both cases limits for v_0 are obtained with a trial-and-error method. Numbers of runs when swarm failed to reach the target are presented in Table 2. One can see that the most effective configurations are: B, AB, BD and ABCD.

The last stage of the experimental research focuses on verification of usefulness of the varying character of particles for escaping from the local optimum trap. Just like in an classic novella by R. L. Stevenson about Dr. Jekyll and Mr. Hyde the particles of selected type change into representatives of another type for a number of iterations, make few moves according to their alternate nature and eventually transform back to their real type. Selection of possible particle types useful for transformations was based on our previous experiences.

The main scheme of JH-strategy for the particle type transformation (named after the main character of Stevenson novella) is as follows: all particles in a swarm turn their types of behaviour into new one for five iterations of the algorithm. The time of five iterations should be devoted exploration necessary to escape from the local optimum trap. Then, the particles transform back and continue their regular "life" following newly discovered attractors and locations. Again, using a trial-and-error method we selected the type C for the

Table 3. Saddle crossing problem: numbers of runs when *JH*-swarm failed to reach the target within series of 1000 runs (for the case of swarm consisting of particles of the C type the strategy of switching into the C type has had to be modified).

Profile vs. size	4	8	12	16	20	24	28	56
A	5	0	0	0	0	0	0	0
B	78	16	5	1	1	0	0	0
C*	717	518	381	289	214	170	116	10
D	77	7	0	0	0	0	0	0
AB	0	0	0	0	0	0	0	0
AC	24	1	0	0	0	0	0	0
AD	13	0	0	0	0	0	0	0
BC	196	77	19	12	4	1	1	0
BD	26	0	0	0	0	0	0	0
CD	145	4	1	0	0	0	0	0
ABCD	4	0	0	0	0	0	0	0

"alternative personality" as the most promising one. Numbers of runs when swarm failed to reach the target are presented in Table 3. The symbol of a star at the letter C in the first column denotes, that for the case of swarm consisting of C-particles the strategy of switching has had to be modified: for the time of the first five iterations instead of turning the particles into the C type (which would be pointless), the particles are turned into the B type. Even brief comparison of Tables 2 and 3 shows that application of the *JH*-transformation significantly improves effectiveness of the swarm (obviously, except from the case of the swarm with C type particles).

For the set of the effective swarm configurations: B, AB, BD and ABCD distributions of the numbers of f.f.c. are depicted in Figs. 12 and 13. Figure 12 presents distributions of the numbers of f.f.c. for selected regular swarms for which in Table 2 one can find also numbers of fails in reaching the target. Respectively, numbers of fails in reaching the target for the *JH*-swarms depicted in Fig. 13 can be found in Table 3.

In the group of tests with regular swarms the computational cost represented as a median of the numbers of f.f.c. for the series of 1000 experiments is least for ABCD-swarms. When the JH-strategy is applied, the cost obtained for all studied JH-swarm configurations decreased and is less than the best cost obtained for ABCD regular swarms. For JH-swarms the values of median are similar for all the cases presented in Fig. 13, however, the best stability measured as a difference between maximum and minimum number of f.f.c. can be observed for BD-swarms.

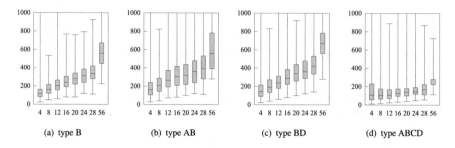

(a) type B   (b) type AB   (c) type BD   (d) type ABCD

Fig. 12. Distributions of the numbers of f.f.c. for a series of 1000 runs of the regular swarms.

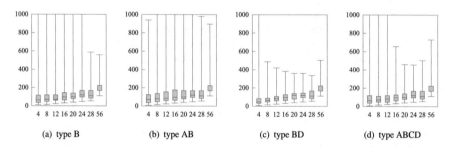

(a) type B   (b) type AB   (c) type BD   (d) type ABCD

Fig. 13. Distributions of the numbers of f.f.c. for a series of 1000 runs of the JH-swarms.

6   Summary

A new measure of particle convergence time (pct) is proposed and studied in this research. For the deterministic model of PSO with inertia weight an upper bound formula of (pct) is derived. These tools are applied to identify four types of particle behaviour depending on parameters ϕ and w. Additionally, two parameters: an angle coefficient for the major axis (a_{ma}) and an eccentricity (ε) are proposed for effective classification of particle behaviour.

In the experimental part it is shown that the proposed classification of particle types can be useful in construction of effective heterogenous swarms. Two test-beds are used: a simple unimodal problem and a saddle crossing problem. A new strategy of run-time particle type control, the JH-strategy is proposed.

Acknowledgements. Authors would like to thank Krzysztof Jura from Cardinal Stefan Wyszyński University in Warsaw, Poland, for his assistance in computer simulations.

References

1. Kennedy, J., Eberhart, R.C.: Particle swarm optimization. In: Proceedings of the IEEE International Conference on Neural Networks, Piscataway, NJ, IEEE 1942–1948 (1995)
2. Shi, Y., Eberhart, R.C.: A modified particle swarm optimizer. In: Proceedings of the IEEE Congress on Evolutionary Computation 1998, pp. 69–73. IEEE Press (1998)
3. Clerc, M., Kennedy, J.: The particle swarm-explosion, stability, and convergence in a multidimensional complex space. IEEE Trans. Evol. Comput. **6**, 58–73 (2002)
4. Trelea, I.C.: The particle swarm optimization algorithm: convergence analysis and parameter selection. Inform. Process. Lett. **85**, 317–325 (2003)
5. van den Bergh, F., Engelbrecht, A.P.: A study of particle swarm optimization particle trajectories. Inf. Sci. **176**, 937–971 (2006)
6. Poli, R.: Mean and variance of the sampling distribution of particle swarm optimizers during stagnation. IEEE Trans. Evol. Comput. **13**, 712–721 (2009)
7. Liu, Q.: Order-2 stability analysis of particle swarm optimization. Evol. Comput. **23**, 187–216 (2015)
8. Bonyadi, M.R., Michalewicz, Z.: Analysis of stability, local convergence, and transformation sensitivity of a variant of particle swarm optimization algorithm. IEEE Trans. Evol. Comput. **20**, 370–385 (2015)
9. Cleghorn, C.W., Engelbrecht, A.P.: Particle swarm convergence: an empirical investigation. In: 2014 IEEE Congress on Evolutionary Computation (CEC), pp. 2524–2530. IEEE Press (2014)
10. Kadirkamanathan, V., Selvarajah, K., Fleming, P.J.: Stability analysis of the particle dynamics in particle swarm optimizer. IEEE Trans. Evol. Comput. **10**, 245–255 (2006)
11. Gazi, V.: Stochastic stability analysis of the particle dynamics in the PSO algorithm. In: 2012 IEEE International Symposium on Intelligent Control (ISIC), pp. 708–713. IEEE Press (2012)
12. Cleghorn, C.W., Engelbrecht, A.P.: A generalized theoretical deterministic particle swarm model. Swarm Intell. **8**, 35–59 (2014)
13. Graham, R.L.: An efficient algorithm for determining the convex hull of a finite planar set. Inf. Process. Lett. **1**, 132–133 (1972)
14. Eberhart, R.C., Shi, Y.: Comparing inertia weights and constriction factors in particle swarm optimization. In: Proceedings of the 2000 Congress on Evolutionary Computation, pp. 84–88. IEEE Service Center, Piscataway (2000)

Fuzzy Computation Theory and Applications

Human Mortality Modeling with a Fuzzy Approach Based on Singular Value Decomposition Technique

Duygun Fatih Demirel$^{(\boxtimes)}$ and Melek Basak

Department of Industrial and Systems Engineering,
Yeditepe University, Istanbul, Turkey
duygunfe@gmail.com, melek.basak@yeditepe.edu.tr

Abstract. Modeling and forecasting human mortality are significant research topics in several disciplines because mortality rates are fundamental in planning and policy decisions. Among various techniques, Lee Carter (LC) model is one of the most popular stochastic method in human mortality modeling. The original LC model was fuzzified to eliminate the assumptions related with homoscedasticity. The existing fuzzy model makes use of ordinary least squares (OLS) technique, which prevents the model to capture the existing fluctuations in data. In this study, a revised version of fuzzy LC model utilizing singular value decomposition (SVD) technique is proposed to overcome this issue. After modeling the mortality rates, their future values are forecasted by a modified first order fuzzy time series technique. For illustration purposes, proposed method is applied to mortality data of Finland. Numerical outputs show that proposed method is statistically better in modeling mortality compared to the existing fuzzy method. In addition, the modified fuzzy time series technique generates better forecasts than the original version.

Keywords: Fuzzy modeling · Lee Carter method · Singular value decomposition · Fuzzy regression · Fuzzy time series · Human mortality modeling

1 Introduction

It is known that human mortality modeling and forecasting play significant roles in strategy development and decision making in diverse sectors. Mortality modeling finds application areas in projecting and forecasting life expectancies, age distributions, unemployment rates, labor force compositions, household consumptions and etc. Together with fertility and migration rates, mortality rates constitute the vital demographic indicators of population dynamics [1]. Age-specific population estimates of immediate future or long-term forecasts based on these vital demographic indicators shape the policies in allocating the resources among public and private investments and the future population [2]. The outputs of the models obtained from fertility, mortality and migration elements form the basis for medium or long term planning in various areas such as labor market [3], public financing [4], insurance and pensions sector [5, 6], education system [2], healthcare services [7], and etc.

© Springer International Publishing AG 2017
J.J. Merelo et al. (eds.), *Computational Intelligence*, Studies in Computational Intelligence 669,
DOI 10.1007/978-3-319-48506-5_11

Population modeling and estimations are performed via diverse methodologies which can basically be grouped as population projection methods and population forecasting methods. The projection methods simply rely on deterministic scenarios for different components of mortality, fertility and migration value combinations [8]. Setting the values of these components generally requires formation of a group of experts. In contrast, population forecasting makes use of historical data to obtain a future population estimate using a stochastic approach which takes component uncertainties into account. In fact, stochastic modeling methods have a significant area in demographic forecasting since they provide estimations for the vital demographic indicators together with forecast intervals for them [8]. Time series methods are major extrapolative stochastic methods used for mortality forecasting based only on historic data [9–12]. Thus, these methods do not involve any exogenous factor such as a disaster or a technological development.

In mortality modeling, Lee-Carter (LC) model is one the most extensively studied stochastic method. This popular method basically computes time varying mortality indices for each age and sex cohort through a matrix decomposition technique. According to Lee and Carter [9], mortality can be modeled as:

$$\ln(\mathbf{m}_{x,t}) = \mathbf{a}_x + \mathbf{b}_x\mathbf{k}_t + \varepsilon_{x,t} \tag{1}$$

where $\mathbf{m}_{x,t}$ is the central death rate for age x at time t, $\mathbf{a}_x$ and $\mathbf{b}_x$ are age-specific constants and $\mathbf{k}_t$ is time-variant mortality index. Here, the error term $\varepsilon_{x,t}$, which represents the random effects that are not reflected by the model, is assumed to be normally distributed with mean 0 and a small constant variance σ_ε^2.

The model in Eq. (1) cannot be fitted by using ordinary least squares (OLS) method since the time-variant mortality indices, $\mathbf{k}_t$'s (independent variables in Eq. (1)) are unknown. Hence, Lee and Carter use singular value decomposition method (SVD) to give estimates for the unknown parameters $\mathbf{a}_x$, $\mathbf{b}_x$ and $\mathbf{k}_t$ of Eq. (1).

In literature, many improvements to the LC model have been suggested. For example, a double bilinear predictor structure can be embedded in the original model to include the effects of age differences [13], whereas a Poisson regression model can be constructed to fit mortality rates at each age group [14]. Moreover, the problems related with outliers in historic data are handled by several parametric and nonparametric smoothing techniques [2, 12, 15–17]. Further developments in LC model can be found in [5, 18].

1.1   Fuzzy LC Model

The reason why LC model is a commonly method is due to its simplicity and ability to represent the decreasing tendencies in mortality rates for most of the developed countries [19]. However, existence of random fluctuations due to external factors causes the original LC model to lose its capability of reflecting the mortality rates [20]. Additionally, for cases with small sample sizes, the assumptions related with error terms cannot be satisfied. Original LC model make use of SVD technique in which error terms are assumed to be normally distributed with a small constant variance, σ_ε^2.

This strict homoscedasticity assumption prevents the model to be applied to the cases where precise and enough number of past data are not available. It is also supposed that the magnitude of this variance has to be small but there is an obvious ambiguity in how small it should be [21]. Considering these shortcomings, Koissi and Shapiro reformulated the original LC model by incorporating fuzziness into it [22]. The fuzzy LC method is based on minimum fuzziness criterion, which is designed as a part of fuzzy least squares regression approach [23].

Koissi and Shapiro's fuzzy LC model is formulated as:

$$\tilde{Y}_{x,t} = \tilde{A}_x \underset{T_W}{\oplus} \tilde{B}_x \underset{T_W}{\otimes} \tilde{K}_t \quad \text{for } x = x_1, \ldots x_N, \, t = t_1, t_1 + 1, \ldots, t_1 + T - 1 \tag{2}$$

Here, $\tilde{Y}_{x,t}$ is the fuzzy ln-mortality rate of age cohort x at time t whose values are known beforehand, $\tilde{A}_x$ and $\tilde{B}_x$ are the unknown fuzzy age-specific parameters, and $\tilde{K}_t$ is the unknown fuzzy time-variant mortality index. $\tilde{A}_x$, $\tilde{B}_x$, and $\tilde{K}_t$ are expressed with fuzzy symmetric triangular numbers as $\tilde{A}_x = (a_x, \alpha_x)$, $\tilde{B}_x = (b_x, \beta_x)$, and $\tilde{K}_t = (k_t, \delta_t)$, where a_x, b_x, and k_t denote the centers and α_x, β_x, and δ_t reflect the spreads of the corresponding fuzzy numbers. With Eq. (2), Koissi and Shapiro consider the ln-mortality rate for age cohort x at time t as a confidence interval by fuzzifying it instead of treating it as a crisp number. Expressing ln-mortality rates with fuzzy numbers may be reasonable as exact values of mortality rates are seldom known because of miscalculations or errors in data collection and recording.

1.2   Motivation for a New Fuzzy LC Method

Fuzzy formulation of LC model in Eq. (2) requires the fuzzification of observed crisp $Y_{x,t}$ values. To meet this requirement, Koissi and Shapiro employ fuzzy least squares regression based on minimum fuzziness criterion [23, 24]. That is, the task is to find $\tilde{A}_0 = (c_{0x}, s_{0x})$, $\tilde{A}_1 = (c_{1x}, s_{1x})$, and $\tilde{Y}_{x,t} = (y_{x,t}, e_{x,t})$ with centers c_{0x}, c_{1x}, and $y_{x,t}$, and spreads s_{0x}, s_{1x}, and $e_{x,t}$, so that:

$$(y_{x,t}, e_{x,t}) = (c_{0x}, s_{0x}) + (c_{1x}, s_{1x}) \times t \tag{3}$$

for each age cohort x. Here, first, ordinary least squares regression (OLS) is applied to obtain center values such that

$$Y_{x,t} = c_{0x} + c_{1x} \times t \tag{4}$$

Then, the spreads are determined by solving linear programming (LP) models for each age cohort x based on minimum fuzziness criterion.

In Eq. (4) time t is considered as the explanatory variable. However, t, the independent variable in Eq. (3), is monotonically increasing, hence, the centers and spreads of ln-mortality rates take linear form, which is not able to capture the existing fluctuations. In this study, to overcome this issue, a modification in fuzzification of crisp $Y_{x,t}$ values based on singular value decomposition (SVD) technique is proposed.

With the proposed modification, the fluctuations in *ln*-mortality rates can be captured by the model. The proposed fuzzy LC model also eliminates the homoscedasticity assumptions and assumptions related to the magnitude of error term variances. Moreover, like Koissi and Shapiro's method, the proposed model can be used in cases where there are concerns about the ambiguity related to data and when the number of data is not sufficient to use the original LC or other stochastic methods.

The proposed method also forecasts the future mortality rates via a modified first order fuzzy time series method. The forecasting operation uses the estimated time-variant fuzzy mortality indices $\tilde{K}_t$'s. $\tilde{K}_t$'s are the input data to forecast *ln*-mortality rates for $t = t_1 + T, t_1 + T + 1, \ldots$ The modified fuzzy time series approach is based on Song and Chissom's method [25].

The rest of the paper is designed as follows: Sect. 2 is dedicated to the proposed methodology while the numerical findings obtained from the application of the proposed method to Finland mortality data are given in Sect. 3. The study is concluded with Sect. 4.

2   Methodology

The proposed fuzzy LC method first deals with modeling human mortality, in which the fuzzy parameters in Eq. (2) are estimated. Next, a first order fuzzy time series approach is utilized to forecast future mortality indices, hence, future mortality rates.

2.1   Modeling Human Mortality

Mortality modeling part of the proposed method can be examined in two phases. Phase I deals with fuzzification of observed $Y_{x,t}$ values, where $Y_{x,t}$ denotes the natural logarithm of mortality rate for age cohort x at time t, for $x = x_1, \ldots x_N$, and $t = t_1, t_1 + 1, \ldots, t_1 + T - 1$. Phase II is devoted to finding the model parameters for estimating the mortality rates. The proposed modifications are mainly in Phase I.

Phase I: Fuzzification of Observed Rates. This phase deals with the fuzzification of observed $Y_{x,t}$ (*ln*-mortality), values. That is, given the natural logarithms of mortality rates $Y_{x,t}$'s, the aim is to find $\tilde{A}_0 = (c_{0x}, s_{0x})$, $\tilde{A}_1 = (c_{1x}, s_{1x})$, and $\tilde{Y}_{x,t} = (y_{x,t}, e_{x,t})$ with centers c_{0x}, c_{1x}, and $y_{x,t}$, and spreads s_{0x}, s_{1x}, and $e_{x,t}$, such that:

$$(y_{x,t}, e_{x,t}) = (c_{0x}, s_{0x}) + (c_{1x}, s_{1x}) \times f_t \tag{5}$$

for each age cohort x, where f_t can be viewed an unknown time-variant fuzzification index. f_t can be expressed as $f_t = g_t(\vec{m}_{xt})$, where g_t is a time dependent function which maps $\vec{m}_{xt}$ to fuzzification index f_t for each time t, and $\vec{m}_{xt}$ is a vector composed of mortality rates $m_{x_1 t}, m_{x_2 t}, \ldots, m_{x_N t}$ for each time t and age cohort $x_i = x_1, \ldots x_N$. Indeed, Eq. (5) employs f_t as the explanatory variable whose value is unknown. This fuzzification index is assumed to be capable of capturing the fluctuations in *ln*-mortality rates. The fuzzification index is different from the independent variable, t, in Eq. (3). This is because t is a monotonically increasing variable, hence the centers and spreads of ln-mortality rates obtained through Koissi and Shapiro's method result in linear pattern. In contrast, the fuzzification index f_t, which consists of the aggregated age cohort mortality rates, does not necessarily show a linear trend. As a result, a better fitting model can be generated via Eq. (5).

With this modification in mind, in Phase I of the proposed method, first the center values, then the spreads in Eq. (5) are found, so the observed crisp ln-mortality rates are transformed into fuzzy numbers. The details of these operations are given in the following two sub-sections.

Finding Center Values. The necessary regression equation for finding the center values in fuzzification of observed ln-mortality rates can be extracted from Eq. (5) as:

$$y_{x,t} = c_{0x} + c_{1x} \times f_t \tag{6}$$

In fact, Eq. (6) is a modified version of Eq. (4), in which the independent variable is replaced with the fuzzification index f_t. Since f_t values are not readily known, rather than using OLS method, SVD technique is utilized to compute the unknown fuzzification indices. SVD transforms the space in which a set of data is defined into a lower dimensional space by putting emphasis on the underlying principle of the original data [26]. Mathematically, the method decomposes an $m \times n$ rectangular matrix A into product of three matrices as:

$$A = FEH^T \tag{7}$$

where F is an $m \times m$ orthogonal matrix whose columns are orthonormal eigenvectors of AA^T, H is an $n \times n$ orthogonal matrix whose columns are orthonormal eigenvectors of $A^T A$, and E is an $m \times n$ diagonal matrix composed of the square roots of eigenvalues from F or H in descending order. Since the diagonal matrix E is comprised of eigenvalues from its left and right eigenvectors, it is possible to reflect the main characteristics of matrix A by SVD technique.

SVD method aims to reorient the dimensional space in which matrix A is defined so that the new coordinate axes are adjusted to follow a similar pattern with the data points of matrix A. That is, the orthogonal vectors formed by the columns of matrices F or H are appointed to be the new coordinate axes for the vector space of original matrix A. Figure 1 illustrates the geometric interpretation of the method as an example. Here, the coordinate plane x_1–x_2 that a hypothetic matrix A with six data points (data points are shown as "x") is defined in can be reoriented into coordinate plane v_1–v_2 with the help of SVD method.

Hence, the nature of SVD technique allows Eq. (6) to capture the fluctuations in ln-mortality rates. Implementing SVD method in Eq. (6) to estimate the unknown parameters c_{0x}, c_{1x}, and f_t requires the following routine [9]:

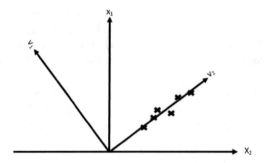

Fig. 1. Geometric interpretation of SVD method for a matrix A.

```
for whole observation set
  normalize f_t 's to sum up to 0
  normalize c_1x 's to sum up to 1
for all time t
  for each age cohort x
    let c_0x be the average of y_x,t
for all age cohorts x
  for each time t
    let f_t be sum of (y_x,t - c_0x)
for each age cohorts x
  find c_1x by regressing (y_x,t - c_0x) on f_t without a constant
end.
```

Finding the Spreads. At this point, using the estimated parameters c_{0x}, c_{1x}, and f_t, the spread optimization part of Koissi and Shapiro's model is rewritten as:

$$\text{minimize} \quad Ts_{0x} + s_{1x} \sum_{t=t_0}^{t_0+T-1} |f_t| \tag{8}$$

subject to:

$$c_{0x} + c_{1x}f_t + (1-h)[s_{0x} + s_{1x}|f_t|] \geq y_{x,t}, \text{ for } \forall t = t_0, t_0+1, \ldots, t_0+T-1 \tag{9}$$

$$c_{0x} + c_{1x}f_t - (1-h)[s_{0x} + s_{1x}|f_t|] \leq y_{x,t}, \text{ for } \forall t = t_0, t_0+1, \ldots, t_0+T-1 \tag{10}$$

$$s_{0x}, s_{1x} \geq 0 \tag{11}$$

Here, the objective is to minimize the total spread. Equations (9) and (10) guarantee that each *ln*-mortality rate $Y_{x,t}$ falls within the fuzzified $\tilde{Y}_{x,t}$ at a level h, which is a predetermined small nonnegative parameter. h can be viewed as the α-cuts in fuzzy set theory. Thus, in the context of fuzzy least squares regression, h is defined to be 0 to prohibit large spreads [22].

Phase II: Estimation of Demographic Rates. Once Phase I is completed and the observed *ln*-mortality rates are fuzzified, the next step is to estimate the mortality rates by finding the unknown fuzzy parameters $\tilde{A}_x$, $\tilde{B}_x$ and $\tilde{K}_t$ in Eq. (2), where $\tilde{A}_x = (a_x, \alpha_x)$, $\tilde{B}_x = (b_x, \beta_x)$, and $\tilde{K}_t = (k_t, \delta_t)$. Actually, the corresponding model in phase II of the proposed methodology is the same as the fuzzy model parameter estimation step of Koissi and Shapiro's method except the solution procedure.

Before presenting Phase II model, it might be practical to state that the characteristics of the fuzzy numbers are not preserved with multiplication of triangular fuzzy numbers. To retain the shape of LR-type fuzzy numbers, weakest triangular norm (T_W) based multiplication and addition can be used [27]. For two symmetric triangular fuzzy numbers $\tilde{A} = (a, l_A)$ and $\tilde{B} = (b, l_B)$, the shape preserving T_W-based multiplication and addition are defined as [22]:

$$\tilde{A} \underset{T_W}{\oplus} \tilde{B} = (a + b, \, \max(l_A, l_B)) \tag{12}$$

$$\tilde{A} \underset{T_W}{\otimes} \tilde{B} = (ab, \, \max(l_A|b|, l_B|a|)) \tag{13}$$

Considering Eqs. (12) and (13), Eq. (2) becomes:

$$\tilde{Y}_{x,t} = (a_x + b_x k_t, \, \max(\alpha_x, |b_x|\delta_t, \beta_x|k_t|)) \tag{14}$$

The unknown parameters a_x, b_x, k_t, α_x, β_x, and δ_t are estimated by minimizing the total squared distance between $\tilde{A}_x \underset{T_W}{\oplus} \tilde{B}_x \underset{T_W}{\otimes} \tilde{K}_t$ and $\tilde{Y}_{x,t}$. Koissi and Shapiro, employ Diamond distance as the fuzzy distance metric in minimization model. Diamond distance between two symmetric triangular fuzzy numbers $\tilde{A}_1 = (a_1, \alpha_1)$ and $\tilde{A}_2 = (a_2, \alpha_2)$ is defined as [28]:

$$D_{LR}(\tilde{A}_1, \tilde{A}_2) = (a_1 - a_2)^2 + [(a_1 - \alpha_1) - (a_2 - \alpha_2)]^2 + [(a_1 + \alpha_1) - (a_2 + \alpha_2)]^2 \tag{15}$$

Minimizing total Diamond distance results in an unconstrained nonlinear optimization problem which is formulated for all age cohorts x and time t as:

$$\text{Minimize} \sum_x \sum_t D_{LR}[\tilde{A}_x \underset{T_W}{\oplus} (\tilde{B}_x \underset{T_W}{\otimes} \tilde{K}_t), \tilde{Y}_{x,t}]^2 \tag{16}$$

where

$$
\begin{aligned}
D_{LR}[\tilde{A}_x \underset{T_w}{\oplus} (\tilde{B}_x \underset{T_w}{\otimes} \tilde{K}_t), \tilde{Y}_{x,t}]^2 &= (a_x + b_x k_t - y_{x,t})^2 + [a_x + b_x k_t - \max\{\alpha_x, |b_x|\delta_t, \beta_x|k_t|\} \\
&\quad - (y_{x,t} - e_{x,t})]^2 + [a_x + b_x k_t + \max\{\alpha_x, |b_x|\delta_t, \beta_x|k_t|\} - (y_{x,t} + e_{x,t})]^2
\end{aligned}
\tag{17}
$$

Applying SVD, a_x can be easily computed as:

$$
a_x = \frac{1}{T} \sum_t y_{x,t}
\tag{18}
$$

Estimating parameters b_x, k_t, α_x, β_x, and δ_t is less forthright, because, the mathematical structure of Eq. (16) does not permit using a derivative based solution approach. At this point, *fminsearch* tool of MATLAB optimization application for unconstrained optimization problems can be used to obtain estimates for the unknown parameters. Indeed, *fminsearch* is a derivative free method for unconstrained nonlinear optimization problems based on Nelder-Mead simplex algorithm (see [29] for details of Nelder and Mead algorithm). Finally, with the computed parameters, fuzzy *ln*-mortality rates can be estimated for $x = x_1, \ldots, x_N$, and $t = t_1, t_1 + 1, \ldots, t_1 + T - 1$ by using Eq. (2).

2.2   Forecasting Future Mortality Rates

In this part of the study, future mortality rates are forecasted by fuzzy time-variant mortality indices $\tilde{K}_t$'s for $t = t_1 + T, t_1 + T + 1, \ldots$ based on the model outputs generated in Sect. 2.1. To forecast future mortality indices, a modified first order fuzzy time series (FFTS) technique is proposed. The proposed FFTS technique is based on the conventional fuzzy time series method of Song and Chissom [25]. Since fuzzy time series approaches divide the observed data into intervals, Song and Chissom identify a fuzzy set as the union of the membership degrees of the intervals to that set. However, in our approach, normalized membership degrees are aggregated to identify a fuzzy set. Based on proposed modified FFTS method, future mortality rates are computed with the forecasted future mortality indices using Eq. (2).

First Order Fuzzy Time Series (FFTS): Conventional Approaches. In general, let $U = \{u_1, u_2, \ldots, u_n\}$ be the universe of discourse. Here, u_k's, $k = 1, \ldots, n$, are the intervals whose union covers the whole dataset. A fuzzy set A_i in U is defined as:

$$
A_i = \frac{f_{A_i}(u_1)}{u_1} + \frac{f_{A_i}(u_2)}{u_2} + \ldots + \frac{f_{A_i}(u_n)}{u_n}
\tag{19}
$$

where f_{A_i} is the membership function of the fuzzy set A_i which maps U to the interval $[0,1]$, $f_{A_i}(u_k) : u_k \to [0, 1]$ stands for the membership degree of u_k in the fuzzy set A_i, $1 \leq k \leq n$ [30].

Consider a time variant universe of discourse $Y(t)$ (t = ..., 0, 1, 2,...), which is composed of intervals representing fuzzy sets $f_i(t)$, $t = ...,$ 0, 1, 2,... Moreover, if $F(t)$ denotes a collection of $f_i(t)$ then $F(t)$ is defined as a fuzzy time series on $Y(t)$. Also, assume that there is a fuzzy relationship $R(t-1, t)$ such that

$$F(t) = F(t-1) \circ R(t-1, t) \tag{20}$$

where "$\circ$" represents a composition operator. Equation (20) indicates that $F(t)$ is caused by $F(t-1)$ and this causality is represented by a fuzzy logical relationship. If $F(t-1) = A_i$ and $F(t) = A_j$, then the fuzzy logical relationship between $F(t-1)$ and $F(t)$ is expressed as:

$$A_i \rightarrow A_j \tag{21}$$

Equation (21) states that the fuzzy relationship R is a first order model of $F(t)$. If there exists some fuzzy logical relationships such that $A_i \rightarrow A_{j_1}$, $A_i \rightarrow A_{j_2}$, ..., $A_i \rightarrow A_{j_i}$, then, these A_{j_k}'s are clustered into a fuzzy logical relationship group $A_i \rightarrow A_{j_1}, A_{j_2}, ... A_{j_k}$.

Proposed FFTS Method for Forecasting Mortality Rates. The existing fuzzy time series methods include fuzzification of crisp observed data. This is done by utilizing fuzzy subsets that correspond to a linguistic expression such that a fuzzy set A_i is defined as in Eq. (19). The conventional fuzzy time series methods express a fuzzy set A_i by the interval u_k^* with the maximum membership (that is, $f_{A_i}(u_k^*)$ is the largest $f_{A_i}(u_k)$) [25].

In this study, fuzzy estimates for mortality indices are generated using the approach discussed in Sect. 2.1. Thus, a modified FFTS method for forecasting mortality indices that uses fuzzy inputs representing crisp data is proposed. The method is composed of seven steps:

Step 1: A universe of discourse U is defined to cover fuzzy mortality indices $\tilde{K}_t$'s to include $\min(k_i - \delta_t)$ and $\max(k_i - \delta_t)$ for all $t - t_1, t_1 + 1, ..., t_1 \mid T$ 1. Then U is partitioned into intervals u_k's, $k = 1, ..., n$. The number of intervals depends on the lengths of the intervals; and the lengths of the intervals are identified according to average based [31], distribution based [31] or ratio based [32] interval length construction approaches.

Step 2: For each interval u_k, midpoint p_k is computed.

Step 3: For each estimated fuzzy mortality index $\tilde{K}_t$, the membership of interval u_k to $\tilde{K}_t$ is calculated by:

$$\mu_{k,t} = \begin{cases} 0, & \text{if } p_k < k_t - \delta_t \\ \frac{p_k - (k_t - \delta_t)}{\delta_t}, & \text{if } k_t - \delta_t < p_k < k_t \\ \frac{(k_t + \delta_t) - p_k}{\delta_t}, & \text{if } k_t < p_k < k_t + \delta_t \\ 0, & \text{if } p_k > k_t - \delta_t \end{cases} \tag{22}$$

Then, for each t, $\mu_{k,t}$'s are normalized to sum up to 1. The normalized $\mu_{k,t}$ is denoted by $norm(\mu_{k,t})$.

Step 4: For each $\tilde{K}_t$, a fuzzy set A_i in U is defined. The index i is determined by the interval u_i into which center k_t takes place. A_i is expressed as:

$$A_i = \frac{\mu_{1,t}}{u_1} + \frac{\mu_{2,t}}{u_2} + \ldots + \frac{\mu_{n,t}}{u_n} \tag{23}$$

Step 5: Fuzzy relationship $A_i \rightarrow A_j$ is identified for each consecutive time periods t -1 and t. One-to-one and one-to-many fuzzy relationships are established as mentioned previously.

Step 6: Forecasted center $\hat{k}_t$ of fuzzy mortality index $\tilde{K}_t$ is derived from $\tilde{K}_{t-1}$ with the fuzzy relationships identified in step 5. Let c_i be the indicator representing fuzzy set A_i corresponding to a fuzzy mortality index $\tilde{K}_t$. c_i is computed by:

$$c_i = \sum_{k=1}^{n} (norm(\mu_{k,t}) * p_k) \tag{24}$$

If there exist different fuzzy sets having the same index i, an average indicator c_i is assigned. After computing the indicators, $\hat{k}_t$ is calculated based on the following principles:

(i)   If there exists a one-to-one relationship $A_i \rightarrow A_j$ between $\tilde{K}_{t-1}$ and $\tilde{K}_t$, then $\hat{k}_t = c_j$

(ii) If $A_j = \emptyset$, that is $A_i \rightarrow \emptyset$, then $\hat{k}_t = c_i$

(iii) If the relationship between $\tilde{K}_{t-1}$ and $\tilde{K}_t$ is a one to many relationship such as $A_i \rightarrow A_{j_1}, A_{j_2}, \ldots, A_{j_n}$, then $\hat{k}_t = \sum_{v=1}^{n} c_{j_v}/n$.

Following this procedure, $\hat{k}_{t+1}$ is forecasted using fuzzy relationships established for $\tilde{K}_t$. Thus, ln-mortality rate $Y_{x,t+1}$ for age cohort x can be computed with

$$\hat{Y}_{x,t+1} = a_x + b_x \hat{k}_{t+1} \tag{25}$$

Hence the six consecutive steps of the proposed method are used in forecasting future ln-mortality rates.

3   Application

The proposed method is applied to mortality data for Finland. Koissi and Shapiro also apply their method on Finland mortality dataset; therefore the reason in selecting Finland as the country for application in this study is that the proposed method can be compared with the existing fuzzy approach. Furthermore, the proposed method claims to capture the fluctuations in data; and the demographic rates in Finland show some fluctuations as a result of some external factors like World War II. In addition to mortality modeling, future mortality rates for Finland are forecasted using the modified FFTS approach described in Sect. 2.2. The numerical findings obtained for mortality modeling and forecasting are given in two sub-sections as follows.

3.1   Mortality Modeling Findings

The mortality data for Finland are taken from "Human Mortality Database" (www. mortality.org). Data is composed of mortality rates (for both genders) of seventeen consecutive five-year-periods 1925–1929, 1930–1934 ..., 2005–2009, and twenty two age cohorts of [0, 1), [1–5), [5, 10), ..., [100, 105) (making 374 data points in total).

The outcomes for fuzzification of observed ln-mortality rates (ln-$M_{x,t}$) which is expressed as Phase I of the proposed method, are depicted in Figs. 2 and 3. Randomly selected age cohorts are [25, 30) and [45–50) respectively in Figs. 2 and 3.

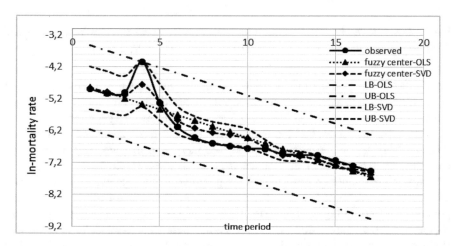

Fig. 2. Observed ln-$M_{x,t}$, fuzzy ln-$M_{x,t}$ with its spreads by Koissi and Shapiro's method, and fuzzy ln-$M_{x,t}$ with its spreads by the proposed method for age group [25, 30).

In both figures, the horizontal axis stand for time periods (1 = 1925–1929, ..., 17 = 2005–2009), whereas the vertical axis depicts the ln-mortality rates. In these figures, observed ln-$M_{x,t}$ (observed); the center of the fuzzified ln-mortality rate (fuzzy center-OLS) together with its corresponding right tail (UB-OLS) and left tail (LB-OLS) obtained via Koissi and Shapiro's method; and the center of the fuzzified ln-mortality rate (fuzzy center-SVD) together with its corresponding right tail (UB-SVD) and left tail (LB-SVD) obtained by the proposed method are provided. Both figures show that the fuzzification of ln-$M_{x,t}$ results in better fits with smaller spread ranges via the proposed method.

To compare the magnitude of spreads generated during fuzzification phase via two methods, two example time periods are given in Tables 1 and 2. These tables illustrate the spreads of fuzzified ln-$M_{x,t}$ values for the first (1925–1929), and the last

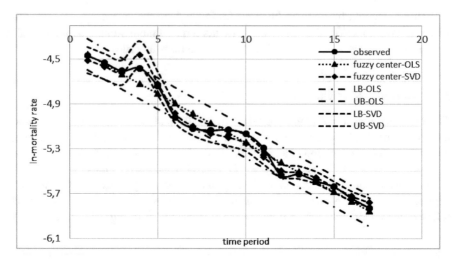

Fig. 3. Observed ln-$M_{x,t}$, fuzzy ln-$M_{x,t}$ with its spreads by Koissi and Shapiro's method, and fuzzy ln-$M_{x,t}$ with its spreads by the proposed method for age group [45, 50).

(2005–2009) time periods that are included in the application respectively. In each table, spreads generated via Koissi and Shapiro's fuzzified LC model (spread$_{OLS}$) and the proposed fuzzy LC model (spread$_{SVD}$) are shown for each age cohort.

Tables 1 and 2 show that the number of smaller spreads generated during fuzzification of ln-$M_{x,t}$ by the proposed method are increasing by time. This trend can be explained by the advances in accurate data approaches which result in vagueness reduction, thus smaller spreads. In fact the proposed method generates larger spreads mostly for the youngest and eldest age cohorts due to extreme (low or high) demographic rates and during time periods in which data collection and recording systems should be treated with caution such as the starting periods or World War II years. This tendency can be seen in Fig. 4, in which number of cases that proposed method

Table 1. Spreads of fuzzified ln-mortality values for Finland, 1925–1929.

Age group	Spread$_{OLS}$	Spread$_{SVD}$	Age group	Spread$_{OLS}$	Spread$_{SVD}$
[0, 1)	0.3220	0.4419	[50, 55)	0.0750	0.1560
[1, 5)	0.4920	0.3556	[55, 60)	0.1250	0.1533
[5, 10)	0.5300	0.1723	[60, 65)	0.1540	0.1837
[10, 15)	0.4890	0.2333	[65, 70)	0.1860	0.2686
[15, 20)	0.9170	0.4520	[70, 75)	0.1920	0.3123
[20, 25)	1.6470	0.9371	[75, 80)	0.2137	0.2960
[25, 30)	1.3170	0.6741	[80, 85)	0.1970	0.2603
[30, 35)	1.0320	0.4829	[85, 90)	0.2110	0.2300
[35, 40)	0.7380	0.3062	[90, 95)	0.2340	0.2473
[40, 45)	0.3860	0.1300	[95,100)	0.2340	0.2556
[45, 50)	0.1380	0.1173	[100,105)	0.3750	0.4252

Table 2. Spreads of fuzzified *ln*-mortality values for Finland, 2005-2009.

Age group	Spread$_{OLS}$	Spread$_{SVD}$	Age group	Spread$_{OLS}$	Spread$_{SVD}$
[0, 1)	0.4180	0.2102	(50, 55)	0.0750	0.0401
[1, 5)	0.4926	0.0852	(55, 60)	0.1250	0.1468
[5, 10)	0.5300	0.0951	(60, 65)	0.1540	0.1450
[10, 15)	0.4890	0.1561	(65, 70)	0.1860	0.1141
[15, 20)	0.9170	0.4520	(70, 75)	0.2240	0.1192
[20, 25)	1.6470	0.0487	(75, 80)	0.2251	0.1801
[25, 30)	1.3170	0.0175	(80, 85)	0.1970	0.1831
[30, 35)	1.0320	0.0194	[85, 90)	0.2110	0.2300
[35, 40)	0.7380	0.0028	[90, 95)	0.2340	0.0542
[40, 45)	0.3860	0.1300	[95,100)	0.2340	0.1011
[45, 50)	0.1380	0.0401	[100,105)	0.3750	0.0003

generates smaller spreads compared to Koissi and Shapiro's method during fuzzification of *ln*-$M_{x,t}$ values. It can be seen from Fig. 4 that proposed method generates smaller spreads for most of the age cohorts for the last time periods (e.g. in 20 cases out of 22 for 2005–2009 period) in which dealing with complex data approaches becomes easier due to technological advancements.

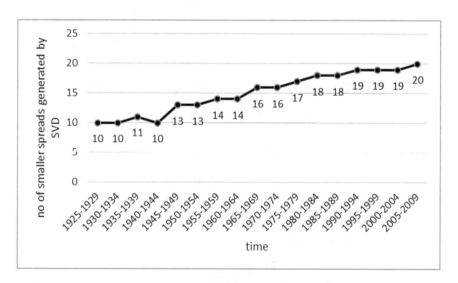

Fig. 4. Number of smaller spreads generated by proposed method during Phase I.

When the whole dataset is taken into account, paired *t*-test results show that the proposed method is statistically superior to Koissi and Shapiro's method in terms of smaller spread generation (*t*-value = 13.53, *p*-value = 0.000), and smaller absolute distances between observed *ln*-$M_{x,t}$ and center values of fuzzified *ln*-$M_{x,t}$ (*t*-value = 5.07, *p*-value = 0.000) during the fuzzification of *ln*-mortality rates.

To display the outputs obtained during Phase II, once again the results for two age cohorts, [25–30) and [45–50), are selected as examples. Figures 5 and 6 illustrate the observed ln-$M_{x,t}$, and centers of estimated fuzzy ln-$M_{x,t}$ with Koissi and Shapiro's and proposed methods for age groups [25, 30) and [45–50) respectively. In both figures, the horizontal axis stand for time periods (1 = 1925–1929, …, 17 = 2005–2009), whereas the vertical axis depicts the ln-mortality rates. The numerical findings show that the proposed method displays better similarity between observed and estimated ln-mortality rates compared to Koissi and Shapiro's method. In fact, paired t-test results show that the proposed method is better in generating smaller spreads (t-value = 13.97, p-value = 0.000) and smaller absolute distances between observed ln-$M_{x,t}$ and centers of fuzzy estimations (t-value = 2.69, p-value = 0.004) during estimation of ln-mortality rates.

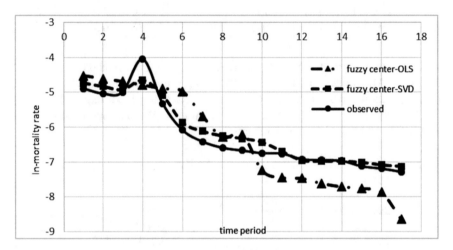

Fig. 5. Comparison of outputs for Phase II in estimating ln-$M_{x,t}$ for age cohort (25–30).

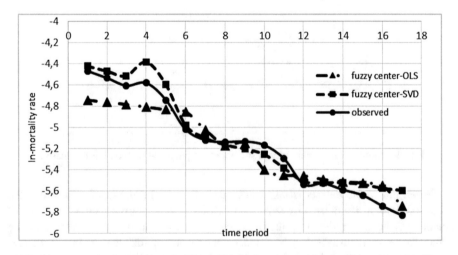

Fig. 6. Comparison of outputs for Phase II in estimating ln-$M_{x,t}$ for age cohort (45–50).

The t-test results can be supported with the amount of errors between observed ln-$M_{x,t}$ and centers of estimated fuzzy ones for the two methods. Mean absolute percentage error (MAPE) is 4.85 % for Koissi and Shapiro's method, while it is found to be 3.13 % for the proposed method. The reason why the proposed method gives better fits is mainly due to the utilization of SVD in fuzzification of observed ln-$M_{x,t}$'s. The usage of OLS in Koissi and Shapiro's method results in centers with linear trends, hence, the fluctuations in data cannot be covered properly.

3.2 Mortality Forecasting Results

In Sect. 3.1, mortality rates up to 2005–2009 period (period 17) have been estimated. Now, the task is to forecast mortality rates for the next period 2010–2014 (period 18) using the estimated mortality indices k_t's (t = 1,...,17) generated. The proposed modified FFTS method given in Sect. 2.2 is applied on k_t's (t = 1,...,17) to forecast k_{18}.

The fuzzy interval lengths are determined by three different approaches: ratio based interval, distribution based interval, and average based interval. In literature, these approaches are reported as the most efficient interval length determination methods [31, 32]. Among them, ratio based interval length setting approach is ranked as the best. Also, the conventional fuzzy time series technique is applied to the same k_t values with the above mentioned interval length determination approaches for comparison purposes, and the results are shown in Tables 3 and 4.

Table 3. Comparison of FFTS methods with 15 data points.

Method	No of intervals	MAPE (%) k_t	MAPE(%) $Y_{x,16}$
Ratio based interval (Con. FTS)	22	6.9269	2.6246
Distribution based interval (Con. FTS)	14	21.6465	2.5964
Average based interval (Con. FTS)	28	7.9715	2.6793
Ratio based interval (Mod. FFTS)	34	8.9307	2.6209
Distribution based interval (Mod. FFTS)	21	9.8947	2.6943
Average based interval (Mod. FFTS)	41	1.6794	2.5742

Table 4. Comparison of FFTS methods with 17 data points.

Method	No of intervals	MAPE (%)
Ratio based interval (Con. FTS)	22	6.2979
Distribution based interval (Con. FTS)	14	14.8642
Average based interval (Con. FTS)	29	8.8718
Ratio based interval (Mod. FFTS)	34	8.7717
Distribution based interval (Mod. FFTS)	21	10.8586
Average based interval (Mod. FFTS)	41	3.2377

Tables 3 and 4 display MAPE between k_t obtained in modeling part and the estimated $\hat{k}_t$ via a time series method using 15 data points (t = 1,...,15) and 17 data points (t = 1,...17) respectively. In these tables, number of fuzzy intervals generated via conventional fuzzy time series method (Con. FTS) and the modified FFTS method (Mod. FFTS) with the three different interval length determination approaches are given in second columns. Both tables show that the modified FFTS method with averaged based interval lengths gives better fits in estimating mortality indices with both sample sizes (Their corresponding MAPEs are 1.6794 % and 3.2377 %). The additional column (MAPE(%) $Y_{x,\ 16}$) in Table 3 displays MAPE in forecasting ln-mortality rates for 1995–1999 time period (t = 16) when 15 data points (k_t obtained from Sect. 3.1, $t = 1,...,15$) are considered. The values in this column show that, modified FFTS method with average based interval lengths slightly outperforms the other methods.

As it is stated in literature, ratio based interval length determination approach outperforms the other techniques for the conventional fuzzy time series method. However, this approach is not the best one when the modified FFTS method is applied. This is because modified FFTS method employs the spreads of $\tilde{K}_t$ together with its center value, whereas the conventional fuzzy time series considers only the center values.

The mortality indices $\hat{k}_t$'s estimated via the six fuzzy time series methods using 17 data points are given in Table 5. k_t values obtained in Sect. 3.1 are given second

Table 5. k_t values estimated via fuzzy time series methods for 17 data points.

t	Modeled k_t	Con.FTS			Mod.FFTS		
		Ratio based interval	Distribution based interval	Average based interval	Ratio based interval	Distribution based interval	Average based interval
1	1.680672						
2	1.571868	1.523127	1.4	1.55	1.43926	1.402006	1.571172
3	1.470318	1.627699	1.6	1.45	1.43926	1.623022	1.470716
4	1.770998	1.627699	1.6	1.75	0.874861	1.623022	1.771359
5	1.283117	1.523127	1.4	1.25	1.43926	1.402006	1.283142
6	0.411846	0.434077	0.5	0.45	0.874861	0.409934	0.412158
7	0.132236	0.153153	0.1	0.15	0.047021	0.133921	0.131434
8	−0.01749	0.025095	−0.1	−0.05	0.003388	−0.06073	−0.01886
9	−0.10348	−0.09535	−0.2	−0.15	−0.02146	−0.14514	−0.10399
10	−0.2234	−0.20863	−0.2	−0.25	−0.06392	−0.14514	−0.22237
11	−0.51925	−0.50962	−0.5	−0.55	−0.41923	−0.51961	−0.52098
12	−0.80825	−0.83379	−0.9	−0.9	−0.91896	−0.90329	−0.89182
13	−0.83351	−0.86847	−0.9	−0.9	−0.91896	−0.90329	−0.89182
14	−0.83351	−0.86847	−0.9	−0.9	−0.91896	−0.90329	−0.89182
15	−0.87163	−0.86847	−0.9	−0.9	−0.91896	−0.90329	−0.89182
16	−0.95437	−0.96838	−0.9	−0.9	−0.91896	−0.90329	−0.89182
17	−0.99782	−0.96838	−0.9	−0.95	−0.91896	−0.90329	−0.97592
18		−0.96838	−0.9	−0.95	−0.99863	−0.90329	−0.97592

column, while the following three columns are $\hat{k}_t$ values estimated via conventional fuzzy time series method (Con.FTS); and the last three columns the modified FFTS method (Mod.FFTS). In Table 5, the superiority of modified FFTS with average based interval length determination approach in estimating k_t can clearly be seen (When columns 2 and 3 for conventional fuzzy time series and columns 2 and 8 for the modified FFTS are compared). The last row of Table 5 is dedicated to k_{18} forecasts, which are utilized in computing the forecasted *ln*-mortality rates.

4 Conclusion

In this study, a modified fuzzy LC method for modeling human mortality is proposed. The proposed method employs SVD technique for fuzzification of observed *ln*-mortality rates, so that the existing fluctuations in data can be captured. Moreover, a modified first order fuzzy time series technique is developed to provide future forecasts for mortality rates. The proposed modified FFTS method uses fuzzy inputs in contrast to conventional time series approaches which employ crisp data.

The proposed method is applied to Finland mortality dataset and the outputs are compared with the existing fuzzy LC approach. Numerical findings show that the proposed method statistically outperforms the existing fuzzy method in terms of better fits with smaller spreads in mortality rate estimations. Based on the estimated fuzzy mortality indices achieved from mortality modeling part of the proposed method, future mortality rates for Finland are forecasted. The forecasts obtained via the modified fuzzy time series approach gives smaller mean absolute errors compared to the existing fuzzy time series method.

Proposed method generates larger spreads for the starting time periods, and youngest and eldest age cohorts. This sounds realistic as data collection and recording techniques may be questionable to some degree for the early periods and there exist extreme values in data for the youngest and eldest age cohorts. It is worth mentioning that the fuzzy range estimated for demographic rates belonging to recent periods are getting smaller with time.

As a future study, an aggregated fuzzy population modeling method can be developed based on the proposed technique of this study. To accomplish this, the proposed method should be extended to cover modeling human fertility and migration as well. Moreover, higher order fuzzy time series models with different interval length determination procedures can be incorporated into the modified fuzzy time series approach to improve the forecast efficiencies.

References

1. Keyfitz, N.: Applied Mathematical Demography. Wiley, New York (1977)
2. Hyndman, R.J., Ullah, M.S.: Robust forecasting of mortality and fertility rates: a functional data approach. Comput. Stat. Data Anal. **51**, 4942–4956 (2007)

3. United Nations: Manuals on methods of estimating population - Manual VIII: Methods for Projections of Urban and Rural Population, Department of Economic and Social Affairs, Population Studies, No. 55, New York (1974)
4. Lindh, T.: Demography as a forecasting tool. Futures **35**, 37–48 (2003)
5. Ahmadi, S.S., Li, J.S.: Coherent mortality forecasting with generalized linear models: a modified time-transformation approach. Insur. Math. Econ. **59**, 194–221 (2014)
6. Danesi, I.L., Haberman, S., Millossovich, P.: Forecasting mortality in subpopulations using Lee-Carter type models: a comparison. Insur. Math. Econ. **62**, 151–161 (2015)
7. French, D.: International mortality modelling - an economic perspective. Econ. Lett. **122**, 182–186 (2014)
8. Booth, H.: Demographic forecasting: 1980 to 2005 in review. Int. J. Forecast. **22**, 547–581 (2006)
9. Lee, R.D., Carter, L.R.: Modelling and forecasting US mortality. J. Am. Stat. Assoc. **87**, 659–671 (1992)
10. Lee, R.D., Tuljapurkar, S.: Stochastic population projections for the U.S.: beyond high, medium and low. J. Am. Stat. Assoc. **89**(428), 1175–1189 (1994)
11. Li, S.H., Chan, W.S.: Outlier analysis and mortality forecasting: the United Kingdom and Scandinavian countries. Scand. Actuarial J. **3**, 187–211 (2005)
12. De Jong, P., Tickle, L.: Extending Lee-Carter mortality forecasting. Math. Popul. Stud. **13**, 1–18 (2006)
13. Renshaw, A.E., Haberman, S.: Lee-Carter mortality forecasting with age specific enhancement. Insur. Math. Econ. **33**(2), 255–272 (2003)
14. Brouhns, N., Denuit, M., Vermunt, J.K.: A Poisson log-bilinear approach to the construction of projected life tables. Insur. Math. Econ. **31**(3), 373–393 (2002)
15. Currie, I.D., Durban, M., Eilers, P.H.C.: Smoothing and forecasting mortality rates. Stat. Model. **4**(4), 279–298 (2004)
16. Lazar, D., Denuit, M.: A multivariate time series approach to projected life tables. Appl. Stochast. Models Bus. Ind. **25**(6), 806–823 (2009)
17. Hatzopoulos, P., Haberman, S.: A dynamic parameterization modeling for the age–period–cohort mortality. Insur. Math. Econ. **49**, 155–174 (2011)
18. Giacometti, R., Bertocchi, M., Rachev, S.T., Fabozzi, F.J.: A comparison of the Lee-Carter model and AR–ARCH model for forecasting mortality rates. Insur. Math. Econ. **50**, 85–93 (2012)
19. Christiansen, M.C., Niemeyer, A., Teigiszerová, L.: Modeling and forecasting duration-dependent mortality rates. Comput. Stat. Data Anal. **83**, 65–81 (2015)
20. Ahcan, A., Medved, D., Olivieri, A., Pitacco, E.: Forecasting mortality for small populations by mixing mortality data. Insur. Math. Econ. **54**, 12–27 (2014)
21. Lee, R.D.: The Lee-Carter method for forecasting mortality, with various extensions and applications. North Am. Actuarial J. **1**(4), 80–91 (2000)
22. Koissi, M.C., Shapiro, A.F.: Fuzzy formulation of Lee-Carter model for mortality forecasting. Insur. Math. Econ. **39**, 287–309 (2006)
23. Tanaka, H., Uejima, S., Asai, K.: Linear regression analysis with fuzzy model. IEEE Trans. Syst. Man Cybern. Syst. **2**, 903–907 (1982)
24. Chang, Y.H.O., Ayyub, B.M.: Fuzzy regression methods – A comparative assessment. Fuzzy Sets Syst. **119**, 187–203 (2001)
25. Song, Q., Chissom, B.S.: Forecasting enrollments with fuzzy time series – part I. Fuzzy Sets Syst. **54**, 1–9 (1993)
26. Mandel, J.: Use of the singular value decomposition in regression analysis. Am. Stat. **36**(1), 15–24 (1982)
27. Meisar, R.: Shape preserving additions of fuzzy intervals. Fuzzy Sets Syst. **86**, 73–78 (1997)

28. Diamond, P.: Fuzzy least squares. Inf. Sci. **46**, 141–157 (1988)
29. Nelder, J.A., Mead, R.: A simplex method for function minimization. Comput. J. **7**(4), 308–313 (1965)
30. Qiu, W., Liu, X., Li, H.: A generalized method for forecasting based on fuzzy time series. Expert Syst. Appl. **38**, 10446–10453 (2011)
31. Huarng, K.: Effective lengths of intervals to improve forecasting in fuzzy time series. Fuzzy Sets Syst. **123**, 387–394 (2001)
32. Huarng, K., Yu, T.H.-K.: Ratio-based lengths of intervals to improve fuzzy time series forecasting. IEEE Trans. Syst. Man Cybern. Part B Cybern. **36**(2), 328–340 (2006)

Synchronization of Chaotic Systems with Unknown Parameters Using Predictive Fuzzy PID Control

Zakaria Driss[✉] and Noura Mansouri

Laboratory of Automatics and Robotic, Faculty of Engineer Sciences,
Department of Electronics, University of Constantine 1, 25000 Constantine, Algeria
drisszakaria1@gmail.com, nor_mansouri@yahoo.fr

Abstract. In this paper, we consider the synchronization of uncertain chaotic systems using predictive fuzzy PID control. The main aim of the study is to show the role of prediction terms as a function of the sort of the controller used to solve the optimization problem. Therefore, two controllers, fuzzy PI+D and fuzzy PD+I controllers, are used in order to compare their abilities concerning the synchronization of chaotic systems in presence and absence of the prediction terms. This survey reveals that the role of the prediction terms depends on the type of the controller used to optimize the cost function. In the case of the fuzzy PD+I controller, the prediction terms seem to be very useful; on the other hand, in the case of the fuzzy PI+D, they restrict the ability of the controller, which leads to reduce its accuracy. Synchronization of two uncertain Lorenz systems is used to show the differences between the two cases.

Keywords: GPC · Predictive fuzzy PID control · Chaotic systems · Synchronization

1 Introduction

Synchronization of chaotic systems has been widely investigated in the last decades. The focal point of chaos synchronization is secure communication, where two systems must be defined, the master or driver system, and the slave or response system. In order to generate synchronization signals, some control methods can be used. However, uncertainties on the parameters cause an obstruction to many classical control approaches. Therefore, some advanced control approaches and improved schemes such as fuzzy logic control (FLC) [1], neural network (NN) [2], adaptive control strategy [3], have been used to resolve this problem.

Chaotic systems have special features, which make them behave in a very interesting way. The common feature between normal nonlinear systems and chaotic systems is that both are deterministic systems. However, chaotic systems exhibit random behavior, which makes them unpredictable. The unpredictability of chaotic systems comes from a phenomenon called the butterfly effect which

© Springer International Publishing AG 2017
J.J. Merelo et al. (eds.), *Computational Intelligence*, Studies in Computational Intelligence 669,
DOI 10.1007/978-3-319-48506-5_12

means the highly sensitive to initial conditions. On the other hand, predictive control needs an accurate model in order to predict the future variation of the controlled system, which make the study of predictive control with unpredictable systems like chaotic systems is a very interesting topic.

Model predictive control (MPC) [4] is a control approach which consists in using a model of a system to predict its output over an extended horizon. In the presence of uncertainties, self-tuning and model-reference adaptive control (MRAC) were used with MPC to solve many problems such as an open-loop unstable plant, a nonminimum-phase plant, a plant with variable or unknown dead-time and a plant with unknown order. However, there was not a general algorithm to solve all these problems at once until the establishment of a general algorithm by D.W. Clarke [5] in 1985 called generalized predictive control (GPC).

The drawback of GPC is the number of mathematical steps the algorithm requests. In order to fix this problem, several advanced control approaches have been involved in GPC such as fuzzy-model-based approach [1], Neural-network [6], and PSO-based model predictive control [7]. One of the most interesting approaches [8] is by involving fuzzy PID controllers to minimize the cost function and to ensure the convergence between the controlled system and the reference trajectory.

In this paper, we consider the performance of predictive fuzzy PID control [8] for the synchronization of uncertain chaotic systems. Fuzzy PI+D and fuzzy PD+I controllers are successively used to check the performance of the proposed control method in the presence and absence of the prediction terms. For the prediction of the future variation of the master and the slave system, an ARX model is used. Lyapunov's second method is used with particle swarm optimization (PSO) algorithm to ensure the stability. To verify the performance of the above predictive approach, we apply it for the synchronization of two uncertain Lorenz systems.

The rest of the paper is arranged as follows: Section 2 presents synchronization of chaotic systems. An overview of predictive fuzzy PID control is introduced in Sect. 3. The main steps of the design of predictive fuzzy PID control and stability analysis are given in Sect. 4. Simulation results are given in Sect. 5. Conclusions are given in Sect. 6.

2   Synchronization of Uncertain Chaotic Systems

Let's consider two n-dimensional chaotic systems, one is designed as the master system:

$$\dot{x}_m = g_m(x, t) \quad 1 \leq m \leq n,$$
$$x = [x_1, x_2, ..., x_n] \in \Re^n \tag{1}$$

and the second is the controlled slave system:

$$\dot{y}_m = f_m(y, t) + u_m(t) \quad 1 \leq m \leq n,$$
$$y = [y_1, y_2, ..., y_n] \in \Re^n \tag{2}$$

where f and g represent unknown nonlinear functions, and $u \in \Re^n$ is the control input.

The Synchronization problem can be considered as a control problem which consists in the design of an appropriate control law $u(t)$ such that:

$$\lim_{t \to \infty} \|y(t) - x(t)\| \to 0. \tag{3}$$

The error states between the two systems are given by:

$$\dot{e}_m = g_m(y, t) - f_m(x, t) + u_m(t) \quad 1 \le m \le n, \tag{4}$$

and the objective is how to design an efficient control law $u_m(t)$ such that the error states converge to zero when the time goes further.

3   An Overview of Predictive Fuzzy PID Control

Predictive fuzzy PID control can be considered as GPC algorithm based on fuzzy PID controllers. GPC algorithm consists mainly in minimizing a cost function that contains the predicted values. There have been many attempts to reduce the complexity of the algorithm by involving some advanced control approaches. To avoid the tedious mathematical steps, fuzzy PID controllers can be used [8]. For the synchronization, the following criterion is used:

$$J_m(k) = \sum_{i=-1}^{N} [x_m(k-i) - y_m(k-i)]^2$$
$$+ \lambda \sum_{j=0}^{N_c} [\Delta u_m(k-j)]^2$$
$$J_m(k) = \sum_{i=-1}^{N} [e_m(k-i)]^2 + \lambda \sum_{j=0}^{N_c} [\Delta u_m(k-j)]^2, \tag{5}$$

where N is the prediction horizon, N_c is the control increment horizon, Δu_m is the incremental output of a controller, $\lambda \ge 0$ is a control increment weight. Figure 1 represents the main structure of predictive fuzzy PID control for synchronization of uncertain chaotic systems.

To get the predicted values of both systems, we use ARX model. For the slave system, the model is given by:

$$\hat{y}_m(k+1) = a_1 y_m(k) + a_2 y_m(k-1) + a_3 y_m(k-2)$$
$$+ a_4 y_m(k-3) + b_1 u_m(k-1), \tag{6}$$

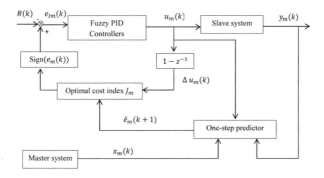

Fig. 1. Block diagram of predictive fuzzy PID control.

while the model of the master is given by:

$$\widehat{x}_m(k+1) = a_1 x_m(k) + a_2 x_m(k-1) + a_3 x_m(k-2)$$
$$+ a_4 x_m(k-3), \qquad (7)$$

where $x_m(k)$, $y_m(k)$ and $u_m(k)$ are the outputs of the master system, the outputs of the slave system and the control inputs respectively; a_1, a_2, a_3, a_4, b_1 are constant parameters.

Thus, the one-step ahead predictor of the error states is given as:

$$\widehat{e}_m(k+1) = a_1 e_m(k) + a_2 e_m(k-1) + a_3 e_m(k-2)$$
$$+ a_4 e_m(k-3) + b_1 u_m(k-1). \qquad (8)$$

Nature inspired optimization algorithms, such as PSO algorithm [9,10], can be used in order to facilitate the adjustment of the parameters of the one-step ahead predictor. PSO algorithm uses the behavior of bird flock, which is called swarm. The swarm has a number of particles. Each particle has a memory, and the movement of a particle within the search region is given by a shift value, called velocity, which is related to its memory and the position of the best particle in the swarm.

The swarm of H particles is defined as:

$$S = \{s_1, s_2, ..., s_H\},$$

and each one is given as:

$$s_w = \left(s_{w1}, s_{w2}, ..., s_{wh}\right)^T, \quad w = 1, 2, ..., H, \quad q = 1, 2, ..., h.$$

Algorithm 1. PSO algorithm.

Step 1. Set $l = 1$.

Step 2. Initial a random swarm S of particles within the search space, and set $P = S$.

Step 3. Evaluate the cost function of each particle of S and P, and get the index g of the best position.

Step 4. while $l < OI$ **do**

Step 5. Update S using (9) and (10).

Step 6. Evaluate S.

Step 7. Update P and redefine index g.

Step 8. $l = l + 1$;.

Step 9. end while.

Step 10. Print best position found.

Particles are moved within search space by using a proper position shift, called velocity, and denoted as:

$$v_w = (v_{w1}, v_{w2}, ..., v_{wh})^T.$$

During the search process, each particle stores the best position which it has ever visited in a memory set:

$$P = \{p_1, p_2, ..., p_H\},$$

where

$$p_w = (p_{w1}, p_{w2}, ..., p_{wh})^T.$$

The velocity of each particle is given as follows:

$$v_{wq}(l+1) = v_{wq}(l) + c_1 R_1 (p_{wq} - s_{wq}(l)) \\ + c_2 R_2 (p_{gq}(l) - s_{wq}(l)), \tag{9}$$

$$s_{wq}(l+1) = s_{wq}(l) + v_{wq}, \tag{10}$$

where R_1, R_2 are random variables between $[0, 1]$; c_1 and c_2 are weighting factors; g is the index of the best position in memory set P.

The optimization algorithm based on PSO is described by the following steps:

Algorithm 1 is used to adjust off-line the parameters of the one -step ahead predictors, where the particles are defined as

$$s_w = (a_{w1}, a_{w2}, a_{w3}, a_{w4}, b_{w1})^T,$$

and the fitness of each particle of the swarm is obtained using the fitness algorithm as described below:

Algorithm 2. Fitness algorithm.

Step 1. Set $k = 1$.

Step 2. While$(k \leq I)$ **do**

Step 3. Calculate the output of the one-step predictor of the slave system using (6).

Step 4. Calculate the cost function value using (5).

Step 5. Obtain $u(k)$ using a fuzzy PID controller.

Step 6. Calculate the output of the slave system.

Step 7. Calculate the error between the slave system and the one-step predictor by $E(k) = |y_m(k) - \hat{y}_m(k)|$.

Step 8. $k = k + 1$;.

Step 9. end while.

Where I is the number of iterations. The fitness of a particle is obtained by calculating the maximum value of E in the permanent case.

4   Design of Predictive Fuzzy PID Control

Fuzzy PID controllers have been categorized among the most successful control approaches in industrial applications. This set of controllers needs only the output of the controlled system in order to find the right control action (free model approach), which make them better than many classical control approaches where an accurate mathematical model must be used. Fuzzy PID controllers are derived from conventional PID controllers, in which fuzzy logic control is involved in traditional PID controllers in order to increase their abilities to handle the complexity of nonlinear systems with uncertainties and time-delay. Therefore, Lu et al. [8] took the advantage of the simplicity of fuzzy PID controllers in order to reduce the complexity of classical GPC algorithm. They proposed to use Fuzzy PD+I controller in order to solve the optimization problem and reach asymptotic stability. This approach doesn't need an accurate model of the controlled system, which makes it suitable to deal with the hypersensibility of chaotic systems during the synchronization process. The idea of predictive fuzzy PID control is to feed the fuzzy PID controllers with the cost function variation J_m and the rate of change of the cost function variation $\dot{J}_m$ instead of the error signal e_m and the rate of change of the error signal $\dot{e}_m$. In this paper, fuzzy PI+D and fuzzy PD+I controllers are used to perform two tasks: drive the slave system to track the output of the master system, make the cost function, J_m, as small as possible.

4.1   Design of Fuzzy PI+D Controller

Fuzzy PI+D controller can be considered as the sum of two sub-controllers [11,12], the fuzzy PI and Fuzzy D controllers. The fuzzy PI controller is derived from conventional PI control, which is given as follows:

$$\begin{cases} u_{PIm}(t) = K_{pm}e_{Jm}(t) + K_{im}\displaystyle\int e_{Jm}(t)dt \\ e_{Jm}(t) = J_m(t) \times Sign(e_m(t)) - R_m(t), \end{cases} \tag{11}$$

where $R_m(t)$ is the reference for the optimal cost index; K_{pm} is the constant proportional gain; K_{im} is integral gain. $e_{Jm}(t)$ and $e_m(t)$ are the error signal from the optimal index J_m, the error between the master and the slave system, respectively.

The convention analog PI controller is given in the frequency s-domain as follows:

$$u_{PIm}(s) = (K_{pm}^c + \frac{K_{im}^c}{s})E_{Jm}(s). \tag{12}$$

To get the digital version, the bilinear transform is applied $s = (2/T)$ $(z-1)/(z+1)$, where $T > 0$, is the sampling time, which leads to the following form:

$$u_{PIm}(z) = (K_{pm}^c - \frac{K_{im}^c T}{2} + \frac{K_{im}^c T}{1 - z^{-1}})E_{Jm}(z). \tag{13}$$

Letting

$$K_{pm} = K_{pm}^c - \frac{K_{im}^c T}{2} \quad \text{and} \quad K_{im} = K_{im}^c T$$

and using the inverse z-transform, we get the digital form of the controller:

$$\begin{aligned} u_{PIm}(kT) - u_{PIm}(kT - T) &= K_{pm}[e_{Jm}(kT) \\ &- e_{Jm}(kT - T)] + K_{im}e_{Jm}(kT). \end{aligned} \tag{14}$$

Dividing (14) by T, we obtain

$$\Delta u_{PIm}(kT) = K_{pm}e_{vm}(kT) + K_{im}e_{pm}(kT), \tag{15}$$

where

$$\Delta u_{PIm}(kT) = \frac{u_{PIm}(kT) - u_{PIm}(kT - T)}{T}, \tag{16}$$

$$e_{vm}(kT) = \frac{e_{Jm}(kT) - e_{Jm}(kT - T)}{T}, \tag{17}$$

$$e_{pm}(kT) = e_{Jm}(kT), \tag{18}$$

$\Delta u_{PIm}(kT)$ is the incremental control output of the PI controller, $e_{pm}(kT)$ the error between the master and the slave system, and $e_{vm}(kT)$ is the error rate. Equation (16) can be written as the following form:

$$u_{PIm}(kT) = u_{PIm}(kT - T) + T\Delta u_{PIm}(kT). \tag{19}$$

To get the fuzzy PI controller, the increment control input $T\Delta u_{PIm}(kT)$ will be replaced by a fuzzy control term $K_{uPIm}\Delta u_{PIm}(kT)$, so that:

$$u_{PIm}(kT) = u_{PIm}(kT - T) + K_{uPIm}\Delta u_{PIm}(kT), \tag{20}$$

where K_{uPIm} is a fuzzy control gain.

The second part of fuzzy PI+D controller is derived from conventional D controller that is given in s-domain, as follows:

$$u_{Dm}(s) = sK_{dm}^c Y_m(s), \tag{21}$$

where $Y_m(s)$ is the outputs of the slave system and K_{dm}^c is the control gain.

Using the bilinear transformation we get:

$$u_{Dm}(z) = K_{dm}^c \frac{2}{T} \frac{1 - z^{-1}}{1 + z^{-1}} Y_m(z), \tag{22}$$

and then taking the inverse z-transform, we get the discrete version of the D controller

$$u_{Dm}(kT) + u_{Dm}(kT - T) = \frac{2K_{dm}^c}{T}[y_m(kT) - y_m(kT - T)]. \tag{23}$$

Letting

$$K_{dm} = \frac{2K_{dm}^c}{T},$$

then, dividing (23) by T, we get

$$\Delta u_{Dm}(kT) = K_{dm}\Delta y_m(kT), \tag{24}$$

where

$$\Delta u_{Dm}(kT) = \frac{u_{Dm}(kT) + u_{Dm}(kT - T)}{T}, \tag{25}$$

and

$$\Delta y_m(kT) = \frac{y_m(kT) - y_m(kT - T)}{T}, \tag{26}$$

We can notice from (24) that there is only one input of the D controller, $\Delta y_m(kT)$, which is not enough to give the right information about the position of the output in the design of the fuzzy rules (below the setpoint or above). Therefore, another signal must be used in (24), which becomes

$$\Delta u_{Dm}(kT) = K_{dm}\Delta y_m(kT) + K_m y_{dm}(kT), \tag{27}$$

where $K_m = 1$ and $y_{dm}(kT) = -e_{Jm}(kT)$.

Thus,

$$u_{Dm}(kT) = -u_{Dm}(kT - T) + T\Delta u_{Dm}(kT). \tag{28}$$

In order to get the fuzzy D controller, the term $T\Delta u_{Dm}(kT)$ is replaced by $K_{uDm}\Delta u_{Dm}(kT)$

$$u_{Dm}(kT) = -u_{Dm}(kT - T) + K_{uDm}\Delta u_{Dm}(kT). \tag{29}$$

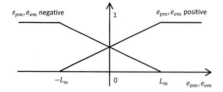

Fig. 2. PI input membership functions.

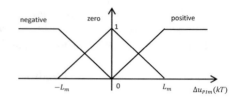

Fig. 3. PI output membership functions.

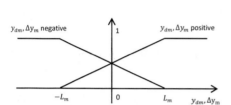

Fig. 4. D input membership functions.

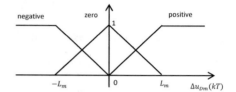

Fig. 5. D output membership functions.

The fuzzy PI+D is a combination of the two controllers, the fuzzy PI controller (20) and the fuzzy D controller (29). Hence, the overall fuzzy PI+D controller is given as follows.

$$
\begin{aligned}
u_{PIDm} &= u_{PIm}(kT) - u_{Dm}(kT) \\
&= u_{PIm}(kT - T) + K_{uPIm}\Delta u_{PIm}(kT) + u_{Dm}(kT - T) \\
&\quad - K_{uDm}\Delta u_{Dm}(kT).
\end{aligned} \tag{30}
$$

The inputs of the Fuzzy PI+D controller are the error $e_{pm}(kT)$, the error rate $e_{vm}(kT)$, the rate of change of the output $\Delta y_m(kT)$ and $y_dm(kT)$. On the other hand, the fuzzy PI+D controller has only one output, the control output $u_{PIDm}(kT)$, which is used as input to the slave system. The design of the fuzzy PI+D controller needs three parts: fuzzification, control rule base, and defuzzification. Figures 2 and 4 give the membership functions for the fuzzification of the inputs, whereas Figs. 3 and 5 give the membership functions of the outputs.

The global stability can be reached by adjusting the fuzzy control rules of both controllers, Fuzzy PI and Fuzzy D, separately. The fuzzy control rules are assigned according to the structure of the controller and the position of the outputs comparing to the reference trajectories.

By using the above membership functions, the following rules can be assigned for the fuzzy PI controller:

(R1) IF $e_{pm} = e_{pm}.n$ AND $e_{vm} = e_{vm}.n$ THEN PI-output $= o.n$,
(R2) IF $e_{pm} = e_{pm}.n$ AND $e_{vm} = e_{vm}.p$ THEN PI-output $= o.z$,
(R3) IF $e_{pm} = e_{pm}.p$ AND $e_{vm} = e_{vm}.n$ THEN PI-output $= o.z$,
(R4) IF $e_{pm} = e_{pm}.p$ AND $e_{vm} = e_{vm}.p$ THEN PI-output $= o.p$,

while the rules of the fuzzy D controller are given as follows:

(R5) IF $y_{dm} = y_{dm}.p$ AND $\Delta y_{vm} = \Delta y_{vm}.p$ THEN D-output $= o.z$,
(R6) IF $y_{dm} = y_{dm}.p$ AND $\Delta y_{vm} = \Delta y_{vm}.n$ THEN D-output $= o.p$,
(R7) IF $y_{dm} = y_{dm}.n$ AND $\Delta y_{vm} = \Delta y_{vm}.p$ THEN D-output $= o.n$,
(R8) IF $y_{dm} = y_{dm}.n$ AND $\Delta y_{vm} = \Delta y_{vm}.n$ THEN D-output $= o.z$.

From the rules, the control action coming from the fuzzy PI+D alternates between the output of the fuzzy PI controller and fuzzy D controller. More precisely, when the fuzzy PI control gets an action, the fuzzy D must set to zero and vice versa. From (11), we can note that only the sing of the errors e_m are used in order to locate the position of the outputs (below or above the reference trajectories), the variation of the cost function cannot be used because it is always positive. We suppose that for each step that x_m is constant. Thus, $e_{pm} = x_m - y_m$ and $e_{vm} = \dot{e}_{pm} = 0 - \dot{y}_m$ are use used to design the rules. For instance, if e_{pm} is negative ($e_{pm}.n$) means that the error is above the setpoint, and if the error rate is negative ($e_{pm}.n$) implies the controller at the previous step is driving the system output upward; the control output, $\Delta u_{Pim}(kT)$, must be set to be negative (R1), and the output of the fuzzy D controller must be set at zero (R5). On the other hand, if e_{pm} is negative ($e_{pm}.n$) means that the error is above the setpoint, and if the error rate is positive ($e_{pm}.n$) implies the controller at the previous step is driving the system output downward; the control output, $\Delta u_{Pim}(kT)$, must be set to be zero (R2), and the role of the fuzzy D controller is to add a positive value in order to make the output down faster (R6). The rest of the rules can be interpreted by the same way. Usually in the defuzzification step, center of mass formula is used to get the increment control outputs of both fuzzy PI and fuzzy D controllers:

$$\Delta u_m = \frac{\sum \text{MVI} \times \text{MVO}}{\sum \text{membership value of input}}, \tag{31}$$

where MVI is membership value of input and MVO is output corresponding to the membership value of input. The intersections between the inputs are divided into 20 adjacent input combination (IC) regions, as shown in Figs. 6 and 7.

Each selected region has specific conditions; for example, in the region IC 1 the flowing conditions must be held: $0 < K_{im}.e_{pm}(kT) < L_m$, $-L_m < K_{pm}.e_{vm}(kT) < 0$ and $K_{pm}.e_{vm}(kT) + K_{im}.e_{pm}(kT) > 0$. By using the rules, the defuzzification formula, and the following equations:

$$e_{pm}.p = \frac{K_{im}e_{pm}(kT) + L_m}{2L_m}, \quad e_{pm}.n = \frac{-K_{im}e_{pm}(kT) + L_m}{2L_m},$$

$$e_{vm}.p = \frac{K_{pm}e_{vm}(kT) + L_m}{2L_m}, \quad e_{vm}.n = \frac{-K_{pm}e_{vm}(kT) + L_m}{2L_m},$$

$$y_{dm}.p = \frac{Ky_{dm}(kT) + L_m}{2L_m}, \quad y_{dm}.n = \frac{-K_m y_{dm}(kT) + L_m}{2L_m},$$

$$\Delta y_m.p = \frac{K_{dm}\Delta y_m(kT) + L_m}{2L_m}, \quad \Delta y_m.n = \frac{-K_{dm}\Delta y_m(kT) + L_m}{2L_m},$$

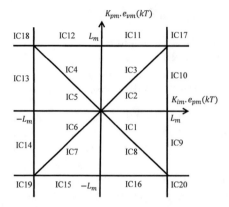

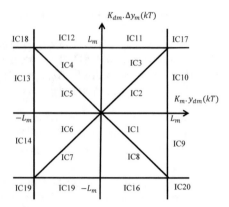

Fig. 6. Regions of the fuzzy PI controller.

Fig. 7. Regions of the fuzzy D controller.

we obtain the defuzzification formulas for all 20 regions of the fuzzy PI:

$$\Delta u_{PIm}(kT) = \frac{L_m[K_{im}.e_{pm}(kT) + K_{pm}.e_{vm}(kT)]}{2(2L_m - K_{im}.|e_{pm}(kT)|)}, \text{ (in IC } 1,2,5,6) \tag{32}$$

$$\Delta u_{PIm}(kT) = \frac{L_m[K_{im}.e_{pm}(kT) + K_{pm}.e_{vm}(kT)]}{2(2L_m - K_{pm}.|e_{vm}(kT)|)}, \text{ (in IC } 3,4,7,8) \tag{33}$$

$$\Delta u_{PIm}(kT) = 1/2[K_{pm}e_{vm}(kT) + L_m], \text{ (in IC } 9,10) \tag{34}$$

$$\Delta u_{PIm}(kT) = 1/2[K_{im}e_{pm}(kT) + L_m], \text{ (in IC } 11,12) \tag{35}$$

$$\Delta u_{PIm}(kT) = 1/2[K_{pm}e_{vm}(kT) - L_m], \text{ (in IC } 13,14) \tag{36}$$

$$\Delta u_{PIm}(kT) = 1/2[K_{im}e_{pm}(kT) - L_m], \text{ (in IC } 15,16) \tag{37}$$

$$\Delta u_{PIm}(kT) = 0, \text{ (in IC } 18,20) \tag{38}$$

$$\Delta u_{PIm}(kT) = L_m, \text{ (in IC } 17) \tag{39}$$

$$\Delta u_{PIm}(kT) = -L_m, \text{ (in IC } 19) \tag{40}$$

whereas for the fuzzy D controller we obtain the following formulas:

$$\Delta u_{Dm}(kT) = \frac{L_m[K_m.y_{dm}(kT) - K_{dm}.\Delta y_m(kT)]}{2(2L_m - K_m.|y_{dm}(kT)|)}, \text{ (in IC } 1,2,5,6) \tag{41}$$

$$\Delta u_{Dm}(kT) = \frac{L_m[K_m.y_{dm}(kT) - K_{dm}.\Delta y_m(kT)]}{2(2L_m - K_{dm}.|\Delta y_m(kT)|)}, \text{ (in IC } 3,4,7,8) \tag{42}$$

$$\Delta u_{Dm}(kT) = 1/2[-K_{dm}\Delta y_m(kT) + L_m], \text{ (in IC } 9,10) \tag{43}$$

$$\Delta u_{Dm}(kT) = 1/2[K_m y_{dm}(kT) - L_m], \text{ (in IC } 11,12) \tag{44}$$

$$\Delta u_{Dm}(kT) = 1/2[-K_{dm}\Delta y_m(kT) - L_m], \text{ (in IC } 13,14) \tag{45}$$

$$\Delta u_{Dm}(kT) = 1/2[K_m y_{dm}(kT) + L_m], \text{ (in IC } 15, 16) \tag{46}$$

$$\Delta u_{Dm}(kT) = 0, \text{ (in IC } 17, 19) \tag{47}$$

$$\Delta u_{Dm}(kT) = -L_m, \text{ (in IC } 18) \tag{48}$$

$$\Delta u_{Dm}(kT) = L_m. \text{ (in IC } 20) \tag{49}$$

4.2   Design of Fuzzy PD+I Controller

The design of the fuzzy PD+I controller passes through the same steps as the fuzzy PI+D [8, 13]. All sorts of fuzzy PID controllers derive from the conventional forms, then the incremental outputs are replaced with the fuzzy terms. The fuzzy core is designed using three steps: fuzzification, control rule base, and defuzzification. The main part of the fuzzy core is the control rule base, which is designed according to the structure of the controller, and the positions of the inputs. In this section, we summarize the principal steps of the design of the fuzzy PD+I controller.

The conventional form of the linear PD+I controller in s-domain is

$$u_{PIDm}(s) = u_{PDm}(s) + u_{Im}(s),$$

where

$$u_{PDm}(s) = (K_{pm}^c + sK_{dm}^c)E_{Jm}(s), \text{ and } u_{Im}(s) = \frac{K_{im}^c}{s}E_{Jm}(s) \Longrightarrow$$

$$u_{PDm}(z) = (K_{pm} + K_{dm}\frac{1-z^{-1}}{1+z^{-1}})E_{Jm}(z), \text{ and } u_{Im}(z) = K_{im}\frac{T}{2}\frac{1+z^{-1}}{1-z^{-1}};$$

$$K_{pm} = K_{pm}^c, K_{dm} = \frac{2}{T}K_{dm}^c, K_{im} = K_{im}^c.$$

The discrete forms of the PD and the I controllers are given by using the inverse z-transform as follows:

$$\Delta u_{PDm}(kT) = K_{pm}d_m(k) + K_{dm}r_m(k), \tag{50}$$

$$\Delta u_{Im}(kT) = K_m r_m(k) + K_{im}e_{Jm}(k-1), \tag{51}$$

where

$$r_m(kT) = \frac{e_{Jm}(k) - e_{Jm}(kT-T)}{T}, \tag{52}$$

$$d_m(kT) = \frac{e_{Jm}(kT) + e_{Jm}(kT-T)}{T}, \tag{53}$$

$$\Delta u_{PDm}(kT) = \frac{u_{PDm}(kT) + u_{PDm}(kT-T)}{T}, \tag{54}$$

$$\Delta u_{Im}(kT) = \frac{u_{Im}(k) - u_{Im}(kT-T)}{T}. \tag{55}$$

Equations (50) and (51) can be rewritten as follows

$$u_{PDm}(kT) = -u_{PDm}(kT-T) + T\Delta u_{PDm}(kT), \tag{56}$$

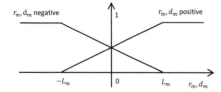

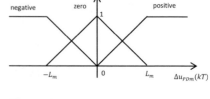

Fig. 8. PD input membership functions.

Fig. 9. PD output membership functions.

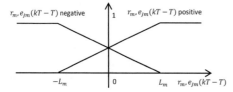

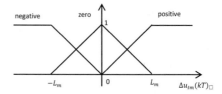

Fig. 10. I input membership functions.

Fig. 11. I output membership functions.

$$u_{Im}(kT) = u_{Im}(kT - T) + T\Delta u_{Im}(kT). \tag{57}$$

To get the fuzzy version of both controllers, the increment control inputs, $T\Delta u_{Im}$ and $T\Delta u_{PDm}$, are replaced by the fuzzy terms, (56) and (57) become:

$$u_{PDm}(kT) = -u_{PDm}(kT - T) + K_{uPDm}\Delta u_{PDm}(kT), \tag{58}$$

$$u_{Im}(kT) = u_{Im}(kT - T) + K_{uIm}\Delta u_{Im}(kT). \tag{59}$$

Figures 8 and 10 give the membership functions for the fuzzification of the inputs, whereas Figs. 9 and 11 give the membership functions of the outputs.

Using the above membership functions, the following rules can be assigned for the fuzzy PD controller:

(R1) IF $d_m = d_m.n$ AND $r_m = r_m.n$ THEN PD-output $= o.z$,
(R2) IF $d_m = d_m.n$ AND $r_m = r_m.p$ THEN PD-output $= o.n$,
(R3) IF $d_m = d_m.p$ AND $r_m = r_m.n$ THEN PD-output $= o.p$,
(R4) IF $d_m = d_m.p$ AND $r_m = r_m.p$ THEN PD-output $= o.z$,

while the rules of the fuzzy I controller are given as follows:

(R5) IF $e_{Jm}(kT - T) = e_{Jm}.p$ AND $r_m = r_m.p$ THEN I-output $= o.p$,
(R6) IF $e_{Jm}(kT - T) = e_{Jm}.p$ AND $r_m = r_m.n$ THEN I-output $= o.z$,
(R7) IF $e_{Jm}(kT - T) = e_{Jm}.n$ AND $r_m = r_m.p$ THEN I-output $= o.z$,
(R8) IF $e_{Jm}(kT - T) = e_{Jm}.n$ AND $r_m = r_m.n$ THEN I-output $= o.n$.

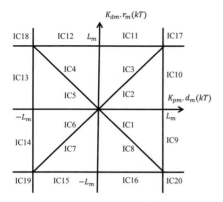

Fig. 12. Regions of the fuzzy PD controller.

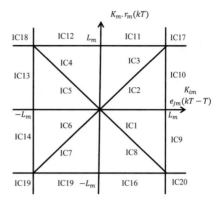

Fig. 13. Regions of the fuzzy I controller.

Twenty adjacent input combination (IC) regions are used for the defuzzification, as shown in Figs. 12 and 13. By using the rules, the defuzzification formula (31), and the following equations:

$$d_m.p = \frac{K_{pm}d_m(kT) + L_m}{2L_m}, \quad d_m.n = \frac{-K_{pm}d_m(kT-T) + L_m}{2L_m},$$

$$r_m.p = \frac{K_{dm}r_m(kT) + L_m}{2L_m}, \quad r_m.n = \frac{-K_{dm}r_m(kT) + L_m}{2L_m},$$

$$e_{Jm}.p = \frac{K_{im}e_{Jm}(kT-T) + L_m}{2L_m}, \quad e_{Jm}.n = \frac{-K_{im}e_{Jm}(kT-T) + L_m}{2L_m},$$

$$r_m.p = \frac{K_m r_m(kT) + L_m}{2L_m}, \quad r_m.n = \frac{-K_m r_m(kT) + L_m}{2L_m},$$

we obtain the defuzzification formulas for all 20 regions of the fuzzy PD.

$$\Delta u_{PDm}(kT) = \frac{L_m[K_{pm}.d_m(kT) - K_{dm}.r_m(kT)]}{2(2L_m - K_{pm}.|d_m(kT)|)}, \text{ (in IC } 1,2,5,6) \tag{60}$$

$$\Delta u_{PDm}(kT) = \frac{L_m[K_{pm}.d_m(kT) - K_{dm}.r_m(kT)]}{2(2L_m - K_{dm}.|r_m(kT)|)}, \text{ (in IC } 3,4,7,8) \tag{61}$$

$$\Delta u_{PDm}(kT) = 1/2[-K_{dm}r_m(kT) + L_m], \text{ (in IC } 9,10) \tag{62}$$

$$\Delta u_{PDm}(kT) = 1/2[K_{pm}d_m(kT) - L_m], \text{ (in IC } 11,12) \tag{63}$$

$$\Delta u_{PDm}(kT) = 1/2[-K_{dm}r_m(kT) - L_m], \text{ (in IC } 13,14) \tag{64}$$

$$\Delta u_{PDm}(kT) = 1/2[K_{pm}d_m(kT) + L_m], \text{ (in IC } 15,16) \tag{65}$$

$$\Delta u_{PDm}(kT) = 0, \text{ (in IC } 19,17) \tag{66}$$

$$\Delta u_{PDm}(kT) = -L_m, \text{ (in IC } 18) \tag{67}$$

$$\Delta u_{PDm}(kT) = L_m, \text{ (in IC 20)} \tag{68}$$

whereas for the fuzzy I controller, we obtain the following formulas:

$$\Delta u_{Im}(kT) = \frac{L_m[K_{im}.e_{Jm}(kT-T) + K_m.r_m(kT)]}{2(2L_m - K_{im}.|e_{Jm}(kT-T)|)}, \text{ (in IC 1,2,5,6)} \tag{69}$$

$$\Delta u_{Im}(kT) = \frac{L_m[K_{im}.e_{Jm}(kT-T) + K_m.r_m(kT)]}{2(2L_m - K_m.|r_m(kT)|)}, \text{ (in IC 3,4,7,8)} \tag{70}$$

$$\Delta u_{Im}(kT) = 1/2[K_m r_m(kT) + L_m], \text{ (in IC 9,10)} \tag{71}$$

$$\Delta u_{Im}(kT) = 1/2[K_{im} e_{Jm}(kT-T) + L_m], \text{ (in IC 11,12)} \tag{72}$$

$$\Delta u_{Im}(kT) = 1/2[K_m r_m(kT) - L_m], \text{ (in IC 13,14)} \tag{73}$$

$$\Delta u_{Im}(kT) = 1/2[K_{im} e_{Jm}(kT-T) - L_m], \text{ (in IC 15,16)} \tag{74}$$

$$\Delta u_{Im}(kT) = 0, \text{ (in IC 20,18)} \tag{75}$$

$$\Delta u_{Im}(kT) = -L_m, \text{ (in IC 19)} \tag{76}$$

$$\Delta u_{Im}(kT) = L_m. \text{ (in IC 17)} \tag{77}$$

4.3   Stability Analysis

The general stability condition can be derived by using Lyapunov's second method [8]. The Lyapunov function is chosen as follows:

$$V = e_{Jm}^2/2 > 0, \tag{78}$$

where $e_{Jm} = J_m(k)$. The time derivation of V is

$$
\begin{aligned}
\dot{V} &= e_{Jm}\dot{e}_{Jm} \approx J_m(k)\left[J_m(k) - J_m(k-1)\right] \\
&= \left[\sum_{i=-1}^{N} e_m(k-i)^2 + \lambda\sum_{j=0}^{N_c}(\Delta u_m(k-j))^2\right] \\
&\quad \times \left[\sum_{i=-1}^{N} e_m(k-i)^2 + \lambda\sum_{j=0}^{N_c}(\Delta u_m(k-j))^2\right. \\
&\quad \left. - \sum_{i=-1}^{N} e_m(k-1-i)^2 - \lambda\sum_{j=0}^{N_c}(\Delta u_m(k-1-j))^2\right] \\
&= \left[\sum_{i=-1}^{N} e_m(k-i)^2 + \lambda\sum_{j=0}^{N_c}(\Delta u_m(k-j))^2\right] \\
&\quad \times [e_m(k+1) - e_m(k-1-N) + \Delta u_m(k) - \Delta u_m(k-1-N_c)].
\end{aligned}
$$

Thus, if the two conditions $e_m(k) - e_m(k-1) < 0$ and $\Delta u_m(k) - \Delta u_m(k-1)$
< 0 are held at every step k then $\dot{V}$ will be less than zero.

The stability conditions of the fuzzy PI+D are given as follows:

$$
\begin{aligned}
C_1 &= \dot{e}_m \approx e_m(k) - e_m(k-1) \\
&= g_m(k) - f_m(k) + u_m(k) \\
&= g_m(k) - f_m(k) + K_{uPIm}\left[K_{im}e_{pm}(kT)\right. \\
&\quad + lK_{pm}\frac{J_m(kT) - J_m(kT-T)}{T}\right] \\
&\quad - K_{uDm}\left[-J_m(kT) + K_{dm}\frac{y_m(kT) - y_m(kT-T)}{T}\right] + u_{PIm}(kT-T) \\
&\quad + u_{Dm}(kT-T) < 0,
\end{aligned}
\tag{79}
$$

$$
\begin{aligned}
C_2 &= \Delta u_m(kT) - \Delta u_m(kT-T) \\
&= u_m(k) - 2u_m(k-1) + u_m(k-2) \\
&= K_{uPIm}K_{im}\left[J_m(kT) - 2J_m(kT-T) + J_m(kT-2T)\right] \\
&\quad + K_{uPIm}K_{pm}\left[\frac{J_m(kT) - 3J_m(kT-T) + 3J_m(kT-2T) - J_m(kT-3T)}{T}\right] \\
&\quad + K_{uDm}\left[J_m(kT) - 2J_m(kT-T) + J_m(kT-2T)\right] \\
&\quad + K_{uDm}K_{dm}\left[\frac{-y_m(kT) + 3y_m(kT-T) - 3y_m(kT-2T) + y_m(kT-3T)}{T}\right] \\
&\quad + u_{PIm}(kT-T) + u_{Dm}(kT-T) \\
&\quad - 2u_{PIm}(kT-2T) - 2u_{Dm}(kT-2T) \\
&\quad + u_{PIm}(kT-3T) + u_{Dm}(kT-3T) < 0,
\end{aligned}
\tag{80}
$$

whereas the stability conditions of the fuzzy PD+I are given as follows:

$$
\begin{aligned}
C_3 &= \dot{e}_m \approx e_m(k) - e_m(k-1) \\
&= g_m(k) - f_m(k) + u_m(k) \\
&= g_m(k) - f_m(k) \\
&\quad + K_{uPDm}\left[K_{pm}\frac{J_m(kT) + J_m(kT-T)}{T} + K_{dm}\frac{J_m(kT) - J_m(kT-T)}{T}\right] \\
&\quad + K_{uIm}\left[K_{im}\frac{J_m(kT) - J_m(kT-T)}{T} + K_{im}J_m(kT-T)\right] - u_{PDm}(kT-T) \\
&\quad + u_{Im}(kT-T) < 0,
\end{aligned}
\tag{81}
$$

$$
\begin{aligned}
C_4 &= \Delta u_m(kT) - \Delta u_m(kT-T) \\
&= u_m(k) - 2u_m(k-1) + u_m(k-2) \\
&= K_{uPDm}K_{pm}\left[\frac{J_m(kT) - J_m(kT-T) - J_m(kT-2T) + J_m(kT-3T)}{T}\right] \\
&\quad + K_{uPDm}K_{dm}\left[\frac{J_m(kT) - 3J_m(kT-T) + 3J_m(kT-2T) - J_m(kT-3T)}{T}\right] \\
&\quad + K_{uIm}K_{im}\left[\frac{J_m(kT) - 3J_m(kT-T) + 3J_m(kT-2T) - J_m(kT-3T)}{T}\right. \\
&\quad + J_m(kT-T) - 2J_m(kT-2T) + J_m(kT-3T)\right] \\
&\quad - u_{PDm}(kT-T) + 2u_{PDm}(kT-2T) - u_{PDm}(kT-3T) \\
&\quad + u_{Im}(kT-T) - 2u_{Im}(kT-3T) + u_{Im}(kT-3T) < 0.
\end{aligned}
\tag{82}
$$

PSO algorithm, described in Algorithm 1, can be used in order to find on-line the gains of both controllers. In the case of the fuzzy PI+D controller, particles are defined as

$$s_w = (K_{wim}, K_{wpm}, K_{wdm}, K_{wuPIm}, K_{wuDm})^T,$$

whereas in the case of the fuzzy PD+I controller, particles are defined as

$$s_w = (K_{wim}, K_{wpm}, K_{wdm}, K_{wuPDm}, K_{wuIm})^T.$$

The fitness of the particles is defined using the stability conditions. For the fuzzy PI+D, the fitness is given as follows:

$$F_{PI+D} = min\{C_1, C_2\},$$

while the fitness of the particles in the case of fuzzy PD+I is given

$$F_{PD+I} = min\{C_3, C_4\}.$$

Moreover, the initial swarm S is chosen using the following formula:

$$S = e_m(kT)rand(sup\,|e_m(kT)|),$$

where $rand(sup\,|e_m(kT)|)$ is a random matrix with random values between $-sup\,|e_m(kT)|)$ and $sup\,|e_m(kT)|)$.

In Lyapunov sense, the minimization of the function $\dot{V}$ using the stability conditions means that the PSO algorithm must find an optimal solution in order to make $\dot{V}$ converge to zero. Therefore, if the best minimum found so far is above zero, the algorithm must keep looking for the minimum. On the other hand, if the minimum is below zero, the algorithm must look for the maximum at the next iteration.

5   Simulation Results

The performance of the proposed algorithm is checked for the synchronization of two uncertain Lorenz systems.

The master is defined by:

$$\begin{cases} \dot{x}_1 = \alpha_1(x_2 - x_1) \\ \dot{x}_2 = (-x_1x_3 + \rho_1x_1 - x_2), \\ \dot{x}_3 = x_1x_2 - \beta_1x_3 \end{cases} \tag{83}$$

where x_1, x_2, x_3 are the state variables and $\alpha_1, \rho_1, \beta_1$ are positive uncertain parameters of the system.

And the slave by:

$$\begin{cases} \dot{y}_1 = \alpha_2(y_2 - y_1) + u_1 \\ \dot{y}_2 = (-y_1y_3 + \rho_2y_1 - y_2) + u_2, \\ \dot{y}_3 = y_1y_2 - \beta_2y_3 + u_3 \end{cases} \tag{84}$$

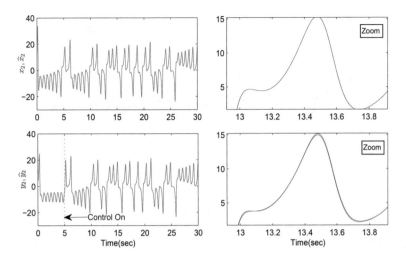

Fig. 14. One step predictors of the states x_2, y_2 using PI+D controller.

where y_1, y_2, y_3 are the state variables, $\alpha_2, \rho_2, \beta_2$ are positive uncertain parameters and u_1, u_2, u_3 are the outputs of the controllers.

The synchronization errors are defined as:

$$\begin{cases} e_1 = x_1 - y_1 \\ e_2 = x_2 - y_2, \\ e_3 = x_3 - y_3 \end{cases} \tag{85}$$

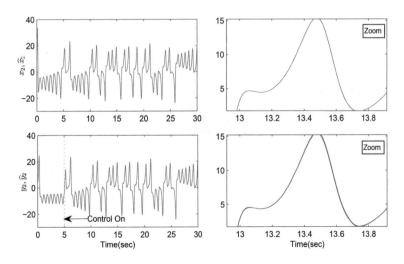

Fig. 15. One step predictors of the states x_2, y_2 using PD+I controller.

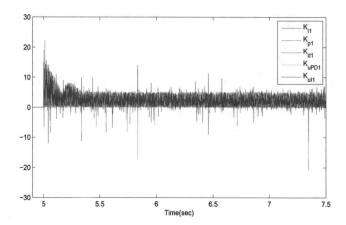

Fig. 16. A sample of the variation of the parameters of the fuzzy PD+I controller.

and the error states as:

$$\begin{cases} \dot{e}_1 = -\alpha_2(y_2 - y_1) + \alpha_1(x_2 - x_1) + u_1 \\ \dot{e}_2 = -\rho_2 y_1 + y_2 + y_1 y_3 + \rho_1 x_1 - x_2 - x_1 x_3 + u_2. \\ \dot{e}_3 = -y_1 y_2 + \beta_2 y_3 + x_1 x_2 - \beta_1 x_3 + u_3 \end{cases} \qquad (86)$$

To synchronize these chaotic systems, we chose the following optimal index:

$$J_m(k) = \sum_{i=-1}^{3} \left[e_m(k-i) \right]^2 + \lambda \sum_{j=0}^{3} \left[\Delta u_m(k-j) \right]^2, \qquad (87)$$

where $\lambda = 0.001$.

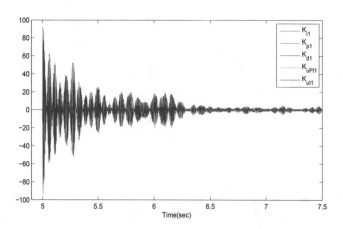

Fig. 17. A sample of the variation of the parameters of the fuzzy PI+D controller.

The one-step predictor of the error states is given as:

$$\begin{aligned}
e_m(k+1) = {}& 0.9497e_m(k) + 0.0141e_m(k-1) \\
& + 0.6806e_m(k-2) + 0.6440e_m(k-3) \\
& + 0.051u_m(k-1).
\end{aligned} \tag{88}$$

Figures 14 and 15 show the one step predictors of the states x_2, y_2, and their prediction errors.

For the numerical simulation, the parameters of the master and the slave systems are chosen respectively as:
$\alpha_1 = 10, \rho_1 = 28, \beta_1 = 8/3$,
$\alpha_2 = 10.5, \rho_2 = 25, \beta_2 = 8/3 + 0.2$.

The initial conditions of the master and the slave systems are taken as:
$x_1(0) = 2, x_2(0) = 10, x_3(0) = -6$,
$y_1(0) = -2, y_2(0) = 5, y_3(0) = 1$.

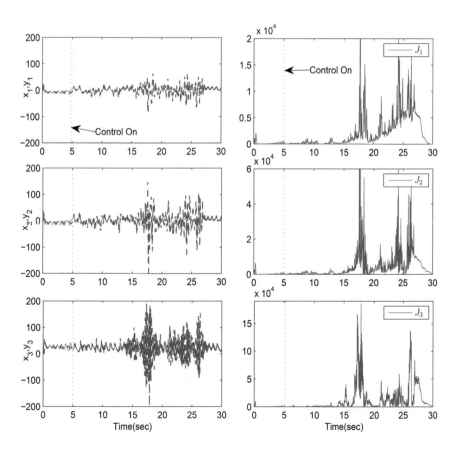

Fig. 18. Synchronization of the Lorenz systems and cost functions variations without prediction terms using fuzzy PD+I controller.

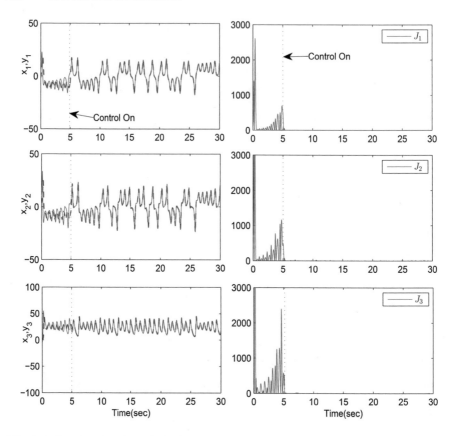

Fig. 19. Synchronization of the Lorenz systems and cost functions variations with prediction terms using fuzzy PD+I controller.

In the first part of the simulation, we present two results obtained using fuzzy PD+I controller. One with prediction terms and the other without. The parameters of the fuzzy PD+I are chosen as: $K_m = 1, L_m = 1$. The rest of the parameters are assigned on-line using the PSO algorithm with the stability conditions as shown in Fig. 16.

In the second part, fuzzy PI+D controller is used instead of fuzzy PD+I controller. The variation of the parameters of the fuzzy PI+D controller are shown in Fig. 17.

Figures 18 and 19 show the results of the synchronization of the two systems and the variations of the cost functions without prediction terms and with prediction terms respectively for the first case, while Figs. 20 and 21 give the results for the second case.

For the first case, we can notice that in the absence of prediction terms, the synchronization between the two systems is destroyed, and the cost functions take huge values. However, in the presence of the prediction terms, the synchronization is achieved and the cost functions converge to zero.

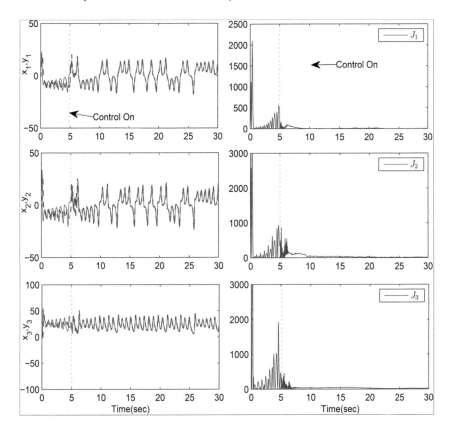

Fig. 20. Synchronization of the Lorenz systems and cost functions variations without prediction terms using fuzzy PI+D controller.

For the second case, the synchronization between the two systems is achieved with and without prediction terms.

Table 1 summarizes all the results obtained by the two controllers with and without prediction terms. In the case of fuzzy PD+I controller, one-step prediction terms ensure the synchronization between the two systems. However, with the fuzzy PI+D controller, they make a noise and reduce the performance of the controller. Moreover, the performance of fuzzy PI+D controller is better than the fuzzy PD+I controller in the two cases. The table shows also that the prediction terms worsen the results, and this can be explained by: the modeling error which is considered as perturbation terms added to the cost functions, the unpredictability behavior of chaotic systems, or the structure of the proposed control method which may need improvements. Although the structure of the algorithm is simpler than many others [14–18], the role of the prediction is still questionable.

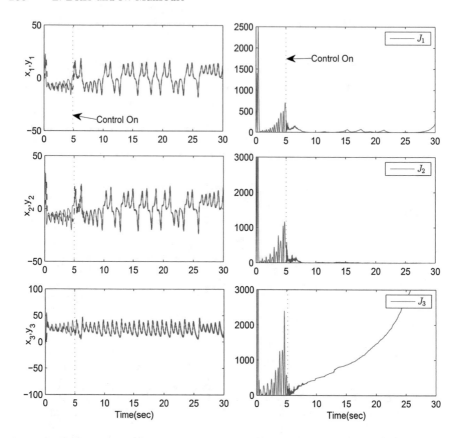

Fig. 21. Synchronization of the Lorenz systems and cost functions variations with prediction terms using fuzzy PI+D controller.

Table 1. Synchronization and cost function errors.

	Without prediction terms		With prediction terms	
	PD+I	PI+D	PD+I	PI+D
$\sum_{m=1}^{3} e_m$	$1.1 \ 10^5$	60.8	1092.9	89.09
$\sum_{m=1}^{3} e_{Jm}$	$2.15 \ 10^7$	285.11	6486.7	3699.6

6   Conclusion

This paper investigated the performance of predictive fuzzy PID control concerning synchronization of uncertain chaotic systems. Using Lyapunov's second method and PSO optimization algorithm the synchronization between the two systems is achieved. However, the performance of the two controllers can be affected by the presence or the absence of the prediction terms. In the case of the fuzzy PD+I controller, the prediction terms give a helping hand to the controller in order to reach the synchronization. On the other hand, in the case of the fuzzy PI+D controller, the prediction terms cause problems and reduce the ability of the controller. Some numerical results are given to demonstrate the role of the prediction terms for synchronization of two uncertain Lorenz systems.

References

1. Lam, H., Leung, F.F.: Synchronization of uncertain chaotic systems based on the fuzzy-model-based approach. Int. J. Bifurcat. Chaos **16**, 1435–1444 (2006)
2. Lam, H.K., Seneviratne, L.D.: Synchronization of chaotic systems using neural-network-based controller. Int. J. Bifurcat. Chaos **17**, 2117–2125 (2007)
3. Sun, Z., Zhu, W., Si, G., Ge, Y., Zhang, Y.: Adaptive synchronization design for uncertain chaotic systems in the presence of unknown system parameters: a revisit. Nonlinear Dyn. **72**, 729–749 (2013)
4. Dumur, D., Boucher, P.: Predictive control application in the machine tool field. In: Clarke, D. (ed.) Advances in Model-Based Predictive Control, pp. 471–482. Oxford University Press, Oxford (1994)
5. Clarke, D.W., Mohtadi, C., Tuffs, P.: Generalized predictive control—Part I. The basic algorithm. Automatica **23**, 137–148 (1987)
6. Jin-quan, H., Lewis, F.L.: Neural-network predictive control for nonlinear dynamic systems with time-delay. IEEE Trans. Neural Netw. **14**, 377–389 (2003)
7. Wang, X., Xiao, J.: PSO-based model predictive control for nonlinear processes. In: Wang, L., Chen, K., Ong, Y.S. (eds.) ICNC 2005. LNCS, vol. 3611, pp. 196–203. Springer, Heidelberg (2005). doi:10.1007/11539117_30
8. Lu, J., Chen, G., Ying, H.: Predictive fuzzy PID control: theory, design and simulation. Inf. Sci. **137**, 157–187 (2001)
9. Parsopoulos, K.E., Vrahatis, M.N., Global, I.: Particle Swarm Optimization and Intelligence: Advances and Applications. Information Science Reference, Hershey (2010)
10. Eberhart, R.C., Kennedy, J.: A new optimizer using particle swarm theory. In: Proceedings of the Sixth International Symposium on Micro Machine and Human Science, vol. 1, pp. 39–43. ACM, New York (1995)
11. Tang, K., Man, K.F., Chen, G., Kwong, S.: An optimal fuzzy PID controller. IEEE Trans. Ind. Electron. **48**, 757–765 (2001)
12. Misir, D., Malki, H.A., Chen, G.: Design and analysis of a fuzzy proportional-integral-derivative controller. Fuzzy Sets Syst. **79**, 297–314 (1996)
13. Malki, H., Li, H., Chen, G., et al.: New design and stability analysis of fuzzy proportional-derivative control systems. IEEE Trans. Fuzzy Syst. **2**, 245–254 (1994)
14. Yan, Z., Wang, J.: Model predictive control of nonlinear systems with unmodeled dynamics based on feedforward and recurrent neural networks. IEEE Trans. Ind. Inform. **8**, 746–756 (2012)
15. Wang, Z., Sun, Y.: Generalized predictive control based on particle swarm optimization for linear/nonlinear process with constraints. In: 2010 Second International Conference on Computational Intelligence and Natural Computing Proceedings (CINC), vol. 1, pp. 303–306. IEEE (2010)
16. Song, Y., Chen, Z., Yuan, Z.: New chaotic PSO-based neural network predictive control for nonlinear process. IEEE Trans. Neural Netw. **18**, 595–601 (2007)

17. Causa, J., Karer, G., Núnez, A., Sáez, D., Škrjanc, I., Zupančič, B.: Hybrid fuzzy predictive control based on genetic algorithms for the temperature control of a batch reactor. Comput. Chem. Eng. **32**, 3254–3263 (2008)
18. Mercieca, J., Fabri, S.: Particle swarm optimization for nonlinear model predictive control. In: The Fifth International Conference on Advanced Engineering Computing and Applications in Sciences, ADVCOMP 2011, pp. 88–93 (2011)

Expanding Gödel Logic with Truth Constants and the Equality, Strict Order, Delta Operators

Dušan Guller$^{(\boxtimes)}$

Department of Applied Informatics, Comenius University,
Mlynská dolina, 842 48 Bratislava, Slovakia
guller@fmph.uniba.sk

Abstract. Concerning the three fundamental first-order fuzzy logics, the set of logically valid formulae is Π_2-complete for Łukasiewicz logic, Π_2-hard for Product logic, and Σ_1-complete for Gödel logic, as with classical first-order logic. Among these fuzzy logics, only Gödel logic is recursively axiomatisable. Hence, it was necessary to provide a hyperresolution-based proof method suitable for automated deduction, as one has done for classical logic. As another step, we can incorporate a countable set of intermediate truth constants of the form $\bar{c}$, $c \in (0,1)$, together with the equality, $\approx$, strict order, $\prec$, projection, Δ, operators in Gödel logic; and modify the hyperresolution calculus, inferring over so-called order clauses. We shall investigate the deduction problem of a formula from a countable theory in this expansion and the so-called canonical standard completeness, where the semantics of Gödel logic is given by the standard G-algebra as well as truth constants are interpreted by 'themselves'. The hyperresolution calculus is refutation sound and complete for a countable order clausal theory under a certain condition for the set of truth constants occurring in the theory. We get an affirmative solution to the open problem of recursive enumerability of unsatisfiable formulae in this expansion of Gödel logic.

Keywords: Gödel logic · Resolution · Many-valued logics · Automated deduction

1 Introduction

In many real-world applications, one may be interested in representation and inference with explicit partial truth; besides the truth constants *0, 1*, intermediate truth constants are involved in. In the literature, two main approaches to expansions with truth constants, are described. Historically, the first one has been introduced in [1], where the propositional Łukasiewicz logic is augmented by truth constants $\bar{r}$, $r \in [0,1]$, Pavelka's logic (*PL*). A formula of the form $\bar{r} \to \phi$ evaluated to 1 expresses that the truth value of ϕ is greater than or equal to r. In [2], further development of evaluated formulae, and in [3], Rational

Dušan Guller—Partially supported by VEGA Grant 1/0592/14.

J.J. Merelo et al. (eds.), *Computational Intelligence*, Studies in Computational Intelligence 669,
DOI 10.1007/978-3-319-48506-5_13

Pavelka's logic (RPL) - a simplification of PL exploiting book-keeping axioms, are described. Another approach relies on traditional algebraic semantics. Various completeness results for expansions of t-norm based logics with countably many truth constants are investigated, among others, in [4–10].

In [11–15], we have generalised the well-known hyperresolution principle to the first-order Gödel logic for the general case. Our approach is based on translation of a formula of Gödel logic to an equivalent satisfiable finite order clausal theory, consisting of order clauses. We have introduced a notion of quantified atom: a formula a is a quantified atom if $a = Qx\,p(t_0, \ldots, t_\tau)$ where Q is a quantifier $(\forall, \exists)$; $p(t_0, \ldots, t_\tau)$ is an atom; x is a variable occurring in $p(t_0, \ldots, t_\tau)$; for all $i \leq \tau$, either $t_i = x$ or x does not occur in t_i (t_i is a free term in the quantified atom). The notion of quantified atom is all important. It permits us to extend classical unification to quantified atoms without any additional computational cost. Two quantified atoms $Qx\,p(t_0, \ldots, t_\tau)$ and $Q'x'\,p'(t'_0, \ldots, t'_\tau)$ are unifiable if $Q = Q'$, $x = x'$, $p = p'$, and the left-right sequence of free terms of $Qx\,p(t_0, \ldots, t_\tau)$ is unifiable with the left-right sequence of free terms of $Q'x'\,p'(t'_0, \ldots, t'_\tau)$ in the standard manner. An order clause is a finite set of order literals of the form $\varepsilon_1 \diamond \varepsilon_2$ where ε_i is an atom or a quantified atom, and $\diamond$ is the connective $=$ or $\prec$. $=$ and $\prec$ are interpreted by the equality and standard strict linear order on $[0, 1]$, respectively. On the basis of the hyperresolution principle, a calculus operating over order clausal theories, has been devised. The calculus is proved to be refutation sound and complete for the countable case with respect to the standard G-algebra $G = ([0, 1], \leq, \vee, \wedge, \Rightarrow, \bar{\ }, =, \prec, \Delta, 0, 1)$ augmented by the binary operators $=$, $\prec$ for $=$, $\prec$, respectively, and by the unary operator Δ for the projection connective Δ. As another step, one may incorporate a countable set of intermediate truth constants of the form $\bar{c}$, $c \in (0, 1)$, to get a modification of the hyperresolution calculus suitable for automated deduction with explicit partial truth [13]. We shall investigate the so-called canonical standard completeness, where the semantics of Gödel logic is given by the standard G-algebra G and truth constants are interpreted by 'themselves'. Note that the Hilbert-style calculus for Gödel logic introduced in [3], is not suitable for expansion with intermediate truth constants. We have $\phi \vdash \psi$ if and only if $\phi \models \psi$ (wrt. G). However, that cannot be preserved after adding intermediate truth constants. Let $c \in (0, 1)$ and a be an atom different from a constant. Then $\bar{c} \models a$ ($\bar{c}$ is unsatisfiable) but $\not\models \bar{c} \to a$, $\not\vdash \bar{c} \to a$, $\bar{c} \not\vdash a$ (from the soundness and the deduction-detachment theorem for this calculus). So, we cannot achieve a strict canonical standard completeness after expansion with intermediate truth constants. On the other side, such a completeness can be feasible for our hyperresolution calculus under a certain condition. We say that a set $\{0, 1\} \subseteq X$ of truth constants is admissible with respect to suprema and infima if, for all $\emptyset \neq Y_1, Y_2 \subseteq X$ and $\bigvee Y_1 = \bigwedge Y_2$, $\bigvee Y_1 \in Y_1, \bigwedge Y_2 \in Y_2$ (truth constants are interpreted by 'themselves'). Then the hyperresolution calculus is refutation sound and complete for a countable order clausal theory if the set of truth constants occurring in the theory, is admissible with respect to suprema and infima. This condition obviously covers the case of finite order clausal theories. We solve the deduction problem of a formula from a countable theory. As an interesting consequence, we get an affirmative solution to the open problem of recursive enumerability of unsatisfiable formulae in Gödel logic with truth constants and the equality, $\mathbf{=}$, strict order, $\prec$, projection, Δ, operators, which strengthens a similar result for prenex formulae of Gödel logic with Δ [16,17]. Some applications of our hyperresolution calculus may lead to computational linguistics, to design and analysis of scientific (natural) language processing systems [18,19].

The paper is organised as follows. Section 2 gives the basic notions and notation concerning the first-order Gödel logic. Section 3 deals with clause form translation. In Sect. 4, we describe the hyperresolution calculus and show its refutational soundness, completeness. Section 5 provides an example for translation and deduction. Section 6 brings conclusions.

2   First-Order Gödel Logic

Throughout the paper, we shall use the common notions and notation of first-order logic. $\mathbb{N} \mid \mathbb{Z}$ designates the set of natural | integer numbers and $\leq \mid <$ the standard order | strict order on $\mathbb{N} \mid \mathbb{Z}$. By $\mathcal{L}$ we denote a first-order language. $Var_{\mathcal{L}} \mid Func_{\mathcal{L}} \mid Pred_{\mathcal{L}} \mid Term_{\mathcal{L}} \mid GTerm_{\mathcal{L}} \mid Atom_{\mathcal{L}} \mid GAtom_{\mathcal{L}}$ denotes the set of all variables | function symbols | predicate symbols | terms | ground terms | atoms | ground atoms of $\mathcal{L}$. $ar_{\mathcal{L}} : Func_{\mathcal{L}} \cup Pred_{\mathcal{L}} \longrightarrow \mathbb{N}$ denotes the mapping assigning an arity to every function and predicate symbol of $\mathcal{L}$. We assume truth constants - nullary predicate symbols $0,\ 1 \in Pred_{\mathcal{L}}$, $ar_{\mathcal{L}}(0) = ar_{\mathcal{L}}(1) = 0$; 0 denotes the false and 1 the true in $\mathcal{L}$. Let $\mathbb{C}_{\mathcal{L}} \subseteq (0,1)$ be countable. In addition, we assume a countable set of nullary predicate symbols $\overline{C}_{\mathcal{L}} = \{\bar{c} \mid \bar{c} \in Pred_{\mathcal{L}}, ar_{\mathcal{L}}(\bar{c}) = 0, c \in \mathbb{C}_{\mathcal{L}}\} \subseteq Pred_{\mathcal{L}}$; $\{0\}, \{1\}, \overline{C}_{\mathcal{L}}$ are pairwise disjoint. $0,\ 1,\ \bar{c} \in \overline{C}_{\mathcal{L}}$ are called truth constants. We denote $Tcons_{\mathcal{L}} = \{0,\ 1\} \cup \overline{C}_{\mathcal{L}} \subseteq Pred_{\mathcal{L}}$. Let $X \subseteq Tcons_{\mathcal{L}}$. We denote $\overline{X} = \{0 \mid 0 \in X\} \cup \{1 \mid 1 \in X\} \cup \{c \mid \bar{c} \in X \cap \overline{C}_{\mathcal{L}}\} \subseteq [0,1]$. We introduce a new unary connective Δ, Delta, and binary connectives $\mathbf{=}$, equality, $\prec$, strict order. By $OrdForm_{\mathcal{L}}$ we designate the set of all so-called order formulae of $\mathcal{L}$ built up from $Atom_{\mathcal{L}}$ and $Var_{\mathcal{L}}$ using the connectives: $\neg$, negation, Δ, $\wedge$, conjunction, $\vee$, disjunction, $\rightarrow$, implication, $\leftrightarrow$, equivalence, $\mathbf{=}$, $\prec$, and the quantifiers: $\forall$, the universal one, $\exists$, the existential one.[1] In the paper, we shall assume that $\mathcal{L}$ is a countable first-order language; hence, all the above mentioned sets of symbols and expressions are countable. Let $\varepsilon \mid \varepsilon_i, 1 \leq i \leq m \mid v_i, 1 \leq i \leq n$, be either an expression or a set of expressions or a set of sets of expressions of $\mathcal{L}$, in general. By $vars(\varepsilon_1, \ldots, \varepsilon_m) \subseteq Var_{\mathcal{L}} \mid freevars(\varepsilon_1, \ldots, \varepsilon_m) \subseteq Var_{\mathcal{L}} \mid boundvars(\varepsilon_1, \ldots, \varepsilon_m) \subseteq Var_{\mathcal{L}} \mid funcs(\varepsilon_1, \ldots, \varepsilon_m) \subseteq Func_{\mathcal{L}} \mid preds(\varepsilon_1, \ldots, \varepsilon_m) \subseteq Pred_{\mathcal{L}} \mid atoms(\varepsilon_1, \ldots, \varepsilon_m) \subseteq Atom_{\mathcal{L}}$ we denote the set of all variables | free variables | bound variables | function symbols | predicate symbols | atoms of $\mathcal{L}$ occurring in $\varepsilon_1, \ldots, \varepsilon_m$. ε is closed iff $freevars(\varepsilon) = \emptyset$. By ℓ we denote the empty sequence. By $|\varepsilon_1, \ldots, \varepsilon_m| = m$

---

[1] We assume a decreasing connective and quantifier precedence: $\forall, \exists, \neg, \Delta, \mathbf{=}, \prec, \wedge,$ $\vee, \rightarrow, \leftrightarrow$.

we denote the length of the sequence $\varepsilon_1, \ldots, \varepsilon_m$. We define the concatenation of the sequences $\varepsilon_1, \ldots, \varepsilon_m$ and $v_1, \ldots, v_n$ as $(\varepsilon_1, \ldots, \varepsilon_m), (v_1, \ldots, v_n) = \varepsilon_1, \ldots, \varepsilon_m, v_1, \ldots, v_n$. Note that concatenation of sequences is associative.

Let X, Y, Z be sets, $Z \subseteq X$; $f : X \longrightarrow Y$ be a mapping. By $\|X\|$ we denote the set-theoretic cardinality of X. X being a finite subset of Y is denoted as $X \subseteq_{\mathcal{F}} Y$. We designate $\mathcal{P}(X) = \{x \mid x \subseteq X\}$; $\mathcal{P}(X)$ is the power set of X; $\mathcal{P}_{\mathcal{F}}(X) = \{x \mid x \subseteq_{\mathcal{F}} X\}$; $\mathcal{P}_{\mathcal{F}}(X)$ is the set of all finite subsets of X; $f[Z] = \{f(z) \mid z \in Z\}$; $f[Z]$ is the image of Z under f; $f|_Z = \{(z, f(z)) \mid z \in Z\}$; $f|_Z$ is the restriction of f onto Z. Let $\gamma \leq \omega$. A sequence δ of X is a bijection $\delta : \gamma \longrightarrow X$. Recall that X is countable if and only if there exists a sequence of X. Let I be a set and $S_i \neq \emptyset$, $i \in I$, be sets. A selector $\mathcal{S}$ over $\{S_i \mid i \in I\}$ is a mapping $\mathcal{S} : I \longrightarrow \bigcup \{S_i \mid i \in I\}$ such that for all $i \in I$, $\mathcal{S}(i) \in S_i$. We denote $Sel(\{S_i \mid i \in I\}) = \{\mathcal{S} \mid \mathcal{S} \text{ is a selector over } \{S_i \mid i \in I\}\}$. $\mathbb{R}$ designates the set of real numbers and $\leq \mid <$ the standard order $\mid$ strict order on $\mathbb{R}$. We denote $\mathbb{R}_0^+ = \{c \mid 0 \leq c \in \mathbb{R}\}$, $\mathbb{R}^+ = \{c \mid 0 < c \in \mathbb{R}\}$; $[0, 1] = \{c \mid 0 \leq c \leq 1, c \in \mathbb{R}\}$; $[0, 1]$ is the unit interval. Let $c \in \mathbb{R}^+$. $\log c$ denotes the binary logarithm of c. Let $f, g : \mathbb{N} \longrightarrow \mathbb{R}_0^+$. f is of the order of g, in symbols $f \in O(g)$, iff there exist $n_0 \in \mathbb{N}$ and $c^* \in \mathbb{R}_0^+$ such that for all $n \geq n_0$, $f(n) \leq c^* \cdot g(n)$.

We define the size of term of $\mathcal{L}$ $|t| : Term_{\mathcal{L}} \longrightarrow \mathbb{N}$ by recursion on the structure of t:

$$|t| = \begin{cases} 1 & \text{if } t \in Var_{\mathcal{L}}, \\ 1 + \sum_{i=1}^{\tau} |t_i| & \text{if } t = f(t_1, \ldots, t_\tau). \end{cases}$$

Subsequently, we define the size of order formula of $\mathcal{L}$ $|\phi| : OrdForm_{\mathcal{L}} \longrightarrow \mathbb{N}$ by recursion on the structure of ϕ:

$$|\phi| = \begin{cases} 1 + \sum_{i=1}^{\tau} |t_i| & \text{if } \phi = p(t_1, \ldots, t_\tau) \in Atom_{\mathcal{L}}, \\ 1 + |\phi_1| & \text{if } \phi = \diamond\phi_1, \\ 1 + |\phi_1| + |\phi_2| & \text{if } \phi = \phi_1 \diamond \phi_2, \\ 2 + |\phi_1| & \text{if } \phi = Qx\,\phi_1. \end{cases}$$

Let $T \subseteq_{\mathcal{F}} OrdForm_{\mathcal{L}}$. We define the size of T as $|T| = \sum_{\phi \in T} |\phi|$. By $varseq(\phi)$, $vars(varseq(\phi)) \subseteq Var_{\mathcal{L}}$, we denote the sequence of all variables of $\mathcal{L}$ occurring in ϕ which is built up via the left-right preorder traversal of ϕ. For example, $varseq(\exists w\,(\forall x\,p(x, x, z) \vee \exists y\,q(x, y, z))) = w, x, x, x, z, y, x, y, z$ and $|w, x, x, x, z, y, x, y, z| = 9$. A sequence of variables will often be denoted as $\bar{x}, \bar{y}, \bar{z}$, etc. Let $Q \in \{\forall, \exists\}$ and $\bar{x} = x_1, \ldots, x_n$ be a sequence of variables of $\mathcal{L}$. By $Q\bar{x}\,\phi$ we denote $Qx_1 \ldots Qx_n\,\phi$.

Gödel logic is interpreted by the standard $\boldsymbol{G}$-algebra augmented by the operators $=\!\!=$, $\prec$, $\boldsymbol{\Delta}$ for the connectives $=\!\!=$, $\prec$, Δ, respectively.

$$\boldsymbol{G} = ([0, 1], \leq, \vee, \wedge, \Rightarrow, \bar{}, =\!\!=, \prec, \boldsymbol{\Delta}, 0, 1)$$

where $\vee \mid \wedge$ denotes the supremum $\mid$ infimum operator on $[0, 1]$;

$$a \Rightarrow b = \begin{cases} 1 & \text{if } a \leq b, \\ b & \text{else}; \end{cases} \qquad \bar{a} = \begin{cases} 1 & \text{if } a = 0, \\ 0 & \text{else}; \end{cases} \qquad a =\!\!= b = \begin{cases} 1 & \text{if } a = b, \\ 0 & \text{else}; \end{cases} \qquad a \prec b = \begin{cases} 1 & \text{if } a < b, \\ 0 & \text{else}; \end{cases}$$

$$\boldsymbol{\Delta} a = \begin{cases} 1 & \text{if } a = 1, \\ 0 & \text{else}. \end{cases}$$

Recall that $\mathbf{G}$ is a complete linearly ordered lattice algebra; $\vee \mid \wedge$ is commutative, associative, idempotent, monotone; $0 \mid 1$ is its neutral element;[2] the residuum operator $\Rightarrow$ of $\wedge$ satisfies the condition of residuation:

$$\text{for all } a, b, c \in \mathbf{G}, a \wedge b \le c \Longleftrightarrow a \le b \Rightarrow c; \tag{1}$$

Gödel negation $^{-}$ satisfies the condition:

$$\text{for all } a \in \mathbf{G}, \bar{a} = a \Rightarrow 0; \tag{2}$$

the following properties, which will be exploited later, hold:[3]

for all $a, b, c \in \mathbf{G}$,

$$a \vee b \wedge c = (a \vee b) \wedge (a \vee c), \quad \text{(distributivity of } \vee \text{ over } \wedge) \tag{3}$$
$$a \wedge (b \vee c) = a \wedge b \vee a \wedge c, \quad \text{(distributivity of } \wedge \text{ over } \vee) \tag{4}$$
$$a \Rightarrow b \vee c = (a \Rightarrow b) \vee (a \Rightarrow c), \tag{5}$$
$$a \Rightarrow b \wedge c = (a \Rightarrow b) \wedge (a \Rightarrow c), \tag{6}$$
$$a \vee b \Rightarrow c = (a \Rightarrow c) \wedge (b \Rightarrow c), \tag{7}$$
$$a \wedge b \Rightarrow c = (a \Rightarrow c) \vee (b \Rightarrow c), \tag{8}$$
$$a \Rightarrow (b \Rightarrow c) = a \wedge b \Rightarrow c, \tag{9}$$
$$((a \Rightarrow b) \Rightarrow b) \Rightarrow b = a \Rightarrow b, \tag{10}$$
$$(a \Rightarrow b) \Rightarrow c = ((a \Rightarrow b) \Rightarrow b) \wedge (b \Rightarrow c) \vee c, \tag{11}$$
$$(a \Rightarrow b) \Rightarrow 0 = ((a \Rightarrow 0) \Rightarrow 0) \wedge (b \Rightarrow 0), \tag{12}$$
$$\Delta a = a \rightleftharpoons 1. \tag{13}$$

An interpretation $\mathcal{I}$ for $\mathcal{L}$ is a triple $\left(\mathcal{U}_{\mathcal{I}}, \{f^{\mathcal{I}} \mid f \in Func_{\mathcal{L}}\}, \{p^{\mathcal{I}} \mid p \in Pred_{\mathcal{L}}\}\right)$ defined as follows: $\mathcal{U}_{\mathcal{I}} \neq \emptyset$ is the universum of $\mathcal{I}$; every $f \in Func_{\mathcal{L}}$ is interpreted as a function $f^{\mathcal{I}} : \mathcal{U}_{\mathcal{I}}^{ar_{\mathcal{L}}(f)} \longrightarrow \mathcal{U}_{\mathcal{I}}$; every $p \in Pred_{\mathcal{L}}$ is interpreted as a $[0, 1]$-relation $p^{\mathcal{I}} : \mathcal{U}_{\mathcal{I}}^{ar_{\mathcal{L}}(p)} \longrightarrow [0, 1]$. A variable assignment in $\mathcal{I}$ is a mapping $Var_{\mathcal{L}} \longrightarrow \mathcal{U}_{\mathcal{I}}$. We denote the set of all variable assignments in $\mathcal{I}$ as $\mathcal{S}_{\mathcal{I}}$. Let $e \in \mathcal{S}_{\mathcal{I}}$ and $u \in \mathcal{U}_{\mathcal{I}}$. A variant $e[x/u] \in \mathcal{S}_{\mathcal{I}}$ of e with respect to x and u is defined as

$$e[x/u](z) = \begin{cases} u & \text{if } z = x, \\ e(z) & \text{else.} \end{cases}$$

Let $t \in Term_{\mathcal{L}}$, $\bar{x}$ be a sequence of variables of $\mathcal{L}$, $\phi \in OrdForm_{\mathcal{L}}$. In $\mathcal{I}$ with respect to e, we define the value $\|t\|_e^{\mathcal{I}} \in \mathcal{U}_{\mathcal{I}}$ of t by recursion on the structure of t, the value $\|\bar{x}\|_e^{\mathcal{I}} \in \mathcal{U}_{\mathcal{I}}^{|\bar{x}|}$ of $\bar{x}$, the truth value $\|\phi\|_e^{\mathcal{I}} \in [0, 1]$ of ϕ by recursion on the structure of ϕ, as follows:

---

[2] Using the commutativity, associativity, idempotence, monotonicity, neutral element of $\vee \mid \wedge$ will not explicitly be referred to.

[3] We assume a decreasing operator precedence: $^{-}, \Delta, \rightleftharpoons, \prec, \wedge, \vee, \Rightarrow$.

$$t \in Var_{\mathcal{L}}, \qquad \|t\|_e^{\mathcal{I}} = e(t);$$

$$t = f(t_1, \ldots, t_\tau), \ \|t\|_e^{\mathcal{I}} = f^{\mathcal{I}}(\|t_1\|_e^{\mathcal{I}}, \ldots, \|t_\tau\|_e^{\mathcal{I}});$$

$$\bar{x} = x_1, \ldots, x_{|\bar{x}|}, \ \|\bar{x}\|_e^{\mathcal{I}} = e(x_1), \ldots, e(x_{|\bar{x}|});$$

$$\phi = \mathit{0}, \qquad \|\phi\|_e^{\mathcal{I}} = 0;$$

$$\phi = \mathit{1}, \qquad \|\phi\|_e^{\mathcal{I}} = 1;$$

$$\phi = \bar{c}, \qquad \|\phi\|_e^{\mathcal{I}} = c;$$

$$\phi = p(t_1, \ldots, t_\tau), \ \|\phi\|_e^{\mathcal{I}} = p^{\mathcal{I}}(\|t_1\|_e^{\mathcal{I}}, \ldots, \|t_\tau\|_e^{\mathcal{I}});$$

$$\phi = \neg \phi_1, \qquad \|\phi\|_e^{\mathcal{I}} = \overline{\|\phi_1\|_e^{\mathcal{I}}};$$

$$\phi = \phi_1 \wedge \phi_2, \qquad \|\phi\|_e^{\mathcal{I}} = \|\phi_1\|_e^{\mathcal{I}} \wedge \|\phi_2\|_e^{\mathcal{I}};$$

$$\phi = \phi_1 \vee \phi_2, \qquad \|\phi\|_e^{\mathcal{I}} = \|\phi_1\|_e^{\mathcal{I}} \vee \|\phi_2\|_e^{\mathcal{I}};$$

$$\phi = \phi_1 \rightarrow \phi_2, \qquad \|\phi\|_e^{\mathcal{I}} = \|\phi_1\|_e^{\mathcal{I}} \Rightarrow \|\phi_2\|_e^{\mathcal{I}};$$

$$\phi = \phi_1 \leftrightarrow \phi_2, \qquad \|\phi\|_e^{\mathcal{I}} = (\|\phi_1\|_e^{\mathcal{I}} \Rightarrow \|\phi_2\|_e^{\mathcal{I}}) \wedge (\|\phi_2\|_e^{\mathcal{I}} \Rightarrow \|\phi_1\|_e^{\mathcal{I}});$$

$$\phi = \phi_1 \,\bar{=}\, \phi_2, \qquad \|\phi\|_e^{\mathcal{I}} = \|\phi_1\|_e^{\mathcal{I}} \,\bar{=}\, \|\phi_2\|_e^{\mathcal{I}};$$

$$\phi = \phi_1 \prec \phi_2, \qquad \|\phi\|_e^{\mathcal{I}} = \|\phi_1\|_e^{\mathcal{I}} \prec \|\phi_2\|_e^{\mathcal{I}};$$

$$\phi = \forall x\, \phi_1, \qquad \|\phi\|_e^{\mathcal{I}} = \bigwedge_{u \in \mathcal{U}_{\mathcal{I}}} \|\phi_1\|_{e[x/u]}^{\mathcal{I}};$$

$$\phi = \exists x\, \phi_1, \qquad \|\phi\|_e^{\mathcal{I}} = \bigvee_{u \in \mathcal{U}_{\mathcal{I}}} \|\phi_1\|_{e[x/u]}^{\mathcal{I}}.$$

Let ϕ be closed. Then, for all $e, e' \in \mathcal{S}_{\mathcal{I}}$, $\|\phi\|_e^{\mathcal{I}} = \|\phi\|_{e'}^{\mathcal{I}}$. Let $e \in \mathcal{S}_{\mathcal{I}} \neq \emptyset$. We denote $\|\phi\|^{\mathcal{I}} = \|\phi\|_e^{\mathcal{I}}$.

Let $\mathcal{L} \mid \mathcal{L}'$ be a first-order language and $\mathcal{I} \mid \mathcal{I}'$ be an interpretation for $\mathcal{L} \mid \mathcal{L}'$. $\mathcal{L}'$ is an expansion of $\mathcal{L}$ iff $Func_{\mathcal{L}'} \supseteq Func_{\mathcal{L}}$ and $Pred_{\mathcal{L}'} \supseteq Pred_{\mathcal{L}}$; on the other side, we say $\mathcal{L}$ is a reduct of $\mathcal{L}'$. $\mathcal{I}'$ is an expansion of $\mathcal{I}$ to $\mathcal{L}'$ iff $\mathcal{L}'$ is an expansion of $\mathcal{L}$, $\mathcal{U}_{\mathcal{I}'} = \mathcal{U}_{\mathcal{I}}$, for all $f \in Func_{\mathcal{L}}$, $f^{\mathcal{I}'} = f^{\mathcal{I}}$, for all $p \in Pred_{\mathcal{L}}$, $p^{\mathcal{I}'} = p^{\mathcal{I}}$; on the other side, we say $\mathcal{I}$ is a reduct of $\mathcal{I}'$ to $\mathcal{L}$, in symbols $\mathcal{I} = \mathcal{I}'|_{\mathcal{L}}$.

An order theory of $\mathcal{L}$ is a set of order formulae of $\mathcal{L}$. Let $\phi, \phi' \in OrdForm_{\mathcal{L}}$, $T \subseteq OrdForm_{\mathcal{L}}$, $e \in \mathcal{S}_{\mathcal{I}}$. ϕ is true in $\mathcal{I}$ with respect to e, written as $\mathcal{I} \models_e \phi$, iff $\|\phi\|_e^{\mathcal{I}} = 1$. $\mathcal{I}$ is a model of ϕ, in symbols $\mathcal{I} \models \phi$, iff, for all $e \in \mathcal{S}_{\mathcal{I}}$, $\mathcal{I} \models_e \phi$. $\mathcal{I}$ is a model of T, in symbols $\mathcal{I} \models T$, iff, for all $\phi \in T$, $\mathcal{I} \models \phi$. ϕ is a logically valid formula iff, for every interpretation $\mathcal{I}$ for $\mathcal{L}$, $\mathcal{I} \models \phi$. ϕ is equivalent to ϕ', in symbols $\phi \equiv \phi'$, iff, for every interpretation $\mathcal{I}$ for $\mathcal{L}$ and $e \in \mathcal{S}_{\mathcal{I}}$, $\|\phi\|_e^{\mathcal{I}} = \|\phi'\|_e^{\mathcal{I}}$. We denote $tcons(\phi) = \{\mathit{0},\ \mathit{1}\} \cup (preds(\phi) \cap \overline{C}_{\mathcal{L}}) \subseteq Tcons_{\mathcal{L}}$ and $tcons(T) = \{\mathit{0},\ \mathit{1}\} \cup (preds(T) \cap \overline{C}_{\mathcal{L}}) \subseteq Tcons_{\mathcal{L}}$.

3   Translation to Clausal Form

In the propositional case [20], we have proposed some translation of a formula to an equivalent *CNF* containing literals of the form either a or $a \rightarrow b$ or

$(a \rightarrow b) \rightarrow b$ where a is a propositional atom and b is either a propositional atom or the propositional constant 0. An output equivalent CNF may be of exponential size with respect to the input formula; we had laid no restrictions on use of the distributivity law (3) during translation to conjunctive normal form. To avoid this disadvantage, we have devised translation to CNF via interpolation using new atoms, which produces an output CNF of linear size at the cost of being only equisatisfiable to the input formula. A similar approach exploiting the renaming subformulae technique can be found in [21–25]. A CNF is further translated to a finite set of order clauses. An order clause is a finite set of order literals of the form $\varepsilon_1 \diamond \varepsilon_2$ where ε_i is either a propositional atom or a propositional constant, 0, 1, and $\diamond \in \{=, \prec\}$.

We now describe some generalisation of the mentioned translation to the first-order case. At first, we introduce a notion of quantified atom. Let $a \in OrdForm_{\mathcal{L}}$. a is a quantified atom of $\mathcal{L}$ iff $a = Qx\, p(t_0, \ldots, t_\tau)$ where $p(t_0, \ldots, t_\tau) \in Atom_{\mathcal{L}}$, $x \in vars(p(t_0, \ldots, t_\tau))$, either $t_i = x$ or $x \notin vars(t_i)$. $QAtom_{\mathcal{L}} \subseteq OrdForm_{\mathcal{L}}$ denotes the set of all quantified atoms of $\mathcal{L}$. $QAtom_{\mathcal{L}}^Q \subseteq QAtom_{\mathcal{L}}$, $Q \in \{\forall, \exists\}$, denotes the set of all quantified atoms of $\mathcal{L}$ of the form $Qx\, a$. Let ε_i, $1 \leq i \leq m$, be either an expression or a set of expressions or a set of sets of expressions of $\mathcal{L}$, in general. By $qatoms(\varepsilon_1, \ldots, \varepsilon_m) \subseteq QAtom_{\mathcal{L}}$ we denote the set of all quantified atoms of $\mathcal{L}$ occurring in $\varepsilon_1, \ldots, \varepsilon_m$. We denote $qatoms^Q(\varepsilon_1, \ldots, \varepsilon_m) = qatoms(\varepsilon_1, \ldots, \varepsilon_m) \cap QAtom_{\mathcal{L}}^Q$, $Q \in \{\forall, \exists\}$. Let $Qx\, p(t_0, \ldots, t_\tau) \in QAtom_{\mathcal{L}}$ and $p(t'_0, \ldots, t'_\tau) \in Atom_{\mathcal{L}}$. We denote

$$boundindset(Qx\, p(t_0, \ldots, t_\tau)) = \{i \,|\, i \leq \tau, t_i = x\} \neq \emptyset.$$

Let $I = \{i \,|\, i \leq \tau, x \notin vars(t_i)\}$ and $r_1, \ldots, r_k$, $r_i \leq \tau$, $k \leq \tau$, for all $1 \leq i < i' \leq k$, $r_i < r_{i'}$, be a sequence such that $\{r_i \,|\, 1 \leq i \leq k\} = I$. We denote

$$freetermseq(Qx\, p(t_0, \ldots, t_\tau)) = t_{r_1}, \ldots, t_{r_k},$$
$$freetermseq(p(t'_0, \ldots, t'_\tau)) = t'_0, \ldots, t'_\tau.$$

We further introduce order clauses in Gödel logic. Let $l \in OrdForm_{\mathcal{L}}$. l is an order literal of $\mathcal{L}$ iff $l = \varepsilon_1 \diamond \varepsilon_2$, $\varepsilon_i \in Atom_{\mathcal{L}} \cup QAtom_{\mathcal{L}}$, $\diamond \in \{=, \prec\}$. The set of all order literals of $\mathcal{L}$ is designated as $OrdLit_{\mathcal{L}} \subseteq OrdForm_{\mathcal{L}}$. An order clause of $\mathcal{L}$ is a finite set of order literals of $\mathcal{L}$; since $=$ is commutative, for all $\varepsilon_1 = \varepsilon_2 \in OrdLit_{\mathcal{L}}$, we identify $\varepsilon_1 = \varepsilon_2$ and $\varepsilon_2 = \varepsilon_1 \in OrdLit_{\mathcal{L}}$ with respect to order clauses. An order clause $\{l_1, \ldots, l_n\}$ is written in the form $l_1 \vee \cdots \vee l_n$. The order clause $\emptyset$ is called the empty order clause and denoted as $\square$. An order clause $\{l\}$ is called a unit order clause and denoted as l; if it does not cause the ambiguity with the denotation of the single order literal l in given context. We designate the set of all order clauses of $\mathcal{L}$ as $OrdCl_{\mathcal{L}}$. Let $l, l_0, \ldots, l_n \in OrdLit_{\mathcal{L}}$ and $C, C' \in OrdCl_{\mathcal{L}}$. We define the size of C as $|C| = \sum_{l \in C} |l|$. By $l \vee C$ we denote $\{l\} \cup C$ where $l \notin C$. Analogously, by $l_0 \vee \cdots \vee l_n \vee C$ we denote $\{l_0\} \cup \cdots \cup \{l_n\} \cup C$ where, for all $i, i' \leq n$ and $i \neq i'$, $l_i \notin C$, $l_i \neq l_{i'}$. By $C \vee C'$ we denote $C \cup C'$. C is a subclause of C', in symbols $C \sqsubseteq C'$, iff $C \subseteq C'$. An order clausal theory of

$\mathcal{L}$ is a set of order clauses of $\mathcal{L}$. A unit order clausal theory is a set of unit order clauses.

Let $\phi, \phi' \in OrdForm_{\mathcal{L}}$, $T, T' \subseteq OrdForm_{\mathcal{L}}$, $S, S' \subseteq OrdCl_{\mathcal{L}}$, $\mathcal{I}$ be an interpretation for $\mathcal{L}$, $e \in \mathcal{S}_{\mathcal{I}}$. Note that $\mathcal{I} \models_e l$ if and only if either $l = \varepsilon_1 \approx \varepsilon_2$, $\|\varepsilon_1 \approx \varepsilon_2\|_e^{\mathcal{I}} = 1$, $\|\varepsilon_1\|_e^{\mathcal{I}} = \|\varepsilon_2\|_e^{\mathcal{I}}$; or $l = \varepsilon_1 \prec \varepsilon_2$, $\|\varepsilon_1 \prec \varepsilon_2\|_e^{\mathcal{I}} = 1$, $\|\varepsilon_1\|_e^{\mathcal{I}} < \|\varepsilon_2\|_e^{\mathcal{I}}$. C is true in $\mathcal{I}$ with respect to e, written as $\mathcal{I} \models_e C$, iff there exists $l^* \in C$ such that $\mathcal{I} \models_e l^*$. $\mathcal{I}$ is a model of C, in symbols $\mathcal{I} \models C$, iff, for all $e \in \mathcal{S}_{\mathcal{I}}$, $\mathcal{I} \models_e C$. $\mathcal{I}$ is a model of S, in symbols $\mathcal{I} \models S$, iff, for all $C \in S$, $\mathcal{I} \models C$. $\phi' \mid T' \mid C' \mid S'$ is a logical consequence of $\phi \mid T \mid C \mid S$, in symbols $\phi \mid T \mid C \mid S \models \phi' \mid T' \mid C' \mid S'$, iff, for every model $\mathcal{I}$ of $\phi \mid T \mid C \mid S$ for $\mathcal{L}$, $\mathcal{I} \models \phi' \mid T' \mid C' \mid S'$. $\phi \mid T \mid C \mid S$ is satisfiable iff there exists a model of $\phi \mid T \mid C \mid S$ for $\mathcal{L}$. Note that both $\square$ and $\square \in S$ are unsatisfiable. $\phi \mid T \mid C \mid S$ is equisatisfiable to $\phi' \mid T' \mid C' \mid S'$ iff $\phi \mid T \mid C \mid S$ is satisfiable if and only if $\phi' \mid T' \mid C' \mid S'$ is satisfiable. We denote $tcons(S) = \{0,\, 1\} \cup (preds(S) \cap \overline{C}_{\mathcal{L}}) \subseteq Tcons_{\mathcal{L}}$. Let $S \subseteq_{\mathcal{F}} OrdCl_{\mathcal{L}}$. We define the size of S as $|S| = \sum_{C \in S} |C|$. l is a simplified order literal of $\mathcal{L}$ iff $l = \varepsilon_1 \diamond \varepsilon_2$, $\{\varepsilon_1, \varepsilon_2\} \not\subseteq Tcons_{\mathcal{L}}$, $\{\varepsilon_1, \varepsilon_2\} \not\subseteq QAtom_{\mathcal{L}}$. The set of all simplified order literals of $\mathcal{L}$ is designated as $SimOrdLit_{\mathcal{L}} \subseteq OrdLit_{\mathcal{L}}$. We denote $SimOrdCl_{\mathcal{L}} = \{C \mid C \in OrdCl_{\mathcal{L}}, C \subseteq SimOrdLit_{\mathcal{L}}\} \subseteq OrdCl_{\mathcal{L}}$. Let $\tilde{f}_0 \notin Func_{\mathcal{L}}$; $\tilde{f}_0$ is a new function symbol. Let $\mathbb{I} = \mathbb{N} \times \mathbb{N}$; $\mathbb{I}$ is an infinite countable set of indices. Let $\tilde{\mathbb{P}} = \{\tilde{p}_{\mathtt{i}} \mid \mathtt{i} \in \mathbb{I}\}$ such that $\tilde{\mathbb{P}} \cap Pred_{\mathcal{L}} = \emptyset$; $\tilde{\mathbb{P}}$ is an infinite countable set of new predicate symbols.

From a computational point of view, the worst case time and space complexity will be estimated using the logarithmic cost measurement. Let $\mathcal{A}$ be an algorithm. $\#\mathcal{O}_{\mathcal{A}}(In) \geq 1$ denotes the number of all elementary operations executed by $\mathcal{A}$ on an input In.

3.1   Substitutions

We assume the reader to be familiar with the standard notions and notation of substitutions. We introduce a few definitions and denotations; some of them are slightly different from the standard ones, but found to be more convenient. Let $X = \{x_i \mid 1 \leq i \leq n\} \subseteq Var_{\mathcal{L}}$. A substitution ϑ of $\mathcal{L}$ is a mapping ϑ : $X \longrightarrow Term_{\mathcal{L}}$. ϑ may be written in the form $x_1/\vartheta(x_1), \ldots, x_n/\vartheta(x_n)$. We denote $dom(\vartheta) = X \subseteq_{\mathcal{F}} Var_{\mathcal{L}}$ and $range(\vartheta) = \bigcup_{x \in X} vars(\vartheta(x)) \subseteq_{\mathcal{F}} Var_{\mathcal{L}}$. The set of all substitutions of $\mathcal{L}$ is designated as $Subst_{\mathcal{L}}$. Let $Qx\,a \in QAtom_{\mathcal{L}}$. ϑ is applicable to $Qx\,a$ iff $dom(\vartheta) \supseteq freevars(Qx\,a)$ and $x \notin range(\vartheta|_{freevars(Qx\,a)})$. We define the application of ϑ to $Qx\,a$ as $(Qx\,a)\vartheta = Qx\,a(\vartheta|_{freevars(Qx\,a) \cup x/x}) \in QAtom_{\mathcal{L}}$. Let ε and ε' be expressions. ε' is an instance of ε of $\mathcal{L}$ iff there exists $\vartheta^* \in Subst_{\mathcal{L}}$ such that $\varepsilon' = \varepsilon\vartheta^*$. ε' is a variant of ε of $\mathcal{L}$ iff there exists a variable renaming $\rho^* \in Subst_{\mathcal{L}}$ such that $\varepsilon' = \varepsilon\rho^*$. Let $C \in OrdCl_{\mathcal{L}}$ and $S \subseteq OrdCl_{\mathcal{L}}$. C is an instance $\mid$ a variant of S of $\mathcal{L}$ iff there exists $C^* \in S$ such that C is an instance $\mid$ a variant of C^* of $\mathcal{L}$. We denote $Inst_{\mathcal{L}}(S) = \{C \mid C \text{ is an instance of } S \text{ of } \mathcal{L}\} \subseteq OrdCl_{\mathcal{L}}$ and $Vrnt_{\mathcal{L}}(S) = \{C \mid C \text{ is a variant of } S \text{ of } \mathcal{L}\} \subseteq OrdCl_{\mathcal{L}}$.

Let E be a set of expressions. ϑ is a unifier of $\mathcal{L}$ for E iff $E\vartheta$ is a singleton set. Let $\theta \in Subst_{\mathcal{L}}$. θ is a most general unifier of $\mathcal{L}$ for E iff θ is a unifier of $\mathcal{L}$ for E, and for every unifier ϑ of $\mathcal{L}$ for E, there exists $\gamma^* \in Subst_{\mathcal{L}}$ such that $\vartheta|_{freevars(E)} = \theta|_{freevars(E)} \circ \gamma^*$. By $mgu_{\mathcal{L}}(E) \subseteq Subst_{\mathcal{L}}$ we denote the set of all most general unifiers of $\mathcal{L}$ for E. Let $\overline{E} = E_0, \ldots, E_n$, $E_i \subseteq A_i$, either $A_i = Term_{\mathcal{L}}$ or $A_i = Atom_{\mathcal{L}}$ or $A_i = QAtom_{\mathcal{L}}$ or $A_i = OrdLit_{\mathcal{L}}$. ϑ is a unifier of $\mathcal{L}$ for $\overline{E}$ iff, for all $i \leq n$, ϑ is a unifier of $\mathcal{L}$ for E_i. θ is a most general unifier of $\mathcal{L}$ for $\overline{E}$ iff θ is a unifier of $\mathcal{L}$ for $\overline{E}$, and for every unifier ϑ of $\mathcal{L}$ for $\overline{E}$, there exists $\gamma^* \in Subst_{\mathcal{L}}$ such that $\vartheta|_{freevars(\overline{E})} = \theta|_{freevars(\overline{E})} \circ \gamma^*$. By $mgu_{\mathcal{L}}(\overline{E}) \subseteq Subst_{\mathcal{L}}$ we denote the set of all most general unifiers of $\mathcal{L}$ for $\overline{E}$.

Theorem 1 (Unification Theorem). *Let* $\overline{E} = E_0, \ldots, E_n$, *either* $E_i \subseteq_{\mathcal{F}} Term_{\mathcal{L}}$ *or* $E_i \subseteq_{\mathcal{F}} Atom_{\mathcal{L}}$. *If there exists a unifier of* $\mathcal{L}$ *for* $\overline{E}$, *then there exists* $\theta^* \in mgu_{\mathcal{L}}(\overline{E})$ *such that* $range(\theta^*|_{vars(\overline{E})}) \subseteq vars(\overline{E})$.

Proof. By induction on $\|vars(\overline{E})\|$; a modification of the proof of Theorem 2.3 (Unification Theorem) in [26], Sect. 2.4, pp. 5–6. □

Theorem 2 (Extended Unification Theorem). *Let* $\overline{E} = E_0, \ldots, E_n$, *either* $E_i \subseteq_{\mathcal{F}} Term_{\mathcal{L}}$ *or* $E_i \subseteq_{\mathcal{F}} Atom_{\mathcal{L}}$ *or* $E_i \subseteq_{\mathcal{F}} QAtom_{\mathcal{L}}$ *or* $E_i \subseteq_{\mathcal{F}} OrdLit_{\mathcal{L}}$, *and* $boundvars(\overline{E}) \subseteq V \subseteq_{\mathcal{F}} Var_{\mathcal{L}}$. *If there exists a unifier of* $\mathcal{L}$ *for* $\overline{E}$, *then there exists* $\theta^* \in mgu_{\mathcal{L}}(\overline{E})$ *such that* $range(\theta^*|_{freevars(\overline{E})}) \cap V = \emptyset$.

Proof. A straightforward consequence of Theorem 1. □

3.2 A Formal Treatment

Translation of an order formula or theory to clausal form, is based on the following lemma:

Lemma 1. *Let* $n_{\phi}, n_0 \in \mathbb{N}$, $\phi \in OrdForm_{\mathcal{L}}$, $T \subseteq OrdForm_{\mathcal{L}}$.

(I) There exist either $J_{\phi} = \emptyset$ *or* $J_{\phi} = \{(n_{\phi}, j) \mid j \leq n_{J_{\phi}}\}$, $J_{\phi} \subseteq \{(n_{\phi}, j) \mid j \in \mathbb{N}\}$, *and* $S_{\phi} \subseteq_{\mathcal{F}} SimOrdCl_{\mathcal{L} \cup \{\tilde{p}_j \mid j \in J_{\phi}\}}$ *such that*

(a) $\|J_{\phi}\| \leq 2 \cdot |\phi|$;

(b) either $J_{\phi} = \emptyset$, $S_{\phi} = \{\Box\}$ *or* $J_{\phi} = S_{\phi} = \emptyset$ *or* $J_{\phi} \neq \emptyset$, $\Box \notin S_{\phi} \neq \emptyset$;

(c) there exists an interpretation $\mathfrak{A}$ *for* $\mathcal{L}$ *and* $\mathfrak{A} \models \phi$ *if and only if there exists an interpretation* $\mathfrak{A}'$ *for* $\mathcal{L} \cup \{\tilde{p}_j \mid j \in J_{\phi}\}$ *and* $\mathfrak{A}' \models S_{\phi}$, *satisfying* $\mathfrak{A} = \mathfrak{A}'|_{\mathcal{L}}$;

(d) $|S_{\phi}| \in O(|\phi|^2)$; *the number of all elementary operations of the translation of* ϕ *to* S_{ϕ}, *is in* $O(|\phi|^2)$; *the time and space complexity of the translation of* ϕ *to* S_{ϕ}, *is in* $O(|\phi|^2 \cdot (\log(1 + n_{\phi}) + \log|\phi|))$;

(e) if $S_{\phi} \neq \emptyset, \{\Box\}$, *then* $J_{\phi} \neq \emptyset$, *for all* $C \in S_{\phi}$, $\emptyset \neq preds(C) \cap \tilde{\mathbb{P}} \subseteq \{\tilde{p}_j \mid j \in J_{\phi}\}$;

(f) for all $a \in qatoms(S_{\phi})$, *there exists* $j^* \in J_{\phi}$ *and* $preds(a) = \{\tilde{p}_{j^*}\}$;

(g) for all $j \in J_{\phi}$, *there exists a sequence* $\bar{x}$ *of variables of* $\mathcal{L}$ *and* $\tilde{p}_j(\bar{x}) \in atoms(S_{\phi})$ *satisfying, for all* $a \in atoms(S_{\phi})$ *and* $preds(a) = \{\tilde{p}_j\}$, $a = \tilde{p}_j(\bar{x})$; *if there exists* $a^* \in qatoms(S_{\phi})$ *and* $preds(a^*) = \{\tilde{p}_j\}$, *then there exists* $Qx\,\tilde{p}_j(\bar{x}) \in qatoms(S_{\phi})$ *satisfying, for all* $a \in qatoms(S_{\phi})$ *and* $preds(a) = \{\tilde{p}_j\}$, $a = Qx\,\tilde{p}_j(\bar{x})$;

(h) $tcons(S_\phi) \subseteq tcons(\phi)$.

(II) There exist $J_T \subseteq \{(i,j) \mid i \geq n_0\}$ and $S_T \subseteq SimOrdCl_{\mathcal{L} \cup \{\tilde{p}_j \mid j \in J_T\}}$ such that

(a) either $J_T = \emptyset$, $S_T = \{\Box\}$ or $J_T = S_T = \emptyset$ or $J_T \neq \emptyset$, $\Box \notin S_T \neq \emptyset$;

(b) there exists an interpretation $\mathfrak{A}$ for $\mathcal{L}$ and $\mathfrak{A} \models T$ if and only if there exists an interpretation $\mathfrak{A}'$ for $\mathcal{L} \cup \{\tilde{p}_j \mid j \in J_T\}$ and $\mathfrak{A}' \models S_T$, satisfying $\mathfrak{A} = \mathfrak{A}'|_{\mathcal{L}}$;

(c) if $T \subseteq_{\mathcal{F}} OrdForm_{\mathcal{L}}$, then $J_T \subseteq_{\mathcal{F}} \{(i,j) \mid i \geq n_0\}$, $\|J_T\| \leq 2 \cdot |T|$, $S_T \subseteq_{\mathcal{F}} SimOrdCl_{\mathcal{L} \cup \{\tilde{p}_j \mid j \in J_T\}}$, $|S_T| \in O(|T|^2)$; the number of all elementary operations of the translation of T to S_T, is in $O(|T|^2)$; the time and space complexity of the translation of T to S_T, is in $O(|T|^2 \cdot \log(1 + n_0 + |T|))$;

(d) if $S_T \neq \emptyset, \{\Box\}$, then $J_T \neq \emptyset$, for all $C \in S_T$, $\emptyset \neq preds(C) \cap \tilde{\mathbb{P}} \subseteq \{\tilde{p}_j \mid j \in J_T\}$;

(e) for all $a \in qatoms(S_T)$, there exists $j^* \in J_T$ and $preds(a) = \{\tilde{p}_{j^*}\}$;

(f) for all $j \in J_T$, there exists a sequence $\bar{x}$ of variables of $\mathcal{L}$ and $\tilde{p}_j(\bar{x}) \in atoms(S_T)$ satisfying, for all $a \in atoms(S_T)$ and $preds(a) = \{\tilde{p}_j\}$, $a = \tilde{p}_j(\bar{x})$; if there exists $a^* \in qatoms(S_T)$ and $preds(a^*) = \{\tilde{p}_j\}$, then there exists $Qx\,\tilde{p}_j(\bar{x}) \in qatoms(S_T)$ satisfying, for all $a \in qatoms(S_T)$ and $preds(a) = \{\tilde{p}_j\}$, $a = Qx\,\tilde{p}_j(\bar{x})$;

(g) $tcons(S_T) \subseteq tcons(T)$.

Proof. Technical, using interpolation. It is straightforward to prove the following statements:

Let $n_\theta \in \mathbb{N}$ and $\theta \in OrdForm_{\mathcal{L}}$. There exists $\theta' \in OrdForm_{\mathcal{L}}$ such that   (14)

(a) $\theta' \equiv \theta$;

(b) $|\theta'| \leq 2 \cdot |\theta|$; θ' can be built up from θ via a postorder traversal of θ with $\#\mathcal{O}(\theta) \in O(|\theta|)$ and the time, space complexity in $O(|\theta| \cdot (\log(1 + n_\theta) + \log|\theta|))$;

(c) θ' does not contain $\neg$ and Δ;

(d) $\theta' \in Tcons_{\mathcal{L}}$; or for every subformula of θ' of the form $\varepsilon_1 \diamond \varepsilon_2$, $\diamond \in \{\wedge, \vee, \leftrightarrow\}$, $\varepsilon_i \neq 0, 1$, $\{\varepsilon_1, \varepsilon_2\} \not\subseteq Tcons_{\mathcal{L}}$; for every subformula of θ' of the form $\varepsilon_1 \to \varepsilon_2$, $\varepsilon_1 \neq 0, 1$, $\varepsilon_2 \neq 1$, $\{\varepsilon_1, \varepsilon_2\} \not\subseteq Tcons_{\mathcal{L}}$; for every subformula of θ' of the form $\varepsilon_1 = \varepsilon_2$, $\{\varepsilon_1, \varepsilon_2\} \not\subseteq Tcons_{\mathcal{L}}$; for every subformula of θ' of the form $\varepsilon_1 \prec \varepsilon_2$, $\varepsilon_1 \neq 1$, $\varepsilon_2 \neq 0$, $\{\varepsilon_1, \varepsilon_2\} \not\subseteq Tcons_{\mathcal{L}}$; for every subformula of θ' of the form $Qx\,\varepsilon_1$, $Q \in \{\forall, \exists\}$, $\varepsilon_1 \notin Tcons_{\mathcal{L}}$;

(e) $tcons(\theta') \subseteq tcons(\theta)$.

The proof is by induction on the structure of θ.

Let $n_\theta \in \mathbb{N}$, $\theta \in OrdForm_{\mathcal{L}} - \{0, 1\}$, (14c, d) hold for θ; $\bar{x}$ be a sequence   (15) of variables, $vars(\theta) \subseteq vars(\bar{x}) \subseteq Var_{\mathcal{L}}$; $i = (n_\theta, j_i) \in \{(n_\theta, j) \mid j \in \mathbb{N}\}$, $\tilde{p}_i \in \tilde{\mathbb{P}}$, $ar(\tilde{p}_i) = |\bar{x}|$. There exist $J = \{(n_\theta, j) \mid j_i + 1 \leq j \leq n_J\} \subseteq \{(n_\theta, j) \mid j \in \mathbb{N}\}$, $j_i \leq n_J$, $i \notin J$, and $S \subseteq_{\mathcal{F}} SimOrdCl_{\mathcal{L} \cup \{\tilde{p}_i\} \cup \{\tilde{p}_j \mid j \in J\}}$ such that

(a) $\|J\| \leq |\theta| - 1$;

(b) there exists an interpretation $\mathfrak{A}$ for $\mathcal{L} \cup \{\tilde{p}_{\mathrm{i}}\}$ and $\mathfrak{A} \models \tilde{p}_{\mathrm{i}}(\bar{x}) \leftrightarrow \theta \in OrdForm_{\mathcal{L} \cup \{\tilde{p}_{\mathrm{i}}\}}$ if and only if there exists an interpretation $\mathfrak{A}'$ for $\mathcal{L} \cup \{\tilde{p}_{\mathrm{i}}\} \cup \{\tilde{p}_{\mathrm{j}} \,|\, \mathrm{j} \in J\}$ and $\mathfrak{A}' \models S$, satisfying $\mathfrak{A} = \mathfrak{A}'|_{\mathcal{L} \cup \{\tilde{p}_{\mathrm{i}}\}}$;

(c) $|S| \leq 27 \cdot |\theta| \cdot (1 + |\bar{x}|)$, S can be built up from θ and $\tilde{f}_0(\bar{x})$ via a preorder traversal of θ with $\#\mathcal{O}(\theta, \tilde{f}_0(\bar{x})) \in O(|\theta| \cdot (1 + |\bar{x}|))$;

(d) for all $C \in S$, $\emptyset \neq preds(C) \cap \tilde{\mathbb{P}} \subseteq \{\tilde{p}_{\mathrm{i}}\} \cup \{\tilde{p}_{\mathrm{j}} \,|\, \mathrm{j} \in J\}$, $\tilde{p}_{\mathrm{i}}(\bar{x}) = 1, \tilde{p}_{\mathrm{i}}(\bar{x}) \prec 1 \notin S$;

(e) for all $a \in qatoms(S)$, there exists $\mathrm{j}^* \in J$ and $preds(a) = \{\tilde{p}_{\mathrm{j}^*}\}$;

(f) for all $\mathrm{j} \in \{\mathrm{i}\} \cup J$, $\tilde{p}_{\mathrm{j}}(\bar{x}) \in atoms(S)$ satisfying, for all $a \in atoms(S)$ and $preds(a) = \{\tilde{p}_{\mathrm{j}}\}$, $a = \tilde{p}_{\mathrm{j}}(\bar{x})$; $\tilde{p}_{\mathrm{i}} \notin preds(qatoms(S))$, for all $\mathrm{j} \in J$, if there exists $a^* \in qatoms(S)$ and $preds(a^*) = \{\tilde{p}_{\mathrm{j}}\}$, then there exists $Qx\,\tilde{p}_{\mathrm{j}}(\bar{x}) \in qatoms(S)$ satisfying, for all $a \in qatoms(S)$ and $preds(a) = \{\tilde{p}_{\mathrm{j}}\}$, $a = Qx\,\tilde{p}_{\mathrm{j}}(\bar{x})$;

(g) $tcons(S) = tcons(\theta)$.

The proof is by induction on the structure of θ using the interpolation rules in Tables 1 and 2.

(I) By (14) for n_ϕ, ϕ, there exists $\phi' \in OrdForm_{\mathcal{L}}$ such that (14a–e) hold for n_ϕ, ϕ, ϕ'. We distinguish three cases for ϕ'. Case 1: $\phi' \in Tcons_{\mathcal{L}} - \{1\}$. We put $J_\phi = \emptyset \subseteq \{(n_\phi, j) \,|\, j \in \mathbb{N}\}$ and $S_\phi = \{\Box\} \subseteq_{\mathcal{F}} SimOrdCl_{\mathcal{L}}$. Case 2: $\phi' = 1$. We put $J_\phi = \emptyset \subseteq \{(n_\phi, j) \,|\, j \in \mathbb{N}\}$ and $S_\phi = \emptyset \subseteq_{\mathcal{F}} SimOrdCl_{\mathcal{L}}$. Case 3: $\phi' \notin Tcons_{\mathcal{L}}$. We put $\bar{x} = varseq(\phi')$, $j_{\mathrm{i}} = 0$, $\mathrm{i} = (n_\phi, j_{\mathrm{i}})$, $ar(\tilde{p}_{\mathrm{i}}) = |\bar{x}|$. We get by (15) for n_ϕ, ϕ', $\bar{x}$, i, $\tilde{p}_{\mathrm{i}}$ that there exist $J = \{(n_\phi, j) \,|\, 1 \leq j \leq n_J\} \subseteq \{(n_\phi, j) \,|\, j \in \mathbb{N}\}$, $j_{\mathrm{i}} \leq n_J$, $\mathrm{i} \notin J$, $S \subseteq_{\mathcal{F}} SimOrdCl_{\mathcal{L} \cup \{\tilde{p}_{\mathrm{i}}\} \cup \{\tilde{p}_{\mathrm{j}} \,|\, \mathrm{j} \in J\}}$, and (15a–g) hold for ϕ', $\bar{x}$, $\tilde{p}_{\mathrm{i}}$, J, S. We put $n_{J_\phi} = n_J$, $J_\phi = \{(n_\phi, j) \,|\, j \leq n_{J_\phi}\} \subseteq \{(n_\phi, j) \,|\, j \in \mathbb{N}\}$, $S_\phi = \{\tilde{p}_{\mathrm{i}}(\bar{x}) = 1\} \cup S \subseteq_{\mathcal{F}} SimOrdCl_{\mathcal{L} \cup \{\tilde{p}_{\mathrm{j}} \,|\, \mathrm{j} \in J_\phi\}}$. (II) straightforwardly follows from (I). The lemma is proved. $\Box$

The described translation produces order clausal theories in some restrictive form, which will be utilised in inference using our order hyperresolution calculus to get shorter deductions in average case, cf. Sect. 5. Let $P \subseteq \tilde{\mathbb{P}}$ and $S \subseteq OrdCl_{\mathcal{L} \cup P}$. S is admissible iff

(a) for all $a \in qatoms(S)$, $preds(a) \subseteq P$;

(b) for all $\tilde{p} \in P$, there exists a sequence $\bar{x}$ of variables of $\mathcal{L}$ and $\tilde{p}(\bar{x}) \in atoms(S)$ satisfying, for all $a \in atoms(S)$ and $preds(a) = \{\tilde{p}\}$, a is an instance of $\tilde{p}(\bar{x})$ of $\mathcal{L} \cup P$; if there exists $a^* \in qatoms(S)$ and $preds(a^*) = \{\tilde{p}\}$, then there exists $Qx\,\tilde{p}(\bar{x}) \in qatoms(S)$ satisfying, for all $a \in qatoms(S)$ and $preds(a) = \{\tilde{p}\}$, a is an instance of $Qx\,\tilde{p}(\bar{x})$ of $\mathcal{L} \cup P$.

(a) and (b) imply that for all $Qx\,a, Q'x'\,a' \in qatoms(S)$, if $preds(a) = preds(a')$, then $Q = Q'$, $x = x'$, $boundindset(Qx\,a) = boundindset(Q'x'\,a')$.

Theorem 3. *Let* $n_0 \in \mathbb{N}$, $\phi \in OrdForm_{\mathcal{L}}$, $T \subseteq OrdForm_{\mathcal{L}}$. *There exist* $J_T^\phi \subseteq \{(i, j) \,|\, i \geq n_0\}$ *and* $S_T^\phi \subseteq SimOrdCl_{\mathcal{L} \cup \{\tilde{p}_{\mathrm{j}} \,|\, \mathrm{j} \in J_T^\phi\}}$ *such that*

Table 1. Binary interpolation rules for $\wedge$, $\vee$, $\rightarrow$, $\leftrightarrow$, $=$, $\prec$.

Case

$\theta = \theta_1 \wedge \theta_2$

$$\frac{\tilde{p}_{\mathbf{i}}(\bar{x}) \leftrightarrow \theta_1 \wedge \theta_2}{\left\{\begin{array}{l} \tilde{p}_{\mathbf{i}_1}(\bar{x}) \prec \tilde{p}_{\mathbf{i}_2}(\bar{x}) \vee \tilde{p}_{\mathbf{i}_1}(\bar{x}) = \tilde{p}_{\mathbf{i}_2}(\bar{x}) \vee \tilde{p}_{\mathbf{i}}(\bar{x}) = \tilde{p}_{\mathbf{i}_2}(\bar{x}), \\ \tilde{p}_{\mathbf{i}_2}(\bar{x}) \prec \tilde{p}_{\mathbf{i}_1}(\bar{x}) \vee \tilde{p}_{\mathbf{i}}(\bar{x}) = \tilde{p}_{\mathbf{i}_1}(\bar{x}), \\ \tilde{p}_{\mathbf{i}_1}(\bar{x}) \leftrightarrow \theta_1, \tilde{p}_{\mathbf{i}_2}(\bar{x}) \leftrightarrow \theta_2 \end{array}\right\}} \tag{16}$$

$|\text{Consequent}| = 15 + 10 \cdot |\bar{x}| + |\tilde{p}_{\mathbf{i}_1}(\bar{x}) \leftrightarrow \theta_1| + |\tilde{p}_{\mathbf{i}_2}(\bar{x}) \leftrightarrow \theta_2| \leq 27 \cdot (1 + |\bar{x}|) + |\tilde{p}_{\mathbf{i}_1}(\bar{x}) \leftrightarrow \theta_1| + |\tilde{p}_{\mathbf{i}_2}(\bar{x}) \leftrightarrow \theta_2|$

$\theta = \theta_1 \vee \theta_2$

$$\frac{\tilde{p}_{\mathbf{i}}(\bar{x}) \leftrightarrow (\theta_1 \vee \theta_2)}{\left\{\begin{array}{l} \tilde{p}_{\mathbf{i}_1}(\bar{x}) \prec \tilde{p}_{\mathbf{i}_2}(\bar{x}) \vee \tilde{p}_{\mathbf{i}_1}(\bar{x}) = \tilde{p}_{\mathbf{i}_2}(\bar{x}) \vee \tilde{p}_{\mathbf{i}}(\bar{x}) = \tilde{p}_{\mathbf{i}_1}(\bar{x}), \\ \tilde{p}_{\mathbf{i}_2}(\bar{x}) \prec \tilde{p}_{\mathbf{i}_1}(\bar{x}) \vee \tilde{p}_{\mathbf{i}}(\bar{x}) = \tilde{p}_{\mathbf{i}_2}(\bar{x}), \\ \tilde{p}_{\mathbf{i}_1}(\bar{x}) \leftrightarrow \theta_1, \tilde{p}_{\mathbf{i}_2}(\bar{x}) \leftrightarrow \theta_2 \end{array}\right\}} \tag{17}$$

$|\text{Consequent}| = 15 + 10 \cdot |\bar{x}| + |\tilde{p}_{\mathbf{i}_1}(\bar{x}) \leftrightarrow \theta_1| + |\tilde{p}_{\mathbf{i}_2}(\bar{x}) \leftrightarrow \theta_2| \leq 27 \cdot (1 + |\bar{x}|) + |\tilde{p}_{\mathbf{i}_1}(\bar{x}) \leftrightarrow \theta_1| + |\tilde{p}_{\mathbf{i}_2}(\bar{x}) \leftrightarrow \theta_2|$

$\theta = \theta_1 \rightarrow \theta_2, \theta_2 \neq 0$

$$\frac{\tilde{p}_{\mathbf{i}}(\bar{x}) \leftrightarrow (\theta_1 \rightarrow \theta_2)}{\left\{\begin{array}{l} \tilde{p}_{\mathbf{i}_1}(\bar{x}) \prec \tilde{p}_{\mathbf{i}_2}(\bar{x}) \vee \tilde{p}_{\mathbf{i}_1}(\bar{x}) = \tilde{p}_{\mathbf{i}_2}(\bar{x}) \vee \tilde{p}_{\mathbf{i}}(\bar{x}) = \tilde{p}_{\mathbf{i}_2}(\bar{x}), \\ \tilde{p}_{\mathbf{i}_2}(\bar{x}) \prec \tilde{p}_{\mathbf{i}_1}(\bar{x}) \vee \tilde{p}_{\mathbf{i}}(\bar{x}) = 1, \tilde{p}_{\mathbf{i}_1}(\bar{x}) \leftrightarrow \theta_1, \tilde{p}_{\mathbf{i}_2}(\bar{x}) \leftrightarrow \theta_2 \end{array}\right\}} \tag{18}$$

$|\text{Consequent}| = 15 + 9 \cdot |\bar{x}| + |\tilde{p}_{\mathbf{i}_1}(\bar{x}) \leftrightarrow \theta_1| + |\tilde{p}_{\mathbf{i}_2}(\bar{x}) \leftrightarrow \theta_2| \leq 27 \cdot (1 + |\bar{x}|) + |\tilde{p}_{\mathbf{i}_1}(\bar{x}) \leftrightarrow \theta_1| + |\tilde{p}_{\mathbf{i}_2}(\bar{x}) \leftrightarrow \theta_2|$

$\theta = \theta_1 \leftrightarrow \theta_2$

$$\frac{\tilde{p}_{\mathbf{i}}(\bar{x}) \leftrightarrow (\theta_1 \leftrightarrow \theta_2)}{\left\{\begin{array}{l} \tilde{p}_{\mathbf{i}_1}(\bar{x}) \prec \tilde{p}_{\mathbf{i}_2}(\bar{x}) \vee \tilde{p}_{\mathbf{i}_1}(\bar{x}) = \tilde{p}_{\mathbf{i}_2}(\bar{x}) \vee \tilde{p}_{\mathbf{i}}(\bar{x}) = \tilde{p}_{\mathbf{i}_2}(\bar{x}), \\ \tilde{p}_{\mathbf{i}_2}(\bar{x}) \prec \tilde{p}_{\mathbf{i}_1}(\bar{x}) \vee \tilde{p}_{\mathbf{i}_2}(\bar{x}) = \tilde{p}_{\mathbf{i}_1}(\bar{x}) \vee \tilde{p}_{\mathbf{i}}(\bar{x}) = \tilde{p}_{\mathbf{i}_1}(\bar{x}), \\ \tilde{p}_{\mathbf{i}_1}(\bar{x}) \prec \tilde{p}_{\mathbf{i}_2}(\bar{x}) \vee \tilde{p}_{\mathbf{i}_2}(\bar{x}) \prec \tilde{p}_{\mathbf{i}_1}(\bar{x}) \vee \tilde{p}_{\mathbf{i}}(\bar{x}) = 1, \\ \tilde{p}_{\mathbf{i}_1}(\bar{x}) \leftrightarrow \theta_1, \tilde{p}_{\mathbf{i}_2}(\bar{x}) \leftrightarrow \theta_2 \end{array}\right\}} \tag{19}$$

$|\text{Consequent}| = 27 + 17 \cdot |\bar{x}| + |\tilde{p}_{\mathbf{i}_1}(\bar{x}) \leftrightarrow \theta_1| + |\tilde{p}_{\mathbf{i}_2}(\bar{x}) \leftrightarrow \theta_2| \leq 27 \cdot (1 + |\bar{x}|) + |\tilde{p}_{\mathbf{i}_1}(\bar{x}) \leftrightarrow \theta_1| + |\tilde{p}_{\mathbf{i}_2}(\bar{x}) \leftrightarrow \theta_2|$

$\theta = \theta_1 = \theta_2, \theta_i \neq 0, 1$

$$\frac{\tilde{p}_{\mathbf{i}}(\bar{x}) \leftrightarrow (\theta_1 = \theta_2)}{\left\{\begin{array}{l} \tilde{p}_{\mathbf{i}_1}(\bar{x}) = \tilde{p}_{\mathbf{i}_2}(\bar{x}) \vee \tilde{p}_{\mathbf{i}}(\bar{x}) = 0, \\ \tilde{p}_{\mathbf{i}_1}(\bar{x}) \prec \tilde{p}_{\mathbf{i}_2}(\bar{x}) \vee \tilde{p}_{\mathbf{i}_2}(\bar{x}) \prec \tilde{p}_{\mathbf{i}_1}(\bar{x}) \vee \tilde{p}_{\mathbf{i}}(\bar{x}) = 1, \tilde{p}_{\mathbf{i}_1}(\bar{x}) \leftrightarrow \theta_1, \tilde{p}_{\mathbf{i}_2}(\bar{x}) \leftrightarrow \theta_2 \end{array}\right\}} \tag{20}$$

$|\text{Consequent}| = 15 + 8 \cdot |\bar{x}| + |\tilde{p}_{\mathbf{i}_1}(\bar{x}) \leftrightarrow \theta_1| + |\tilde{p}_{\mathbf{i}_2}(\bar{x}) \leftrightarrow \theta_2| \leq 27 \cdot (1 + |\bar{x}|) + |\tilde{p}_{\mathbf{i}_1}(\bar{x}) \leftrightarrow \theta_1| + |\tilde{p}_{\mathbf{i}_2}(\bar{x}) \leftrightarrow \theta_2|$

$\theta = \theta_1 \prec \theta_2, \theta_1 \neq 0, \theta_2 \neq 1$

$$\frac{\tilde{p}_{\mathbf{i}}(\bar{x}) \leftrightarrow (\theta_1 \prec \theta_2)}{\left\{\begin{array}{l} \tilde{p}_{\mathbf{i}_1}(\bar{x}) \prec \tilde{p}_{\mathbf{i}_2}(\bar{x}) \vee \tilde{p}_{\mathbf{i}}(\bar{x}) = 0, \\ \tilde{p}_{\mathbf{i}_2}(\bar{x}) \prec \tilde{p}_{\mathbf{i}_1}(\bar{x}) \vee \tilde{p}_{\mathbf{i}_2}(\bar{x}) = \tilde{p}_{\mathbf{i}_1}(\bar{x}) \vee \tilde{p}_{\mathbf{i}}(\bar{x}) = 1, \tilde{p}_{\mathbf{i}_1}(\bar{x}) \leftrightarrow \theta_1, \tilde{p}_{\mathbf{i}_2}(\bar{x}) \leftrightarrow \theta_2 \end{array}\right\}} \tag{21}$$

$|\text{Consequent}| = 15 + 8 \cdot |\bar{x}| + |\tilde{p}_{\mathbf{i}_1}(\bar{x}) \leftrightarrow \theta_1| + |\tilde{p}_{\mathbf{i}_2}(\bar{x}) \leftrightarrow \theta_2| \leq 27 \cdot (1 + |\bar{x}|) + |\tilde{p}_{\mathbf{i}_1}(\bar{x}) \leftrightarrow \theta_1| + |\tilde{p}_{\mathbf{i}_2}(\bar{x}) \leftrightarrow \theta_2|$

Table 2. Unary interpolation rules for $\to$, $\equiv$, $\prec$, $\forall$, $\exists$.

Case

$\theta = \theta_1 \to 0$

$$\frac{\tilde{p}_\mathbf{i}(\bar{x}) \leftrightarrow (\theta_1 \to 0)}{\{\tilde{p}_{\mathbf{i}_1}(\bar{x}) \equiv 0 \vee \tilde{p}_\mathbf{i}(\bar{x}) \equiv 0, 0 \prec \tilde{p}_{\mathbf{i}_1}(\bar{x}) \vee \tilde{p}_\mathbf{i}(\bar{x}) \equiv 1, \tilde{p}_{\mathbf{i}_1}(\bar{x}) \leftrightarrow \theta_1\}} \tag{22}$$

$|\text{Consequent}| = 12 + 4 \cdot |\bar{x}| + |\tilde{p}_{\mathbf{i}_1}(\bar{x}) \leftrightarrow \theta_1| \leq 27 \cdot (1 + |\bar{x}|) + |\tilde{p}_{\mathbf{i}_1}(\bar{x}) \leftrightarrow \theta_1|$

$\theta = \theta_1 \equiv 0$

$$\frac{\tilde{p}_\mathbf{i}(\bar{x}) \leftrightarrow (\theta_1 \equiv 0)}{\{\tilde{p}_{\mathbf{i}_1}(\bar{x}) \equiv 0 \vee \tilde{p}_\mathbf{i}(\bar{x}) \equiv 0, 0 \prec \tilde{p}_{\mathbf{i}_1}(\bar{x}) \vee \tilde{p}_\mathbf{i}(\bar{x}) \equiv 1, \tilde{p}_{\mathbf{i}_1}(\bar{x}) \leftrightarrow \theta_1\}} \tag{23}$$

$|\text{Consequent}| = 12 + 4 \cdot |\bar{x}| + |\tilde{p}_{\mathbf{i}_1}(\bar{x}) \leftrightarrow \theta_1| \leq 27 \cdot (1 + |\bar{x}|) + |\tilde{p}_{\mathbf{i}_1}(\bar{x}) \leftrightarrow \theta_1|$

$\theta = \theta_1 \equiv 1$

$$\frac{\tilde{p}_\mathbf{i}(\bar{x}) \leftrightarrow (\theta_1 \equiv 1)}{\{\tilde{p}_{\mathbf{i}_1}(\bar{x}) \equiv 1 \vee \tilde{p}_\mathbf{i}(\bar{x}) \equiv 0, \tilde{p}_{\mathbf{i}_1}(\bar{x}) \prec 1 \vee \tilde{p}_\mathbf{i}(\bar{x}) \equiv 1, \tilde{p}_{\mathbf{i}_1}(\bar{x}) \leftrightarrow \theta_1\}} \tag{24}$$

$|\text{Consequent}| = 12 + 4 \cdot |\bar{x}| + |\tilde{p}_{\mathbf{i}_1}(\bar{x}) \leftrightarrow \theta_1| \leq 27 \cdot (1 + |\bar{x}|) + |\tilde{p}_{\mathbf{i}_1}(\bar{x}) \leftrightarrow \theta_1|$

$\theta = 0 \prec \theta_1$

$$\frac{\tilde{p}_\mathbf{i}(\bar{x}) \leftrightarrow (0 \prec \theta_1)}{\{0 \prec \tilde{p}_{\mathbf{i}_1}(\bar{x}) \vee \tilde{p}_\mathbf{i}(\bar{x}) \equiv 0, \tilde{p}_{\mathbf{i}_1}(\bar{x}) \equiv 0 \vee \tilde{p}_\mathbf{i}(\bar{x}) \equiv 1, \tilde{p}_{\mathbf{i}_1}(\bar{x}) \leftrightarrow \theta_1\}} \tag{25}$$

$|\text{Consequent}| = 12 + 4 \cdot |\bar{x}| + |\tilde{p}_{\mathbf{i}_1}(\bar{x}) \leftrightarrow \theta_1| \leq 27 \cdot (1 + |\bar{x}|) + |\tilde{p}_{\mathbf{i}_1}(\bar{x}) \leftrightarrow \theta_1|$

$\theta = \theta_1 \prec 1$

$$\frac{\tilde{p}_\mathbf{i}(\bar{x}) \leftrightarrow (\theta_1 \prec 1)}{\{\tilde{p}_{\mathbf{i}_1}(\bar{x}) \prec 1 \vee \tilde{p}_\mathbf{i}(\bar{x}) \equiv 0, \tilde{p}_{\mathbf{i}_1}(\bar{x}) \equiv 1 \vee \tilde{p}_\mathbf{i}(\bar{x}) \equiv 1, \tilde{p}_{\mathbf{i}_1}(\bar{x}) \leftrightarrow \theta_1\}} \tag{26}$$

$|\text{Consequent}| = 12 + 4 \cdot |\bar{x}| + |\tilde{p}_{\mathbf{i}_1}(\bar{x}) \leftrightarrow \theta_1| \leq 27 \cdot (1 + |\bar{x}|) + |\tilde{p}_{\mathbf{i}_1}(\bar{x}) \leftrightarrow \theta_1|$

$\theta = \forall x\, \theta_1$

$$\frac{\tilde{p}_\mathbf{i}(\bar{x}) \leftrightarrow \forall x\, \theta_1}{\{\tilde{p}_\mathbf{i}(\bar{x}) \equiv \forall x\, \tilde{p}_{\mathbf{i}_1}(\bar{x}), \tilde{p}_{\mathbf{i}_1}(\bar{x}) \leftrightarrow \theta_1\}} \tag{27}$$

$|\text{Consequent}| = 5 + 2 \cdot |\bar{x}| + |\tilde{p}_{\mathbf{i}_1}(\bar{x}) \leftrightarrow \theta_1| \leq 27 \cdot (1 + |\bar{x}|) + |\tilde{p}_{\mathbf{i}_1}(\bar{x}) \leftrightarrow \theta_1|$

$\theta = \exists x\, \theta_1$

$$\frac{\tilde{p}_\mathbf{i}(\bar{x}) \leftrightarrow \exists x\, \theta_1}{\{\tilde{p}_\mathbf{i}(\bar{x}) \equiv \exists x\, \tilde{p}_{\mathbf{i}_1}(\bar{x}), \tilde{p}_{\mathbf{i}_1}(\bar{x}) \leftrightarrow \theta_1\}} \tag{28}$$

$|\text{Consequent}| = 5 + 2 \cdot |\bar{x}| + |\tilde{p}_{\mathbf{i}_1}(\bar{x}) \leftrightarrow \theta_1| \leq 27 \cdot (1 + |\bar{x}|) + |\tilde{p}_{\mathbf{i}_1}(\bar{x}) \leftrightarrow \theta_1|$

(i) *there exists an interpretation $\mathfrak{A}$ for $\mathcal{L}$ and $\mathfrak{A} \models T$, $\mathfrak{A} \not\models \phi$ if and only if there exists an interpretation $\mathfrak{A}'$ for $\mathcal{L} \cup \{\tilde{p}_j \mid j \in J_T^\phi\}$ and $\mathfrak{A}' \models S_T^\phi$, satisfying $\mathfrak{A} = \mathfrak{A}'|_{\mathcal{L}}$;*

(ii) *if $T \subseteq_\mathcal{F} OrdForm_\mathcal{L}$, then $J_T^\phi \subseteq_\mathcal{F} \{(i,j) \mid i \geq n_0\}$, $\|J_T^\phi\| \in O(|T| + |\phi|)$, $S_T^\phi \subseteq_\mathcal{F} SimOrdCl_{\mathcal{L} \cup \{\tilde{p}_j \mid j \in J_T^\phi\}}$, $|S_T^\phi| \in O(|T|^2 + |\phi|^2)$; the number of all elementary operations of the translation of T and ϕ to S_T^ϕ, is in $O(|T|^2 + |\phi|^2)$; the time and space complexity of the translation of T and ϕ to S_T^ϕ, is in $O(|T|^2 \cdot \log(1 + n_0 + |T|) + |\phi|^2 \cdot (\log(1 + n_0) + \log |\phi|))$;*

(iii) *S_T^ϕ is admissible;*

(iv) *$tcons(S_T^\phi) \subseteq tcons(\phi) \cup tcons(T)$.*

Proof. Similar to that of Lemma 1(I). We get by Lemma 1(II) for $n_0 + 1$, T that there exist $J_T \subseteq \{(i,j) \mid i \geq n_0 + 1\}$, $S_T \subseteq SimOrdCl_{\mathcal{L} \cup \{\tilde{p}_j \mid j \in J_T\}}$, and Lemma 1(II a–g) hold for $n_0 + 1$, T, J_T, S_T. By (14) for n_0, ϕ, there exists $\phi' \in OrdForm_\mathcal{L}$ such that (14a–e) hold for n_0, ϕ, ϕ'. We distinguish three cases for ϕ'. Case 1: $\phi' \in Tcons_\mathcal{L} - \{1\}$. We put $J_T^\phi = J_T \subseteq \{(i,j) \mid i \geq n_0 + 1\} \subseteq \{(i,j) \mid i \geq n_0\}$ and $S_T^\phi = S_T \subseteq SimOrdCl_{\mathcal{L} \cup \{\tilde{p}_j \mid j \in J_T^\phi\}}$. Case 2: $\phi' = 1$. We put $J_T^\phi = \emptyset \subseteq \{(i,j) \mid i \geq n_0\}$ and $S_T^\phi = \{\Box\} \subseteq SimOrdCl_\mathcal{L}$. Case 3: $\phi' \notin Tcons_\mathcal{L}$. We put $\bar{x} = varseq(\phi')$, $j_i = 0$, $i = (n_0, j_i)$, $ar(\tilde{p}_i) = |\bar{x}|$. We get by (15) for n_0, $\forall \bar{x}\, \phi'$, $\bar{x}$, i, $\tilde{p}_i$ that there exist $J = \{(n_0, j) \mid 1 \leq j \leq n_J\} \subseteq \{(n_0, j) \mid j \in \mathbb{N}\}$, $j_i \leq n_J$, $i \notin J$, $S \subseteq_\mathcal{F} SimOrdCl_{\mathcal{L} \cup \{\tilde{p}_i\} \cup \{\tilde{p}_j \mid j \in J\}}$, and (15a–g) hold for $\forall \bar{x}\, \phi'$, $\bar{x}$, $\tilde{p}_i$, J, S. We put $J_T^\phi = J_T \cup \{i\} \cup J \subseteq \{(i,j) \mid i \geq n_0\}$ and $S_T^\phi = S_T \cup \{\tilde{p}_i(\bar{x}) \prec 1\} \cup S \subseteq SimOrdCl_{\mathcal{L} \cup \{\tilde{p}_j \mid j \in J_T^\phi\}}$. The theorem is proved. $\square$

Corollary 1. *Let $n_0 \in \mathbb{N}$, $\phi \in OrdForm_\mathcal{L}$, $T \subseteq OrdForm_\mathcal{L}$. There exist $J_T^\phi \subseteq \{(i,j) \mid i \geq n_0\}$ and $S_T^\phi \subseteq SimOrdCl_{\mathcal{L} \cup \{\tilde{p}_j \mid j \in J_T^\phi\}}$ such that*

(i) *$T \models \phi$ if and only if S_T^ϕ is unsatisfiable;*

(ii) *if $T \subseteq_\mathcal{F} OrdForm_\mathcal{L}$, then $J_T^\phi \subseteq_\mathcal{F} \{(i,j) \mid i \geq n_0\}$, $\|J_T^\phi\| \in O(|T| + |\phi|)$, $S_T^\phi \subseteq_\mathcal{F} SimOrdCl_{\mathcal{L} \cup \{\tilde{p}_j \mid j \in J_T^\phi\}}$, $|S_T^\phi| \in O(|T|^2 + |\phi|^2)$; the number of all elementary operations of the translation of T and ϕ to S_T^ϕ, is in $O(|T|^2 + |\phi|^2)$; the time and space complexity of the translation of T and ϕ to S_T^ϕ, is in $O(|T|^2 \cdot \log(1 + n_0 + |T|) + |\phi|^2 \cdot (\log(1 + n_0) + \log |\phi|))$;*

(iii) *S_T^ϕ is admissible;*

(iv) *$tcons(S_T^\phi) \subseteq tcons(\phi) \cup tcons(T)$.*

Proof. Let $T \models \phi$. Then, for every interpretation $\mathfrak{A}$ for $\mathcal{L}$, $\mathfrak{A} \not\models T$ or $\mathfrak{A} \models \phi$; by Theorem 3(i), there does not exist an interpretation $\mathfrak{A}'$ for $\mathcal{L} \cup \{\tilde{p}_j \mid j \in J_T^\phi\}$ and $\mathfrak{A}' \models S_T^\phi$; S_T^ϕ is unsatisfiable.

Let S_T^ϕ is unsatisfiable. Then, for every interpretation $\mathfrak{A}'$ for $\mathcal{L} \cup \{\tilde{p}_j \mid j \in J_T^\phi\}$, $\mathfrak{A}' \not\models S_T^\phi$; by Theorem 3(i), there does not exist an interpretation $\mathfrak{A}$ for $\mathcal{L}$ and $\mathfrak{A} \models T$, $\mathfrak{A} \not\models \phi$; for every interpretation $\mathfrak{A}$ for $\mathcal{L}$, $\mathfrak{A} \not\models T$ or $\mathfrak{A} \models \phi$; $T \models \phi$; (i) holds.

(ii–iv) are the same as Theorem 3(ii–iv); (ii–iv) hold. The corollary is proved. □

4 Hyperresolution over Order Clauses

In this section, we propose an order hyperresolution calculus with truth constants operating over order clausal theories, and prove its refutational soundness, completeness.

4.1 Order Hyperresolution Rules

At first, we introduce some basic notions and notation concerning chains of order literals. A chain Ξ of $\mathcal{L}$ is a sequence $\Xi = \varepsilon_0 \diamond_0 v_0, \ldots, \varepsilon_n \diamond_n v_n$, $\varepsilon_i \diamond_i v_i \in OrdLit_{\mathcal{L}}$, such that for all $i < n$, $v_i = \varepsilon_{i+1}$. ε_0 is the beginning element of Ξ and v_n the ending element of Ξ. $\varepsilon_0 \Xi v_n$ denotes Ξ together with its respective beginning and ending element. Let $\Xi = \varepsilon_0 \diamond_0 v_0, \ldots, \varepsilon_n \diamond_n v_n$ be a chain of $\mathcal{L}$. Ξ is an equality chain of $\mathcal{L}$ iff, for all $i \leq n$, $\diamond_i ={=}$. Ξ is an increasing chain of $\mathcal{L}$ iff there exists $i^* \leq n$ such that $\diamond_{i^*} ={\prec}$. Ξ is a contradiction of $\mathcal{L}$ iff Ξ is an increasing chain of $\mathcal{L}$ of the form $\varepsilon_0 \Xi 0$ or $1 \Xi v_n$ or $\varepsilon_0 \Xi \varepsilon_0$. Let $S \subseteq OrdCl_{\mathcal{L}}$ be unit and $\Xi = \varepsilon_0 \diamond_0 v_0, \ldots, \varepsilon_n \diamond_n v_n$ be a chain | an equality chain | an increasing chain | a contradiction of $\mathcal{L}$. Ξ is a chain | an equality chain | an increasing chain | a contradiction of S iff, for all $i \leq n$, $\varepsilon_i \diamond_i v_i \in S$.

Let $\tilde{\mathbb{W}} = \{\tilde{w}_i \,|\, i \in \mathbb{I}\}$ such that $\tilde{\mathbb{W}} \cap (Func_{\mathcal{L}} \cup \{\tilde{f}_0\}) = \emptyset$; $\tilde{\mathbb{W}}$ is an infinite countable set of new function symbols. Let $\mathcal{L}$ contain a constant (nullary function) symbol. Let $P \subseteq \tilde{\mathbb{P}}$ and $S \subseteq OrdCl_{\mathcal{L} \cup P}$. We denote $GOrdCl_{\mathcal{L}} = \{C \,|\, C \in OrdCl_{\mathcal{L}} \text{ is closed}\} \subseteq OrdCl_{\mathcal{L}}$, $GInst_{\mathcal{L}}(S) = \{C \,|\, C \in GOrdCl_{\mathcal{L}} \text{ is an instance of } S \text{ of } \mathcal{L}\} \subseteq GOrdCl_{\mathcal{L}}$, $ordtcons(S) = \{0 \prec 1\} \cup \{0 \prec \bar{c} \,|\, \bar{c} \in tcons(S) \cap \overline{C}_{\mathcal{L}}\} \cup \{\bar{c} \prec 1 \,|\, \bar{c} \in tcons(S) \cap \overline{C}_{\mathcal{L}}\} \cup \{\bar{c}_1 \prec \bar{c}_2 \,|\, \bar{c}_1, \bar{c}_2 \in tcons(S) \cap \overline{C}_{\mathcal{L}}, c_1 < c_2\} \subseteq GOrdCl_{\mathcal{L}}$. A basic order hyperresolution calculus is defined as follows. The first rule is a central order hyperresolution one with obvious intuition.

(Basic order hyperresolution rule) (29)

$$\frac{l_0 \vee C_0, \ldots, l_n \vee C_n \in S_{\kappa-1}}{\bigvee_{i=0}^{n} C_i \in S_{\kappa}};$$

$l_0, \ldots, l_n$ is a contradiction of $\mathcal{L}_{\kappa-1}$.

We say that $\bigvee_{i=0}^{n} C_i$ is a basic order hyperresolvent of $l_0 \vee C_0, \ldots, l_n \vee C_n$. The second and third rules are auxiliary ones that order derived atoms in both the cases $qatoms(S) = \emptyset$ and $qatoms(S) \neq \emptyset$, which is exploited in the proof of the completeness of the calculus, Theorem 4.

(Basic order trichotomy rule) (30)

$$\frac{a, b \in atoms(S_{\kappa-1}), a \in \overline{C}_{\mathcal{L}}, b \notin Tcons_{\mathcal{L}}, qatoms(S) = \emptyset}{a \prec b \vee a = b \vee b \prec a \in S_{\kappa}}.$$

(*Basic order trichotomy rule*) (31)

$$\frac{a, b \in atoms(S_{\kappa-1}) - \{0,\ 1\}, \{a, b\} \nsubseteq Tcons_{\mathcal{L}}, qatoms(S) \neq \emptyset}{a \prec b \vee a \doteq b \vee b \prec a \in S_{\kappa}.}$$

$a \prec b \vee a \doteq b \vee b \prec a$ is a basic order trichotomy resolvent of a and b. The next two rules order a quantified atom and its ground instances.

(*Basic order $\forall$-quantification rule*) (32)

$$\frac{\forall x\, a \in qatoms^{\forall}(S_{\kappa-1})}{\forall x\, a \prec a\gamma \vee \forall x\, a \doteq a\gamma \in S_{\kappa}};$$
$t \in GTerm_{\mathcal{L}_{\kappa-1}}, \gamma = x/t \in Subst_{\mathcal{L}_{\kappa-1}}, dom(\gamma) = \{x\} = vars(a).$

$\forall x\, a \prec a\gamma \vee \forall x\, a \doteq a\gamma$ is a basic order $\forall$-quantification resolvent of $\forall x\, a$.

(*Basic order $\exists$-quantification rule*) (33)

$$\frac{\exists x\, a \in qatoms^{\exists}(S_{\kappa-1})}{a\gamma \prec \exists x\, a \vee a\gamma \doteq \exists x\, a \in S_{\kappa}};$$
$t \in GTerm_{\mathcal{L}_{\kappa-1}}, \gamma = x/t \in Subst_{\mathcal{L}_{\kappa-1}}, dom(\gamma) = \{x\} = vars(a).$

$a\gamma \prec \exists x\, a \vee a\gamma \doteq \exists x\, a$ is a basic order $\exists$-quantification resolvent of $\exists x\, a$. The last two rules introduce a witness with respect to infimum | supremum, as a ground term with a new function symbol, between a derived quantified atom and an atom | a quantified atom. They also ensure a total order over a derived quantified atom and atoms | quantified atoms together with Rules (32) and (33), which is exploited in the proof of the completeness.

(*Basic order $\forall$-witnessing rule*) (34)

$$\frac{\forall x\, a \in qatoms^{\forall}(S_{\kappa-1}), b \in atoms(S_{\kappa-1}) \cup qatoms(S_{\kappa-1})}{a\gamma \prec b \vee b \doteq \forall x\, a \vee b \prec \forall x\, a \in S_{\kappa}};$$
$\tilde{w} \in \tilde{\mathbb{W}} - Func_{\mathcal{L}_{\kappa-1}}, ar(\tilde{w}) = |freetermseq(\forall x\, a), freetermseq(b)|,$
$\gamma = x/\tilde{w}(freetermseq(\forall x\, a), freetermseq(b)) \in Subst_{\mathcal{L}_{\kappa}}, dom(\gamma) = \{x\} = vars(a).$

$a\gamma \prec b \vee b \doteq \forall x\, a \vee b \prec \forall x\, a$ is a basic order $\forall$-witnessing resolvent of $\forall x\, a$ and b.

(*Basic order $\exists$-witnessing rule*) (35)

$$\frac{\exists x\, a \in qatoms^{\exists}(S_{\kappa-1}), b \in atoms(S_{\kappa-1}) \cup qatoms(S_{\kappa-1})}{b \prec a\gamma \vee \exists x\, a \doteq b \vee \exists x\, a \prec b \in S_{\kappa}};$$
$\tilde{w} \in \tilde{\mathbb{W}} - Func_{\mathcal{L}_{\kappa-1}}, ar(\tilde{w}) = |freetermseq(\exists x\, a), freetermseq(b)|,$
$\gamma = x/\tilde{w}(freetermseq(\exists x\, a), freetermseq(b)) \in Subst_{\mathcal{L}_{\kappa}}, dom(\gamma) = \{x\} = vars(a).$

$b \prec a\gamma \vee \exists x\, a \doteq b \vee \exists x\, a \prec b$ is a basic order $\exists$-witnessing resolvent of $\exists x\, a$ and b.

The basic order hyperresolution calculus can be generalised to an order hyperresolution one. Intuition behind rules is similar to that in the basic case.

(Order hyperresolution rule) (36)

$$\frac{\bigvee_{j=0}^{k_0} \varepsilon_j^0 \diamond_j^0 v_j^0 \vee \bigvee_{j=1}^{m_0} l_j^0, \ldots, \bigvee_{j=0}^{k_n} \varepsilon_j^n \diamond_j^n v_j^n \vee \bigvee_{j=1}^{m_n} l_j^n \in S_{\kappa-1}^{Vr}}{\left(\bigvee_{i=0}^{n} \bigvee_{j=1}^{m_i} l_j^i\right)\theta \in S_\kappa};$$

for all $i < i' \leq n$,

$freevars(\bigvee_{j=0}^{k_i} \varepsilon_j^i \diamond_j^i v_j^i \vee \bigvee_{j=1}^{m_i} l_j^i) \cap freevars(\bigvee_{j=0}^{k_{i'}} \varepsilon_j^{i'} \diamond_j^{i'} v_j^{i'} \vee \bigvee_{j=1}^{m_{i'}} l_j^{i'}) = \emptyset,$

$\theta \in mgu_{\mathcal{L}_{\kappa-1}}\left(\bigvee_{j=0}^{k_0} \varepsilon_j^0 \diamond_j^0 v_j^0, l_1^0, \ldots, l_{m_0}^0, \ldots, \bigvee_{j=0}^{k_n} \varepsilon_j^n \diamond_j^n v_j^n, l_1^n, \ldots, l_{m_n}^n,\right.$

$\left. \{v_0^0, \varepsilon_0^1\}, \ldots, \{v_0^{n-1}, \varepsilon_0^n\}, \{a, b\}\right),$

$dom(\theta) = freevars(\{\varepsilon_j^i \diamond_j^i v_j^i \mid j \leq k_i, i \leq n\}, \{l_j^i \mid 1 \leq j \leq m_i, i \leq n\}),$

$a = \varepsilon_0^0, b = 1$ or $a = v_0^n, b = 0$ or $a = \varepsilon_0^0, b = v_0^n,$

there exists $i^* \leq n$ such that $\diamond_0^{i^*} =\prec$.

$\left(\bigvee_{i=0}^{n} \bigvee_{j=1}^{m_i} l_j^i\right)\theta$ is an order hyperresolvent of $\bigvee_{j=0}^{k_0} \varepsilon_j^0 \diamond_j^0 v_j^0 \vee \bigvee_{j=1}^{m_0} l_j^0, \ldots, \bigvee_{j=0}^{k_n} \varepsilon_j^n \diamond_j^n v_j^n \vee \bigvee_{j=1}^{m_n} l_j^n.$

(Order trichotomy rule) (37)

$$\frac{a, b \in atoms(S_{\kappa-1}), a \in \overline{C}_{\mathcal{L}}, b \notin Tcons_{\mathcal{L}}, qatoms(S) = \emptyset}{a \prec b \vee a = b \vee b \prec a \in S_\kappa}.$$

(Order trichotomy rule) (38)

$$\frac{a, b \in atoms(S_{\kappa-1}^{Vr}) - \{0,\ 1\}, \{a, b\} \not\subseteq Tcons_{\mathcal{L}}, qatoms(S) \neq \emptyset}{a \prec b \vee a = b \vee b \prec u \in S_\kappa};$$

$vars(a) \cap vars(b) = \emptyset.$

$a \prec b \vee a = b \vee b \prec a$ is an order trichotomy resolvent of a and b.

(Order $\forall$-quantification rule) (39)

$$\frac{\forall x\, a \in qatoms^\forall(S_{\kappa-1})}{\forall x\, a \prec a \vee \forall x\, a = a \in S_\kappa}.$$

$\forall x\, a \prec a \vee \forall x\, a = a$ is an order $\forall$-quantification resolvent of $\forall x\, a$.

(Order $\exists$-quantification rule) (40)

$$\frac{\exists x\, a \in qatoms^\exists(S_{\kappa-1})}{a \prec \exists x\, a \vee a = \exists x\, a \in S_\kappa}.$$

$a \prec \exists x\, a \vee a \doteq \exists x\, a$ is an order $\exists$-quantification resolvent of $\exists x\, a$.

(*Order $\forall$-witnessing rule*) (41)

$$\frac{\forall x\, a \in qatoms^{\forall}(S_{\kappa-1}^{Vr}), b \in atoms(S_{\kappa-1}^{Vr}) \cup qatoms(S_{\kappa-1}^{Vr})}{a\gamma \prec b \vee b \doteq \forall x\, a \vee b \prec \forall x\, a \in S_{\kappa}};$$

$freevars(\forall x\, a) \cap freevars(b) = \emptyset,$
$\tilde{w} \in \tilde{\mathbb{W}} - Func_{\mathcal{L}_{\kappa-1}}, ar(\tilde{w}) = |freetermseq(\forall x\, a), freetermseq(b)|,$
$\gamma = x/\tilde{w}(freetermseq(\forall x\, a), freetermseq(b)) \cup id|_{vars(a)-\{x\}} \in Subst_{\mathcal{L}_{\kappa}},$
$dom(\gamma) = \{x\} \cup (vars(a) - \{x\}) = vars(a).$

$a\gamma \prec b \vee b \doteq \forall x\, a \vee b \prec \forall x\, a$ is an order $\forall$-witnessing resolvent of $\forall x\, a$ and b.

(*Order $\exists$-witnessing rule*) (42)

$$\frac{\exists x\, a \in qatoms^{\exists}(S_{\kappa-1}^{Vr}), b \in atoms(S_{\kappa-1}^{Vr}) \cup qatoms(S_{\kappa-1}^{Vr})}{b \prec a\gamma \vee \exists x\, a \doteq b \vee \exists x\, a \prec b \in S_{\kappa}};$$

$freevars(\exists x\, a) \cap freevars(b) = \emptyset,$
$\tilde{w} \in \tilde{\mathbb{W}} - Func_{\mathcal{L}_{\kappa-1}}, ar(\tilde{w}) = |freetermseq(\exists x\, a), freetermseq(b)|,$
$\gamma = x/\tilde{w}(freetermseq(\exists x\, a), freetermseq(b)) \cup id|_{vars(a)-\{x\}} \in Subst_{\mathcal{L}_{\kappa}},$
$dom(\gamma) = \{x\} \cup (vars(a) - \{x\}) = vars(a).$

$b \prec a\gamma \vee \exists x\, a \doteq b \vee \exists x\, a \prec b$ is an order $\exists$-witnessing resolvent of $\exists x\, a$ and b.

Let $\mathcal{L}_0 = \mathcal{L} \cup P$, a reduct of $\mathcal{L} \cup \tilde{\mathbb{W}} \cup P$, and $S_0 = \emptyset \subseteq GOrdCl_{\mathcal{L}_0} \mid OrdCl_{\mathcal{L}_0}$. Let $\mathcal{D} = C_1, \ldots, C_n$, $C_{\kappa} \in GOrdCl_{\mathcal{L} \cup \tilde{\mathbb{W}} \cup P} \mid OrdCl_{\mathcal{L} \cup \tilde{\mathbb{W}} \cup P}$, $n \geq 1$. $\mathcal{D}$ is a deduction of C_n from S by basic order hyperresolution iff, for all $1 \leq \kappa \leq n$, $C_{\kappa} \in ordtcons(S) \cup GInst_{\mathcal{L}_{\kappa-1}}(S)$, or there exist $1 \leq j_k^* \leq \kappa - 1$, $k = 1, \ldots, m$, such that C_{κ} is a basic order resolvent of $C_{j_1^*}, \ldots, C_{j_m^*} \in S_{\kappa-1}$ using Rule (29)–(35) with respect to $\mathcal{L}_{\kappa-1}$ and $S_{\kappa-1}$; $\mathcal{D}$ is a deduction of C_n from S by order hyperresolution iff, for all $1 \leq \kappa \leq n$, $C_{\kappa} \in ordtcons(S) \cup S$, or there exist $1 \leq j_k^* \leq \kappa - 1$, $k = 1, \ldots, m$, such that C_{κ} is an order resolvent of $C'_{j_1^*}, \ldots, C'_{j_m^*} \in S_{\kappa-1}^{Vr}$ using Rule (36)–(42) with respect to $\mathcal{L}_{\kappa-1}$ and $S_{\kappa-1}$ where $C'_{j_k^*}$ is a variant of $C_{j_k^*} \in S_{\kappa-1}$ of $\mathcal{L}_{\kappa-1}$; $\mathcal{L}_{\kappa}$ and S_{κ} are defined by recursion on $1 \leq \kappa \leq n$ as follows:

$$\mathcal{L}_{\kappa} = \begin{cases} \mathcal{L}_{\kappa-1} \cup \{\tilde{w}\} & \text{in case of Rule (34), (35) } | \text{ (41), (42)}, \\ \mathcal{L}_{\kappa-1} & \text{else}, \end{cases} \quad \text{a reduct of } \mathcal{L} \cup \tilde{\mathbb{W}} \cup P;$$

$S_{\kappa} = S_{\kappa-1} \cup \{C_{\kappa}\} \subseteq GOrdCl_{\mathcal{L}_{\kappa}} \mid OrdCl_{\mathcal{L}_{\kappa}},$

$S_{\kappa}^{Vr} = Vrnt_{\mathcal{L}_{\kappa}}(S_{\kappa}) \subseteq OrdCl_{\mathcal{L}_{\kappa}}.$

$\mathcal{D}$ is a refutation of S iff $C_n = \square$. We denote

$$clo^{\mathcal{BH}}(S) = \{C \mid \text{there exists a deduction of } C \text{ from } S$$
$$\text{by basic order hyperresolution}\} \subseteq GOrdCl_{\mathcal{L} \cup \tilde{\mathbb{W}} \cup P},$$
$$clo^{\mathcal{H}}(S) = \{C \mid \text{there exists a deduction of } C \text{ from } S$$
$$\text{by order hyperresolution}\} \subseteq OrdCl_{\mathcal{L} \cup \tilde{\mathbb{W}} \cup P}.$$

4.2   Refutational Soundness and Completeness

We are in position to prove the refutational soundness and completeness of the order hyperresolution calculus. At first, we list some auxiliary lemmata.

Lemma 2 (Lifting Lemma). *Let $\mathcal{L}$ contain a constant symbol. Let $P \subseteq \tilde{\mathbb{P}}$ and $S \subseteq OrdCl_{\mathcal{L} \cup P}$. Let $C \in clo^{\mathcal{BH}}(S)$. There exists $C^* \in clo^{\mathcal{H}}(S)$ such that C is an instance of C^* of $\mathcal{L} \cup \tilde{\mathbb{W}} \cup P$.*

Proof. Technical, analogous to the standard one. □

Lemma 3 (Reduction Lemma). *Let $\mathcal{L}$ contain a constant symbol. Let $P \subseteq \tilde{\mathbb{P}}$ and $S \subseteq OrdCl_{\mathcal{L} \cup P}$. Let $\{\bigvee_{j=0}^{k_i} \varepsilon_j^i \diamond_j^i v_j^i \vee C_i \mid i \leq n\} \subseteq clo^{\mathcal{BH}}(S)$ such that for all $\mathcal{S} \in \mathcal{S}el(\{\{j \mid j \leq k_i\}_i \mid i \leq n\})$, there exists a contradiction of $\{\varepsilon_{\mathcal{S}(i)}^i \diamond_{\mathcal{S}(i)}^i v_{\mathcal{S}(i)}^i \mid i \leq n\} \subseteq GOrdCl_{\mathcal{L} \cup \tilde{\mathbb{W}} \cup P}$. There exists $\emptyset \neq I^* \subseteq \{i \mid i \leq n\}$ such that $\bigvee_{i \in I^*} C_i \in clo^{\mathcal{BH}}(S)$.*

Proof. Technical, analogous to the one of Proposition 2, [27]. □

Lemma 4 (Unit Lemma). *Let $\mathcal{L}$ contain a constant symbol. Let $P \subseteq \tilde{\mathbb{P}}$ and $S \subseteq OrdCl_{\mathcal{L} \cup P}$. Let $\square \notin clo^{\mathcal{BH}}(S) = \{\bigvee_{j=0}^{k_\iota} \varepsilon_j^\iota \diamond_j^\iota v_j^\iota \mid \iota < \gamma\}$, $\gamma \leq \omega$. There exists $\mathcal{S}^* \in \mathcal{S}el(\{\{j \mid j \leq k_\iota\}_\iota \mid \iota < \gamma\})$ such that there does not exist a contradiction of $\{\varepsilon_{\mathcal{S}^*(\iota)}^\iota \diamond_{\mathcal{S}^*(\iota)}^\iota v_{\mathcal{S}^*(\iota)}^\iota \mid \iota < \gamma\} \subseteq GOrdCl_{\mathcal{L} \cup \tilde{\mathbb{W}} \cup P}$.*

Proof. Technical, a straightforward consequence of König's Lemma and Lemma 3. □

Let $\{0, 1\} \subseteq X \subseteq [0, 1]$. X is admissible with respect to suprema and infima iff, for all $\emptyset \neq Y_1, Y_2 \subseteq X$ and $\bigvee Y_1 = \bigwedge Y_2$, $\bigvee Y_1 \in Y_1$, $\bigwedge Y_2 \in Y_2$. Let $\{0, 1\} \subseteq Tc \subseteq Tcons_{\mathcal{L}}$. Tc is admissible with respect to suprema and infima iff $\{0, 1\} \subseteq \overline{Tc} \subseteq [0, 1]$ is admissible with respect to suprema and infima.

Theorem 4 (Refutational Soundness and Completeness). *Let $\mathcal{L}$ contain a constant symbol. Let $P \subseteq \tilde{\mathbb{P}}$, $S \subseteq OrdCl_{\mathcal{L} \cup P}$, $tcons(S)$ be admissible with respect to suprema and infima. $\square \in clo^{\mathcal{H}}(S)$ if and only if S is unsatisfiable.*

Proof. ($\Longrightarrow$) Let $\mathfrak{A}$ be a model of S for $\mathcal{L} \cup P$ and $C \in clo^{\mathcal{H}}(S) \subseteq OrdCl_{\mathcal{L} \cup \tilde{\mathbb{W}} \cup P}$. Then there exists an expansion $\mathfrak{A}'$ of $\mathfrak{A}$ to $\mathcal{L} \cup \tilde{\mathbb{W}} \cup P$ such that $\mathfrak{A}' \models C$. The proof is by complete induction on the length of a deduction of C from S by order hyperresolution. Let $\square \in clo^{\mathcal{H}}(S)$ and $\mathfrak{A}$ be a model of S for $\mathcal{L} \cup P$. Hence, there exists an expansion $\mathfrak{A}'$ of $\mathfrak{A}$ to $\mathcal{L} \cup \tilde{\mathbb{W}} \cup P$ such that $\mathfrak{A}' \models \square$, which is a contradiction; S is unsatisfiable.

($\Longleftarrow$) Let $\square \notin clo^{\mathcal{H}}(S)$. Then, by Lemma 2 for S, $\square$, $\square \notin clo^{\mathcal{BH}}(S)$; we have $\mathcal{L}$, $\tilde{\mathbb{P}}$, $\tilde{\mathbb{W}}$ are countable, $P \subseteq \tilde{\mathbb{P}}$, $S \subseteq OrdCl_{\mathcal{L} \cup P}$, $clo^{\mathcal{BH}}(S) \subseteq GOrdCl_{\mathcal{L} \cup \tilde{\mathbb{W}} \cup P}$; P, $\mathcal{L} \cup P$, $OrdCl_{\mathcal{L} \cup P}$, S, $\mathcal{L} \cup \tilde{\mathbb{W}} \cup P$, $GOrdCl_{\mathcal{L} \cup \tilde{\mathbb{W}} \cup P}$, $clo^{\mathcal{BH}}(S)$ are countable; there exists $\gamma_1 \leq \omega$ and $\square \notin clo^{\mathcal{BH}}(S) = \{\bigvee_{j=0}^{k_\iota} \varepsilon_j^\iota \diamond_j^\iota v_j^\iota \mid \iota < \gamma_1\}$; by Lemma 4 for S, there exists $\mathcal{S}^* \in \mathcal{S}el(\{\{j \mid j \leq k_\iota\}_\iota \mid \iota < \gamma_1\})$ and there does not exist a contradiction of $\{\varepsilon^\iota_{\mathcal{S}^*(\iota)} \diamond^\iota_{\mathcal{S}^*(\iota)} \upsilon^\iota_{\mathcal{S}^*(\iota)} \mid \iota < \gamma_1\} \subseteq GOrdCl_{\mathcal{L} \cup \tilde{\mathbb{W}} \cup P}$. We put $\mathbb{S} = \{\varepsilon^\iota_{\mathcal{S}^*(\iota)} \diamond^\iota_{\mathcal{S}^*(\iota)} \upsilon^\iota_{\mathcal{S}^*(\iota)} \mid \iota < \gamma_1\} \subseteq GOrdCl_{\mathcal{L} \cup \tilde{\mathbb{W}} \cup P}$. Then $ordtcons(S) \subseteq clo^{\mathcal{BH}}(S)$, $\mathbb{S} \supseteq ordtcons(S)$ is countable, unit, $(q)atoms(\mathbb{S}) \subseteq (q)atoms(clo^{\mathcal{BH}}(S))$; there does not exist a contradiction of $\mathbb{S}$. We have $\mathcal{L}$ contains a constant symbol. Hence, there exists $cn^* \in Func_{\mathcal{L}}$, $ar_{\mathcal{L}}(cn^*) = 0$. We put $\tilde{\mathbb{W}}^* = funcs(\mathbb{S}) \cap \tilde{\mathbb{W}} \subseteq \tilde{\mathbb{W}}$, $\tilde{\mathbb{W}}^* \cap (Func_{\mathcal{L}} \cup \{\tilde{f}_0\}) \subseteq \tilde{\mathbb{W}} \cap (Func_{\mathcal{L}} \cup \{\tilde{f}_0\}) = \emptyset$,

$$\mathcal{U}_{\mathfrak{A}} = GTerm_{\mathcal{L} \cup \tilde{\mathbb{W}}^* \cup P}, cn^* \in \mathcal{U}_{\mathfrak{A}} \neq \emptyset,$$
$$\mathcal{B} = atoms(\mathbb{S}) \cup qatoms(\mathbb{S}) \subseteq GAtom_{\mathcal{L} \cup \tilde{\mathbb{W}}^* \cup P} \cup QAtom_{\mathcal{L} \cup \tilde{\mathbb{W}}^* \cup P}.$$

We have $\mathbb{S}$ is countable. Then $tcons(S) = atoms(ordtcons(S)) \subseteq atoms(\mathbb{S}) \subseteq \mathcal{B}$, $\mathcal{B} = tcons(S) \cup (\mathcal{B} - tcons(S))$, $tcons(S) \cap (\mathcal{B} - tcons(S)) = \emptyset$, $atoms(\mathbb{S})$, $qatoms(\mathbb{S})$, $\mathcal{B}$, $tcons(S)$, $\mathcal{B} - tcons(S)$ are countable; there exist $\gamma_2 \leq \omega$ and a sequence $\delta_2 : \gamma_2 \longrightarrow \mathcal{B} - tcons(S)$ of $\mathcal{B} - tcons(S)$. Let $\varepsilon_1, \varepsilon_2 \in \mathcal{B}$. $\varepsilon_1 \triangleq \varepsilon_2$ iff there exists an equality chain $\varepsilon_1 \equiv \varepsilon_2$ of $\mathbb{S}$. Note that $\triangleq$ is a binary symmetric transitive relation on $\mathcal{B}$. $\varepsilon_1 \triangleleft \varepsilon_2$ iff there exists an increasing chain $\varepsilon_1 \equiv \varepsilon_2$ of $\mathbb{S}$. Note that $\triangleleft$ is a binary transitive relation on $\mathcal{B}$.

$$0 \not\triangleq 1, 1 \not\triangleq 0, 0 \triangleleft 1, 1 \not\triangleleft 0, \text{ for all } \varepsilon \in \mathcal{B}, \varepsilon \not\triangleleft 0, 1 \not\triangleleft \varepsilon, \varepsilon \not\triangleleft \varepsilon. \tag{43}$$

The proof is straightforward; we have that there does not exist a contradiction of $\mathbb{S}$. Note that $\triangleleft$ is also irreflexive and a partial strict order on $\mathcal{B}$.

Let $tcons(S) \subseteq X \subseteq \mathcal{B}$. A partial valuation $\mathcal{V}$ is a mapping $\mathcal{V} : X \longrightarrow [0, 1]$ such that $\mathcal{V}(0) = 0$, $\mathcal{V}(1) = 1$, for all $\bar{c} \in tcons(S) \cap \overline{C}_{\mathcal{L}}$, $\mathcal{V}(\bar{c}) = c$. We denote $dom(\mathcal{V}) = X$, $tcons(S) \subseteq dom(\mathcal{V}) \subseteq \mathcal{B}$. We define a partial valuation $\mathcal{V}_\alpha$ by recursion on $\alpha \leq \gamma_2$ as follows:

$$\mathcal{V}_0 = \{(0,0), (1,1)\} \cup \{(\bar{c}, c) \mid \bar{c} \in tcons(S) \cap \overline{C}_{\mathcal{L}}\};$$
$$\mathcal{V}_\alpha = \mathcal{V}_{\alpha-1} \cup \{(\delta_2(\alpha-1), \lambda_{\alpha-1})\} \quad (1 \leq \alpha \leq \gamma_2 \text{ is a successor ordinal}),$$
$$\mathbb{E}_{\alpha-1} = \{\mathcal{V}_{\alpha-1}(a) \mid a \triangleq \delta_2(\alpha-1), a \in dom(\mathcal{V}_{\alpha-1})\},$$
$$\mathbb{D}_{\alpha-1} = \{\mathcal{V}_{\alpha-1}(a) \mid a \triangleleft \delta_2(\alpha-1), a \in dom(\mathcal{V}_{\alpha-1})\},$$
$$\mathbb{U}_{\alpha-1} = \{\mathcal{V}_{\alpha-1}(a) \mid \delta_2(\alpha-1) \triangleleft a, a \in dom(\mathcal{V}_{\alpha-1})\},$$
$$\lambda_{\alpha-1} = \begin{cases} \dfrac{\bigvee \mathbb{D}_{\alpha-1} + \bigwedge \mathbb{U}_{\alpha-1}}{2} & \text{if } \mathbb{E}_{\alpha-1} = \emptyset, \\ \bigvee \mathbb{E}_{\alpha-1} & \text{else}; \end{cases}$$
$$\mathcal{V}_{\gamma_2} = \bigcup_{\alpha < \gamma_2} \mathcal{V}_\alpha \; (\gamma_2 \text{ is a limit ordinal}).$$

For all $\alpha \leq \alpha' \leq \gamma_2$, $\mathcal{V}_\alpha$ is a partial valuation, $dom(\mathcal{V}_\alpha) = tcons(S) \cup \delta_2[\alpha]$, $\mathcal{V}_\alpha \subseteq \mathcal{V}_{\alpha'}$. $\tag{44}$

The proof is by induction on $\alpha \leq \gamma_2$.

We list some auxiliary statements without proofs:

$$\text{If } qatoms(S) = \emptyset, \text{ then } qatoms(clo^{\mathcal{BH}}(S)) = \emptyset. \tag{45}$$

$$tcons(S) = tcons(clo^{\mathcal{BH}}(S)). \tag{46}$$

For all $a, b \in atoms(clo^{\mathcal{BH}}(S)) \cup qatoms(clo^{\mathcal{BH}}(S))$, there exist a $\quad$ (47) deduction $C_1, \ldots, C_n$, $n \geq 1$, from S by basic order hyperresolution, associated $\mathcal{L}_n$, S_n, $S_n \subseteq GOrdCl_{\mathcal{L}_n}$, such that $a, b \in atoms(S_n) \cup qatoms(S_n)$.

For all $\emptyset \neq A \subseteq_{\mathcal{F}} atoms(clo^{\mathcal{BH}}(S)) \cup qatoms(clo^{\mathcal{BH}}(S))$, there exist $\quad$ (48) a deduction $C_1, \ldots, C_n$, $n \geq 1$, from S by basic order hyperresolution, associated $\mathcal{L}_n$, S_n, $S_n \subseteq GOrdCl_{\mathcal{L}_n}$, such that $A \subseteq atoms(S_n) \cup qatoms(S_n)$.

For all $a \in tcons(S) \cap \overline{C}_{\mathcal{L}}$, $b \in \mathcal{B} - tcons(S)$, either $a \lhd b$ or $a \triangleq b$ or $\quad$ (49) $b \lhd a$.

Let $qatoms(S) \neq \emptyset$. For all $a, b \in \mathcal{B} - \{0, 1\}$, either $a \lhd b$ or $(a = b$ or $\quad$ (50) $a \triangleq b)$ or $b \lhd a$.

$$\begin{aligned} &\text{For all } \alpha \leq \gamma_2, \text{ for all } a, b \in dom(\mathcal{V}_\alpha), \quad (51)\\ &\quad \text{if } a \triangleq b, \text{ then } \mathcal{V}_\alpha(a) = \mathcal{V}_\alpha(b);\\ &\quad \text{if } a \lhd b, \text{ then } \mathcal{V}_\alpha(a) < \mathcal{V}_\alpha(b);\\ &\quad \text{if } \mathcal{V}_\alpha(a) = 0, \text{ then } a = 0 \text{ or } a \triangleq 0;\\ &\quad \text{if } \mathcal{V}_\alpha(a) = 1, \text{ then } a = 1 \text{ or } a \triangleq 1;\\ &\text{for all } \alpha < \gamma_2,\\ &\quad \mathcal{V}_\alpha[dom(\mathcal{V}_\alpha)] \text{ is admissible with respect to suprema and infima.} \end{aligned}$$

The proof is by induction on $\alpha \leq \gamma_2$ using the assumption that $\overline{tcons(S)}$ is admissible with respect to suprema and infima.

We put $\mathcal{V} = \mathcal{V}_{\gamma_2}$, $dom(\mathcal{V}) \overset{(44)}{=} tcons(S) \cup \delta[\gamma_2] = tcons(S) \cup (\mathcal{B} - tcons(S)) = \mathcal{B}$. We further list some other auxiliary statements without proofs:

$$\begin{aligned} &\text{For all } a, b \in \mathcal{B}, \quad (52)\\ &\quad \text{if } a \triangleq b, \text{ then } \mathcal{V}(a) = \mathcal{V}(b);\\ &\quad \text{if } a \lhd b, \text{ then } \mathcal{V}(a) < \mathcal{V}(b). \end{aligned}$$

For all $Qx \, a \in qatoms(clo^{\mathcal{BH}}(S))$ and $u \in \mathcal{U}_{\mathfrak{A}}$, $\quad$ (53) $a(x/u) \in atoms(clo^{\mathcal{BH}}(S))$.

For all $a \in \mathcal{B}$, (54)
    if $a = \forall x b$, then $\mathcal{V}(a) = \bigwedge_{u \in \mathcal{U}_{\mathfrak{A}}} \mathcal{V}(b(x/u))$;
    if $a = \exists x b$, then $\mathcal{V}(a) = \bigvee_{u \in \mathcal{U}_{\mathfrak{A}}} \mathcal{V}(b(x/u))$.

We put

$$f^{\mathfrak{A}}(u_1, \ldots, u_\tau) = \begin{cases} f(u_1, \ldots, u_\tau) & \text{if } f \in Func_{\mathcal{L} \cup \tilde{\mathcal{W}}^* \cup P}, \\ cn^* & \text{else}, \end{cases} \quad f \in Func_{\mathcal{L} \cup \tilde{\mathcal{W}} \cup P}, u_i \in \mathcal{U}_{\mathfrak{A}};$$

$$p^{\mathfrak{A}}(u_1, \ldots, u_\tau) = \begin{cases} \mathcal{V}(p(u_1, \ldots, u_\tau)) & \text{if } p(u_1, \ldots, u_\tau) \in \mathcal{B}, \\ 0 & \text{else}, \end{cases} \quad p \in Pred_{\mathcal{L} \cup \tilde{\mathcal{W}} \cup P}, u_i \in \mathcal{U}_{\mathfrak{A}};$$

$$\mathfrak{A} = (\mathcal{U}_{\mathfrak{A}}, \{f^{\mathfrak{A}} \mid f \in Func_{\mathcal{L} \cup \tilde{\mathcal{W}} \cup P}\}, \{p^{\mathfrak{A}} \mid p \in Pred_{\mathcal{L} \cup \tilde{\mathcal{W}} \cup P}\}),$$

an interpretation for $\mathcal{L} \cup \tilde{\mathcal{W}} \cup P$.

For all $C \in S$ and $e \in \mathcal{S}_{\mathfrak{A}}$, $C(e|_{freevars(C)}) \in clo^{\mathcal{BH}}(S)$. (55)

It is straightforward to prove that for all $a \in \mathcal{B}$ and $e \in \mathcal{S}_{\mathfrak{A}}$, $\|a\|_e^{\mathfrak{A}} = \mathcal{V}(a)$. Let $l = \varepsilon_1 \doteq \varepsilon_2 \in \mathbb{S}$ and $e \in \mathcal{S}_{\mathfrak{A}}$. Then $\varepsilon_1, \varepsilon_2 \in \mathcal{B}$, $\varepsilon_1 \triangleq \varepsilon_2$, by (52) for $\varepsilon_1, \varepsilon_2$, $\mathcal{V}(\varepsilon_1) = \mathcal{V}(\varepsilon_2)$, $\|l\|_e^{\mathfrak{A}} = \|\varepsilon_1 \doteq \varepsilon_2\|_e^{\mathfrak{A}} = \|\varepsilon_1\|_e^{\mathfrak{A}} \doteq \|\varepsilon_2\|_e^{\mathfrak{A}} = \mathcal{V}(\varepsilon_1) \doteq \mathcal{V}(\varepsilon_2) = 1$. Let $l = \varepsilon_1 \prec \varepsilon_2 \in \mathbb{S}$ and $e \in \mathcal{S}_{\mathfrak{A}}$. Then $\varepsilon_1, \varepsilon_2 \in \mathcal{B}$, $\varepsilon_1 \triangleleft \varepsilon_2$, by (52) for $\varepsilon_1, \varepsilon_2$, $\mathcal{V}(\varepsilon_1) < \mathcal{V}(\varepsilon_2)$, $\|l\|_e^{\mathfrak{A}} = \|\varepsilon_1 \prec \varepsilon_2\|_e^{\mathfrak{A}} = \|\varepsilon_1\|_e^{\mathfrak{A}} \prec \|\varepsilon_2\|_e^{\mathfrak{A}} = \mathcal{V}(\varepsilon_1) \prec \mathcal{V}(\varepsilon_2) = 1$. So, for all $l \in \mathbb{S}$ and $e \in \mathcal{S}_{\mathfrak{A}}$, for both the cases $l = \varepsilon_1 \doteq \varepsilon_2 \in \mathbb{S}$ and $l = \varepsilon_1 \prec \varepsilon_2 \in \mathbb{S}$, $\|l\|_e^{\mathfrak{A}} = 1$; $\|l\|_e^{\mathfrak{A}} = 1$. Let $C \in S \subseteq OrdCl_{\mathcal{L} \cup P}$ and $e \in \mathcal{S}_{\mathfrak{A}}$. Then $e : Var_{\mathcal{L}} \longrightarrow \mathcal{U}_{\mathfrak{A}}$, $freevars(C) \subseteq_{\mathcal{F}} Var_{\mathcal{L}}$, $e|_{freevars(C)} \in Subst_{\mathcal{L} \cup \tilde{\mathcal{W}}^* \cup P}$, $dom(e|_{freevars(C)}) = freevars(C)$, $range(e|_{freevars(C)}) = \emptyset$; $e|_{freevars(C)}$ is applicable to C; by (55) for C, e, $C(e|_{freevars(C)}) \in clo^{\mathcal{BH}}(S)$, there exists $l^* \in C(e|_{freevars(C)})$ and $l^* \in \mathbb{S}$, $\|l^*\|_e^{\mathfrak{A}} = 1$; there exists $l^{**} \in C \in OrdCl_{\mathcal{L} \cup P}$ and $l^{**} \in OrdLit_{\mathcal{L} \cup P} \subseteq OrdLit_{\mathcal{L} \cup \tilde{\mathcal{W}}^* \cup P}$, $freevars(l^{**}) \subseteq freevars(C)$; $e|_{freevars(l^{**})}$ is applicable to l^{**}, $l^{**}(e|_{freevars(l^{**})}) = l^*$; for all $t \in Term_{\mathcal{L} \cup \tilde{\mathcal{W}}^* \cup P}$, $a \in Atom_{\mathcal{L} \cup \tilde{\mathcal{W}}^* \cup P} \cup QAtom_{\mathcal{L} \cup \tilde{\mathcal{W}} \cup P}$, $l \in OrdLit_{\mathcal{L} \cup \tilde{\mathcal{W}}^* \cup P}$, $\|t\|_e^{\mathfrak{A}} = t(e|_{vars(t)}) = \|t(e|_{vars(t)})\|_e^{\mathfrak{A}}$, $\|a\|_e^{\mathfrak{A}} = \|a(e|_{freevars(a)})\|_e^{\mathfrak{A}}$, $\|l\|_e^{\mathfrak{A}} = \|l(e|_{freevars(l)})\|_e^{\mathfrak{A}}$; the proof is by induction on t and by definition; $\|l^{**}\|_e^{\mathfrak{A}} = \|l^{**}(e|_{freevars(l^{**})})\|_e^{\mathfrak{A}} = \|l^*\|_e^{\mathfrak{A}} = 1$; $\mathfrak{A} \models_e C$; $\mathfrak{A} \models S$, $\mathfrak{A}|_{\mathcal{L} \cup P} \models S$; S is satisfiable. The theorem is proved. $\square$

Consider $S = \{0 \prec a\} \cup \{a \prec \frac{1}{n} \mid n \geq 2\} \subseteq OrdCl_{\mathcal{L}}$, $a \in Pred_{\mathcal{L}} - Tcons_{\mathcal{L}}$, $ar_{\mathcal{L}}(a) = 0$. $tcons(S)$ is not admissible with respect to suprema and infima; for $\{0\}$ and $\{\frac{1}{n} \mid n \geq 2\}$, $\bigvee\{0\} = \bigwedge\{\frac{1}{n} \mid n \geq 2\} = 0$, $0 \notin \{\frac{1}{n} \mid n \geq 2\}$. S is unsatisfiable; both the cases $\|a\|^{\mathfrak{A}} = 0$ and $\|a\|^{\mathfrak{A}} > 0$ lead to $\mathfrak{A} \not\models S$ for every interpretation $\mathfrak{A}$ for $\mathcal{L}$. However, $\square \notin clo^{\mathcal{H}}(S) = S \cup \{0 \prec 1\} \cup \{0 \prec \frac{1}{n} \mid n \geq 2\} \cup \{\frac{1}{n} \prec 1 \mid n \geq 2\} \cup \{\frac{1}{n_1} \prec \frac{1}{n_2} \mid n_1 > n_2 \geq 2\} \cup \{\frac{1}{n} \prec a \vee \frac{1}{n} \doteq a \vee a \prec \frac{1}{n} \mid n \geq 2\} \cup \{\frac{1}{n} \doteq a \vee a \prec \frac{1}{n} \mid n \geq 2\} \cup \{\frac{1}{n} \prec a \vee a \prec \frac{1}{n} \mid n \geq 2\}$, using Rules (37) and (36); $clo^{\mathcal{H}}(S)$ contains the order clauses from S, from $ordtcons(S)$, and some superclauses of them. So, the condition on $tcons(S)$ being admissible with respect to suprema and infima, is necessary.

The deduction problem of an order formula from an order theory can be solved as follows:

Corollary 2. *Let $\mathcal{L}$ contain a constant symbol. Let $n_0 \in \mathbb{N}$, $\phi \in OrdForm_{\mathcal{L}}$, $T \subseteq OrdForm_{\mathcal{L}}$, $tcons(T)$ be admissible with respect to suprema and infima. There exist $J_T^{\phi} \subseteq \{(i,j) \mid i \geq n_0\}$ and $S_T^{\phi} \subseteq SimOrdCl_{\mathcal{L} \cup \{\tilde{p}_j \mid j \in J_T^{\phi}\}}$ such that $tcons(S_T^{\phi})$ is admissible with respect to suprema and infima; $T \models \phi$ if and only if $\Box \in clo^{\mathcal{H}}(S_T^{\phi})$.*

Proof. By Corollary 1 for n_0, ϕ, T, there exist

$$J_T^{\phi} \subseteq \{(i,j) \mid i \geq n_0\}, S_T^{\phi} \subseteq SimOrdCl_{\mathcal{L} \cup \{\tilde{p}_j \mid j \in J_T^{\phi}\}}$$

and Corollary 1(i, iv) hold for ϕ, T, S_T^{ϕ}; we have $tcons(T)$ is admissible with respect to suprema and infima, $tcons(S_T^{\phi}) \subseteq tcons(\phi) \cup tcons(T)$; $tcons(\phi) \subseteq_{\mathcal{F}} Tcons_{\mathcal{L}}$, $tcons(S_T^{\phi})$ is admissible with respect to suprema and infima; we have $T \models \phi$ if and only if S_T^{ϕ} is unsatisfiable; by Theorem 4 for $\{\tilde{p}_j \mid j \in J_T^{\phi}\}$, S_T^{ϕ}, S_T^{ϕ} is unsatisfiable if and only if $\Box \in clo^{\mathcal{H}}(S_T^{\phi})$; $T \models \phi$ if and only if $\Box \in clo^{\mathcal{H}}(S_T^{\phi})$. The corollary is proved. □

Corollary 3. *The set of unsatisfiable order formulae of $\mathcal{L}$ is recursively enumerable.*

Proof. Without loss of generality, we may assume that $\mathcal{L}$ contains a constant symbol. Let $\phi \in OrdForm_{\mathcal{L}}$. Then ϕ contains a finite number of truth constants and $tcons(\{\phi\})$ is admissible with respect to suprema and infima. ϕ is unsatisfiable if and only if $\{\phi\} \models 0$. Hence, the problem that ϕ is unsatisfiable can be reduced to the deduction problem $\{\phi\} \models 0$ after a constant number of steps. Let $n_0 \in \mathbb{N}$. By Corollary 2 for n_0, 0, $\{\phi\}$, there exist $J_{\{\phi\}}^{0} \subseteq \{(i,j) \mid i \geq n_0\}$, $S_{\{\phi\}}^{0} \subseteq SimOrdCl_{\mathcal{L} \cup \{\tilde{p}_j \mid j \in J_{\{\phi\}}^{0}\}}$ and $tcons(S_{\{\phi\}}^{0})$ is admissible with respect to suprema and infima, $\{\phi\} \models 0$ if and only if $\Box \in clo^{\mathcal{H}}(S_{\{\phi\}}^{0})$; if $\{\phi\} \models 0$, then $\Box \in clo^{\mathcal{H}}(S_{\{\phi\}}^{0})$ and we can decide it after a finite number of steps. This straightforwardly implies that the set of unsatisfiable order formulae of $\mathcal{L}$ is recursively enumerable. The corollary is proved. □

5 An Example

In this section, we illustrate the solution to the deduction problem by an example. We show that $\phi = \forall x\, (q(x) \prec \overline{0.3}) \rightarrow \exists x\, q(x) \prec \overline{0.3} \vee \exists x\, q(x) = \overline{0.3} \in OrdForm_{\mathcal{L}}$ is logically valid using the proposed translation to clausal form and the order hyperresolution calculus.

$$\phi = \forall x \, (q(x) \prec \overline{0.3}) \to \exists x \, q(x) \prec \overline{0.3} \lor \exists x \, q(x) = \overline{0.3}$$

$$\left\{ \tilde{p}_0(x) \prec 1, \tilde{p}_0(x) \leftrightarrow \Big(\underbrace{\forall x \, (q(x) \prec \overline{0.3})}_{\tilde{p}_1(x)} \to \underbrace{\exists x \, q(x) \prec \overline{0.3} \lor \exists x \, q(x) = \overline{0.3}}_{\tilde{p}_2(x)} \Big) \right\} \tag{18}$$

$$\left\{ \tilde{p}_0(x) \prec 1, \tilde{p}_1(x) \prec \tilde{p}_2(x) \lor \tilde{p}_1(x) = \tilde{p}_2(x) \lor \tilde{p}_0(x) = \tilde{p}_2(x), \right.$$

$$\tilde{p}_2(x) \prec \tilde{p}_1(x) \lor \tilde{p}_0(x) = 1,$$

$$\left. \tilde{p}_1(x) \leftrightarrow \forall x \underbrace{(q(x) \prec \overline{0.3})}_{\tilde{p}_3(x)}, \tilde{p}_2(x) \leftrightarrow \underbrace{\exists x \, q(x) \prec \overline{0.3}}_{\tilde{p}_4(x)} \lor \underbrace{\exists x \, q(x) = \overline{0.3}}_{\tilde{p}_5(x)} \right\} \tag{27}, \tag{17}$$

$$\left\{ \tilde{p}_0(x) \prec 1, \tilde{p}_1(x) \prec \tilde{p}_2(x) \lor \tilde{p}_1(x) = \tilde{p}_2(x) \lor \tilde{p}_0(x) = \tilde{p}_2(x), \right.$$

$$\tilde{p}_2(x) \prec \tilde{p}_1(x) \lor \tilde{p}_0(x) = 1, \tilde{p}_1(x) = \forall x \, \tilde{p}_3(x), \tilde{p}_3(x) \leftrightarrow \underbrace{q(x)}_{\tilde{p}_6(x)} \prec \underbrace{\overline{0.3}}_{\tilde{p}_7(x)},$$

$$\tilde{p}_4(x) \prec \tilde{p}_5(x) \lor \tilde{p}_4(x) = \tilde{p}_5(x) \lor \tilde{p}_2(x) = \tilde{p}_4(x),$$

$$\tilde{p}_5(x) \prec \tilde{p}_4(x) \lor \tilde{p}_2(x) = \tilde{p}_5(x),$$

$$\left. \tilde{p}_4(x) \leftrightarrow \underbrace{\exists x \, q(x) \prec \overline{0.3}}_{\tilde{p}_8(x) \quad \tilde{p}_9(x)}, \tilde{p}_5(x) \leftrightarrow \underbrace{\exists x \, q(x) = \overline{0.3}}_{\tilde{p}_{10}(x) \quad \tilde{p}_{11}(x)} \right\} \tag{21}, \tag{20}$$

$$\left\{ \tilde{p}_0(x) \prec 1, \tilde{p}_1(x) \prec \tilde{p}_2(x) \lor \tilde{p}_1(x) = \tilde{p}_2(x) \lor \tilde{p}_0(x) = \tilde{p}_2(x), \right.$$

$$\tilde{p}_2(x) \prec \tilde{p}_1(x) \lor \tilde{p}_0(x) = 1, \tilde{p}_1(x) = \forall x \, \tilde{p}_3(x), \tilde{p}_6(x) \prec \tilde{p}_7(x) \lor \tilde{p}_3(x) = 0,$$

$$\tilde{p}_7(x) \prec \tilde{p}_6(x) \lor \tilde{p}_7(x) = \tilde{p}_6(x) \lor \tilde{p}_3(x) = 1, \tilde{p}_6(x) = q(x), \tilde{p}_7(x) = \overline{0.3},$$

$$\tilde{p}_4(x) \prec \tilde{p}_5(x) \lor \tilde{p}_4(x) = \tilde{p}_5(x) \lor \tilde{p}_2(x) = \tilde{p}_4(x),$$

$$\tilde{p}_5(x) \prec \tilde{p}_4(x) \lor \tilde{p}_2(x) = \tilde{p}_5(x),$$

$$\tilde{p}_8(x) \prec \tilde{p}_9(x) \lor \tilde{p}_4(x) = 0, \tilde{p}_9(x) \prec \tilde{p}_8(x) \lor \tilde{p}_9(x) = \tilde{p}_8(x) \lor \tilde{p}_4(x) = 1,$$

$$\tilde{p}_8(x) \leftrightarrow \exists x \underbrace{q(x)}_{\tilde{p}_{12}(x)}, \tilde{p}_9(x) = \overline{0.3},$$

$$\tilde{p}_{10}(x) = \tilde{p}_{11}(x) \lor \tilde{p}_5(x) = 0,$$

$$\tilde{p}_{10}(x) \prec \tilde{p}_{11}(x) \lor \tilde{p}_{11}(x) \prec \tilde{p}_{10}(x) \lor \tilde{p}_5(x) = 1,$$

$$\left. \tilde{p}_{10}(x) \leftrightarrow \exists x \underbrace{q(x)}_{\tilde{p}_{13}(x)}, \tilde{p}_{11}(x) = \overline{0.3} \right\} \tag{28}$$

$$S^\phi = \left\{ \boxed{\tilde{p}_0(x) \prec 1} \right. \tag{1}$$

$$\boxed{\tilde{p}_1(x) \prec \tilde{p}_2(x) \vee \tilde{p}_1(x) \doteq \tilde{p}_2(x)} \vee \tilde{p}_0(x) \doteq \tilde{p}_2(x) \tag{2}$$

$$\tilde{p}_2(x) \prec \tilde{p}_1(x) \vee \boxed{\tilde{p}_0(x) \doteq 1} \tag{3}$$

$$\boxed{\tilde{p}_1(x) \doteq \forall x\, \tilde{p}_3(x)} \tag{4}$$

$$\tilde{p}_6(x) \prec \tilde{p}_7(x) \vee \boxed{\tilde{p}_3(x) \doteq 0} \tag{5}$$

$$\tilde{p}_7(x) \prec \tilde{p}_6(x) \vee \tilde{p}_7(x) \doteq \tilde{p}_6(x) \vee \tilde{p}_3(x) \doteq 1 \tag{6}$$

$$\boxed{\tilde{p}_6(x) \doteq q(x)} \tag{7}$$

$$\boxed{\tilde{p}_7(x) \doteq \overline{0.3}} \tag{8}$$

$$\boxed{\tilde{p}_4(x) \prec \tilde{p}_5(x) \vee \tilde{p}_4(x) \doteq \tilde{p}_5(x)} \vee \boxed{\tilde{p}_2(x) \doteq \tilde{p}_4(x)} \tag{9}$$

$$\boxed{\tilde{p}_5(x) \prec \tilde{p}_4(x)} \vee \boxed{\tilde{p}_2(x) \doteq \tilde{p}_5(x)} \tag{10}$$

$$\boxed{\tilde{p}_8(x) \prec \tilde{p}_9(x)} \vee \tilde{p}_4(x) \doteq 0 \tag{11}$$

$$\tilde{p}_9(x) \prec \tilde{p}_8(x) \vee \tilde{p}_9(x) \doteq \tilde{p}_8(x) \vee \boxed{\tilde{p}_4(x) \doteq 1} \tag{12}$$

$$\boxed{\tilde{p}_8(x) \doteq \exists x\, \tilde{p}_{12}(x)} \tag{13}$$

$$\boxed{\tilde{p}_{12}(x) \doteq q(x)} \tag{14}$$

$$\boxed{\tilde{p}_9(x) \doteq \overline{0.3}} \tag{15}$$

$$\boxed{\tilde{p}_{10}(x) \doteq \tilde{p}_{11}(x)} \vee \tilde{p}_5(x) \doteq 0 \tag{16}$$

$$\boxed{\tilde{p}_{10}(x) \prec \tilde{p}_{11}(x) \vee \tilde{p}_{11}(x) \prec \tilde{p}_{10}(x)} \vee \boxed{\tilde{p}_5(x) \doteq 1} \tag{17}$$

$$\boxed{\tilde{p}_{10}(x) \doteq \exists x\, \tilde{p}_{13}(x)} \tag{18}$$

$$\boxed{\tilde{p}_{13}(x) \doteq q(x)} \tag{19}$$

$$\left. \boxed{\tilde{p}_{11}(x) \doteq \overline{0.3}} \right\} \tag{20}$$

Rule (36) : [1][3] :

$$\boxed{\tilde{p}_2(x) \prec \tilde{p}_1(x)} \tag{21}$$

repeatedly **Rule (36)** : [2][21] :

$$\boxed{\tilde{p}_0(x) \doteq \tilde{p}_2(x)} \tag{22}$$

repeatedly **Rule (36)** : [11][12] :

$$\boxed{\tilde{p}_4(x) \doteq 0} \vee \boxed{\tilde{p}_4(x) \doteq 1} \tag{23}$$

repeatedly **Rule (36)** : [16][17] :

$$\boxed{\tilde{p}_5(x) \doteq 0} \vee \boxed{\tilde{p}_5(x) \doteq 1} \tag{24}$$

repeatedly **Rule (36)** : [1][9][10][22][23][24] :

$$\boxed{\tilde{p}_4(x) \doteq 0} \tag{25}$$

repeatedly **Rule (36)** : [1][9][10][22][23][24] :

$$\boxed{\tilde{p}_5(x) = 0} \tag{26}$$

$0 \prec 1 \in ordtcons(S^\phi)$

$$\boxed{0 \prec 1} \tag{27}$$

Rule (36) : [12][25][27] :

$$\boxed{\tilde{p}_9(x) \prec \tilde{p}_8(x)} \vee \tilde{p}_9(x) = \tilde{p}_8(x) \tag{28}$$

Rule (36) : [17][26][27] :

$$\tilde{p}_{10}(x) \prec \tilde{p}_{11}(x) \vee \boxed{\tilde{p}_{11}(x) \prec \tilde{p}_{10}(x)} \tag{29}$$

Rule (39) : $\forall x \, \tilde{p}_3(x)$:

$$\boxed{\forall x \, \tilde{p}_3(x) \prec \tilde{p}_3(x) \vee \forall x \, \tilde{p}_3(x) = \tilde{p}_3(x)} \tag{30}$$

repeatedly **Rule (36)** : [4][5][21][30] :

$$\boxed{\tilde{p}_6(x) \prec \tilde{p}_7(x)} \tag{31}$$

Rule (42) : $\exists x \, \tilde{p}_{12}(x), \overline{0.3}$:

$$\boxed{\overline{0.3} \prec \tilde{p}_{12}(\tilde{w}_{(0,0)})} \vee \exists x \, \tilde{p}_{12}(x) \prec \overline{0.3} \vee \exists x \, \tilde{p}_{12}(x) = \overline{0.3} \tag{32}$$

Rule (42) : $\exists x \, \tilde{p}_{13}(x), \overline{0.3}$:

$$\boxed{\overline{0.3} \prec \tilde{p}_{13}(\tilde{w}_{(1,1)})} \vee \exists x \, \tilde{p}_{13}(x) \prec \overline{0.3} \vee \exists x \, \tilde{p}_{13}(x) = \overline{0.3} \tag{33}$$

Rule (42) : $\exists x \, \tilde{p}_{12}(x), \exists x \, \tilde{p}_{13}(x)$:

$$\boxed{\exists x \, \tilde{p}_{13}(x) \prec \tilde{p}_{12}(\tilde{w}_{(2,2)})} \vee \exists x \, \tilde{p}_{12}(x) \prec \exists x \, \tilde{p}_{13}(x) \vee \exists x \, \tilde{p}_{12}(x) = \exists x \, \tilde{p}_{13}(x) \tag{34}$$

Rule (36) : [7][8][14][31]; $x/\tilde{w}_{(0,0)}$: [32] :

$$\boxed{\exists x \, \tilde{p}_{12}(x) \prec \overline{0.3} \vee \exists x \, \tilde{p}_{12}(x) = \overline{0.3}} \tag{35}$$

Rule (36) : [7][8][19][31]; $x/\tilde{w}_{(1,1)}$: [33] :

$$\boxed{\exists x \, \tilde{p}_{13}(x) \prec \overline{0.3} \vee \exists x \, \tilde{p}_{13}(x) = \overline{0.3}} \tag{36}$$

Rule (40) : $\exists x \, \tilde{p}_{13}(x)$:

$$\boxed{\tilde{p}_{13}(x) \prec \exists x \, \tilde{p}_{13}(x) \vee \tilde{p}_{13}(x) = \exists x \, \tilde{p}_{13}(x)} \tag{37}$$

repeatedly **Rule (36)** : [14][19][37]; $x/\tilde{w}_{(2,2)}$: [34] :

$$\boxed{\exists x \, \tilde{p}_{12}(x) \prec \exists x \, \tilde{p}_{13}(x) \vee \exists x \, \tilde{p}_{12}(x) = \exists x \, \tilde{p}_{13}(x)} \tag{38}$$

repeatedly **Rule (36)** : [13][15][28][35] :

$$\boxed{\tilde{p}_9(x) = \tilde{p}_8(x)} \tag{39}$$

repeatedly **Rule (36)** : [18][20][29][36] :

$$\boxed{\tilde{p}_{10}(x) \prec \tilde{p}_{11}(x)} \tag{40}$$

repeatedly **Rule (36)** : [13][15][18][20][38][39][40] :

$$\square \tag{41}$$

6   Conclusions

In the paper, we have refined the hyperresolution calculus proposed in [11–15], which is suitable for automated deduction in the first-order Gödel logic with explicit partial truth. Gödel logic is expanded by a countable set of intermediate truth constants of the form $\bar{c}$, $c \in (0,1)$. We have modified translation of a formula to an equivalent satisfiable finite order clausal theory, consisting of order clauses. An order clause is a finite set of order literals of the form $\varepsilon_1 \diamond \varepsilon_2$ where ε_i is an atom or a quantified atom, and $\diamond$ is the connective $\approx$ or $\prec$. $\approx$ and $\prec$ are interpreted by the equality and standard strict linear order on $[0,1]$, respectively. We have investigated the so-called canonical standard completeness, where the semantics of Gödel logic is given by the standard G-algebra and truth constants are interpreted by 'themselves'. The refined hyperresolution calculus is refutation sound and complete for a countable order clausal theory if the set of truth constants occurring in the theory, is admissible with respect to suprema and infima. This condition covers the case of finite order clausal theories. We have solved the deduction problem of a formula from a countable theory and got an affirmative solution to the open problem of recursive enumerability of unsatisfiable formulae in Gödel logic with truth constants and the equality, $\approx$, strict order, $\prec$, projection, Δ, operators.

References

1. Pavelka, J.: On fuzzy logic I, II, III. Semantical completeness of some many-valued propositional calculi. Math. Logic Q. **25**, 45–52, 119–134, 447–464 (1979)
2. Novák, V., Perfilieva, I., Močkoř, J.: Mathematical Principles of Fuzzy Logic. The Springer International Series in Engineering and Computer Science. Springer, New York (1999)
3. Hájek, P.: Metamathematics of Fuzzy Logic. Trends in Logic. Springer, Dordrech (2001)
4. Esteva, F., Godo, L., Montagna, F.: The $L\Pi$ and $L\Pi\frac{1}{2}$ logics: two complete fuzzy systems joining Łukasiewicz and product logics. Arch. Math. Log. **40**, 39–67 (2001)
5. Savický, P., Cignoli, R., Esteva, F., Godo, L., Noguera, C.: On product logic with truth-constants. J. Log. Comput. **16**, 205–225 (2006)
6. Esteva, F., Godo, L., Noguera, C.: On completeness results for the expansions with truth-constants of some predicate fuzzy logics. In: Stepnicka, M., Novák, V., Bodenhofer, U. (eds.) Proceedings of the 5th EUSFLAT Conference New Dimensions in Fuzzy Logic and Related Technologies, Ostrava, Czech Republic, September 11–14 2007, vol. 2, pp. 21–26 (2007). Regular Sessions, Universitas Ostraviensis
7. Esteva, F., Gispert, J., Godo, L., Noguera, C.: Adding truth-constants to logics of continuous t-norms: axiomatization and completeness results. Fuzzy Sets Syst. **158**, 597–618 (2007)
8. Esteva, F., Godo, L., Noguera, C.: First-order t-norm based fuzzy logics with truth-constants: distinguished semantics and completeness properties. Ann. Pure Appl. Logic **161**, 185–202 (2009)
9. Esteva, F., Godo, L., Noguera, C.: Expanding the propositional logic of a t-norm with truth-constants: completeness results for rational semantics. Soft Comput. **14**, 273–284 (2010)

10. Esteva, F., Godo, L., Noguera, C.: On expansions of WNM t-norm based logics with truth-constants. Fuzzy Sets Syst. **161**, 347–368 (2010)
11. Guller, D.: An order hyperresolution calculus for Gödel logic - general first-order case. In: Rosa, A.C., Correia, A.D., Madani, K., Filipe, J., Kacprzyk, J. (eds.) IJCCI 2012 - Proceedings of the 4th International Joint Conference on Computational Intelligence, Barcelona, Spain, 5–7 October 2012, pp. 329–342. SciTePress (2012)
12. Guller, D.: A generalisation of the hyperresolution principle to first order Gödel logic. In: Madani, K., Correia, A.D., Rosa, A.C., Filipe, J. (eds.) Computational Intelligence - International Joint Conference, IJCCI 2012, Barcelona, Spain, 5–7 October 2012. Studies in Computational Intelligence, vol. 577, pp. 159–182. Springer (2015). Revised Selected Papers
13. Guller, D.: An order hyperresolution calculus for Gödel logic with truth constants. In: Dourado, A., Cadenas, J.M., Filipe, J. (eds.) FCTA 2014 - Proceedings of the International Conference on Fuzzy Computation Theory and Applications, part of IJCCI 2014, Rome, Italy, 22–24 October 2014, pp. 37–52. SciTePress (2014)
14. Guller, D.: Unsatisfiable formulae of Gödel logic with truth constants and $=$, $\prec$, Δ are recursively enumerable. In: Tan, Y., Shi, Y., Buarque, F., Gelbukh, A., Das, S., Engelbrecht, A. (eds.) ICSI 2015. LNCS, vol. 9142, pp. 242–250. Springer, Heidelberg (2015). doi:10.1007/978-3-319-20469-7_27
15. Guller, D.: An order hyperresolution calculus for Gödel logic with truth constants and equality, strict order, Delta. In: Rosa, A.C., Dourado, A., Correia, K.M., Filipe, J., Kacprzyk, J. (eds.) FCTA 2015 - Proceedings of the 7th International Joint Conference on Computational Intelligence, Lisbon, Portugal, 12–14 November 2015, pp. 31–46. SciTePress (2015)
16. Baaz, M., Fermüller, C.G.: A resolution mechanism for prenex Gödel logic. In: Dawar, A., Veith, H. (eds.) CSL 2010. LNCS, vol. 6247, pp. 67–79. Springer, Heidelberg (2010). doi:10.1007/978-3-642-15205-4_9
17. Baaz, M., Ciabattoni, A., Fermüller, C.G.: Theorem proving for prenex Gödel logic with Delta: checking validity and unsatisfiability. Logical Methods Comput. Sci. **8**(1:20), 1–23 (2012)
18. Mandelíková, L.: Analysis and Interpretation of Scientific Text. TnUni Press, Trenchin (2012)
19. Mandelíková, L.: Sociocultural Connections of the Language. TnUni Press, Trenchin (2014)
20. Guller, D.: A DPLL procedure for the propositional Gödel logic. In: Filipe, J., Kacprzyk, J. (eds.) ICFC-ICNC 2010 - Proceedings of the International Conference on Fuzzy Computation and International Conference on Neural Computation, Parts of the International Joint Conference on Computational Intelligence IJCCI 2010, Valencia, Spain, 24–26 October 2010, pp. 31–42. SciTePress (2010)
21. Plaisted, D.A., Greenbaum, S.: A structure-preserving clause form translation. J. Symb. Comput. **2**, 293–304 (1986)
22. de la Tour, T.B.: An optimality result for clause form translation. J. Symb. Comput. **14**, 283–302 (1992)
23. Hähnle, R.: Short conjunctive normal forms in finitely valued logics. J. Log. Comput. **4**, 905–927 (1994)
24. Nonnengart, A., Rock, G., Weidenbach, C.: On generating small clause normal forms. In: Kirchner, C., Kirchner, H. (eds.) CADE 1998. LNCS, vol. 1421, pp. 397–411. Springer, Heidelberg (1998). doi:10.1007/BFb0054274
25. Sheridan, D.: The optimality of a fast CNF conversion and its use with SAT. In: SAT (2004)

26. Apt, K.R.: Introduction to logic programming. Technical report CS-R8826, Centre for Mathematics and Computer Science, Amsterdam, The Netherlands (1988)
27. Guller, D.: On the refutational completeness of signed binary resolution and hyper-resolution. Fuzzy Sets Syst. **160**, 1162–1176 (2009). Featured Issue: Formal Methods for Fuzzy Mathematics. Approximation and Reasoning, Part II

Improvement of LCC Prediction Modeling Based on Correlated Parameters and Model Structure Uncertainty Propagation

Ahlem Ferchichi[✉], Wadii Boulila, and Imed Riadh Farah

Laboratoire RIADI, Ecole Nationale des Sciences de l'Informatique,
Manouba, Tunisia
ferchichi.ahlem@gmail.com, wadii.boulila@riadi.rnu.tn,
riadh.farah@ensi.rnu.tn

Abstract. Land cover change (LCC) mapping is one of the basic tasks for environmental monitoring and management. The most significant factors in determining the performance of model of LCC prediction are its structure and parameter optimization. However, these factors are generally marred by uncertainties which affect the reliability of decision about changes. The reduction of these uncertainties is deemed as essential elements for LCC prediction modeling. Propagation of uncertainty appears as good alternative for decreasing the uncertainty related to LCC prediction process and therefore obtain more relevant decision. On the other hand, correlation analysis between model parameters is often neglected. This affects the reliability of the model and makes it difficult to better determine the uncertainty related to model parameters. Several studies in literature depicts that evidence theory can be applied to propagate uncertainty associated to LCC prediction models and to solve multidimensional problems. This paper presents an effective optimization scheme for the LCC prediction modeling based on the uncertainty propagation of model parameters and model structure. Uncertainty propagation is analyzed by using evidence theory without and with considering correlations. In this study, change prediction of land cover in Saint-Denis City, Reunion Island of next 5 years (2016) was anticipated using multitemporal Spot-4 satellite images acquired at the dates 2006 and 2011. Results show good performances of the proposed approach in improving prediction of the LCC. Results also demonstrate that the proposed approach is an effective and efficient method due to its adequate degree of accuracy.

Keywords: LCC prediction · Parameter uncertainty · Correlation analysis · Model structure uncertainty · Uncertainty propagation · Evidence theory

1 Introduction

LCC is a topic that has recently received considerable attention in the prospective modeling domain. Predicting changes in forthcoming years may play a

© Springer International Publishing AG 2017
J.J. Merelo et al. (eds.), *Computational Intelligence*, Studies in Computational Intelligence 669,
DOI 10.1007/978-3-319-48506-5_14

significant role in planning and optimal use of resources and harnessing the non-normative changes in the future [7,8]. In literature, several models are proposed to predict LCC such as cellular automata models [1], markov chain model [5], logistic regression models [2], agent-based models [3], data mining models [6], and artificial neural networks [4]. In general, input parameters of each of these models and the prediction model structure are marred by uncertainties which affect the reliability of decision about these changes [7,9,10]. Uncertainty in model parameters is due to natural variability, measurement inaccuracy, and errors in handling and processing data [11,12]. Model structure encloses uncertainty which is due to model assumptions/approximations, hypotheses, and scale effects. Propagation of uncertainty helps improve the change prediction process and decrease the associated uncertainties. More recent applications to propagate uncertainty in remote sensing have been reported, e.g. [7,13–16]. For example, Boulila et al. [7] focused on propagating uncertainty related to input parameters of LCC prediction model without taking into account the uncertainty in model structure. Authors explore how those uncertainties propagate through LCC model responses using probabilistic method. Sexton et al. [13] proposed a model for the input parameters uncertainty propagation from continuous estimates of tree cover to categorical forest cover and change. Cockx et al. [16] developed an approach of quantification and reduction of the uncertainties to improve the reliability of urban growth models in land-use mapping and land-use change model parameter assessment. Others few works have investigated the issue of propagating uncertainty related to model structure. Bastola et al. [17] proposed a probabilistic approach to study the role of the uncertainty of model structure in climate change.

A major limitation of many existing approaches is that they take into account a single uncertainty source (model parameters or model structure) for land cover decision making processes.

Propagation of uncertainty based on a single uncertainty source is prone to statistical bias and underestimation of uncertainty. Ignoring one of uncertainty source could lead to over-confident inferences and decisions that are more risky than one thinks they are.

On the other hand, correlation analysis between system parameters is often neglected when modeling this system. Correlation of parameters often blurs the model uncertainty and makes it difficult to determine parameters uncertainty.

Several studies in literature depict that evidence theory can be applied to propagate uncertainty associated to LCC prediction models and to solve multidimensional problems. The aim of this paper is to propagate the uncertainty associated with both LCC prediction model parameters and model structure using evidence theory without and with considering correlations between parameters. The proposed approach is divided into four main steps: (1) parameter uncertainty identification step is used to identify uncertain input parameters, their types of uncertainty (aleatory and/or epistemic), their sources of uncertainty, their reduction factors, and their correlations, according to study area and used data, (2) parameter uncertainty propagation step is used to propagate input parameters uncertainty, (3) model structure uncertainty identification step is used to identify the LCC prediction model structure, their types of uncertainty, and their sources of uncertainty, and 4) model structure uncertainty propagation step is used to propagate the uncertainty associated to LCC prediction model structure.

This paper is outlined as follows. Section 2 introduces the theory of evidence. Section 3 presents materials and methods for the uncertainty propagation of LCC prediction process using evidence theory. Section 4 depicts experiment results. Finally, conclusions are outlined in Sect. 5.

2   Evidence Theory

Evidence theory, also called as Dempster-Shafer theory, was initially developed by [18] and formalized by [19]. The evidence theory has the potential to quantify aleatory and epistemic uncertainties. This theory is also used for propagating correlated and uncorrelated input parameters through LCC prediction models. In this section, the basic notations of the evidence theory are introduced.

Frame of Discernment (FD). The FD is defined by the finest possible subdivisions of the sets, and the finest possible subdivision is called the elementary proposition.

Basic Probability Assignment (BPA). Let Θ be a finite set of mutually exclusive and exhaustive hypotheses, and 2^{Θ} be the power set of Θ. The fundamental concept for representing imperfection is the BPA, which defines a mapping function (m) of 2^{Θ} to the interval between 0 and 1. The measure m, BPA function, must satisfy the following axioms:

$$m(A) \geq 0, \quad \forall A \subseteq \Theta. \tag{1}$$

$$m(\emptyset) = 0 \tag{2}$$

$$\sum m(A) = 1. \quad \forall A \subseteq \Theta. \tag{3}$$

Belief and Plausibility Functions. The measures of uncertainty provided by evidence theory are known as belief (Bel) and plausibility (Pl), which also lie in the interval $[0, 1]$. Given a body of evidence, the (Bel) and (Pl) can be derived from the BPA by

$$Bel(B) = \sum_{A \subseteq B} m(A). \tag{4}$$

$$Pl(B) = \sum_{B \cap A \neq \emptyset} m(A). \tag{5}$$

The formulas make it easy to see that the belief function, (Bel), is calculated by summing the BPAs that totally agree with the event B, while the plausibility function, (Pl), is calculated by summing BPAs that agree with the event B totally and partially. These two functions can be derived from each other. For example, the belief function can be derived from the plausibility function in the following way:

$$Bel(B) = 1 - Pl(\overline{B}) \tag{6}$$

The relationship between belief and plausibility functions is

$$Bel(B) \leq Pl(B) \tag{7}$$

which shows that as a measure of "event B is true", if $P(B)$ is the true value of the measure of set B is true, then $Pl(B)$ is the upper bound of $P(B)$, and $Bel(B)$ is the lower bound, so

$$Bel(A) \leq P(A) \leq Pl(A) \tag{8}$$

Dempster's Rule of Combining. The Dempster's rule of combination is an operation that plays a central role in the evidence theory. The BPAs induced by several sources are aggregated using this rule in order to yield a global BPA that synthesizes the knowledge of the different sources. Take two BPA structures, m_1 and m_2, for instance, the combined structure m_{12} is calculated in the following manner:

$$m_{12}(A) = \frac{\sum_{B \cap C = A} m_1(B) m_2(C)}{1 - K} \quad when \quad A \neq \emptyset \tag{9}$$

$$m_{12}(\emptyset) = 0, \quad when \quad K = \sum_{B \cap C \neq \emptyset} m_1(B) m_2(C) \tag{10}$$

The coefficient K represents the mass that the combination assigns to $\emptyset$ and reflects the conflict among the sources. The denominator in Dempster's rule, $1 - K$, is a normalization factor, which throws out the opinion of those experts who assert that the object under consideration does not exist.

3   Materials and Proposed Approach

3.1   Study Area and Data

The study area is Reunion Island. It is a French territory of 2500 km^2 located in the Indian Ocean, 200 km South-West of Mauritius and 700 km to the East of Madagascar (Fig. 1). Mean annual temperatures decrease from 24 °C in the lowlands to 12 °C at ca 2000 m. Mean annual precipitation ranges from 3 m on the eastern windward coast, up to 8 m in the mountains and down to 1 m along the south western coast. Vegetation is most clearly structured along gradients of altitude and rainfall [35]. Reunion Island has a strong growth in a limited area with an estimated population of 833,000 in 2010 that will probably be more than 1 million in 2030 [32]. These significant changes put pressure on agricultural and natural areas. The urban areas expanded by 189 % over the period from 1989 to 2002 [33] and available land became a rare and coveted resource. The landscapes are now expected to fulfil multiple functions i.e. urbanisation, agriculture production and ecosystem conservation. This causes conflicts among stakeholders about their planning and management [34]. Among cities on Reunion Island which is affected bu urban sprawl, Saint−Denis. It is the capital of Reunion Island, and the city with the most inhabitants on the island (Fig. 1). It hosts all the important administrative offices, and is also a cultural center with numerous museums. Saint-Denis is also the largest city in all of the French Overseas Departments.

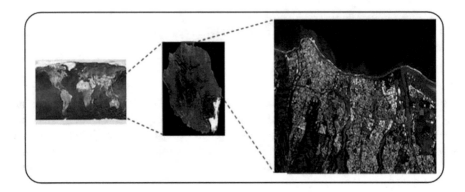

Fig. 1. Location of the study area.

Available remote sensing data for this research include classified images of land over of Saint−Denis from SPOT-4 images for the years 2006 and 2011 (Fig. 2). Selecting these images benefits from advantages such as a broad and integrated view, multispectral images and replicated coverage in different time periods. For this study, satellite data were classified after initial corrections and processing in order to prepare the data for extracting useful information. Spectral, geometric, and atmospheric corrections of images were conducted to make features manifest, increase the quality of images and to eliminate the adverse effects of light and atmosphere. Five types of land cover are determined in the study area such as water, urban, forest, bare soil, and vegetation.

3.2   Proposed Approach

Predicting future changes may play a significant role in planning and optimal use of resources and harnessing the non-normative changes in the future. As we mentioned, several models are proposed in order to predict LCC [1–6]. In this paper, we apply the LCC prediction model described by Boulila et al. in [6] to the Saint-Denis City, Reunion Island. This model exploits data mining concepts to build predictions and decisions for several remote sensing fields. It takes into account uncertainty related to the spatiotemporal mining process in order to provide more accurate and reliable information about LCC in satellite images. The prediction model proposed by Boulila et al. in [6] is divided into three main steps. It starts by a similarity measurement step to find similar states (in the object database) to a query state (representing the query object at a given date).

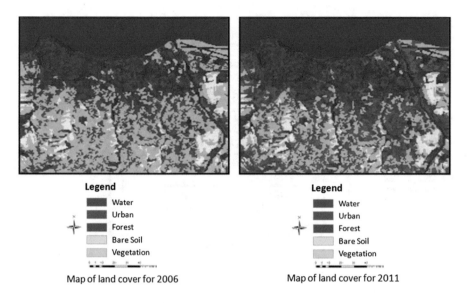

Fig. 2. Land cover maps.

The second step is composed by three substeps: (1) finding the corresponding model for the state, (2) finding all forthcoming states in the model (states having dates superior to the date of the retrieved state), and (3) for each forthcoming date, build the spatiotemporal change tree for the retrieved state. The third step is to construct the spatiotemporal change for the query state. Interested readers can refer to [6,20].

The uncertainty propagation through this model can be carried out by the following steps (Fig. 3): (1) identifying the uncertain parameters and their correlations; (2) propagating parameter uncertainty through the LCC prediction model using evidence theory without and with considering correlation between parameters; (3) identifying types and sources of uncertainty of model structure; (4) propagating the model structure uncertainty using evidence theory.

Parameters of LCC Prediction Model. Input parameters of LCC prediction model describe object features or descriptors extracted from satellite images and which are subject of studying changes. In this study, we extracted different features from SPOT-4 satellite images by using ten spectral, five texture, seven shape, one vegetation, and three climate descriptors. Spectral descriptors are: mean values and standard deviation values of green (MG, SDG), red (MR, SDR), NIR (MN, SDN), SWIR (MS, SDS) and monospectral (MM, SDM) bands for each image object. Texture descriptors are: homogeneity (Hom), contrast (Ctr), entropy (Ent), standard deviation (SD) and correlation (Cor) generated from GLCM (Gray Level Co-occurrence Matrix). Shape and spatial relationships descriptors are: area (A), length/width (LW), shape index (SI), roundness

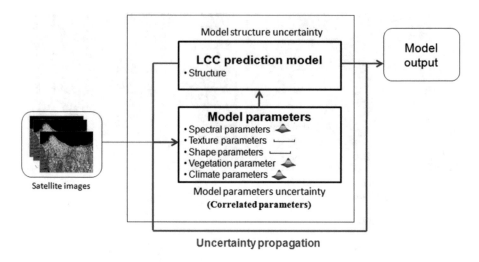

Fig. 3. General modelling proposed framework.

(R), density (D), metric relations (MR) and direction relations (DR). Vegetation descriptor is: NDVI (Normalized Difference Vegetation Index) that is the ratio of the difference between NIR and red reflectance. Finally, Climate descriptors are: temperature (Tem), humidity (Hum), and pressure (Pre). These descriptors were selected based on previous results as reported in [5]. Uncertainties related to these input parameters can be very numerous and affect model outputs. In general, these uncertainties can be of two types: aleatory and epistemic. The former comes from the natural variability of a random event, while the latter represents a lack of knowledge. Aleatory uncertainty is often referred to as irreducible uncertainty because a better understanding of the natural phenomena cannot reduce it. On the contrary, improving our background knowledge can reduce our epistemic uncertainty; therefore, we call it reducible uncertainty. The type of uncertainty of each parameter depends on sources of its uncertainty. Therefore its necessary to identify uncertainty sources that should be considered for processing. Table 1 shows the uncertain input parameters of LCC prediction model, their types, and their sources of uncertainty. In the current work, the spectral, NDVI, and climate parameters are modeled as an aleatory uncertain variables described with normal probability distributions. The texture and shape parameters are modeled as epistemic uncertain variables because their uncertainty originates due to the lack of knowledge in a physical model, and they are represented with intervals with specified bounds. In the literature, several studies have proposed some solutions (reduction factors) of uncertainty sources of the uncertain input parameters. Table 1 shows these reduction factors. According to these factors, we can improve the quality of input parameters before their use in propagating uncertainty step. We act on the distribution of each parameter to reduce their uncertainty sources. On the other hand, input parameters in remote sensing systems are not independent of each other. The value of one parameter can influence the value of another. LCC prediction models usually contain a large number of correlated parameters leading to non-identifiability problems. In this study, the Pearson correlation coefficient is used for analyses of linear correlation between input parameters of LCC prediction model.

Propagation of Parameters Uncertainty. The objective of this step is to propagate aleatory and epistemic input parameters through LCC prediction model without and with considering correlations. Different kinds of theories have been developed to propagate the uncertainty, including probability theory [6], fuzzy sets [21], possibility theory [22], p-box approach [23], fuzzy probabilities [24], and evidence theory [19], etc. Among the mentioned approaches, evidence theory seems to be more general than other modeling techniques. Under different cases, it can provide equivalent formulations to classical probability theory, possibility theory, p-box approach and fuzzy sets, respectively. The proposed approach takes advantages of the evidence theory to combine aleatory and epistemic uncetainty in a very natural way and to consider correlation between input parameters [25] for the LCC prediction process. In addition, evidence theory has been widely used in the remote sensing literature about 20 years ago. The main use of this theory has been in data fusion, image segmentation and classification, climate change [26–28]. In this section, the procedures of propagating the unified structures dealing with both aleatory and epistemic uncertainty and while considering correlation between parameters will be addressed. For the proposed model, first we should identify which type of uncertainty of each parameter. To illustrate the proposed method, we use a simple transfer function which has two uncertain parameters

$$Y = f(E, A). \tag{11}$$

where E represents the epistemic uncertainty parameter, A represents the aleatory uncertainty parameter and Y is the model response of the LCC. For E, the epistemic uncertainty is generally expressed by a series of subsets of the universal set associated with a BPA structure just as $\{[E_1^L, E_1^U]/m(1), [E_2^L, E_2^U]/m(2), ..., [E_k^L, E_k^U]/m(k), ...|k \in (1, 2, ..., M)\}$. Where M is the total number of subintervals of E and $m(k)$ represents the BPA value associated with the $k th$ subinterval $[E_k^L, E_k^U]$. When there are different BPA structures, we can use combining rule to integrate them into a combined BPA structure as $E_j/m(E_j)(j \in [1, 2, , M])$ ultimately, where E_j is also an interval as $[E_j^L, E_j^U]$ and $m(E_j)$ is the BPA value associated with the interval E_j. For A, assuming A is normal distribution $A \sim (\mu, \sigma)$, the distribution scope can be truncated to $[\mu - \xi\sigma, \mu + \xi\sigma]$ approximately and then we can discretize the approximate interval into N subintervals $[A_i^L, A_i^U]$, $i \in [1, 2, , N]$, and for each subinterval the basic probability value is defined as show in Eq. (12)

$$m(A_i) = \int_{A_i^L}^{A_i^U} f(x)dx, i \in [1, 2, ..., N]. \tag{12}$$

where A_i is defined as $\{A_i|x \in [A_i^L, A_i^U]\}$ and $f(x)$ is the probability density distribution function (pdf) of x. Obviously for the random parameter, the equivalent

Table 1. Types, sources, and reductions factors of uncertainty of input parameters of LCC prediction model.

Parameters	Uncertainty type	Uncertainty sources	Reduction factors of uncertainty
Spectral parameters	Aleatory	-Spectral reflectance of the surface -Sensor calibration and sensor noise -Effect of mixed pixels -Effect of a shift in the channel location -Pixel registration between several spectral channels -Atmospheric temperature and moisture profile -Effect of haze particles -Instrument's operation conditions -Atmospheric conditions and clouds -Surface conditions that change the target reflectance -Topography and viewing geometry	-Strict requirements for the instruments design -Envisaging of appropriate procedures for on-board calibration -Choosing appropriate algorithms for radiometric and atmospheric correction -Reducing the wavelength range of the irradiance or spectral response measurement -Reducing the cloud shadows and cloud contamination effects and reducing errors of sensor system itself
Texture parameters	Epistemic	-Spatial interaction between the size of the object in the scene and the spatial resolution of the sensor -Ambiguity in the object/background distinction	-Using high spatial resolution and choosing appropriate methods for segmentation
Shape parameters	Epistemic	-Accounting for the seasonal position of the sun with respect to the Earth -Conditions in which the image was acquired change in the scene's illumination -Atmospheric conditions -Observation geometry	-Improving platforms stability and carriers velocity -Technological enhancement of the sensors themselves -Reducing effects of atmospheric conditions -Improving the overall segmentation quality -Reducing the number of bad pixels and the size of bad areas -Improvement of the uncertainty of pixel response
Vegetation parameter	Aleatory	-Variation in the brightness of soil background -Red and NIR bands -Atmospheric perturbations -Variability in the structure of sub-pixel	-Choosing appropriate algorithms for atmospheric correction -Reducing errors in surface measurements for the NIR and red bands -Reducing temporal variations effects in the solar zenith and azimuth angles -Reducing sun angle effects and noise contamination
Climate parameters	Aleatory	-Atmospheric correction -Sensor noise -Land surface emissivity -Aerosols and other gaseous absorbers -Angular effects -Wavelength uncertainty -Full-width half maximum of the sensor and band-pass effects	-Choosing appropriate algorithms for atmospheric correction -Reducing errors of sensor system itself -Reducing emissivity variations -Reducing sun angle effects and solar heating -Reducing errors of radiometer calibration and the errors of radiation -Reducing errors of spatial and temporal variability of clouds

BPA values within specified intervals are equal to the area under the pdf. After obtaining the BPA structures of all the uncertain parameters, we can integrate them into a joint structure. The joint BPA structure is defined by the Cartesian product, which is synthesized as

$$C = A \times E = \{c_{ij} = A_i \times E_j\} \tag{13}$$

where C denotes the Cartesian set of all the uncertain parameters and c_{ij} is the element of C.

− When the uncertain parameters, E and A, are independent, the joint BPA for c_{ij} is defined by multiplying the BPA of A_i to the BPA of E_j.

$$m(c_{ij}) = m(A_i) \times m(E_j) \tag{14}$$

The focal element c_{ij} is included by the joint FD, and its BPA is just equal to the multiplication of the corresponding marginal BPAs.

− When the uncertain parameters, E and A, are correlated, we will develop a new evidence theory model which takes into account the correlation among parameters based on ellipsoidal model [10]. This model is originally proposed for non-probabilistic uncertainty analysis. Here the ellipsoidal model is extended to deal with the correlated evidence parameters. For this purpose, a multidimensional ellipsoid is constructed by making all possible realizations of the N-dimensional inter-correlated evidence parameters fall into a joint FD:

$$\Omega = \{\mathbf{X} | (\mathbf{X} - \mathbf{X}^c)^T \mathbf{G} (\mathbf{X} - \mathbf{X}^c) \leq 1\} \tag{15}$$

where the ellipsoidal center $\mathbf{X}^c$ is obtained through the marginal FDs:

$$X_m{}^c = \frac{X_m{}^L + X_m{}^R}{2}, m = 1, 2, ..., N \tag{16}$$

where $X_m \in c_{ij}$ are the evidence parameters (aleatory and epistemic parameters).

The symmetric positive-definite characteristic matrix $\mathbf{G}$ determines the size and orientation of the ellipsoid, reflecting the degree and the manner of correlation between the evidence parameters. Obviously, one should assign the belief probabilities only to the elements c_{ij} that are partially or totally falling into the ellipsoid model. Thus, a joint BPA is formulated as

$$m(c_{ij} \cap \Omega) = \frac{m(A_i) \times m(E_j)}{S}, c_{ij} \cap \Omega \neq 0 \tag{17}$$

where S is a normalization factor to make the total BPAs of m equal to 1.0, which is given by

$$S = \sum_{c_{ij} \cap \Omega \neq 0} m(c_{ij}) \tag{18}$$

Then, get the upper and lower CDFs of system response y via evidence reasoning. Let $\Theta_Y = \{d_{ij} : d_{ij} = f(c_{ij}), c_{ij} \subset \Theta_{\mathbf{X}}\}$ denotes the frame of discernment of Y, where d_{ij} is its focal element, f is the LCC prediction model in (11), and $\Theta_{\mathbf{X}}$ is the frame of discernment of $\mathbf{X}$. After determining the sets, c_{ij} and d_{ij}, the belief and plausibility functions are evaluated by checking all propositions of the joint BPA structure, as given in the following equations [14].

$$Bel_Y(d_{ij}) = Bel_{\mathbf{X}}[f^{-1}(d_{ij})] = \sum_{c_{ij} \subset f^{-1}(d_{ij})} m_{\mathbf{X}}(c_{ij}) \tag{19}$$

$$Pl_Y(d_{ij}) = Pl_{\mathbf{X}}[f^{-1}(d_{ij})] = \sum_{c_{ij} \cap f^{-1}(d_{ij}) \neq \emptyset} m_{\mathbf{X}}(c_{ij}) \tag{20}$$

Then

$$Bel_Y(y < v) = Bel_{\mathbf{X}}[f^{-1}(Y_v)] = \sum_{c_{ij} \subset f^{-1}(Y_v)} m_{\mathbf{X}}(c_{ij}) \tag{21}$$

$$Pl_Y(y < v) = Pl_{\mathbf{X}}[f^{-1}(Y_v)] = \sum_{c_{ij} \cap f^{-1}(Y_v) \neq \emptyset} m_{\mathbf{X}}(c_{ij}) \tag{22}$$

$$Y_v = \{y : y < v, y \in \Theta_Y\} \tag{23}$$

From (8),

$$Bel_Y(y < v) \leq P(y < v) \leq Pl_Y(y < v) \tag{24}$$

Obviously, Bel_Y is the lower CDF of the LCC prediction system response Y, and Pl_Y is the upper CDF.

- *Algorithm of the Ellipsoidal Model Construction:* Assuming that there are t experimental samples $X^{(r)}$, $r = 1, 2, ..., t$ for the N evidence parameters and each sample is an N-dimensional vector, the ellipsoidal model can be established as follows:
  1. Take a pair of evidence parameters X_m and X_n $(m \neq n)$ at a time from the uncertain parameter set.
  2. Extract the values of X_m and X_n from the t experimental samples and construct a corresponding bivariant sample set $(X_m^{(r)}, X_n^{(r)})$, $r = 1, 2, ..., t$.
  3. Create a minimum ellipse enveloping the obtained bivariant samples and obtain the corresponding rotation angle θ.
  4. Compute the covariance (Cov) and correlation coefficient (ρ) of the two uncertain parameters X_m and X_n based on the value of θ:
     - $Cov(X_m, X_n) = \frac{tan(\theta)}{1-tan^2(\theta)}((X_m^w)^2 - (X_n^w)^2)$ where $X_m^w = \frac{X_m{}^L + X_m{}^R}{2}$ and $X_n^w = \frac{X_n{}^L + X_n{}^R}{2}$ represent the radii of X_m and X_n, respectively.
     - $\rho_{X_m X_n} = \frac{Cov(X_m, X_n)}{X_m^w X_n^w}$, $-1 \leq \rho_{X_m X_n} \leq 1$.
  5. Repeat the above process for all pairs of uncertain parameters, and obtain a total of $N(N-1)/2$ covariances and correlation coefficients for all parameters.
  6. Create a covariance matrix C based on the calculated covariances.
  7. Finally, an ellipsoidal model can be obtained:

$$\Omega = \{\mathbf{X}|(\mathbf{X} - \mathbf{X}^c)^T \mathbf{C}^{-1}(\mathbf{X} - \mathbf{X}^c) \leq 1\} \tag{25}$$

Model Structure of LCC Prediction. Models of LCC prediction are simplifications of reality; they are theoretical abstractions that represent systems in such a way that essential features crucial to the theory and its application are identified and highlighted [29]. LCC models are tools to support the analysis of the causes and consequences of LCC for a better understanding of the system functionality, and to support land-use planning and policy [30]. Models are useful for simplifying the complex suite of socioeconomic and biophysical forces that influence the rate and spatial pattern of LCC and for estimating the impacts of changes [30]. Generally, each prediction model structure is represented by a number of hypotheses which are decisions or judgments considered by analysts. From a prediction model structure to another, the representations are different. Therefore it can keep uncertain representations. For example, when two hypotheses H_1, H_2 are given by two different experts, then we have two different structural models M_1 and M_2. The uncertainty related to LCC prediction model structure is often neglected in the process of uncertainty treatment although according to Droguett and Mosleh [31] their impact on results was important and sometimes even more important than the impact of input parameters uncertainties. These impacts should therefore be taken into account in the final decision process. In most cases, uncertainty about LCC prediction model structure is a form of epistemic uncertainty because we are unsure whether their constructions are reasonable and complete. It would be aleatory uncertainty only if the structure of the governing model were itself to change over time, across space, or among components in some population.

Propagation of Model Structure Uncertainty. The uncertainty propagation of LCC prediction model structure is implemented in combination with the propagation of the input parameters uncertainty. In this section, as input parameters uncertainty are modeled by evidence theory, we use this technique in this framework. Suppose that a set of alternative models M_k, $1 \leq k \leq K$ represents the uncertainty related to the choice of model. For each model M_k, input parameters uncertainty is propagated through this model. Consequently, the output indicator Y is characterized by a set of uncertainty representations according to each alternative model. Thus, for all alternative models M_k, $1 \leq k \leq K$, we have a set of pairs of belief and plausibility functions for output variable Y, noted $\{[Bel_1(Y), Pl_1(Y)], [Bel_2(Y), Pl_2(Y)], ..., [Bel_K(Y), Pl_K(Y)]\}$. The difference between these representations reflects the variation associated to LCC prediction model structure uncertainty. These different representations $[Bel_i(Y), Pl_i(Y)]$, $1 \leq i \leq K$ can be combined into a single representation. Therefore, the final uncertainty representation of output variable Y can be obtained by the following formulas.

$$Bel^*(Y) = min(Bel_1(Y), Bel_2(Y), ..., Bel_K(Y)) \qquad (26)$$

$$Pl^*(Y) = max(Pl_1(Y), Pl_2(Y), ..., Pl_K(Y)) \qquad (27)$$

The belief and plausibility functions $[Bel^*(Y), Pl^*(Y)]$ take into account both model parameters and structure uncertainty of LCC prediction in the final output result.

4   Experiment Results

The aim of this section is to validate and to evaluate the performance of the proposed approach in propagating uncertainty related to input parameters of LCC prediction model and uncertainty related to model structure. This section is divided into two parts: validation of the uncertainty propagation of LCC prediction model and validation of LCC prediction maps.

4.1   Validation of the Uncertainty Propagation of LCC Prediction Model

The validation section is divided into two main steps: (1) Propagating uncertainty of input parameters, and (2) Propagating uncertainty of input parameters and model structure.

Propagating Uncertainty of Input Parameters. As mentioned already, input parameters of LCC prediction model (M_1) are marred by aleatory and epistemic uncertainty. Ignoring each of these types can affect the results of uncertainty propagation. To illustrate the importance of propagating both uncertainty types in LCC prediction model, the analysis with pure aleatory uncertainty assumption was conducted where all 26 uncertain input parameters were treated as aleatory with normal probability distributions. In this case, the cumulative distribution function (CDF) of output representing only the uncertainty in input parameters is obtained via evidence theory. Figure 4 shows this distribution based on 10,000 samples. The probability distribution of probability results obtained by Monte Carlo method with 10,000 samples is also plotted in Fig. 4. Now, for both aleatory and epistemic uncertainty propagation of only input parameters in LCC prediction model (M_1), the specified bounds were utilized for the epistemic uncertain input parameters and the normal probability distributions were utilized for aleatory uncertain input parameters. The 10,000 samples were selected for both aleatory and epistemic cases. The CDF produced is shown in Fig. 5.

As we have indicated also, input parameters of LCC prediction model (M_1) are highly correlated. Then, it is necessary to study the effect of these correlated input parameters on output response variation and uncertainty propagation results. According to Pearson correlation coefficient, for example we found statistically significant correlations between the shape parameters (Fig. 6). Summarily, the shape index was significantly correlated with the Area, Density, and length/width. Moreover, we found statistically significant correlations between the NDVI parameter and climate parameters (Fig. 7). The NDVI parameter had a highest correlation with the temperature and pressure. Another result is that

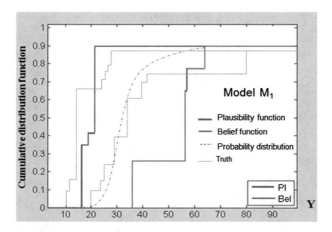

Fig. 4. Belief and plausibility functions of LCC prediction model output: case of aleatory uncertainty.

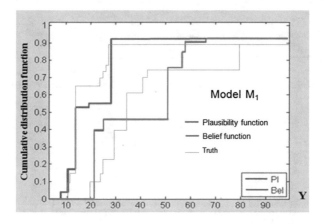

Fig. 5. Belief and plausibility functions of LCC prediction model output: case of aleatory and epistemic uncertainty.

the spectral parameters were correlated with texture parameters. To compare the influence of input parameters, Fig. 8 shows both the distribution of the LCC when the correlation (dependence) between input parameters is considered and the cumulative distribution of LCC when correlation between parameters is not taken into consideration in LCC prediction model (M_1). Note that capturing the true relationship among the input parameters can be crucially important to the accurate computation of the uncertainty in model predictions.

Propagation of Parameters and Model Structure Uncertainty. In order to illustrate the importance of modeling and taking into account uncertainty in the model structure, we used the LCC prediction model described by

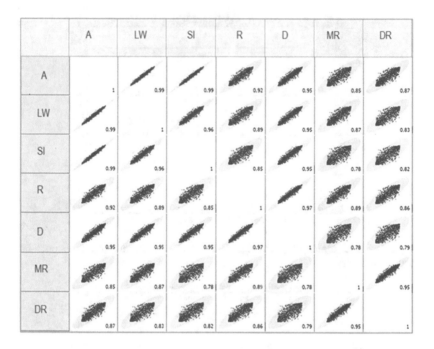

Fig. 6. Scatter plots of shape parameters of satellite image objects.

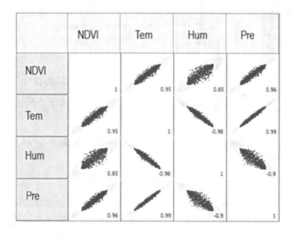

Fig. 7. Scatter plots of NDVI and climate parameters of satellite image objects.

Boulila et *al.* in [6] with three different hypotheses. Then, we have three different models such as M_1/H_1, M_2/H_2, and M_3/H_3. Thus, to take into account the model structure uncertainty in the final result, for each prediction model, the uncertainty associated with the input parameters are first propagated. Figure 5 shows the result of the belief and plausibility functions of the LCC prediction

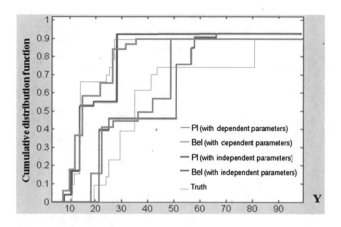

Fig. 8. Comparison between Belief and plausibility functions of LCC prediction model output for (in) dependence analysis.

model where only input parameters uncertainty is propagated. After the input parameters uncertainty propagation of three prediction models such as M_1/H_1, M_2/H_2, and M_3/H_3, we have three uncertainty representations of LCCs, which are shown in Fig. 9. The difference between these three representations presents the impact of uncertainty of LCC prediction model structure. Compared with the result of LCC prediction model (M_1), we can see that, this difference is important. Figure 10 shows the belief and plausibility functions representing the integrated input parameters and model structure uncertainty about the LCCs. Note that the combined effect of model structure and input parameters uncertainty lead to a wider uncertainty bound of the LCC when compared against the input parameters uncertainty case (Fig. 5). Also, the most probable value for the LCC when only input parameters uncertainty is taken into account is considerably lower than in the combined case, indicating an underestimation of the LCC.

4.2   Validation of LCC Prediction Maps

The validation of LCC prediction maps consisted on two phases. First, the 2011 LCC was simulated using the 2006 datasets, which was then compared with the actual LCC in 2011 to evaluate the accuracy and the performance of the proposed approach. Second, forthcoming changes are simulated using the actual 2011 datasets. Figure 11 compares actual and simulated percentages occupied by the different land cover types (water, urban, forest, bare soil, and vegetation) between 2006 and 2011; it shows that the simulated changes generally matched that of the actual changes. These results confirm that the LCC prediction model were reasonable to describe the LCC and the proposed approach can simulate the prediction of LCC with an acceptable accuracy. After the validation of the proposed approach, the next step is to simulate the LCC in 2016, assuming

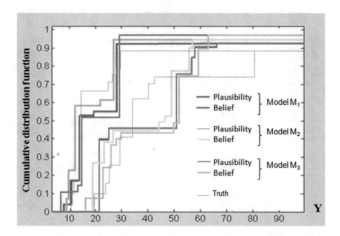

Fig. 9. Belief and plausibility functions of LCCs for three different prediction models.

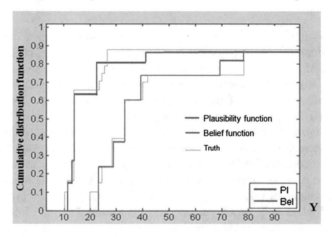

Fig. 10. Belief and plausibility functions of the combined input parameters and model structure uncertainty for LCCs.

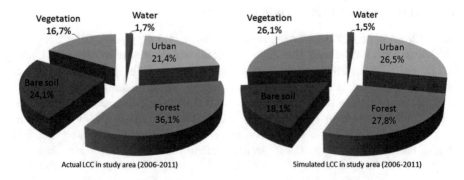

Fig. 11. Categorical distribution of the actual and simulated LCC between 2006 and 2011.

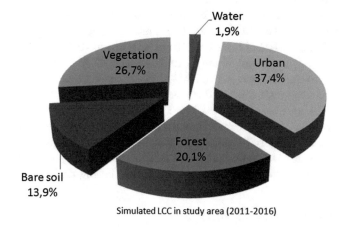

Fig. 12. Categorical distribution of the simulated LCC between 2011 and 2016.

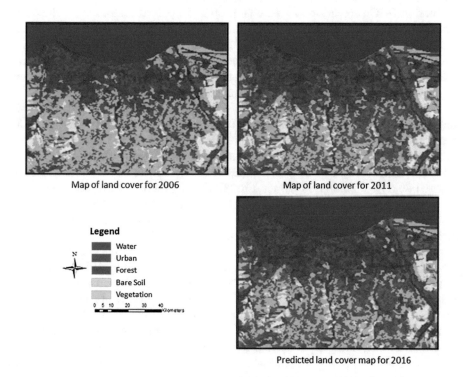

Fig. 13. Comparison between the land cover maps for years 2006 and 2011 and the predicted land cover map for 2016.

the changes between 2006 and 2011 will continue during the next time interval. In this simulation, the LCC and input parameters acquired in 2011 were used as input to simulate the LCC in 2016. Figure 12 shows the simulated changes between 2011 and 2016. There have been significant LCC where urban land covered 26,5 % of simulated changes in 2011 and 37,4 % in 2016. This could be attributed to an increase in population by increased demands for residential land. The resulting effect was the decrease in forest land from 27,8 % of simulated changes in 2011 and to 20,1 % in 2016. From these results, it can be found the replacing the land natural cover (forest land) in the study area by residential land (urban land). Figure 13 maps the simulated future changes compared with land cover maps for the years 2006 and 2011. These results indicate usefulness and applicability of the proposed approach in predicting the LCC.

5   Conclusion and Discussion

Uncertainty propagation through LCC prediction model with both correlated parameters and model structure uncertainty is presented in this paper. This paper proposes the use of an evidence theory as the uncertainty propagation method in LCC prediction model. The proposed approach is based on the identification of uncertainty sources of satellite image object parameters of LCC prediction model. Then, the uncertain input parameters are first propagated with considering mixed aleatory-epistemic uncertainty and correlation between parameters through the LCC prediction model. After that, model structure uncertainty of LCC prediction are also propagated with evidence theory.

The results show the importance to propagate both aleatory and epistemic uncertainty and to consider correlation between input parameters through LCC prediction model. Proposed approach studied changes prediction of land cover in Saint-Denis City, Reunion Island of next 5 years (2016) using multi-temporal Spot-4 satellite images in 2006 and 2011. Results indicated that the urban land covered 26,5 % of simulated changes in 2011 and 37,4 % in 2016 and the forest land covered 27,8 % of simulated changes in 2011 and 20,1 % in 2016. From these results, it can be found the replacing the land natural cover (forest land) in the study area by residential land (urban land).

Results show that the LCC prediction model using 26 input parameters representing both spectral, texture, shape, vegetation and climate characteristics, could simulate the LCC changes with satisfactory degrees of accuracy. This study has also demonstrated the usefulness of propagating uncertainty of input parameters and model structure of LCC prediction in providing land cover maps and change information, which are very valuable for planning and management. Additional work, however, is needed to understand how to reduce the computational cost in the proposed approach. LCC prediction model with a large number of uncertain input parameters is more complex. To optimize, it is important to study the sensitivity of input parameters and also the sensitivity of model structure.

References

1. Ballestores, J.F., Qiu, Z., Nedorezova, B.N., Nedorezov, L.V., Ferrarini, A., Ramathilaga, A., Ackah, M.: An integrated parcel-based land use change model using cellular automata and decision tree. Proc. Int. Acad. Ecol. Environ. Sci. **2**(2), 53–69 (2012)
2. Tayyebi, A., Perry, P.C., Tayyebi, A.H.: Predicting the expansion of an urban boundary using spatial logistic regression and hybrid raster-vector routines with remote sensing and GIS. Int. J. Geogr. Inf. Sci. **28**(4), 1–21 (2013)
3. Ralha, C.G., Abreu, C.G., Coelho, C.G., Zaghetto, A., Macchiavello, B., Machado, R.B.: A multi-agent model system for land-use change simulation. Environ. Model. Softw. **42**, 30–46 (2013)
4. Li, X.: Neural-network-based cellular automata for simulating multiple land use changes using GIS. Int. J. Geogr. Inf. Sci. **16**(4), 323–343 (2002)
5. Razavi, B.S.: Predicting the trend of land use changes using artificial neural network and Markov chain model (Case Study: Kermanshah City). Res. J. Environ. Earth Sci. **6**(4), 215–226 (2014)
6. Boulila, W., Farah, I.R., Ettabaa, K.S., Solaiman, B., Ghezala, H.B.: A data mining based approach to predict Spatio-temporal changes in satellite images. Int. J. Appl. Earth **13**(3), 386–395 (2011)
7. Boulila, W., Bouatay, A., Farah, I.R.: A probabilistic collocation method for the imperfection propagation: application to land cover change prediction. J. Multimedia Process. Technol. **5**(1), 12–32 (2014)
8. Hubert-Moy, L., Corgne, S., Mercier, G., Solaiman, B.: Land use and land cover change prediction with the theory of evidence: a case study in an intensive agricultural region of France. In: Proceedings of the Fifth International Conference on Information Fusion, vol. 1, pp. 114–121 (2002)
9. Ferchichi, A., Boulila, W., Farah, I.R.: Parameter and structural model imperfection propagation using evidence theory in land cover change prediction. In: International Conference Image Processing, Applications and Systems, pp. 1–6 (2014)
10. Liu, R., Sun, J., Wang, J., Li, X.: Study of remote sensing based parameter uncertainty in production Efficiency Models. In: IEEE International Geoscience and Remote Sensing Symposium, pp. 3303–3306 (2010)
11. Peters, R.A.: A new algorithm for image noise reduction using mathematical morphology. IEEE Trans. Image Process. **4**(5), 554–568 (1995)
12. Hulley, G.C., Hughes, C.G.: Quantifying uncertainties in land surface temperature and emissivity retrievals from ASTER and MODIS thermal infrared data. J. Geophys. Res. **117**(D23), 1–18 (2012)
13. Sexton, J.O., Noojipady, P., Anand, A., Song, X.P., McMahon, S., Huang, C., Feng, M., Channan, S., Townshend, J.R.: A model for the propagation of uncertainty from continuous estimates of tree cover to categorical forest cover and change. Remote Sens. Environ. **156**, 418–425 (2015)
14. Shi, W.Z., Ehlers, M.: Determining uncertainties and their propagation in dynamic change detection based on classified remotely-sensed images. Int. J. Remote Sens. **17**(14), 2729–2741 (1996)
15. Crosettoa, M., Ruiz, J.A.M., Crippac, B.: Uncertainty propagation in models driven by remotely sensed data. Remote Sens. Environ. **76**(3), 373–385 (2001)
16. Cockx, K., Van de Voorde, T., Canters, F.: Quantifying uncertainty in remote sensing-based urban land-use mapping. Int. J. Appl. Earth Obs. Geoinf. **31**, 154–166 (2014)

17. Bastola, S., Murphy, C., Sweeney, J.: The role of hydrological modeling uncertainties in climate change impact assessments of Irish river catchments. Adv. Water Resour. **34**(5), 562–576 (2011)
18. Klir, G.J.: On the alleged superiority of probability representation of uncertainty. IEEE Trans. Fuzzy Syst. **2**(1), 27–31 (1994)
19. Shafer, G.A.: Mathematical Theory of Evidence. Princeton University Press, Princeton (1976)
20. Boulila, W., Farah, I.R., Ettabaa, K.S., Solaiman, B.: Combining decision fusion and uncertainty propagation to improve land cover change prediction in satellite image databases. J. Multimedia Process. Technol. **2**(3), 127–139 (2011)
21. Zadeh, L.A.: Fuzzy sets as a basis for a theory of possibility. Fuzzy Sets and Syst. **100**, 9–34 (1999)
22. Mourelatos, Z.P., Zhou, J.: Reliability estimation and design with insufficient data based on possibility theory. AIAA J. **43**(8), 1696–1705 (2005)
23. Zhang, H., Mullen, R.L., Muhanna, R.L.: Safety structural analysis with probability boxes. Int. J. Reliab. Saf. **6**, 110–129 (2012)
24. Beer, M.: Fuzzy probability theory. In: Meyers, R.A. (ed.) Encyclopedia of Complexity and Systems Science, vol. 6, pp. 4047–4059. Springer, New York (2009)
25. Gao, Z.Q., Dennis, O.: The temporal and spatial relationship between NDVI and climatological parameters in Colorado. J. Geog. Sci. **11**(4), 411–419 (2001)
26. Abdallah, N.B., Mouhous-Voyneau, N., Denoeux, T.: Using Dempster-Shafer Theory to model uncertainty in climate change and environmental impact assessments. In: International Conference on Information Fusion, pp. 2117–2124 (2013)
27. Wu, Z., Gao, F.: Image Classification Based on Dempster-Shafer evidence theory and neural network. In: Second WRI Global Congress on Intelligent Systems, vol. 2, pp. 296–298 (2010)
28. Garvey, T.D.: Evidential reasoning for land-use classification. In: Analytical Methods in Remote Sensing for Geographic Information Systems, International Association of Pattern Recognition, Technical Committee 7 Workshop, Paris (1986)
29. Batty, M.: Urban modeling. In: International Encyclopedia of Human Geography, pp. 51–58. Elsevier, Oxford (2009)
30. Verburg, P.H., Schot, P.P., Dijst, M.J., Veldkamp, A.: Land use change modelling: current practice and research priorities. Geojournal **61**, 309–324 (2004)
31. Droguett, E.L., Mosleh, A.: Integrated treatment of model and parameter uncertainties through a Bayesian approach. J. Risk Reliab. **227**(1), 41–54 (2012)
32. INSEE Reunion, Bilan demographique 2009, Rsultats (40) (2011)
33. Durieux, L., Lagabrielle, E., Andrew, N.: A method for monitoring building construction in urban sprawl areas using object-based analysis of Spot 5 images and existing GIS data. ISPRS J. Photogramm. **63**, 399–408 (2008)
34. van der Valk, A.: The Dutch planning experience. Landsc. Urban Plan. **58**(2), 201–210 (2002)
35. Cadet, T.: La vegetation de l'ile de La reunion, etude phytocologique et phytosociologique, Ph.D. thesis. University of Aix Marseille (1980)

In a Quest for Suitable Similarity Measures to Compare Experience-Based Evaluations

Marcelo Loor[1,2]([✉]) and Guy De Tré[1]

[1] Department of Telecommunications and Information Processing, Ghent University,
Sint-Pietersnieuwstraat 41, 9000 Ghent, Belgium
{Marcelo.Loor,Guy.DeTre}@UGent.be
[2] Department of Electrical and Computer Engineering, ESPOL University,
Campus Gustavo Galindo V., Km. 30.5 Via Perimetral, Guayaquil, Ecuador

Abstract. After representing experience-based evaluations as intuition-istic fuzzy sets (IFSs), one might expect that all of the existing similarity measures for IFSs could be used to compare them. However, only some of those measures seem to be suitable to do so according to a psychological perspective which indicates that similarity measures assuming symmetry and transitivity could not reflect properly the perceived similarity. Consequently, to determine empirically their suitability for such comparisons, several similarity measures for IFSs were tested on simulated experience-based evaluations. This paper presents our findings about how each of them reflected the perceived similarity among the simulated experience-based evaluation sets.

Keywords: Experience-based evaluations · Similarity Measures · Intu-itionistic fuzzy sets

1 Introduction

If you ask about comic books suitable for 7-year-old kids, a coworker who does not like slang expressions might judge 'Popeye the Sailor' as a *quite unsuitable* comic book, whereas a coworker who learned eating spinach due to "they are the source of Popeye's super strength" might judge it as a *totally suitable* one. We deem the evaluations resulting from this kind of judgments to be *experience-based evaluations*, which mainly depend on what each person has experienced or understood about a particular concept (e.g., 'comic books suitable for 7-year-old kids').

Imagine that your sister is looking for a proper comic book for your 7-year-old nephew. If you want to know which of your coworkers could choose a comic book on behalf of your sister, you might be interested in measuring the level to which the evaluations given by each coworker are similar to your sister's evaluations. A problem in such similarity comparisons is that those experience-based evaluations are fairly subjective and a "pseudo-matching" between them is possible, i.e., the evaluations could match even though the evaluators have distinct understandings of the evaluated concept [8].

© Springer International Publishing AG 2017
J.J. Merelo et al. (eds.), *Computational Intelligence*, Studies in Computational Intelligence 669,
DOI 10.1007/978-3-319-48506-5_15

Considering that an experience-based evaluation could be imprecise and marked by hesitation, in [8] the authors proposed modeling it as an element of an intuitionistic fuzzy set, or IFS for short [1,2]. However, the authors pointed out that, to compare two IFSs that represent experience-based evaluation sets, the similarity measures based on a metric distance approach such as the studied in [12,15] might not be applicable to this case because of their implicit assumption about symmetry and transitivity, which does not reflect judgments of similarity observed from a psychological perspective [16].

To study empirically which of those similarity measures can be used to compare such IFSs, which is the main purpose of this work, we tested the similarity measures in comparisons between pairs of IFSs resulting from simulations of experience-based evaluation processes. Our motivation for this study is to complement the existing theoretical work within the context of IFSs to find suitable methods that allow us to compare experience-based evaluation sets given from persons that might have different learning experiences. As an extended version of [10], in this paper we tested additional (configurations of) similarity measures, which include among them a novel version of an existing *directional* similarity measure.

To simulate an experience-based evaluation process, we first made use of an algorithm that uses support vector machines [17,18] to learn how a human editor categorizes newswire stories under a given scenario. We then made use of the previous knowledge to evaluate the level to which other stories fit into one of the learned categories and, thus, we obtained the simulated experience-based evaluation sets. Each of the established learning scenarios included a training collection that contains a certain proportion of opposite examples in relation to the original data, which consist of manually categorized newswire stories— by opposite example is meant that, e.g., if a story is assigned to a particular category in the original training collection, the story will not be assigned to the category in the training collection related to the current scenario.

An interesting aspect about testing the similarity measures in that way is that we can observe how they reflect the perceived similarity between two experience-based evaluation sets given from dissimilar learning scenarios. For instance, we could test a similarity measure to observe how it reflects the perceived similarity between the IFSs given by two persons who use training collections having examples that are totally opposite to each other—here, one can anticipate that the resulting level of similarity will be the lowest.

The remainder of this work is structured as follows: Sect. 2 presents the IFS concept as well as the similarity measures that were tested; Sect. 3 describes how the simulated experience-based evaluation sets were obtained; Sect. 4 describes the test procedure that was carried out for each of the chosen similarity measures; Sect. 5 presents the results and our findings during the testing process; and Sect. 6 concludes the paper.

2   Preliminaries

This section presents a brief introduction to the IFS concept and shows how an IFS is used to model an experience-based evaluation set. Additionally, it presents some of the existing similarity measures for IFSs and introduces the formal notation that has been used throughout the paper.

2.1   IFS Concept

In [1,2], an *intuitionistic fuzzy set*, IFS for short, was proposed as an extension of a *fuzzy set* [19] and was defined as follows:

Definition 1 ([1,2]). *Consider an object x in the universe of discourse X and a set $A \subseteq X$. An intuitionistic fuzzy set is a collection*

$$A^* = \{\langle x, \mu_A(x), \nu_A(x)\rangle | (x \in X) \wedge (0 \le \mu_A(x) + \nu_A(x) \le 1)\}, \tag{1}$$

such that the functions $\mu_A : X \mapsto [0,1]$ and $\nu_A : X \mapsto [0,1]$ define the degree of membership *and the* degree of non-membership *of $x \in X$ to the set A respectively.*

In addition, the equation

$$h_A(x) = 1 - \mu_A(x) - \nu_A(x) \tag{2}$$

was proposed in [1] to represent the lack of knowledge (or hesitation) about the membership or non-membership of x to the set A.

Modeling Experience-Based Evaluations. An IFS can be used to model an experience-based evaluation set [8]. For instance, $X = \{$'*Popeye the Sailor*', '*The Avengers*'$\}$ could represent the 'comic books' that you asked your coworkers to evaluate for, and A could represent a set of the 'comic books suitable for 7-year-old kids'. If so, the IFS $A^* = \{\langle$'Popeye the Sailor', 0, 0.8$\rangle$, $\langle$'The Avengers', 0.5, 0.3 $\rangle\}$ might represent the evaluations given by one of your coworkers.

IFS Notation. Even though Definition 1 and the previous example show the difference between the IFS A^* and the set A, as it was suggested in [1] we shall hereafter use A instead of A^* as a notation for an IFS.

2.2   Similarity Measures for IFSs

Let A and B be two IFSs in $X = \{x_1, \cdots, x_n\}$, a similarity measure S is usually defined as a mapping $S : X^2 \mapsto [0,1]$ such that $S(A, B)$ denotes the level to which A is similar to B with 0 and 1 representing the lowest and the highest levels respectively.

Recalling the difference between an IFS P^* and a set P in Definition 1, the IFSs A and B in $S(A, B)$ can correspond to

$$P_{@A}^* = \{\langle x_i, \mu_{P_{@A}}(x_i), \nu_{P_{@A}}(x_i)\rangle | (x_i \in X) \wedge (0 \leq \mu_{P_{@A}}(x_i) + \nu_{P_{@A}}(x_i) \leq 1)\},$$

and

$$P_{@B}^* = \{\langle x_i, \mu_{P_{@B}}(x_i), \nu_{P_{@B}}(x_i)\rangle | (x_i \in X) \wedge (0 \leq \mu_{P_{@B}}(x_i) + \nu_{P_{@B}}(x_i) \leq 1)\},$$

respectively, where $P_{@A}$ and $P_{@B}$ represent the individual understanding of P as seen from the perspectives of the evaluators who provide the IFSs A and B. This means that, in the context of experience-based evaluations, $S(A, B)$ measures the similarity between IFSs A and B with regard to individual understandings of a *common set* P. For instance, if P represents a collection of 'comic books suitable for 7-year-old kids', $S(A, B)$ will measure the similarity between two experience-based evaluation sets taking into account the individual understandings of P that the providers of IFSs A and B might have.

The above clarification is needed because we identify two approaches in the formulation of similarity measures for IFSs: a *symmetric* (or metric distance) approach, which considers that $S(A, B) = S(B, A)$ always holds; and a *directional* approach, which considers that $S(A, B) = S(B, A)$ only holds in situations in which the evaluators who provide the IFSs A and B have the same understandings of the common set behind these IFSs.

Symmetric Similarity Measures. Among others, the following symmetric similarity measures for IFSs have been studied:

$$S_{H3D}(A, B) = 1 - \frac{1}{2n} \sum_{i=1}^{n} \left(|\mu_A(x_i) - \mu_B(x_i)| + |\nu_A(x_i) - \nu_B(x_i)| + |h_A(x_i) - h_B(x_i)| \right)$$
(3)

and

$$S_{H2D}(A, B) = 1 - \frac{1}{2n} \sum_{i=1}^{n} \left(|\mu_A(x_i) - \mu_B(x_i)| + |\nu_A(x_i) - \nu_B(x_i)| \right),$$
(4)

which are based on Hamming distance [13];

$$S_{E3D}(A, B) = 1 - \sqrt{\frac{1}{2n} \sum_{i=1}^{n} \left((\mu_A(x_i) - \mu_B(x_i))^2 + (\nu_A(x_i) - \nu_B(x_i))^2 + (h_A(x_i) - h_B(x_i))^2 \right)}$$
(5)

and

$$S_{E2D}(A, B) = 1 - \sqrt{\frac{1}{2n} \sum_{i=1}^{n} \left((\mu_A(x_i) - \mu_B(x_i))^2 + (\nu_A(x_i) - \nu_B(x_i))^2 \right)}$$
(6)

which are based on Euclidean distance [13]; and

$$S_{COS}(A,B) = \frac{1}{n}\sum_{i=1}^{n}\frac{\mu_A(x_i)\mu_B(x_i)+\nu_A(x_i)\nu_B(x_i)+h_A(x_i)h_B(x_i)}{\sqrt{\mu_A(x_i)^2+\nu_A(x_i)^2+h_A(x_i)^2}\sqrt{\mu_B(x_i)^2+\nu_B(x_i)^2+h_B(x_i)^2}},$$

(7)

which is based on Bhattacharyas's distance [15].

In addition, symmetric similarity measures that include the "notion of complement" in their definitions have been proposed in [14]:

$$S_{SK1}(A,B) = 1 - f\left(l(A,B),l(A,B^c)\right),$$

(8)

$$S_{SK2}(A,B) = \frac{1 - f\left(l(A,B),l(A,B^c)\right)}{1 + f\left(l(A,B),l(A,B^c)\right)},$$

(9)

$$S_{SK3}(A,B) = \frac{(1 - f\left(l(A,B),l(A,B^c)\right))^2}{(1 + f\left(l(A,B),l(A,B^c)\right))^2}$$

(10)

and

$$S_{SK4}(A,B) = \frac{e^{-f(l(A,B),l(A,B^c))} - e^{-1}}{1 - e^{-1}}.$$

(11)

Herein, B^c is the complement of B, i.e.,

$$B^c = \{\langle x_i, \nu_B(x_i), \mu_B(x_i)\rangle | (x_i \in X) \wedge (0 \le \mu_B(x_i) + \nu_B(x_i) \le 1)\},$$

(12)

$l(A,B)$ could be the "3D version" of the Hamming distance between A and B [12,14], i.e.,

$$l(A,B) = \frac{1}{2n}\sum_{i=1}^{n}\left(|\mu_A(x_i)-\mu_B(x_i)|+|\nu_A(x_i)-\nu_B(x_i)|+|h_A(x_i)-h_B(x_i)|\right),$$

(13)

and

$$f\left(l(A,B),l(A,B^c)\right) = \frac{l(A,B)}{l(A,B) + l(A,B^c)}.$$

(14)

Directional Similarity Measures. To the best of our knowledge, only two directional similarity measures for IFSs have been studied. Both are briefly described below.

The first directional similarity measure is defined by the equation

$$S^{\alpha}(A,B) = 1 - \frac{1}{n}\sum_{i=1}^{n}|dif^{\alpha}(\mathbf{a_i},\mathbf{b_i})|,$$

(15)

where $\alpha \in [0,1]$ is called *hesitation splitter*,

$$\mathbf{a_i} = \begin{pmatrix} \mu_A(x_i) + & \alpha h_A(x_i) \\ \nu_A(x_i) + (1-\alpha)h_A(x_i) \end{pmatrix}$$

and

$$\mathbf{b_i} = \begin{pmatrix} \mu_B(x_i) + & \alpha h_B(x_i) \\ \nu_B(x_i) + & (1-\alpha)h_B(x_i) \end{pmatrix}$$

are vector interpretations of the IFS-elements in IFSs A and B related to x_i [9], and

$$dif^\alpha(\mathbf{a_i}, \mathbf{b_i}) = (\mu_A(x_i) - \mu_B(x_i)) + \alpha(h_A(x_i) - h_B(x_i)) \tag{16}$$

is the *spot difference* between the IFS-elements corresponding to x_i in A and B respectively [9].

The second directional similarity measure is defined by

$$S_{@A}^\alpha(A, B) = \Delta_{@A} \cdot S^\alpha(A, B), \tag{17}$$

which is an extension of (15) based on a factor $\Delta_{@A} \in [0, 1]$ that indicates the level to which the understandings of the common set behind IFSs A and B are in alignment. In [8], $\Delta_{@A}$ is conceived as the weight of a *connotation-differential print* (CDP) between A and B as seen from the perspective of the evaluator who provides A—therein a CDP is defined as a sequence that represents any difference in the understandings of the common set behind IFSs A and B. Since such a difference in understandings is deemed to be subjective, the assembling of a CDP will depend on either the perspective of who provides A or the perspective of who provides B (i.e., it is directional); so will do its weight.

3   Simulation

As was mentioned in the Introduction, the aim of this work is to study empirically which of the similarity measures presented in Sect. 2.2 can be used to compare experience-based evaluation sets represented as IFSs. Hence, in this section we describe both the learning and the evaluation processes that were used to obtain the IFSs that represent the *simulated* experience-based evaluation sets.

3.1   Learning Process

In this part we describe the data, scenarios and algorithm that were employed to simulate how a human editor categorizes newswire stories.

Learning Data. We made use of the Reuters Corpora Volume I (RCV1) [11], which is a collection of manually categorized newswire stories provided by Reuters, Ltd. Specifically, we made use of the corrected version RCV1.v2, which is available (and fully described) in [7]. This collection has 804414 newswire stories, each assigned to one or more (sub) categories within three main categories: *Topics, Regions* and *Industries*.

The 23149 stories contained in the training file *lyrl2004_tokens_train.dat* of that collection were used to learn how to categorize newswire stories into one or more of the following categories from *Topics*: *ECAT, E11, E12, GSCI, GSPO, GTOUR, GVIO, CCAT, C12, C13, GCAT, G15, GDEF, GDIP, GDIS, GENT, GENV, GFAS, GHEA* and *GJOB*. The interested reader is referred to [7] for a full description of these categories.

Learning Scenarios. We established the following scenarios to learn how to categorize newswire stories into each of the chosen categories:

- *R0*: All the stories in the training data preserve the assignation of the training category in its original state.
- *R20, R40, R60, R80, R100*: The assignation of the training category is opposite to its original state in the 20, 40, 60, 80 and 100 % of the stories in the training data respectively. The assignation of the training category in the remainder of the stories is preserved. The selection of the stories that do not preserve the original state is made through a simple random sampling.

For instance, consider the story with code 2286, which was assigned to the category *ECAT*. In the scenario *R20*, if the training category is *ECAT* and the story is selected to change its category, the story will be considered as a nonmember of *ECAT*.

Learning Algorithm. We made use of an algorithm based on *support vector machines*, or SVM for short [17,18], which have been successfully used in statistical learning theory. Specifically, we made use of the application of SVMs for the *text categorization problem* proposed in [6], which has demonstrated superior results to deal with such a problem [7].

In the context of the text categorization problem, the words in a newswire story are the features that determine whether the story belongs or not to a category. This follows an intuition in which, according to his/her experience, a person focuses on the words in a document to decide whether it fits or not into a given category.

To use the SVM algorithm, each story must be modeled as a vector whose components are the words in the story. A story might contain words such as 'the', 'of' or 'at' that have a negligible impact on the categorization decision, or words such as 'learning', 'learned' or 'learn' that have a common stem. To simplify the vector representation, such words are usually filtered out and stemmed by using different algorithms. Hence, for the sake of reproducibility of the simulation, we made use of the stories in the training file *lyrl2004_tokens_train.dat* [7], which already have reduced and stemmed words. For example, the story with code 2320 has the following words: *tuesday, stock, york, seat, seat, nys, level, million, million, million, sold, sold, current, off, exchang, exchang, exchang, bid, prev, sale, mln.*

Since the impact of the words on the categorization decision could be different, a weight should be assigned to each word. Thus, to compute the (initial) weight of a word in a story (or document), as it was suggested in [7], we applied the equation

$$weight(f, x) = (1 + \ln n(f, x)) \ln \left(|X_0| / n(f, X_0) \right), \tag{18}$$

which is a variant of tf-idf weighting given in [3] where X_0 is the training collection (i.e., the collection of stories in $lyrl2004\_tokens\_train.dat$), $x \in X_0$ is a story, f is a word in x, $n(f, x)$ is the number of occurrences of f in x, $n(f, X_0)$ is the number of stories in X_0 that contain f, and $|X_0|$ is the number of stories in X_0 (i.e., $|X_0| = 23149$). For example, the weight of the word $exchang$ in the story with code 2320 is given by $weight(exchang, 2320) = (1 + \ln 3) \ln (23149/2485) = 4.6834$.

After computing the weight of each word in a story with code i, say x_i, we represented x_i as a vector $\mathbf{x}_i = \beta_{i,1}\hat{\mathbf{f}}_1 + \cdots + \beta_{i,|F|}\hat{\mathbf{f}}_{|F|}$ such that:

- F is a dictionary having all the distinct words in the training collection X_0;
- $|F|$ is the number of words in F (for the chosen training collection, $|F| = 47152$);
- $\hat{\mathbf{f}}_k$ is a unit vector that represents an axis related to a word $f_k \in F$ (i.e., $\hat{\mathbf{f}}_k$ belongs to a multi-dimensional feature space in which each dimension corresponds to a word $f_k \in F$); and
- $\beta_{i,k} = weight(f_k, x_i)$ is the weight of f_k in x_i (if f_k is not present in the story, $\beta_{i,k}$ will be fixed to 0).

Since the stories may have different number of words, each $\beta_{i,k}$ in $\mathbf{x}_i$ was divided by $\|\mathbf{x}_i\| = \sqrt{\mathbf{x}_i \cdot \mathbf{x}_i}$, i.e., $\mathbf{x}_i$ was transformed to a unit vector [7].

Idea Behind the SVM Algorithm. So far we have described how each story x_i in the training collection X_0 was represented by a vector $\mathbf{x}_i$. To describe how we made use of those vectors (and the resulting ones later on), in what follows we briefly explain the idea behind the SVM algorithm (see [4] for a tutorial about SVM).

In Fig. 1 the vectors corresponding to stories that fit into a given category (i.e., positive examples) are depicted with gray-circle heads, while the vectors that do not fit into the category (i.e., negative examples) are depicted with

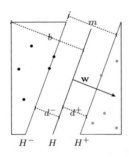

Fig. 1. Idea behind the SVM algorithm.

black-circle heads. The hyperplane H separates the positive from the negative examples—here H is defined by $\mathbf{w} \cdot \mathbf{x} + b = 0$, where $\mathbf{w}$ is a vector perpendicular to H, $\mathbf{x}$ is a point lying on H, and b is the perpendicular distance between H and the origin. The hyperplane H^+ is parallel to H and contains the closest positive example to it. The hyperplane H^- is also parallel to H and contains the closest negative example to it. The margin $m = d^+ + d^-$ between H^+ and H^- is the largest. The *support vectors* are the vectors whose heads lay either on H^- or H^+.

To find the hyperplane H that maximizes the margin between H^+ and H^- the following quadratic programming problem should be solved

$$\Lambda = \sum_{i=1}^{n} \lambda_i - \frac{1}{2} \sum_{i=1,j=1}^{n} \lambda_i \lambda_j y_i y_j \mathbf{x}_i \cdot \mathbf{x}_j, \tag{19}$$

where $\mathbf{x}_i$ and $\mathbf{x}_j$ are the vectors corresponding to stories in the training collection, y_i (or y_j) denotes whether the $\mathbf{x}_i$ (or $\mathbf{x}_j$) fits ($y_i = 1$) or not ($y_i = -1$) into the category, $\lambda_i, \lambda_j \geq 0$, and n is the number of stories in the training collection. The solution is given by both

$$\mathbf{w} = \sum_{k=1}^{n} \lambda_k y_k \mathbf{x}_k \tag{20}$$

and

$$b = y_k - \mathbf{w} \cdot \mathbf{x}_k, \tag{21}$$

for any $\mathbf{x}_k$ such that $\lambda_k > 0$.

To compute both (20) and (21), we made use of the package *SVMLight Version V6.02* [5]. We issued the command "*svm_learn.exe -c 1 svmTrainingFile svmModelFile*", where *svmTrainingFile* is an input file that contains the training vectors for a category under a given scenario, and *svmModelFile* is an output file that contains the solution (or model) of the scenario-category learning process. Using the 6 scenarios and 20 categories described above, we obtained 120 *scenario-category* models during this learning process—hereafter a model will be referred to using the nomenclature *scenario-category*.

3.2   Evaluation Process

Consider a collection of newswire stories X. To evaluate the level to which a newswire story $x \in X$ fits into a category, say *ECAT*, under a given scenario, say *R20*, we use the *R20-ECAT* model, which represents the experience (or knowledge) acquired after the previous learning process. After evaluating all the newswire stories in X, we obtain an evaluation set for X. This evaluation set corresponds to the *simulated* experience-based evaluation set given by a person who learned the concept *ECAT* using the training data specified in the scenario *R20*.

The data and the process that were utilized to generate such *simulated* experience-based evaluation sets are described below.

Evaluation Data. We made use of the first 12500 newswire stories in each of the following files from RCV1.v2 [7]:

- *lyrl2004_tokens_test_pt0.dat,*
- *lyrl2004_tokens_test_pt1.dat,*
- *lyrl2004_tokens_test_pt2.dat* and
- *lyrl2004_tokens_test_pt3.dat.*

With these 50000 stories, we built 1000 50-story collections.

Obtaining an IFS as a Result of an Evaluation Process. Let X_k be one of the 50-story collections that constitute the evaluation data. To evaluate the level to which a story $x_i \in X_k$ fits into a category, say C, under a given (learning) scenario, say LS, we made use of the LS-C model resulting from the previous learning process to obtain an IFS-element $\langle x_i, \mu_C(x_i), \nu_C(x_i) \rangle$ as follows.

First, we represented x_i as a vector $\mathbf{x}_i = \beta_{i,1}\hat{\mathbf{f}}_1 + \cdots + \beta_{i,|F|}\hat{\mathbf{f}}_{|F|}$ according to the procedure described in the previous section, where X_0 corresponds to the training collection in the scenario S.

Then, we made use of $\mathbf{w} = \omega_1\hat{\mathbf{f}}_1 + \cdots + \omega_{|F|}\hat{\mathbf{f}}_{|F|}$ and b in the S-C model to figure out $\mu_C(x_i)$ and $\nu_C(x_i)$ by means of the equations

$$\mu_C(x_i) = \check{\mu}_C(x_i)/\sigma \tag{22}$$

and

$$\nu_C(x_i) = \check{\nu}_C(x_i)/\sigma \tag{23}$$

respectively, where

$$\check{\mu}_C(x_i) = \begin{cases} \frac{\left(\sum_{j=1}^{|F|} \beta_{i,j}\omega_j\right)+|b|}{\|\mathbf{x}_i\|\|\mathbf{w}\|} & : (\beta_{i,j}\omega_j > 0) \wedge (b < 0); \\ \frac{\sum_{j=1}^{|F|} \beta_{i,j}\omega_j}{\|\mathbf{x}_i\|\|\mathbf{w}\|} & : (\beta_{i,j}\omega_j > 0) \wedge (b \geq 0); \\ 0 & : \text{otherwise}; \end{cases} \tag{24}$$

$$\check{\nu}_C(x_i) = \begin{cases} \frac{\left(\sum_{j=1}^{|F|} |\beta_{i,j}\omega_j|\right)+b}{\|\mathbf{x}_i\|\|\mathbf{w}\|} & : (\beta_{i,j}\omega_j < 0) \wedge (b > 0) \\ \frac{\sum_{j=1}^{|F|} |\beta_{i,j}\omega_j|}{\|\mathbf{x}_i\|\|\mathbf{w}\|} & : (\beta_{i,j}\omega_j < 0) \wedge (b \leq 0); \\ 0 & : \text{otherwise}; \end{cases} \tag{25}$$

and

$$\sigma = \max\left(1, \check{\mu}_C(x_i) + \check{\nu}_C(x_i)\right), \forall x_i \in X_k. \tag{26}$$

Finally, after computing all the IFS-elements for each $x_i \in X_k$, we obtained an IFS that represents the simulated experience-based evaluations for the stories in X_k according to what was learned (or experienced) about the category C under the scenario LS.

Since we built 1000 50-story collections, we obtained 1000 IFSs for each scenario-category model. We made use of the notation $C_{@LS}(X_k)$ to denote an

Table 1. IFSs that represent the simulated experience-based evaluations for the stories in each $X_k \in \{X_1, \cdots, X_{1000}\}$ according to what was learned about category $E11$ under the scenarios $R0$, $R20$, $R40$, $R60$ and $R100$ respectively.

$E11$					
Scenario	50-story Collections				
	X_1	$\cdots$	X_k	$\cdots$	X_{1000}
$R0$	$E11_{@R0}(X_1)$	$\cdots$	$E11_{@R0}(X_k)$	$\cdots$	$E11_{@R0}(X_{1000})$
$R20$	$E11_{@R20}(X_1)$	$\cdots$	$E11_{@R20}(X_k)$	$\cdots$	$E11_{@R20}(X_{1000})$
$R40$	$E11_{@R40}(X_1)$	$\cdots$	$E11_{@R40}(X_k)$	$\cdots$	$E11_{@R40}(X_{1000})$
$R60$	$E11_{@R60}(X_1)$	$\cdots$	$E11_{@R60}(X_k)$	$\cdots$	$E11_{@R60}(X_{1000})$
$R80$	$E11_{@R80}(X_1)$	$\cdots$	$E11_{@R80}(X_k)$	$\cdots$	$E11_{@R80}(X_{1000})$
$R100$	$E11_{@R100}(X_1)$	$\cdots$	$E11_{@R100}(X_k)$	$\cdots$	$E11_{@R100}(X_{1000})$

IFS that represents the simulated experience-based evaluations for the stories in X_k according to what was learned about category C under a scenario LS. For example, Table 1 shows the IFSs that represent the simulated experience-based evaluations for the stories in each $X_k \in \{X_1, \cdots, X_{1000}\}$ according to what was learned about category $E11$ under the scenarios $R0$, $R20$, $R40$, $R60$ and $R100$ respectively.

Considering that we chose 20 categories and built 6 scenarios during the learning phase, we obtained a total of 120000 IFSs during this phase.

4   Testing

In this section we describe how the similarly measures presented in Sect. 2.2 were tested with the IFSs that represent simulated experience-based evaluation sets.

4.1   A Point of Reference for the Perceived Similarity

Consider a scenario-category model LS-C represented by both $\mathbf{w}$ and b according to the Eqs. (20) and (21) respectively (see Sect. 3.1). Consider then a story $x_i \in X_k$ represented by $\mathbf{x_i}$, where X_k is one of the 50-story collections in the evaluation data (see Sect. 3.2). Consider finally a collection $Y_k = \{y_i | (y_i = \mathbf{w} \cdot \mathbf{x_i} + b)\}$ such that y_i is the *SVM-based evaluation* of story $x_i \in X_k$ fitting into the category C under the scenario LS. In this context, the decision about the fittingness of the story x_i into the category C under the scenario LS will depend on y_i: when $y_i > 0$, the decision will be "x_i fits into C;" when $y_i < 0$, the decision will be "x_i does not fit into C;" and when $y_i = 0$, no decision will be taken. A visual interpretation of this decision process is observable in Fig. 1: when $y_i > 0$ the head of the vector $\mathbf{x_i}$ corresponding to story x_i will be on the H^+-side, i.e., it will have a gray-circle head; when $y_i < 0$ the head of $\mathbf{x_i}$ will be on the H^--side, i.e., it will have a black-circle head; and when $y_i = 0$ the head of $\mathbf{x_i}$ will be on H

(see [7] for more details about the influence of this decision process in the text categorization problem).

Now consider the collections $Y_{k@L1}$ and $Y_{k@L2}$ having SVM-based evaluations under scenarios $L1$ and $L2$ respectively. Consider also $y_{i@L1} \in Y_{k@L1}$ and $y_{i@L2} \in Y_{k@L2}$. In this situation, when

$$((y_{i@L1} < 0 \wedge y_{i@L2} < 0) \vee (y_{i@L1} > 0 \wedge y_{i@L2} > 0) \vee (y_{i@L1} = 0 \wedge y_{i@L2} = 0))$$

is *true*, an *agreement on decision* about the fittingness of story x_i between the evaluations given under scenarios $L1$ and $L2$ occurs.

We made use of the agreements on decisions between $Y_{k@L1}$ and $Y_{k@L2}$ to obtain an *agreement-on-decision ratio*, AoD for short, which is expressed by

$$AoD(Y_{k@L1}, Y_{k@L2}) = n/N, \tag{27}$$

where n represents the number of agreements on decision between $Y_{k@L1}$ and $Y_{k@L2}$, and N represents the number of stories in X_k. Since the AoD ratio denotes how similar the decisions are, we deemed it to be an indicator of the perceived similarity between the evaluations given by two persons that learned (or experienced) C under $L1$ and $L2$ respectively.

4.2   Testing Procedure and Settings

As was mentioned in the Introduction, an experience-based evaluation mainly depends on what an evaluator has experienced or learned about a particular concept. Thus, one could expect that the level of similarity between the evaluation sets given by two evaluators who learned a concept under the same (learning) scenario will be greater than or equal to the level of similarity between the evaluation sets given by two evaluators who learned the same concept under different scenarios. For instance, consider three evaluators: P, Q and R. While P and Q learned about the category $E11$ under the same scenario $R0$, R learned so under the scenario $R80$. Consider also that the IFSs $E11_{@P}(X_k)$, $E11_{@Q}(X_k)$ and $E11_{@R}(X_k)$ represent the experience-based evaluation sets about the fittingness of the stories in the 50-story collection X_k into category $E11$ given by P, Q and R respectively. In this context, one could expect that the similarity between $E11_{@P}(X_k)$ and $E11_{@Q}(X_k)$ will be greater than the similarity between $E11_{@P}(X_k)$ and $E11_{@R}(X_k)$.

We made use of the above intuition to test the similarity measures presented in Sect. 3.2. Since we chose the AoD ratio as an indicator of the perceived similarity, we first tested it to observe how the agreement on decisions between two SVM-based evaluation sets is affected according to their respective learning scenarios. We then tested the similarity measures, some of them with different configurations.

Testing the Agreement-on-Decision Ratio. Again, one could expect that the AoD ratio between two SVM-based evaluation sets resulting from the same scenario will be greater than the AoD ratio between two SVM-based evaluation sets resulting from distinct scenarios. Thus, we considered the question: *Is there sufficient evidence in the evaluation data to suggest that the mean AoD ratio is different after altering a given percentage of the training data?* To answer this, for each category and for each 50-story collection, we obtained the AoD ratio between the SVM-based evaluation set given under scenario $R0$ (i.e., $R0$ is a *referent* scenario) and each of the SVM-based evaluation sets given under the scenarios $R0, R20, R40, R60, R80$ and $R100$ respectively (see Algorithm 1).

Algorithm 1. Obtaining AoD ratios.

Require: $ChosenCategories$ {see Section 3.1}
Require: $LearningScenarios$ {see Section 3.1}
Require: $50storyCollections$ {see Section 3.2}
Require: $SVMEvals$ {see Section 4.1}
 1: $Z \leftarrow \emptyset$ {resulting ratios}
 2: **for all** $C \in ChosenCategories$ **do**
 3:     **for all** $X_k \in 50storyCollections$ **do**
 4:         $Y_{k@R0} \leftarrow SVMEvals[X_k][R0][C]$
 5:         **for all** $LS \in LearningScenarios$ **do**
 6:             $Y_{k@LS} \leftarrow SVMEvals[X_k][LS][C]$
 7:             $r \leftarrow AoD(Y_{k@R0}, Y_{k@LS})$
 8:             $Z[C][LS][X_k] \leftarrow r$
 9: **return** Z

Testing the Similarity Measures. To test the similarity measures, we computed the level of similarity between the IFS given under scenario $R0$ and each of the IFSs given under the scenarios $R0, R20, R40, R60, R80$ and $R100$ respectively by means of each of the established similarity measures. We did so through the steps described in Algorithm 2. As could be noticed, the computation was performed for each category, for each 50-story collection, for each scenario and for each similarity measure. For readability, hereafter we shall use the acronym placed as subscript in each of the given similarity measures to refer to each of them. For instance, we shall use $H2D$ to refer to S_{H2D} (see Eq. 6).

Since two of the similarity measures presented in Sect. 2.2, namely S^{α} (see Eq. 15) and its extended version (see Eq. 17), needed to be configured, we applied to them the configurations described below before the test.

The similarity measure S^{α} was configured with hesitation splitters $\alpha = 0, 0.5$ and 1—we shall use the label $VB\text{-}\alpha$ to refer to each of the possible configurations of this measure.

With respect to the extended version of S^{α}, two different methods were applied to compute the $\Delta_{@A}$ factor, i.e., two forms of this measure were used during the test.

Algorithm 2. Testing similarity measures.

Require: $SimMeasures$ {see Sections 2.2 and 4.2}
Require: $ChosenCategories$ {see Section 3.1}
Require: $LearningScenarios$ {see Section 3.1}
Require: $50storyCollections$ {see Section 3.2}
Require: $IFSEvals$ {see Section 3.2}
1: $Z \leftarrow \emptyset$ {resulting levels}
2: **for all** $C \in ChosenCategories$ **do**
3:      **for all** $X_k \in 50storyCollections$ **do**
4:          $C_{@R0}(X_k) \leftarrow IFSEvals[X_k][R0][C]$
5:          **for all** $LS \in LearningScenarios$ **do**
6:              $C_{@LS}(X_k) \leftarrow IFSEvals[X_k][LS][C]$
7:              **for all** $S \in SimMeasures$ **do**
8:                  $l \leftarrow S(C_{@R0}(X_k), C_{@LS}(X_k))$
9:                  $Z[C][LS][X_k][S] \leftarrow l$
10: **return** Z

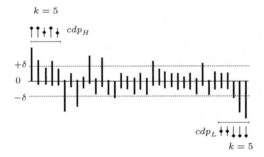

Fig. 2. Obtaining a CDP and its weight. The bars represent the spot differences between the elements of IFSs A and B. The CDPs for the k-highest and the k-lowest IFS-elements according to A's perspective are denoted by cdp_H and cdp_L respectively.

In the first form, labeled XVB-α-w, $\Delta_{@A}$ was computed by means of the method $weightCDP(A, B, \alpha, w)$, in which A and B are the IFSs in the comparison, α is the hesitation splitter, and $w \in [0, 1]$ is a value that allows us to obtain a CDP (see Sect. 3.2) between A and B according to the *wide* of the average gap between the membership and non-membership values as seen from the perspective of who provides A. This method involves the following steps:

1. Obtain $\delta \in [0, 1]$ for IFS A through

$$\delta = \frac{w}{n} \sum_{i=1}^{n} (\mu_A(x_i) + \nu_A(x_i)). \tag{28}$$

2. Compute the spot differences among the IFS-elements in A and B using (16).
3. Order the IFS-elements in A by descending membership values and then by ascending non-membership values.

4. Fix $k = 0.1n$ (i.e., $k = 5$) and obtain the connotation-differential mark-ers (i.e., ⟐, ⇑ and ⇓ [9]) for the k-highest and the k-lowest IFS-elements in the arranged IFS A (see Fig. 2). For a spot difference s, the marker will be: ⟐ when $|s| \le \delta$; ⇑ when $s > \delta$; and ⇓ when $s < -\delta$.

5. Build the CDPs cdp_H and cdp_L with the markers corresponding to k-highest and the k-lowest IFS-elements respectively (see Fig. 2).

6. Fix $v[⟐] = 1$, $v[⇑] = 0.01$ and $v[⇓] = 0.01$, and compute $\Delta_{@A}$ by means of

$$\Delta_{@A} = \max\left(\frac{1}{k}\sum_{m \in cdp_H} v[m], \frac{1}{k}\sum_{m \in cdp_L} v[m]\right) \qquad (29)$$

In the second form, labeled $XVBr$-α, $\Delta_{@A}$ was computed through a novel method, called $spotRatios$, which involves the following steps:

1. Fix $k = 0.1n$ (i.e., $k = 5$).
2. Order the IFS-elements in A by descending membership values and then by ascending non-membership values. After that, put the top k into a collection H and the bottom k in a collection L.
3. Compute $\Delta_{@A}$ by means of

$$\Delta_{@A} = \frac{1}{2k}\left(\sum_{\mathbf{a}_i \in H}^{k} spotRatio(\mathbf{a}_i, \mathbf{b}_i, \alpha) + \sum_{\mathbf{a}_i \in L}^{k} spotRatio(\mathbf{a}_i, \mathbf{b}_i, \alpha)\right), \qquad (30)$$

where $\mathbf{a}_i$ and $\mathbf{b}_i$ are vector interpretations of the IFS-elements in A and B related to x_i (see Sect. 2.2), α is the hesitation splitter, and $spotRatio$ is defined by Algorithm 3.

5   Results and Discussion

This section presents the results after following the test conditions described in the previous section.

5.1   Agreement-on-Decision Ratio as an Indicator of the Perceived Similarity

To answer the question *is there sufficient evidence in the evaluation data to suggest that the mean AoD ratio is different after altering a given percentage of the training data?*, we first made use of the collection resulting of Algorithm 1 to compute the averages of the AoD ratios per scenario-category. We then ran the t-test for the null hypothesis *"the average of the AoD ratio is the same after altering the r% of the training data"* in contrast to the alternative one *"the average of the AoD ratio is different after altering the r% of the training data"* according to r given in each scenario (see Table 2).

Algorithm 3. $spotRatio(\mathbf{a}_i, \mathbf{b}_i, \alpha)$.

1: $r \leftarrow 0.5$ {default value}
2: $a_\mu \leftarrow \mu_A(x_i) + \alpha h_A(x_i)$
3: $a_\nu \leftarrow \nu_A(x_i) + (1 - \alpha)h_A(x_i)$
4: $b_\mu \leftarrow \mu_B(x_i) + \alpha h_B(x_i)$
5: $b_\nu \leftarrow \nu_B(x_i) + (1 - \alpha)h_B(x_i)$
6: $o_\mu \leftarrow \nu_A(x_i) + \alpha h_A(x_i)$ {$\mathbf{o}_i$ is the vector representation of the complement of the
    IFS-element $\langle x_i, \mu_A(x_i), \nu_A(x_i) \rangle$ represented by $\mathbf{a}_i$ }
7: $o_\nu \leftarrow \mu_A(x_i) + (1 - \alpha)h_A(x_i)$
8: $A_{ao} \leftarrow a_\mu o_\nu - a_\nu o_\mu$ {A_{ao} is the area of the paralellogram formed by $\mathbf{a}_i$ and $\mathbf{o}_i$}
9: $A_{bo} \leftarrow b_\mu o_\nu - b_\nu o_\mu$ {A_{bo} is the area of the paralellogram formed by $\mathbf{b}_i$ and $\mathbf{o}_i$}
10: **if** $|A_{ao}| > 0$ **then**
11:      $r \leftarrow A_{bo}/A_{ao}$
12:      **if** $r > 0$ **then**
13:          $r \leftarrow \min(1, r)$
14:      **else**
15:          $r \leftarrow 0$
16: **return** r

The results in Table 2 show that, for the scenarios *R20, R40, R60, R80* and *R100*, the t-values were statistically significant ($p < 0.05$). Consequently, we can say that there is sufficient evidence in the evaluation data to suggest that the average of the AoD ratio is different after altering the 20, 40, 60, 80 or 100 % of the training data.

Recalling that we deemed the AoD ratio to be an indicator of the perceived similarity, we can confidently expect that it will be affected by the different learning scenarios established in the simulation. This can be observed in the bivariate plot depicted in Fig. 3, which shows a strongly negative (or inverse) relationship ($R = -0.9741$) between the averages of the AoD ratios and the percentage of opposites included in the learning scenarios.

5.2   How Each Similarity Measure Reflects the Perceived Similarity

To observe how each of the configurations of similarity measures given in Sect. 4.2 reflects the perceived similarity between the simulated IFSs, we first made use of the collection resulting of Algorithm 2 to compute the averages of the levels of similarity per scenario-category. Then, we obtained linear models for the relationships between each one of those averages and the percentage of opposites considered in each scenario. After that, each of the resulting models was contrasted with the linear model corresponding to the AoD ratio. As an indicator of how well a similarity measure reflects the perceived similarity, we computed a *manifest* index, which is defined by

$$m = (a_{SM}/a_{AoD})(b_{SM}/b_{AoD})(R^2_{SM}/R^2_{AoD}), \qquad (31)$$

where a_{SM} and a_{AoD} are the slopes, b_{SM} and b_{AoD} are the intercepts, and R^2_{SM} and R^2_{AoD} are the R-statistics in the linear models corresponding to the

Table 2. Averages of the AoD ratios per scenario-category, and t-test for the null hypothesis "the average of the AoD ratio is the same after altering the $r\%$ of the training data" according to r given in each scenario (e.g., $r = 20$ in scenario $R20$), where $R0$ $(r = 0)$ is the referent scenario.

Category	Scenarios				
	R20	R40	R60	R80	R100
C12	0.7292	0.5900	0.4757	0.2897	0.0001
C13	0.7385	0.6091	0.4281	0.2766	0.0002
CCAT	0.9372	0.7505	0.2711	0.0663	0.0001
E11	0.6431	0.5740	0.4722	0.3519	0.0003
E12	0.7156	0.5792	0.4796	0.3080	0.0001
ECAT	0.8187	0.6273	0.4052	0.1853	0.0002
G15	0.7186	0.5954	0.4781	0.3039	0.0002
GCAT	0.9314	0.7472	0.2690	0.0661	0
GDEF	0.6515	0.5717	0.4668	0.3602	0.0002
GDIP	0.7433	0.5990	0.4587	0.2672	0.0002
GDIS	0.7229	0.5951	0.4729	0.3022	0.0002
GENT	0.7066	0.5796	0.4802	0.3209	0.0002
GENV	0.6941	0.6009	0.4812	0.3248	0.0004
GFAS	0.6763	0.5787	0.4882	0.3457	0.0002
GHEA	0.7016	0.5850	0.4636	0.3451	0.0003
GJOB	0.7359	0.5883	0.4542	0.2886	0.0003
GSCI	0.6899	0.5854	0.4844	0.3424	0.0004
GSPO	0.8208	0.6508	0.4130	0.1962	0
GTOUR	0.5383	0.5197	0.4866	0.4663	0.0005
GVIO	0.7551	0.6368	0.4749	0.2796	0.0002
Mean	0.7334	0.6082	0.4452	0.2844	0.0002
stdDev	0.0913	0.0551	0.0643	0.0951	0.0001
N	20	20	20	20	20
df	19	19	19	19	19
t-value	13.06	31.80	38.59	33.67	34025.85
p-value	0	0	0	0	0

similarity measure SM and the AoD ratio respectively. For readability, we shall use hereafter SM-vs.-OP to denote the relationship between the averages of the levels (of similarity) resulting from the (configuration of) similarity measure SM and the percentage of opposites OP.

The results in Table 3 show that, in contrast to what happens with the AoD ratio, the averages of the levels of $H2D$, $H3D$, $E2D$, $E3D$, COS, VB-0, VB-0.5 and VB-1 are hardly affected by the variation of the percentage of

opposites—notice the broad difference among the slopes of the linear models
corresponding to these similarity measures and the slope of the linear model
corresponding to the AoD ratio. By way of illustration, if we use the resulting
model for COS (i.e., $y = -0.0004x + 0.9831$) to compute the level to which the
average of evaluations given under the scenarios $R0$ and $R100$ are similar, we
will obtain $y = 0.9827$ as a result—since $R100$ contains the 100 % of opposite
training examples in relation to $R0$, we fix $x = 1$ to make this computation. As
noticed, the computed level differs markedly from the result obtained for AoD
(i.e., $y = 0.0469$). Such a remarkable difference is reflected by the lowest manifest index (i.e., $m = 0$) and is observable in Fig. 4(c), which shows the average
of the similarity levels per scenario vs the percentage of opposites included in
each scenario.

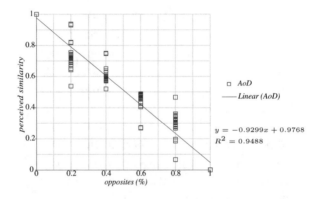

Fig. 3. Bivariate plot between the averages of the AoD ratios and the percentage of
opposites included in the learning scenarios. The relationship is represented by means of
a linear model and described by the statistic R (Pearson Product Moment Correlation).

Regarding the averages of the levels of $SK1$, $SK2$, $SK3$ and $SK4$, the results
suggest that the levels computed with $SK1$, $SK2$ and $SK4$ are fairly affected by
the variation of the percentage of opposites. Notice that the correlation indices
for $SK1$-vs.-OP, $SK2$-vs.-OP and $SK4$-vs.-OP (i.e., $R = -0.8632$, $R = -0.8319$
and $R = -0.8463$ respectively) denote fairly strong negative relationships that
are roughly comparable with the strongly negative relationship ($R = -0.9741$)
in AoD-vs.-OP. Moreover, Figs. 4(e), (f) and (h) show that $SK1$, $SK2$ and $SK4$
reflect properly the perceived similarity between the evaluations given under the
scenarios $R0$ and $R100$ in contrast to, e.g., $H3D$, COS or VB-0.5 (see Figs. 4(b),
(c) and (d)). However, these similarity measures seem to reflect more or less
properly the perceived similarity between the evaluations given under the scenarios $R0$ and, e.g., $R20$ or $R80$, which affects the values of the m-indices related
to their linear models.

With respect to the averages of the levels of the form XVB-α-w of (17), the
results in Table 3 show that the levels computed with two of them, namely XVB-
0-0.05 and XVB-0-0.1, are fairly affected by the variation of the percentage of

Table 3. Linear models and m-indices for each SM-vs.-OP representing the relationship between the averages levels that result from the (configuration of) similarity measure SM and the percentage of opposites OP.

SM	SM-vs.-OP (linear model: $y = ax + b$)			m-index
	slope (a)	intercept (b)	R^2	
H2D	−0.0139	0.9939	0.4128	0.0066
H3D	−0.0138	0.9852	0.1442	0.0023
E2D	−0.0171	0.9920	0.4189	0.0082
E3D	−0.0167	0.9853	0.2034	0.0039
COS	−0.0004	0.9831	0	0
SK1	**−0.7302**	**0.8666**	**0.7451**	**0.5471**
SK2	−0.7287	0.7547	0.6920	0.4416
SK3	−0.7242	0.6041	0.5242	0.2661
SK4	−0.7298	0.7848	0.7163	0.4761
VB-0	−0.0133	0.9955	0.4527	0.0070
VB-0.5	−0.0133	0.9955	0.4528	0.0070
VB-1	−0.0144	0.9922	0.3170	0.0053
XVB-0-0.05	−0.7318	0.7831	0.6871	0.4569
XVB-0.5-0.05	−0.6738	0.6999	0.5974	0.3269
XVB-1-0.05	−0.6388	0.6185	0.4666	0.2139
XVB-0-0.1	−0.6587	0.9307	0.6560	0.4666
XVB-0.5-0.1	−0.6240	0.8358	0.6878	0.4162
XVB-1-0.1	−0.5727	0.6978	0.4805	0.2228
XVB-0-0.2	−0.4218	1.0241	0.4575	0.2293
XVB-0.5-0.2	−0.4657	1.0029	0.5944	0.3221
XVB-1-0.2	−0.4335	0.8321	0.4414	0.1847
XVBr-0	−0.7368	0.8069	0.6740	0.4650
XVBr-0.5	**−0.7971**	**0.8984**	**0.8499**	**0.7062**
XVBr-1	−0.7368	0.8069	0.6740	0.4650
AoD	**−0.9299**	**0.9768**	**0.9488**	**1**

opposites as well. In contrast to $SK1$, $SK2$ and $SK4$, Figs. 4(i) and (j) suggest that XVB-0-0.05 and XVB-0-0.1 reflect more properly the perceived similarity between the evaluations given under the scenario $R0$ and the evaluations given under the scenarios $R20$, $R40$ or $R60$. However, the figures also suggest that both measures do not reflect so properly the perceived similarity between the evaluations given under the scenarios $R0$ and $R80$ or $R100$—notice that the average of the similarity levels between $R0$ and $R80$ is greater that the average of the similarity levels between $R0$ and $R60$.

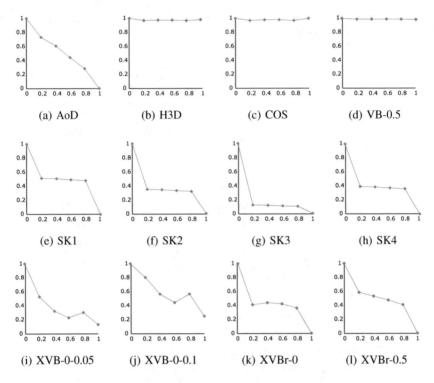

Fig. 4. Averages of the similarity levels per scenario versus the percentage of opposites included in each scenario.

Since the form XVB-α-w is based on the weight of a CDP and the computation of this weight was based on the w-parameter in our testing procedure (see Sect. 4.2), we performed additional tests to observe the influence of this parameter on the quality of the results of this similarity measure. In such additional tests, we configured (17) with $\alpha = 0$ and $w = 0.05, 0.1, 0.15, \cdots, 1$ and used the same nomenclature (i.e., XVB-α-w) to label each configuration. Figure 5 shows how the m-index corresponding to the linear model for each XVB-0-w-vs-OP relationship is affected by the w-parameter. As noticed, the peak m-index is reached at $w = 0.1$ and is projected to decline after that point. Recalling from Sect. 4.2, the w-parameter determines the *wide* of the average gap between the membership and non-membership values, which is then used to build a CDP for the IFSs in the similarity comparison as seen from the perspective of the person who provides the referent IFS. This means that, in this scenario, a spot difference with a magnitude less than or equal to the 10 % of the average gap between the membership and non-membership values (see Sects. 2.2 and 4.2) will roughly reflect a similar understanding (or knowledge) of the evaluate concept. This result seems to support the idea behind a CDP, which suggests that *"a difference in understanding of a concept could be marked by a difference in one or more evaluations"* [8].

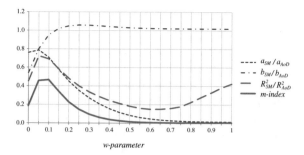

Fig. 5. Influence of the w-parameter on the quality of the m-index for XVB-0-w.

Concerning the averages of the levels of the form $XVBr$-α of (17), the results reported in Table 3 show that the levels computed by $XVBr$-0.5 are strongly affected by the variation of the percentage of opposites. As noticed, the correlation index for $XVBr$-0.5-$vs.$-OP ($R = -0.9219$) denote a very strong inverse relationship that is comparable with the correlation index for AoD-$vs.$-OP ($R = -0.9741$). In addition, Fig. 4(l) shows that the averages of the similarity levels between the evaluations given under scenario $R0$ and the given under the other scenarios are well reflected by $XVBr$-0.5, which is indicated by the best m-index reported (i.e., $m = 0.7062$). However, with respect to the levels computed by $XVBr$-0 and $XVBr$-1, the results suggest that such levels are affected by the variation of the percentage of opposites but not as strong as $XVBr$-0.5. A potential weakness of $XVBr$-0 when compared to $XVBr$-0.5 is shown in Fig. 4(k). Notice that, in contrast to $XVBr$-0.5, the average of the similarity levels computed by $XVBr$-0 between $R0$ and $R20$ is a little less than the average computed between $R0$ and $R40$.

5.3   Discussion

The results obtained suggest that one of the configurations of the similarity measure (17), namely $XVBr$-0.5, overcomes the other (configurations of) similarity measures when dealing with similarity comparisons among the simulated IFSs. However, it was found that other (configurations of) similarity measures such as XVB-0-0.1 or $SK1$ can be (partially) effective in comparisons between IFSs resulting from particular scenarios. For instance, $SK1$ seems to reflect properly the perceived similarity between the evaluations resulting from completely opposite understandings but it reflects in a lesser extent the perceived similarity among the evaluations resulting from roughly opposite (or slightly similar) understandings.

A possible explanation for those results might be that, by means of the factor $\Delta_{@A}$, the configurations of the similarity measure (17) take into account what is understood as a qualitative difference between two IFS-elements from the perspective of the evaluator who provides the IFS A. This situation is observable in the two evaluated forms of this measure: in the form XVB-α-w, when both the magnitude and the direction of a spot difference (see Sect. 2.2), as well as the average gap between the membership and non-membership components of the IFS-elements in A are used in the computation of $\Delta_{@A}$ (see Eq. 29); and in the form $XVBr$-α, when both the magnitude and the direction of a vector product $\mathbf{a}_i \times \mathbf{o}_i$, in which $\mathbf{a}_i$ is a vector representing an IFS-element in A and $\mathbf{o}_i$ is a vector representing the complement of that IFS-element, are used as points of reference in the (internal) computation of $\Delta_{@A}$ (see Algorithm 3). Even though both forms try to detect and quantify any qualitative difference between two IFS-elements, the results suggest that the form $XVBr$-α applies a more effective method. Notice that, in contrast to the form XVB-α-w, the form $XVBr$-α does not need a threshold value to quantify a difference (or similarity) in understandings, i.e., the parameter w is not necessary. Notice also that the method applied by $XVBr$-α seems to agree in some extent with the "notion of complement" used in the definitions of $SK1$, $SK2$, $SK3$ and $SK4$.

Another possible explanation for the results might be that a gap between the membership and non-membership components is contextually related to the categorization decision (see Sects. 3.2 and 4.1), which is deemed to be a point of reference for the perceived similarity through the agreement on decision ratio. Hence, a similarity measure such as (17) that takes into account the aforesaid gap could reflect more adequately the similarity perceived from the perspective of who makes the categorization decision.

Even though these results are based on simulated IFSs that use a manually categorized newswire stories, they need to be interpreted with caution because of the dependency of the IFSs with the learning algorithm and the (text categorization) context that were chosen for the simulations. Consequently, conducting simulations with other learning algorithms and experiments with real evaluators is recommended and subject to further study.

6   Conclusions

In this work we have conducted an empirical study that aims to determine which similarity measures are suitable to compare experience-based evaluations (XBEs) represented as intuitionistic fuzzy sets (IFSs) [1,2]. Herein by *'experience-based evaluation'* we mean a judgment that depends on what a person has experienced or understood about a particular concept or topic.

During the study, several similarity measures for IFSs were used in comparisons of simulated XBEs, which were obtained after learning through a support vector learning algorithm [17,18] how human editors categorize newswire stories under different scenarios. This made it possible to assess the level to which each similarity measure reflects the perceived similarity in comparisons of XBEs that might be given by persons with different backgrounds.

Taken together, the results obtained suggest that the studied similarity measures could be categorized as *suitable*, *partially suitable* and *unsuitable* while comparing XBEs.

The first category includes an improved version of the similarity measure proposed in [8], which uses a new method to quantify what is understood as a qualitative difference between two IFS-elements. A configuration of this measure seem to reflect well the perceived similarity among the simulated XBEs and, moreover, it overcomes the other tested similarity measures.

The second category is constituted by the similarity measures including the "notion of complement" in their definitions [12,14], as well as by some configurations of the original version of the similarity measure proposed in [8]. These measures seem to be (partially) effective in comparisons between XBEs resulting from particular scenarios.

The last category consists of similarity measures such as the proposed in [9] that could not reflect the perceived similarity between XBEs resulting from opposite scenarios.

Despite the results seem to be significant for choosing a proper similarity measure to compare human XBEs represented as IFSs, they need to be interpreted with caution because of the dependency of the simulated XBEs with the learning algorithm and the context that were chosen for the simulations. Hence, further studies with evaluations provided by human evaluators are recommended.

References

1. Atanassov, K.T.: Intuitionistic fuzzy sets. Fuzzy Sets Syst. **20**(1), 87–96 (1986)
2. Atanassov, K.T.: On Intuitionistic Fuzzy Sets Theory: Studies in Fuzziness and Soft Computing, vol. 283. Springer, Heidelberg (2012)
3. Buckley, C., Salton, G., Allan, J.: The effect of adding relevance information in a relevance feedback environment. In: Croft, B.W., van Rijsbergen, C.J. (eds.) SIGIR 1994, pp. 292–300. Springer, London (1994). http://dl.acm.org/citation.cfm?id=188490.188586
4. Burges, C.J.: A tutorial on support vector machines for pattern recognition. Data Min. Knowl. Discov. **2**(2), 121–167 (1998)
5. Joachims, T.: Making large-scale SVM learning practical. In: Schölkopf, B., Burges, C., Smola, A. (eds.) Advances in Kernel Methods - Support Vector Learning, chap. 11, pp. 169–184. MIT Press, Cambridge, MA (1999)
6. Joachims, T.. Text categorization with support vector machines: learning with many relevant features. In: Nédellec, C., Rouveirol, C. (eds.) ECML 1998. LNCS, vol. 1398, pp. 137–142. Springer, Heidelberg (1998). doi:10.1007/BFb0026683
7. Lewis, D.D., Yang, Y., Rose, T.G., Li, F.: RCV1: a new benchmark collection for text categorization research. J. Mach. Learn. Res. **5**, 361–397 (2004)
8. Loor, M., De Tré, G.: Connotation-differential prints - comparing what is connoted through (fuzzy) evaluations. In: Proceedings of the International Conference on Fuzzy Computation Theory and Applications, pp. 127–136 (2014)
9. Loor, M., De Tré, G.: Vector based similarity measure for intuitionistic fuzzy sets. In: Atanassov, K.T., Baczyński, M., Drewniak, J., Kacprzyk,J., Krawczak, M., Szmidt, E., Wygralak, M., Zadrożny, S. (eds.) Modern Approaches in Fuzzy Sets, Intuitionistic Fuzzy Sets, Generalized Nets and Related Topics: Volume I: Foundations, SRI-PAS, pp. 105–127 (2014)
10. Loor, M., De Tré, G.: Choosing suitable similarity measures to compare intuitionistic fuzzy sets that represent experience-based evaluation sets. In: Proceedings of the 7th International Joint Conference on Computational Intelligence (IJCCI 2015) - Volume 2: FCT, pp. 57—68 (2015)

11. Rose, T., Stevenson, M., Whitehead, M.: The reuters corpus volume 1-from yesterday's news to tomorrow's language resources. In: LREC, vol. 2, pp. 827–832 (2002)
12. Szmidt, E.: Similarity measures between intuitionistic fuzzy sets. Distances and Similarities in Intuitionistic Fuzzy Sets. Studies in Fuzziness and Soft Computing, vol. 307, pp. 87–129. Springer, Cham (2014)
13. Szmidt, E., Kacprzyk, J.: Distances between intuitionistic fuzzy sets. Fuzzy Sets Syst. **114**(3), 505–518 (2000)
14. Szmidt, E., Kacprzyk, J.: A concept of similarity for intuitionistic fuzzy sets and its use in group decision making. In: IEEE International Conference on Fuzzy Systems, pp. 1129–1134 (2004)
15. Szmidt, E., Kacprzyk, J.: Geometric similarity measures for the intuitionistic fuzzy sets. In: 8th conference of the European Society for Fuzzy Logic and Technology (EUSFLAT-13), pp. 840–847. Atlantis Press (2013)
16. Tversky, A.: Features of similarity. Psychol. Rev. **84**(4), 327 (1977)
17. Vapnik, V.N.: The Nature of Statistical Learning Theory. Springer, New York (1995)
18. Vapnik, V.N., Vapnik, V.: Statistical Learning Theory, 1st edn. Wiley, New York (1998)
19. Zadeh, L.: Fuzzy sets. Inf. Control **8**(3), 338–353 (1965)

Implementing Adaptive Vectorial Centroid in Bayesian Logistic Regression for Interval Type-2 Fuzzy Sets

Ku Muhammad Naim Ku Khalif and Alexander Gegov[✉]

School of Computing, University of Portsmouth,
Buckingham Building, Lion Terrace, Portsmouth PO1 3HE, UK
{muhammad.khalif,alexander.gegov}@port.ac.uk

Abstract. A prior distributions in standard Bayesian knowledge are assumed to be classical probability distribution. It is required to representable those probabilities of fuzzy events based on Bayesian knowledge. Propelled by such real applications, in this research study, the theoretical foundations of Vectorial Centroid of interval type-2 fuzzy set with Bayesian logistic regression is introduced. As opposed of utilising type-1 fuzzy set, type-2 fuzzy set is recommended based on the involvement of uncertainty quantity. It additionally highlights the association of fuzzy sets with Bayesian logistic regression permits the use of fuzzy attributes by considering the need of human intuition in data analysis. It may be worth including here that this proposed methodology then applied for BUPA liver-disorder dataset and validated theoretically and empirically.

Keywords: Machine learning · Bayesian logistic regression · Interval type-2 fuzzy set · Defuzzification · Vectorial centroid · Human intuition · Uncertainty

1 Introduction

Through the most recent decade, uncertainty problems are common issue in complex systems. In describing uncertainty, a lot of techniques have drawn the attentions of researchers and applied scientist over last decade. Decisions are made based on information given which known as data. However, information about decision is always uncertain. In real-world phenomena, the uncertain information may consist of randomness, vagueness and fuzziness. Machine learning has always been considered as an integral part of the field of artificial intelligence. In artificial intelligence research area, the main problems that always arise are: how to reason uncertain information precisely and: how to reason using uncertain information [1]. Machine learning is certainly one of the most significant subfields of modern artificial intelligence. In recent years, machine learning systems have been adopted standard framework to deal with imprecision in data analysis.

In dealing with imprecise data, type-1 fuzzy set is used as a unique tool to erase these imprecision appropriately. Uncertainty is closely connected with probability, which establishes the formal framework in machine learning systems. Uncertainty and

© Springer International Publishing AG 2017
J.J. Merelo et al. (eds.), *Computational Intelligence*, Studies in Computational Intelligence 669,
DOI 10.1007/978-3-319-48506-5_16

fuzziness are very prominent phenomena in science and engineering applications, where most of researchers nowadays are often used type-1 fuzzy for their case studies. Some of the input data sets, we can't describe straight away or objectively because they have different interpretations and very subjective. Even, type-1 fuzzy set can't tackle the uncertainty component completely because the degree of membership grade of type-1 fuzzy set is focusing on imprecision only. Type-2 fuzzy set is capable to deal with uncertainty or approximate reasoning, especially for the machine learning systems with a mathematical model that is difficult to derive. [2] claim that type-1 fuzzy set only describe imprecise not uncertainty. On particular motivation for the further interest in type-1 fuzzy set that its' provide a better scope for modelling uncertainty than type-1 fuzzy set [3].

Zadeh [4] introduced fuzzy set theory in representing vagueness or imprecision in a mathematical approach. In order to do so, the foremost motivation of using fuzzy set shows its ability in appropriately dealing with imprecise numerical quantities and subjective preferences of decision makers [5]. Fuzzy numbers are represented as possibility distribution where most of the real-world phenomena exist in nature are fuzzy rather than probabilistic or deterministic [6]. It was specifically designed to mathematically represent to randomness and also provide formalised tools for dealing with imprecision essential to many real problems nowadays. Technologies nowadays have been developed in fuzzy set that have potential to support all of the steps that encompass a process of model orientation and knowledge discovery. In particular, fuzzy set theory can be used in data analysis to model vague data in terms of fuzzy set. These are some contributions that fuzzy set can assist machine learning which are: (1) graduality; (2) granulity; (3) interpretability; (4) robustness; (5) representation of uncertainty; (6) incorporation of background knowledge and; (7) aggregation, combination and information fusion [7]. In particular, fuzzy set theory can already be used in the data selection or data processing. For analysing the fuzzy data, there are two different ways: (1) fuzzifying the mapping from data to model and: (2) embed the data into more complex mathematical spaces, like fuzzy metric spaces.

Type-2 fuzzy set notion was introduced by Zadeh [8] as an extension of the type-1 fuzzy set. In accordance to [9], type-2 fuzzy set can be considered as fuzzy membership function where the membership value for each element in type-2 is a fuzzy set in $[0, 1]$, different with type-1 fuzzy set where the membership value is in crisp condition between $[0, 1]$. The interval type-2 fuzzy set is extensively used in type-2 fuzzy set family in many practical science and engineering areas [10]. The participation of higher level uncertainty of type-2 fuzzy set compared to type-1, provide additional degrees of freedom to represent the uncertainty and the fuzziness of real-world problems. Uncertainty can be divided into two types which are inter and intra personal uncertainties, in improvising the representation of type-1 fuzzy set in the literature of fuzzy sets. This is also supported by [11] where there are supposedly two kinds of uncertainties that are related to linguistic characteristics namely intra-personal uncertainty and inter-personal uncertainty, In fact, a lot of experts have applied interval type-2 fuzzy set in machine learning systems analysis. Due to implementing interval type-2 fuzzy set in real-world problems, the way to handle is different and much more complex compared to type-1 fuzzy set. The contribution of centroid of type-2 fuzzy set still now commonly used uncertainty measure for modelling problems. Interval type-2 fuzzy set offer an approach to handle knowledge uncertainty in machine learning systems.

Defuzzification plays an important role in the performance of fuzzy system's modelling techniques [12]. Generally, defuzzification process is guided by the output fuzzy subset that one value would be the selected as a single crisp value as the system output. There are variation defuzzification methods have largely developed. Nevertheless, they have difference performances in difference applications and there is a general method can satisfactory performance in all conditions [13]. The centroid defuzzification methods of fuzzy numbers have been explored for the last few decades that commonly used and have been applied in various discipline areas. The computation complexity of type-2 fuzzy set is very problematic to handle into practical applications because of characterised by their footprint of uncertainty (FOU) [14]. Two typical directions in computing type-2 fuzzy set intensively: (1) type-reduction [9, 14, 15] and; (2) direct defuzzification [16]. Most experts applied type-reduction methods in handling the complexity of type-2 fuzzy set by finding the equivalent type-1 fuzzy set. Though, direct defuzzification computational for type-2 fuzzy set is still under study.

The possibility mean value for interval sets was introduced by [17] where the notations of lower possibilistic and upper possibilistic mean values are defined the interval-valued possibilistic mean. From probabilistic viewpoint, the possibility mean value of fuzzy sets can be represented as expected values which is same function as direct defuzzification method where it does not need type-reduction or conversion stage into type-1 fuzzy to get the outputs. [16] extend the concept of [17] about possibility mean value of type-1 fuzzy set which introduce the lower and upper possibility mean value for interval type-2 fuzzy set. In this paper, the comparative simulation results between the proposed of the extension of Vectorial Centroid [18] and possibility mean value proposed by [16] method for interval type-2 fuzzy set, where in some cases it will give illogical and irrational results that inconsistent with human intuition. This method can't cater all possible cases of interval type-2 fuzzy set properly since some of the results are dispersed far away from the closed interval bounded by the expectations calculated from its upper and lower distribution functions.

Due to growths in computational capability and technology development, data are being generated for understanding details real-world problems in health nowadays that associated with clinic test, diseases, disorder, genetic cases and so forth [19]. Yet, with the availability of large datasets become the essential challenges of a new methods of statistical analysis and modelling. Logistic regression model is one of machine learning systems that used in handling these problems with high-dimensional data. The dataset that represents binary dependent attribute where it uses logit transform to predict probabilities directly. Logistic regression is a model-based approach to mapping observers' distribution. When applied within Bayesian setting, logistic regression provides a useful platform for integrating expert knowledge, in the form of a prior, with empirical data [20]. Probability is a complete with parametric models that let us characterised random uncertainty [10]. Prior knowledge can be amalgamated into Bayesian logistic regression and the method is computationally efficient.

Issues with respects to representation capability of fuzzy sets in machine learning systems on uncertainty become one of the significant problems in decision making environments. The aim of the present paper is to illustrate the extension of Vectorial

Centroid [18] method for interval type-2 fuzzy set that consider the illustration of Bayesian algorithm about the parameters of a logistic regression model. Aiming at the problems pointed out above, new centroid defuzzification for interval type-2 fuzzy set is proposed that easy to understand, more flexible and more intelligent compared to existing methods. The proposed method also considers the need of human intuition and gives logical results while dealing with machine learning systems. In this research study, classification dataset with binary dependent attribute is used. The observations in this dataset, we worked on "BUPA liver-disorder" that were sampled by BUPA Medical Research Ltd. There are 7 attributes that consist of six independent attributes and one binary dependent attribute. The BUPA liver-disorder dataset represents blood tests indicating a property of liver disorders that may increase from excessive alcohol consumption.

The remainder of this paper is organised as follows: Sect. 2 introduces the concepts of type-2 fuzzy set, interval type-2 fuzzy set, centroid method that proposed by [16] and Bayesian logistic regression. Section 3 views the proposed new centroid method for interval type-2 fuzzy sets using Vectorial Centroid method. Section 4 illustrates the implementation of proposed method with Bayesian logistic regression in BUPA liver-disorder and compares the results with [16] method. Section 5 summarises the main results and draws conclusion.

2 Preliminaries

2.1 Interval Type-2 Fuzzy Set

Definition 1: A type-2 fuzzy set $\tilde{\tilde{A}}$ in the universe of discourse X represented by the type-2 membership function μ. If all $\mu_{\tilde{\tilde{A}}}(x, u) = 1$, then $\tilde{\tilde{A}}$ is called an interval type-2 fuzzy set. An interval type-2 fuzzy set can be considered as a special case type-2 fuzzy set, denoted as follows [5]:

$$\tilde{\tilde{A}} = \int_{x \in X} \int_{u \in J_x} 1/(x, u), \text{where } J_x \subseteq [0, 1] \tag{1}$$

Definition 2: The upper and lower membership function of an interval type-2 fuzzy set are type-1 fuzzy set membership functions, respectively. A trapezoidal interval type-2 fuzzy set can be represented by $\tilde{\tilde{A}}_i = (\tilde{A}_i^U, \tilde{A}_i^L) = ((a_{i1}^U, a_{i2}^U, a_{i3}^U, a_{i4}^U; H_1(\tilde{A}_i^U), \tilde{H}_2(\tilde{A}_i^U)),$ $(a_{i1}^L, a_{i2}^L, a_{i3}^L, a_{i4}^L; H_1(\tilde{A}_i^L), H_2(\tilde{A}_i^L)))$ where can be shown in Fig. 1 [16]. The $\tilde{A}_i^U$ and $\tilde{A}_i^L$ are type-1 fuzzy sets, $a_{i1}^U, a_{i2}^U, a_{i3}^U, a_{i4}^U, a_{i1}^L, a_{i2}^L, a_{i3}^L$ and a_{i4}^L are the reference points of the interval type-2 fuzzy set $\tilde{\tilde{A}}$, $H_j(\tilde{A}_i^U)$ denote the membership value of the element $a_{i(j+1)}^U$ in the upper trapezoidal membership function $\tilde{A}_i^U$, $1 \leq j \leq 2$, $H_j(\tilde{A}_i^L)$ denotes the membership value of the element $a_{i(j+1)}^L$ in the lower trapezoidal membership function

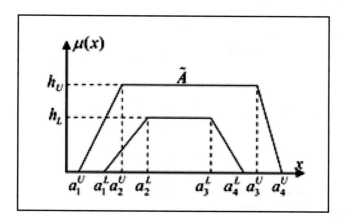

Fig. 1. The upper trapezoidal membership function $\tilde{A}_i^U$ and lower trapezoidal membership function $\tilde{A}_i^L$ of interval type-2 fuzzy set.

$\tilde{A}_i^L, 1 \leq j \leq 2$, and for $H_1(\tilde{A}_i^U) \in [0,1], H_2(\tilde{A}_i^U) \in [0,1], H_1(\tilde{A}_i^L) \in [0,1]$ and $1 \leq i \leq n$, $H_2(\tilde{A}_i^U) \in [0,1]$ [5].

2.2 Bayesian Logistic Regression

The principal of Bayesian inference for logistic regression analyses follows the typical pattern for Bayesian analysis [21]:

1. Write down the likelihood function of the data
2. Form a prior distribution over all unidentified parameters
3. Find posterior distribution using Bayes theorem over all parameters

Likelihood Function: The likelihood contribution from the ith subject is binomial

$$likelihood_i = \pi(x_i)^{y_i}(1 - \pi(x_i))^{(1-y_i)} \tag{2}$$

where $\pi(x_i)$ represents the probability of the event for subject i, which has covariate vector x_i and y_i specifies the liver-disorder $y_i = 1$, or liver-normal $y_i = 2$ of the event for the subject. Logistic regression is denoted as

$$\pi(x) = \frac{e^{\beta_0 + \beta_1 X_i + \dots + \beta_p X_p}}{1 + e^{\beta_0 + \beta_1 X_i + \dots + \beta_p X_p}} \tag{3}$$

So the likelihood from the ith subject is

$$likelihood_i = \left(\frac{e^{\beta_0 + \beta_1 X_{i1} + \dots + \beta_p X_{ip}}}{1 + e^{\beta_0 + \beta_1 X_{i1} + \dots + \beta_p X_{ip}}} \right)^{y_i} \left(1 - \frac{e^{\beta_0 + \beta_1 X_{i1} + \dots + \beta_p X_{ip}}}{1 + e^{\beta_0 + \beta_1 X_{i1} + \dots + \beta_p X_{ip}}} \right)^{(1-y_i)}$$

$$likelihood_i = \prod_{i=1}^{n}\left[\left(\frac{e^{\beta_0+\beta_1X_{i1}+\dots+\beta_pX_{ip}}}{1+e^{\beta_0+\beta_1X_{i1}+\dots+\beta_pX_{ip}}}\right)^{y_i}\left(1-\frac{e^{\beta_0+\beta_1X_{i1}+\dots+\beta_pX_{ip}}}{1+e^{\beta_0+\beta_1X_{i1}+\dots+\beta_pX_{ip}}}\right)^{(1-y_i)}\right]$$

(4)

Prior Distribution: In general, any prior distribution can be used, depending on the available prior information.

$$\beta_j \sim Normal(c_j, \sigma_j^2)$$

(5)

The most common choice for c is zero, and σ is usually chosen to be large enough to be considered as non-informative in the range from $\sigma = 10$ to $\sigma = 100$.

Posterior Distribution via Bayes Theorem: The posterior distribution is divided by multiplying the prior distribution over all parameter by the full likelihood function, so that

$$Posterior = \prod_{i=1}^{n}\left[\left(\frac{e^{\beta_0+\beta_1X_{i1}+\dots+\beta_pX_{ip}}}{1+e^{\beta_0+\beta_1X_{i1}+\dots+\beta_pX_{ip}}}\right)^{y_i}\left(1-\frac{e^{\beta_0+\beta_1X_{i1}+\dots+\beta_pX_{ip}}}{1+e^{\beta_0+\beta_1X_{i1}+\dots+\beta_pX_{ip}}}\right)^{(1-y_i)}\right]$$
$$\times \prod_{j=0}^{p}\frac{1}{\sqrt{2\pi}\sigma_j}\exp\left\{-\frac{1}{2}\left(\frac{\beta_j-c_j}{\sigma_j}\right)^2\right\}$$

(6)

The latter part of the above expression being recognised as normal distribution for the β parameters. For liver-disorder classification problem, $p(y=1|\beta_p x_p)$, will be an estimate of the probability that the pth document belongs to the category. The decision of whether to assign the category can be based on comparing the probability estimate with a threshold or by computing which decision gives optimal expected utility.

2.3   Interval-Valued Possibility Mean Value

The concept of interval-valued possibility mean value are divided into two parts which are lower and upper possibility mean value. The lower $\underline{M}(\widetilde{\widetilde{A}})$ and upper $\overline{M}(\widetilde{\widetilde{A}})$ possibility mean value for interval type-2 fuzzy set are denoted as follow [16]:

$$\underline{M}(\widetilde{\widetilde{A}}) = \int_0^{h_U}\alpha\left(a_1^U + \frac{a_2^U - a_1^U}{h_u}\alpha\right)d\alpha + \int_0^{h_L}\beta\left(a_1^L + \frac{a_2^L - a_1^L}{h_L}\beta\right)d\beta$$

(7)

$$\overline{M}(\widetilde{\widetilde{A}}) = \int_0^{h_U}\alpha\left(a_4^U + \frac{a_3^U - a_4^U}{h_u}\alpha\right)d\alpha + \int_0^{h_L}\beta\left(a_4^L + \frac{a_3^L - a_4^L}{h_L}\beta\right)d\beta$$

(8)

For crisp value, we can compute by using the average of lower and upper possibility mean value above that denoted as follows

$$M(\overset{\approx}{A}) = \frac{M + \overline{M}}{2} \tag{9}$$

In this paper, the numerical analysis for proposed methodology is compared with interval-valued possibility mean value that proposed by [16].

3 Proposed Method

As noted in the introduction, the useful of interval type-2 fuzzy sets nowadays are widely applied in many research areas in dealing with uncertainty in data analysis which consistent with human intuition at the same time. A lot of researchers attempt to eliminate the need of human intuition in data analysis processes. Human intuition is strictly can't be eliminated because it can drive us towards uncertainty problems. This study simplify the concept of attributes to $\mu_{\overset{\approx}{A}} \in [0, 1]$ for fuzzy events. The values of attributes correspond to interval type-2 fuzzy sets. This study proposed a new centroid defuzzification method for Bayesian logistic regression algorithm. The methodology consist of two stages here namely:

A. *Stage one*
   The development of an extension of the Vectorial Centroid defuzzification for interval type-2 fuzzy set.

Let consider by $\overset{\approx}{A}_i = (\tilde{A}_i^U, \tilde{A}_i^L) = ((a_{i1}^U, a_{i2}^U, a_{i3}^U, a_{i4}^U; H_1(\tilde{A}_i^U), \tilde{H}_2(\tilde{A}_i^L)), (a_{i1}^L, a_{i2}^L, a_{i3}^L, a_{i4}^L; H_1(\tilde{A}_i^L), H_2(\tilde{A}_i^L)))$ as the interval type-2 fuzzy set. The complete method process of Vectorial Centroid is signified as follows:

Step 1: Compute the centroid points of the three parts of α, β and γ in interval type-2 fuzzy set representation as shown in Fig. 2.

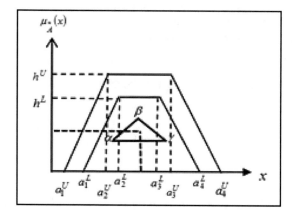

Fig. 2. Vectorial Centroid plane representation.

$$\alpha_{\overline{\alpha},\underline{\alpha}}(x,y) = \left(\frac{1}{3}(a_1^U + \frac{1}{2}a_2^U) + \frac{1}{3}(a_1^L + \frac{1}{2}a_2^L), \frac{1}{6}(h^U + h^L)\right) \qquad (10)$$

$$\beta_{\overline{\beta},\underline{\beta}}(x,y) = \left(\frac{1}{4}(a_2^U + a_3^U + a_2^L + a_3^L), \frac{1}{4}(h^U + h^L)\right) \qquad (11)$$

$$\gamma_{\overline{\gamma},\underline{\gamma}}(x,y) = \left(\frac{1}{3}(a_3^U + \frac{1}{2}a_4^U) + \frac{1}{3}(a_3^L + \frac{1}{2}a_4^L), \frac{1}{6}(h^U + h^L)\right) \qquad (12)$$

Step 2: Connect all vertices centroids points of α, β and γ each other, where it will generate another triangular plane inside of trapezoid plane.

Step 3: The centroid index of Vectorial Centroid of $(\tilde{x},\tilde{y})$ with vertices α, β and γ can be calculated as:

$$VC_{\underset{A}{\approx}}(\tilde{x},\tilde{y}) = \left(\frac{\alpha_{\overline{\alpha},\underline{\alpha}}(x,y) + \beta_{\overline{\beta},\underline{\beta}}(x,y) + \gamma_{\overline{\gamma},\underline{\gamma}}(x,y)}{3}, \beta_{\overline{\beta},\underline{\beta}}(x,y)\right.$$
$$\left. + \left[\frac{2}{3}\left(\frac{\alpha_{\overline{\alpha},\underline{\alpha}}(x,y) + \gamma_{\overline{\gamma},\underline{\gamma}}(x,y)}{2} - \beta_{\overline{\beta},\underline{\beta}}(x,y)\right)\right]\right) \qquad (13)$$

Vectorial Centroid can be summarised as

$$VC_{\underset{A}{\approx}}(\tilde{x},\tilde{y}) = \left(\frac{1}{9}\left[a_1^U + a_1^L + \frac{5}{4}(a_2^U + a_2^L) + \frac{7}{4}(a_3^U + a_3^L)\right.\right.$$
$$\left.\left. + \frac{1}{2}(a_4^U + a_4^L)\right], \frac{11}{36}(h^U + h^L)\right) \qquad (14)$$

where

α: the centroid coordinate of first triangle plane
β: the centroid coordinate of rectangle plane
γ: the centroid coordinate of second triangle plane

$(\tilde{x},\tilde{y})$: the centroid coordinate of fuzzy number $\widetilde{\widetilde{A}}$

Centroid index of Vectorial Centroid can be generated using Euclidean distance by [23]:

$$R(\widetilde{\widetilde{A}}) = \sqrt{\tilde{x}^2 + \tilde{y}^2} \qquad (15)$$

B. *Stage two*
   The implementation of Vectorial Centroid in Bayesian logistic regression.

Integrating fuzzy sets with Bayesian logistic regression in fuzzy states of nature, where if there is fuzzy dataset, defuzzification process is needed in converting into crisp values where at the same time the fuzzy nature is not lost. Reinterpretation of degree $\mu_{\tilde{\approx}} \in [0, 1]$ using Vectorial Centroid to the $P(y = 1|\beta_p X_p)$ is developed as follows:

Step 1: Lift the reintergration of the fuzzy values membership function using trapezoidal interval type-2 fuzzy sets. Vectorial Centroid formulation are applied for trapezoidal interval type-2 fuzzy set rule formula. The $\mu_{\tilde{\approx}}$ represents as $\tilde{\tilde{A}}_1 = (\tilde{A}_1^U, \tilde{A}_1^L) = ((a_{11}^U, a_{12}^U, a_{13}^U, a_{14}^U; H_1(\tilde{A}_1^U), H_2(\tilde{A}_1^U)), (a_{11}^L, a_{12}^L, a_{14}^L; H_1(\tilde{A}_1^L), H_2(\tilde{A}_1^L)))$ in calculation to avoid cluttering.

Step 2: The centroid index of Vectorial Centroid, $R(\tilde{\tilde{A}})$ is inserted into Bayesian logistic regression rule as

$$R(\tilde{\tilde{A}}) = \sqrt{\tilde{x}^2 + \tilde{y}^2} = \mu(\tilde{\tilde{A}})$$

The computational process of likelihood and posterior distribution of fuzzy Bayesian logistic regression using Vectorial Centroid are denoted as

$$likelihood_i = \prod_{i=1}^{n} \left[\left(\frac{e^{\beta_0 + \beta_1 \mu(\tilde{\tilde{A}})_{i1} + \ldots + \beta_p \mu(\tilde{\tilde{A}})_{ip}}}{1 + e^{\beta_0 + \beta_1 \mu(\tilde{\tilde{A}})_{i1} + \ldots + \beta_p \mu(\tilde{\tilde{A}})_{ip}}} \right)^{y_i} \left(1 - \frac{e^{\beta_0 + \beta_1 \mu(\tilde{\tilde{A}})_{i1} + \ldots + \beta_p \mu(\tilde{\tilde{A}})_{ip}}}{1 + e^{\beta_0 + \beta_1 \mu(\tilde{\tilde{A}})_{i1} + \ldots + \beta_p \mu(\tilde{\tilde{A}})_{ip}}} \right)^{(1-y_i)} \right]$$

$$(16)$$

$$Posterior = \prod_{i=1}^{n} \left[\left(\frac{e^{\beta_0 + \beta_1 \mu(\tilde{\tilde{A}})_{i1} + \ldots + \beta_p \mu(\tilde{\tilde{A}})_{ip}}}{1 + e^{\beta_0 + \beta_1 \mu(\tilde{\tilde{A}})_{i1} + \ldots + \beta_p \mu(\tilde{\tilde{A}})_{ip}}} \right)^{y_i} \left(1 - \frac{e^{\beta_0 + \beta_1 \mu(\tilde{\tilde{A}})_{i1} + \ldots + \beta_p \mu(\tilde{\tilde{A}})_{ip}}}{1 + e^{\beta_0 + \beta_1 \mu(\tilde{\tilde{A}})_{i1} + \ldots + \beta_p \mu(\tilde{\tilde{A}})_{ip}}} \right)^{(1-y_i)} \right]$$
$$\times \prod_{j=0}^{p} \frac{1}{\sqrt{2\pi}\sigma_j} \exp\left\{ -\frac{1}{2} \left(\frac{\beta_j - c_j}{\sigma_j} \right)^2 \right\}$$

$$(17)$$

4 Experimental Settings

The experiment is conducted using 10-fold cross validation on BUPA liver-disorder dataset from UCI machine learning repository [22] is used where donated by BUPA Medical Research Ltd. This liver-disorder classification dataset has 345 examples, 7 attributes and binary classes for dependent attribute. The first 5 attributes are measurements taken by blood tests that are thought to be sensitive to liver-disorders and might arise from excessive alcohol consumption. The sixth attribute is a sort of selector attribute where the subjects are single male individuals. The seventh attribute shows a selector on the dataset which being used to split into two categories that indicating the class identity. The attributes include:

a. Mean corpuscular volume,
b. Alkaline phosphatase,
c. Aspartate aminotransferase,
d. Gamma-glutamyl transpeptidase,
e. Alamine aminotransferase,
f. Number if half-pint equivalents of alcoholic beverage drunk per day, and
g. Output attributes either liver disorder or liver normal

Among all the people, there are 145 belonging to the liver-disorder group and 200 belonging to the liver-normal group. These attributes are selected with the aid of experts. The original dataset are fuzzified randomly in interval type-2 fuzzy sets form in operating centroid methods. Suppose that attribute alkaline phosphatase, aspartate aminotransferase, gamma-glutamyl transpeptidase and alamine aminotransferase are fuzzy events, $\mu_{\beta_i x_i}$. Below describes the example of interval type-2 fuzzy sets are used in this research study:

Example: If the trapezoidal interval type-2 fuzzy set $\tilde{\tilde{A}}_i = (\tilde{A}_i^U, \tilde{A}_i^L) = ((15.35, 16.68, 18.06, 20.51; 1), (16, 17, 18, 19; 0.9))$, then the centre points are computed using proposed of extension Vectorial Centroid and interval-valued possibility mean value [16] formulation respectively as follows:

Vectorial Centroid:
$VC(x) = 17.3678$ and $VC(y) = 0.58056$

Centroid index Vectorial Centroid, $VC(\tilde{\tilde{R}}) = 17.3775$
Interval-Valued Possibility Mean Value:

$$M(\tilde{\tilde{A}}) = \left[M_*(\tilde{\tilde{A}}), M^*(\tilde{\tilde{A}})\right] = [14.8683, 16.8633]$$

Crisp possibility mean value, $M(\tilde{\tilde{A}}) = 15.8658$

5   Simulation Results

This section illustrates the validation process of the methodology in theoretically and empirically. Therefore, the theoretical of Vectorial Centroid validation process are as follow:

A. *Stage one*
  The relevant properties considered for qualifying the applicability of centroid for interval type-2 fuzzy set, where they depend on the practicality within the area of research however, they are not considered as complete. Therefore, without loss of generality, the relevant properties of the centroid are as follow:

Let $\tilde{\tilde{A}}$ and $\tilde{\tilde{B}}$ are be trapezoidal and triangular interval type-2 fuzzy set respectively, while $VC_{\tilde{\tilde{A}}}(\tilde{x}, \tilde{y})$ and $VC_{\tilde{\tilde{B}}}(\tilde{x}, \tilde{y})$ be centroid points for $\tilde{\tilde{A}}$ and $\tilde{\tilde{B}}$ respectively. Centroid index of Vectorial Centroid, (R) shows the crisp value of centroid point that is denoted as $R(\tilde{\tilde{A}}) = \sqrt{\tilde{x}^2 + \tilde{y}^2}$.

Property 1: *If $\widetilde{\widetilde{A}}$ and $\widetilde{\widetilde{B}}$ are embedded and symmetry, then $R(\widetilde{\widetilde{A}}) > R(\widetilde{\widetilde{B}})$.*

Proof: Since $\widetilde{\widetilde{A}}$ and $\widetilde{\widetilde{B}}$ are embedded and symmetry, hence from Eq. (15) we have $\sqrt{\tilde{x}_{\widetilde{A}}^2 + \tilde{y}_{\widetilde{A}}^2} > \sqrt{\tilde{x}_{\widetilde{B}}^2 + \tilde{y}_{\widetilde{B}}^2}$. Therefore, $R(\widetilde{\widetilde{A}}) > R(\widetilde{\widetilde{B}})$.

Property 2: *If $\widetilde{\widetilde{A}}$ and $\widetilde{\widetilde{B}}$ are embedded with $(h^U, h^L)_{\widetilde{A}} > (h^U, h^L)_{\widetilde{B}}$, then $R(\tilde{A}) > R(\tilde{B})$.*

Proof: Since $\widetilde{\widetilde{A}}$ and $\widetilde{\widetilde{B}}$ are embedded and *with* $(h^U, h^L)_{\widetilde{A}} \approx (h^U, h^L)_{\widetilde{B}}$, hence we know that $\tilde{y}_{\tilde{A}} > \tilde{y}_{\tilde{B}}$.

Then, from Eq. (15) we have $\sqrt{\tilde{x}_{\widetilde{A}}^2 + \tilde{y}_{\widetilde{A}}^2} > \sqrt{\tilde{x}_{\widetilde{B}}^2 + \tilde{y}_{\widetilde{B}}^2}$. Therefore, $R(\widetilde{\widetilde{A}}) > R(\widetilde{\widetilde{B}})$.

Property 3: *If $\widetilde{\widetilde{A}}$ is singleton fuzzy number, then $R(\widetilde{\widetilde{A}}) = \sqrt{\tilde{x}_{\widetilde{A}}^2 + \tilde{y}_{\widetilde{A}}^2}$.*

Proof: For any crisp (real) interval type-2 fuzzy set, we know that $a_1^U = a_2^U = a_3^U = a_4^U = a_1^L = a_2^L = a_3^L = a_4^L = \tilde{x}_{\widetilde{A}}$ which are equivalent to Eq. (14). Therefore, $R(\widetilde{\widetilde{A}}) = \sqrt{\tilde{x}_{\widetilde{A}}^2 + \tilde{y}_{\widetilde{A}}^2}$.

Property 4: *If $\widetilde{\widetilde{A}}$ is any symmetrical or asymmetrical interval type-2 fuzzy number, then $a_1^U < R(\widetilde{\widetilde{A}}) < a_4^U$.*

Proof: Since any symmetrical or asymmetrical interval type-2 fuzzy set has $a_1^U \leq a_2^U \leq a_3^U \leq a_4^U$, hence $a_1^U < VC_{\widetilde{A}}(\tilde{x}, \tilde{y}) < a_4^U$. Therefore, $a_1^U < R(\widetilde{\widetilde{A}}) < a_4^U$.

B. *Stage two*

Aforementioned, the empirical validation is implemented where the BUPA liver-disorder data set is used in conducting Bayesian Logistic Regression data analysis.

Note that this study is considered all type of possible interval type-2 fuzzy set for attributes randomly as Figs. 3, 4, 5, 6, 7, 8, 9, 10, 11 and 12.

In simulation analysis, the data set is randomly partitioned into 10 equal sized partitions. Then, one of the partitions is used to test, while the rest of the partitions is dedicated to train the base learner. This procedure is repeated ten times so that each partition is used for the test exactly one time. Here, a mean accuracy of the individual results is combined. Table 1 presents a comparative results between classical Bayesian logistic regression (BLR-Classic), Bayesian logistic regression using possibility mean value [16] method (BLR-PMV), and Bayesian logistic regression using the extension of Vectorial Centroid (BLR-VC). The comparison results are based on accuracy, precision, sensitivity, specificity, Kappa statistic, and some error terms: Mean Absolute Error (MAE), Root Mean Square Error (RMSE), Relative Absolute Error (RAE) and Root Relative Square Error (RRSE).

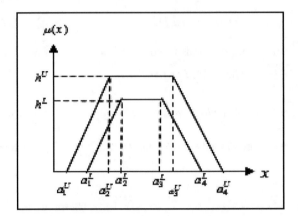

Fig. 3. Trapezoidal non-normal symmetry.

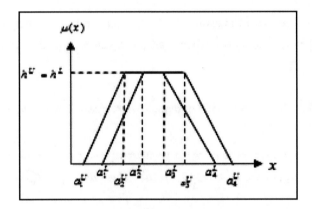

Fig. 4. Trapezoidal normal symmetry.

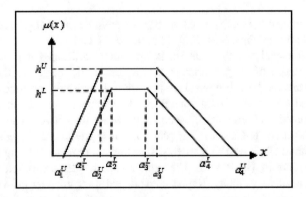

Fig. 5. Trapezoidal non-normal asymmetry.

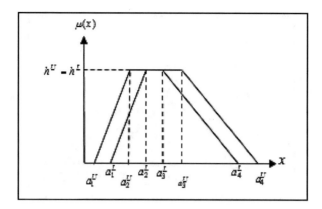

Fig. 6. Trapezoidal normal asymmetry.

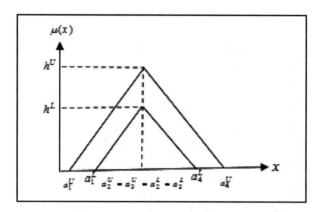

Fig. 7. Triangular non-normal symmetry.

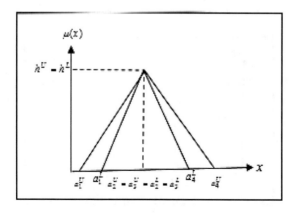

Fig. 8. Triangular normal symmetry.

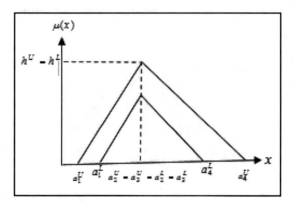

Fig. 9. Triangular non-normal asymmetry.

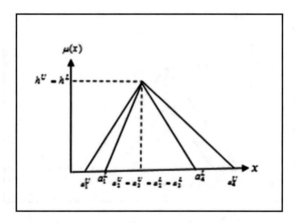

Fig. 10. Triangular normal asymmetry.

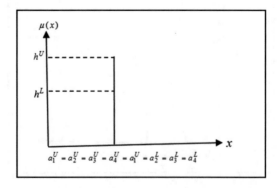

Fig. 11. Singleton non-normal.

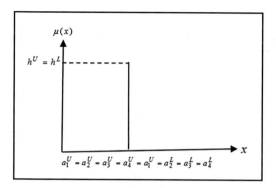

Fig. 12. Singleton Normal.

Table 1. Accuracy, precision, sensitivity, specificity Kappa statistic and error terms.

Method	BLR-Classic	BLR-PMV	BLR-VC
Accuracy	67.2464 %	58.5507 %	68.1159 %
Precision	17.67 %	1.4 %	30.34 %
Sensitivity	82 %	66.67 %	83.02 %
Specificity	64.75 %	58.41 %	65.41 %
Kappa Statistic	0.2613	0.0203	0.2832
Error:			
MAE	0.3275	0.4145	0.3188
RMSE	0.5723	0.6438 %	0.5647
RAE	67.2025 %	85.0438 %	65.4183 %
RRSE	115.9404 %	130.4259 %	114.391 %

The accuracy and precision of a measurement system plays important role in quantifying the actual measure value. It is commonly used as metric for evaluation of machine learning systems. In addition to classification, accuracy results obtained by the algorithm. The precision is dependent of accuracy where the model can be very precise but inaccurate. The higher the value of accuracy and precision, the better classification prediction is made. In this research study, Table 1 shows the classification accuracy results that show the correctness of a model classifies the dataset in each class. Below shows the formulation of accuracy and precision:

$$Accuracy : \frac{TotalPositive + TotalNegative}{Positive + Negative} \tag{17}$$

$$Precision : \frac{TotalPositive}{TotalPositive + FalsePositive} \tag{18}$$

The classification accuracy results of BLR-Classic, BLR-PMV and BLR-VC are 67.2464 %, 58.5507 % and 68.1159 % respectively. It shows that the proposed methodology is significantly more accurate and very promising compared to others.

The highest precision in this case study is BLR-VC with 30.34 %, followed by BLR-Classic with 17.67 % and BLR-PMV with 1.4 %. Precision discusses the closeness of two or more measurements to each other.

The sensitivity test mentions to the ability of the test to correctly identify those observers with positive predictive value. A high sensitivity is clearly imperative where the test is used to identify the correct class. But, specificity test is inversely proportional to sensitivity where it has the ability of the test to correctly identify those observers with negative predictive value [24]. Below are formulation to calculate sensitivity and specificity:

$$Sensitivity : \frac{TotalPositive}{TotalPositive + FalseNegative} \tag{19}$$

$$Specificity : \frac{TotalNegative}{FalsePositive + TotalNegative} \tag{20}$$

It is interesting to observe that the proposed method, BLR-VC produces the highest sensitivity and specificity value with 83.02 % and 65.41 % respectively. The results for BLR-PMV shows the lowest results for sensitivity and specificity with 66.67 % and 58.41 % respectively. It depicts that the goodness of prediction of both tests for BLR-PMV is lesser than BLR-Classic and BLR-VC.

Kappa statistic technique is used to measure the agreement of two classifiers and estimate the probability of two classifiers agree simply by chance [25]. Known as chance-corrected measure of agreement between classification and the true classes, it is an evaluation metric which is based on the difference between the actual agreement in the error matrix and the chance agreement. The values for Kappa range from 0 to 1 and the higher the value of Kappa statistic, the stronger the strength of agreement between two classifiers by chance.

$$KappaStat, k = \frac{p_o - p_e}{1 - p_e} \tag{21}$$

where

p_o is relative observed agreement among raters,
p_e is the hypothetical probability of chance agreement.

BLR-VC shows the highest value of Kappa statistic with 0.2832 followed by BLR-Classic and BLR-PMV with 0.2613 and 0.0203 respectively. It seems that, the proposed methodology produced strong agreement between two classifiers by chance.

The final part in Table 1 depicts the errors for the experiment carried out. The errors are computed using Mean Absolute Error (MAE), Root Mean Square Error (RMSE), Relative Absolute Error (RAE) and Root Relative Square Error (RRSE). All the statistic error terms compare true values to theirs estimates, but do it in a slightly different way. Below depict the formulation in calculating MAE, RMSE, RAE and RRSE:

$$MAE = \frac{1}{N}\sum_{i=1}^{N}\left|\hat{\theta}_i - \theta_i\right| \tag{22}$$

$$RMSE = \sqrt{\frac{1}{N}\sum_{i=1}^{N}(\hat{\theta}_i - \theta_i)^2} \tag{23}$$

$$RAE = \frac{\sum_{i=1}^{N}\left|\hat{\theta}_i - \theta_i\right|}{\sum_{i=1}^{N}\left|\bar{\theta}_i - \theta_i\right|} \tag{24}$$

$$RRSE = \sqrt{\frac{\sum_{i=1}^{N}(\hat{\theta}_i - \theta_i)^2}{\sum_{i=1}^{N}(\bar{\theta}_i - \theta_i)^2}} \tag{25}$$

These error terms demonstrates how disperse away the estimated values from the true value of θ. MAE and RMSE calculate the average difference between those two values. For every data point, the distance is take vertically from the point of the corresponding estimated value on the curve fit and square the value. RMSE is directly interpretable in terms of measurement units that measure of goodness of fit. In RAE and RRSE, we divide those differences by the variation of θ where they have a scale from 0 to 1, then we would multiply those value by 100 to get the similarity in 0–100 scale. In this simulation results, the proposed methodology, BLR-MC performs better in error terms where all of these errors are less than BLR-Classic and BLR-PMV.

6 Conclusion

The usefulness of fuzzy Bayesian knowledge in understanding and modelling complex uncertainty associated with real-world problems is presented in this paper. This research study has carried out an extension based Vectorial Centroid for interval type-2 fuzzy set with Bayesian logistic regression. Bayesian logistic regression algorithm that takes into account the need of fuzzy events in attributes. This work consist of two stages which are: (1) The development of Vectorial Centroid defuzzification method for interval type-2 fuzzy set and: (2) The implementation of Vectorial Centroid in Bayesian logistic regression. For the primary stage, the development of new centroid method can cater all the possible cases of interval type-2 fuzzy set precisely that matching for human intuition. The implementation in Bayesian logistic regression using proposed method on stage two is easily capable constructed and handled in data analysis when dealing with fuzzy data sets. The contribution of the paper can be summarised as follows. First, the development of new defuzzification method which can cater all possible cases in interval type-2 fuzzy sets and considering human judgment or intuition. Second, the presented hybrid intelligent classification model utilised the consistency-based feature selection between new Vectorial Centroid defuzzification method and Bayesian logistic regression model. The experimental results on Table 1 indicates that the proposed hybrid classification model can obtain very promising results in terms of accuracy, precision, sensitivity, specificity, Kappa statistics and error terms.

Although the model proposed in this paper is relatively simple conceptually, some drawbacks may exist in this research study. First, the proposed hybrid classification model for interval type-2 fuzzy numbers was developed and tested on BUPA liver-disorder dataset from WEKA (Waikato Environment for Knowledge Analysis) software. The useful of interval type-2 fuzzy set for attributes are randomly computed. Hence, the implementation or development of special linguistic terms for attributes are needed for remarkable outputs. Second, the scope of this research study is focused to be automated diagnosis liver-disorder. Still, more experimental work should be enthusiastic to obtain a medical classification model with a better ability of generalization under fuzzy environment. Finally, the proposed Vectorial Centroid only applied for Bayesian logistic regression. It should be applied and compared with other machine learning systems in the future work that would make research much more convincing.

Despite the above drawbacks, this study can be profitable alternatively in the set of existing Bayesian logistic regression algorithms for various problems in machine learning such as inference, classification, clustering, regression and so forth. There are four relevant properties for centroid development are constructed and well proved in theoretical validation, where corresponding with all possible interval type-2 fuzzy set representation. Several tests for validation have been done and the results have been studied in-depth using BUPA liver-disorder classification dataset from UCI machine learning repository. With this promising preliminary simulation results, the proposed research study presents more effective in dealing with fuzzy events empirically. To conclude, the main focus of this research study can be continued in order to make some contributions by considering real case study drawn for diverse fields crossing ecology, health, genetics, finance and so forth. To take care of the hybrid of fuzzy machine learning systems, a more general concepts of general fuzzy numbers and fuzzy vectors along the characterising function must be applied in capturing the imprecision and uncertainty in data analysis.

References

1. Tang, Y., Pan, H., Xu, Y.: Fuzzy Naïve Bayes classifier base on fuzzy clustering. In: IEEE International Conference on Systems, Mans and Cybernetics, vol. 5, pp. 6pp. IEEE Press (2002)
2. Klir, G., Yuan, B.: Fuzzy Sets and Fuzzy Logic: Theory and Application. Prentice Hall, Upper Saddle River (1995)
3. Wagner, C., Hargas, H.: Uncertainty and type-2 fuzzy sets systems. In: IEEE UK Workshop on Computational Intelligent (UKCI) (2010)
4. Zadeh, L.A.: Fuzzy sets. J. Inf. Control **8**, 338–353 (1965)
5. Deng, H.: Comparing and ranking fuzzy numbers using ideal solutions. J. Appl. Math. Model. **38**, 1638–1646 (2013)

6. Zimmerman, H.J.: An application – oriented view of modelling uncertainty. Eur. J. Oper. Res. **122**, 190–198 (2000)
7. Hullermeire, E.: Fuzzy sets in machine learning and data mining. J. Appl. Soft Comput. **11**, 1493–1505 (2011)
8. Zadeh, L.A.: The concept of a linguistic variable and its application to approximate reasoning. J. Inf. Sci. **8**, 199–249 (1975)
9. Karnik, N.N., Mendel, J.M.: Centroid of type-2 fuzzy set. J. Inf. Sci. **132**, 195–220 (2001)
10. Mendel, J.M., John, R.I., Liu, F.L.: Interval type-2 fuzzy logical systems made simple. J. IEEE Trans. Fuzzy Syst. **14**, 808–821 (2006)
11. Wallsten, T.S., Budescu, D.V.: A review of human linguistic probability processing general principles and empirical evidences. J. Knowl. Eng. Rev. **10**(1), 43–62 (1995)
12. Yager, R.R., Filev, D.P.: Essential of Fuzzy Modelling and Control. Wiley, New York (1994)
13. Mogharreban, N., Dilalla, L.F.: Comparison of defuzzification techniques for analysis of non-interval data. In: Annual Meeting of the North American on Fuzzy Information Processing Society, NAFIPS 2006, pp. 257–260 (2006)
14. Mendel, J.M.: Uncertain Rule-Based Fuzzy Logic Systems: Introduction and New Directions. Prentice-Hall, Upper Saddle River (2001)
15. Liu, F.: An effect centroid type-reduction strategy for general type-2 fuzzy logic system. J. Inf. Sci. **178**, 2224–2236 (2008)
16. Gong, Y., Hu, N., Zhang, J., Liu, G., Deng, J.: Multi-attribute group decision making method based on geometric Bonferroni mean operator of trapezoidal interval type-2 fuzzy numbers. J. Comput. Ind. Eng. **81**, 167–176 (2015)
17. Carlsson, C., Fuller, R.: On possibility mean value and variance of fuzzy numbers. J. Fuzzy Sets Syst. **122**, 315–326 (2001)
18. Khalif, K.M.N.K, Gegov, A.: Generalised fuzzy Bayesian network with adaptive vectorial centroid. In: 16th World Congress of the International Fuzzy Systems Association (IFSA) & 9th Conference of European Society for Fuzzy Logic and Technology (EUSFLAT), pp. 757–764. Atlantis Press, Gijon (2015)
19. Chen, C.C.M., Schwender, H., Keith, J., Nunkesser, R., Mengersen, K., Macrossan, P.: Method of identifying SNP interactions: a review on variations of logic regression, random forest and bayesian logistic regression. IEEE/ACM Trans. Comput. Biol. Bioinf. **8**, 1580–1591 (2011)
20. Choy, S.L.: Priors: Silent or Active Partners of Bayesian Inference? Case Studies in Bayesian Statistical Modelling and Analysis. Wiley, Sussex (2013)
21. Bayesian Inference for Logistic Regression Parameters. http://www.medicine.mcgill.ca/epidemiology/joseph/courses/EPIB-621/bayeslogit.pdf
22. Liver Disorder Data Set, UCI Machine Learning Repository. https://archive.ics.uci.edu/ml/datasets/Liver+Disorders
23. Cheng, C.H.: A new approach for ranking fuzzy numbers by distance method. J. Fuzzy Set Syst. **95**, 307–317 (1998)
24. Clinical Tests: Sensitivity and Specificity. http://ceaccp.oxfordjournals.org/content/8/6/221.full
25. Jeong, D., Kang, D., Won, S.: feature selection for steel defects classification. In: International Conference on Control Automation and Systems (ICCAS), pp. 338–341. IEEE Press, Gyeonngi-do (2010)

Neural Computation Theory and Applications

High Capacity Content Addressable Memory with Mixed Order Hyper Networks

Kevin Swingler[(⊠)]

Computing Science and Maths, University of Stirling,
Stirling FK9 4LA, Scotland, UK
kms@cs.stir.ac.uk

Abstract. A mixed order hyper network (MOHN) is a neural network in which weights can connect any number of neurons, rather than the usual two. MOHNs can be used as content addressable memories (CAMs) with higher capacity than standard Hopfield networks. They can also be used for regression learning of functions in $f : \{-1, 1\}^n \to \mathbb{R}$ in which the turning points are equivalent to memories in a CAM. This paper presents a number of methods for learning an energy function from data that can act as either a CAM or a regression model and presents the advantages of using such an approach.

1 Introduction

For a long time, the multi layer perceptron (MLP) [10] has been a very popular choice for performing non-linear regression on functions of multiple inputs. It has the advantage of being easy to apply to problems where there is little or no knowledge of the structure of the function to be learned, particularly concerning the interactions between input variables. The weight learning algorithm (back propagation of error, for example) simultaneously discovers features (interactions between inputs) in the function that underlies the training data and the correct values for the regression coefficients, given those features. This leads to two significant and well known disadvantages of the MLP, namely the so called 'black box problem' that means it is very difficult for a human analyst to learn much about the structure of the underlying function from the structure of the network and the problem of local minima in the error function that are a result of the hidden units failing to encode the correct interactions between input variables.

Another type of neural network, the Hopfield network (HN) [6], is used as a content addressable memory. Unlike MLPs, HNs do not have neurons that are distinguished as either inputs or outputs. All neurons can take external input and all can be updated via weights from other neurons. Their new outputs can be read as if they were output neurons. HNs can be trained to store a set of patterns in $\{-1, 1\}^n$ and their dynamics are such that when their neurons are set to a pattern that has not been stored, updating the neuron values causes the network to move to represent a pattern that is in its memory. Such networks suffer from two main short comings. They have a low capacity for storing memories compared to the

© Springer International Publishing AG 2017
J.J. Merelo et al. (eds.), *Computational Intelligence*, Studies in Computational Intelligence 669,
DOI 10.1007/978-3-319-48506-5_17

number of neurons they contain and they often contain many spurious attractors - memories that were not loaded but are recalled if the starting pattern is closer to them than to a stored memory.

The problems of both MLPs and HNs described above are addressed by Mixed Order Hyper Networks (MOHNs) [16], which make the structure of the function explicit, meaning that human readability is greatly improved. There are no local minima in the error function and they have a far higher memory capacity than HNs. This paper presents and compares a number of methods for calculating the correct weight values for a MOHN of fixed structure. Different learning rules have different strengths and weaknesses. Some, for example may be carried out in an on line mode, meaning that the data need not be all stored in memory at one time. On line learning also allows partially learned networks to be updated in light of new data or as part of an algorithm to discover the correct connection structure. Standard regression techniques such as ordinary least squares (OLS) and Least Absolute Shrinkage and Selection Operator (LASSO) may be applied when on line learning is not required. LASSO also has the advantage that weights that do not contribute to the function end up with values equal to zero.

MOHNs have been shown to be useful as fitness function models if used as part of a metaheuristic constraint satisfaction (or combinatorial optimisation) algorithm [13,17]. In such cases, it is not always necessary to learn the whole function space correctly, but sufficient to build a model where the attractors in the energy function correspond to turning points (local optima) in the fitness function. A simpler learning rule is sufficient to build such models, and is presented here.

2   Mixed Order Hyper Networks

A Mixed Order Hyper Network is a neural network in which weights can connect any number of neurons. A MOHN has a fixed number of n neurons and $\leq 2^n$ weights, which may be added or removed dynamically during learning. Each neuron can be in one of two states: $u_i \in \{-1, 1\}$. The state of the MOHN is determined by the values of the vector, $\mathbf{u} = u_0 \ldots u_{n-1}$. The structure of a MOHN is defined by a set, $\mathbf{W}$ of real valued weights, each connecting $0 \leq k \leq n$ neurons. The weights define a hyper graph connecting the elements of $\mathbf{u}$. Each weight, w_j has an integer index that is determined by the indices of the neurons it connects:

$$\mathbf{W} \subseteq \{w_j : j = 0 \ldots 2^{n-1}\} \quad w_j \in \mathbb{R} \tag{1}$$

The weights each have an associated order, defined by the number of neurons they connect. There is a single zero-order weight, which connects no neurons, but has a weight all the same. There are n first order weights, which are the equivalent of bias inputs in a standard neural network. In general, there are $\binom{n}{k}$ possible weights of order k in a network of size n. For convenience of notation, the set of k neurons connected to weight w_j is denoted Q_j, meaning that the index j defines a neuron subset. This is done by creating an n bit binary number,

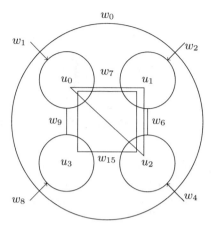

Fig. 1. A four neuron MOHN with some of the weights shown. w_7 is the triangle and w_{15} is the square.

where bit i is set to one to indicate that neuron i is part of the subset and zero otherwise. The resulting binary number, stated in base 10 becomes the weight index. For example, the weight connecting neurons $\{0, 1, 2\}$ is w_7 as setting the bits 0,1,2 in a four bit number gives 0111, which is 7 in base 10. Consequently, we can write $Q_7 = \{u_0, u_1, u_2\}$. Figure 1 shows an example MOHN where $n = 4$.

2.1   Using a MOHN

MOHNs can be applied to a number of different computational intelligence tasks such as building a content addressable memory, performing regression, clustering, classification, probability distribution estimation and as fitness function models for use in heuristic optimisation. These different tasks involve different methods of use and require different approaches to estimating the values on the weights.

3   The Weight Estimation Rules

This section presents the different methods for estimating the weights needed to allow a MOHN to perform a particular task. Although, in theory the weights can be designed by hand, all of the methods described here are based on learning from data. In what follows, a single training example consists of a vector of input variables and a real valued output denoted $(\mathbf{x}, y)$. The training data as a whole is denoted D.

3.1   Hebbian Learning

The simplest of the MOHN learning rules is an extension to the Hebbian rule employed by a Hopfield network to allow it to work with higher order weights.

In this case, the training data consists only of input patterns, $\mathbf{x}$ and no function output is specified. Learning involves setting $u_i = x_i$ for each neuron and the weight update is then:

$$w_j = w_j + \prod_{u \in Q_j} u \tag{2}$$

The Hebbian learning rule allows the MOHN to be used as a content addressable memory (CAM). The CAM learning algorithm is given in Algorithm 1.

Algorithm 1. Loading Pattern $\mathbf{x}$ into a MOHN CAM.

$u_i = x_i \; \forall i$	▷ Set the neuron outputs to equal the pattern to be learned
$w_j = w_j + \prod_{u \in Q_j} u \; \forall w_j \in W$	▷ Update the weights according to equation 2

For a network that is fully connected at order two, Algorithm 1 is the same as loading patterns into a standard Hopfield network. When the MOHN contains higher order weights, the capacity of the network is increased. Patterns are recalled as they are in a Hopfield network, by setting the neuron values to a noisy or degraded pattern and allowing the network to settle using a neuron update rule that first calculates an activation value for each neuron, a_i using Eq. 3 and then applies a threshold using Eq. 4.

$$a_i = \sum_{j: u_i \in Q_j} \left(w_j \prod_{k \in Q_j \setminus i} u_k \right) \tag{3}$$

where $j : u_i \in Q_j$ makes j enumerate the index of each weight that connects to u_i and $k \in Q_j \setminus i$ indicates the index of every member of Q_j, except neuron i itself. A neuron's output is then calculated using the threshold function in Eq. 4.

$$u_i = \begin{cases} 1, & \text{if } a_i > 0 \\ -1, & \text{otherwise} \end{cases} \tag{4}$$

An attractor state is a pattern across $\mathbf{u}$ from which the application of Eqs. 3 and 4 results in no change to any of the neuron outputs. A trained MOHN settles to an attractor point by repeated application of the activation rules 3 and 4, choosing neurons in random order. Algorithm 2 describes the algorithm for settling from a pattern to an attractor.

The dynamics of Algorithm 2 have an underlying Lyapunov function, just as they do in a standard HNN and will always settle to a local minimum of the associated energy function. [19] report a capacity for binary valued order k networks of the order of $n^k / \ln n$, a figure that is also reported by [7]. In fact, as described below, such networks are capable of representing any arbitrary Lyapunov function and therefore a network with the right structure will be able to represent any possible number of turning points.

Algorithm 2. Settling a trained MOHN to an attractor point.

---

repeat
    $ch \leftarrow FALSE$ ▷ Keep track of whether or not a change has been made
    $visited \leftarrow \{\}$ ▷ Keep track of which neurons have been visited
    repeat
        $i \leftarrow rand(i : i \notin visited)$ ▷ Pick a random unset neuron
        $temp \leftarrow u_i$ ▷ Make a note of its value for later comparison
        $\text{Update}(u_i)$ ▷ Update the neuron's output using equations 3 and 4
        if $u_i \neq temp$ **then**
            $ch \leftarrow TRUE$
        end if   ▷ If a change was made to the neuron's output, note the fact
        $visited \leftarrow \{visited \cup i\}$ ▷ Add the neuron's index to the visited set
    until $\|visited\| = n$ ▷ Loop until all neurons have been updated
until $!ch$ ▷ Loop if any neuron value has changed

3.2 Weighted Hebbian Learning

Let $f(\mathbf{x})$ be a multi-modal function where each local maximum represents a pattern of interest. These patterns might be local optima in an optimisation task, archetypes in a clustering task or examples of a satisfaction of multiple constraints, for example. A MOHN can be trained as a CAM in which the attractors are the local maxima of the function. The learning rule is a weighted version of the Hebbian rule:

$$w_j = \sum_{\mathbf{x} \in D} \frac{1}{|D|} f(\mathbf{x}) \prod_{u \in Q_i} u \qquad (5)$$

Previous work [17] has shown that the weighted Hebbian rule is capable of learning the local maxima of a function from samples of $\mathbf{x}, f(\mathbf{x})$ and that the capacity of the resulting networks for storing such attractors was equal to the capacity of a CAM trained using Eq. 2. The difference between Eqs. 2 and 5 is that the target patterns are known in the first case, but unknown in the second, where they are local maxima of y in a function that is learned from a sample of $(\mathbf{x}, y)$ pairs. Note also that experiments have shown that the training data need not contain a single example of any of the attractor patterns for the method to work.

Parity Count Learning. When the inputs (both single variables and products of variable subsets) are uncorrelated (i.e. orthogonal) and each input has an even distribution of values, the weighted Hebbian rule produces the correct weight values in a single pass of the data. When the distribution of values across each variable is uneven, a better estimate of weight values may be made by taking into account how often the input product on each weight is positive or negative during learning. Each weight is set to equal the difference between the average of the output y when the weight's input is positive and when it is negative.

Let D_j^+ be the set of sub-patterns learned by w_j that contain values whose product is positive and D_j^- be the set of sub-patterns learned by w_j that contain values whose product is negative. Now let $\langle y^+ \rangle$ be the average value of y associated with the members of D_j^+ and $\langle y^- \rangle$ be the average value of y associated with the members of D_j^-:

$$\langle y^+ \rangle = \frac{1}{|D_j^+|} \sum_{\mathbf{x} \in D_j^+} f(\mathbf{x}) \tag{6}$$

Similarly, $\langle y^- \rangle$ is calculated as a sum over $\mathbf{x} \in D_j^-$. The weight calculation is simply

$$w_j = \frac{1}{2}(\langle y^+ \rangle - \langle y^- \rangle) \tag{7}$$

The averages may be maintained online so that the weight values are always correct at any time during learning (rather than summing and dividing at the end of a defined training set). W_0 is set in a similar way. The Weighted Hebbian calculation, $w_0 = \langle y \rangle$ means that w_0 is just the average of the output, y across the training sample. This can be improved by taking into account the distribution of patterns across each input.

$$w_0 = \langle y \rangle - \sum_{j:w_j \in W} w_j \Big\langle \prod_{x \in w_j} x \Big\rangle \tag{8}$$

where $x \in w_j$ indicates the neurons connected to weight w_j and $\langle \prod_{x \in w_j} x \rangle$ is the average value across all input patterns of the product of the values of x connected to w_j.

3.3   Regression Rules

The weighted Hebbian update rule is capable of capturing the turning points in a function, but cannot accurately reproduce the output of the function itself across all of the input space. Such networks have an energy function[1] and this can be used as a regression function for estimating $\hat{y} = f(\mathbf{x})$ in the form

$$\hat{y} = \sum_i w_j \prod_{u \in Q_i} u \tag{9}$$

The weight values for the regression may be calculated either in a single off line calculation or using an on line weight update rule.

Off Line Regression. To use ordinary least squares (OLS) [3] to estimate the weights offline, a matrix X is constructed where each row represents a training example and each column represents a weight. The first column represents W_0 and always contains a 1. The remaining columns contain the product of the values of the inputs connected by the column's weight, $\prod_{x \in Q_i} x$. A vector Y takes the

---

[1] The regression Eq. 9 is actually the negative of the energy function, which is minimised by applying the settling Algorithm 2.

output values associated with each of the input rows and the parameters are calculated using singular value decomposition:

$$\beta = (X^T X)^{-1} X^T y \tag{10}$$

where X^T is the transpose of X, X^{-1} is the inverse of X and β becomes a vector from which the weights of the MOHN may be directly read so that $w_0 = \beta_0$ and the remaining weights take values from β in the same sequence as they were inserted into the matrix X.

LASSO Learning Rule. The LASSO algorithm [18] may also be used to learn the values on the weights of the MOHN. Each input vector is set up in the same way as described for OLS, by calculating the product of the input values connected to each weight and the coefficients generated by LASSO are read back into the weights of the MOHN in the same order. LASSO performs regression with an additional constraint on the L^1 norm of the weight vector. The learning algorithm minimises the sum:

$$\sum_{\mathbf{x} \in D} (f(\mathbf{x}) - \hat{f}(\mathbf{x})) + \lambda \sum_{w \in W} |w| \tag{11}$$

where λ controls the degree of regularisation. When $\lambda = 0$, the LASSO solution becomes the OLS solution. With $\lambda > 0$ the regularisation causes the sum of the absolute weight values to shrink such that weights with the least contribution to error reduction take a value of zero. This not only allows LASSO to reject input variables that contribute little, but also to reject higher order weights that are not needed. LASSO can be used as a simple method for choosing network structure by over-connecting a network and then removing all the zero valued weights after LASSO regression has been performed.

On Line Learning. The weights of a MOHN can also be estimated on line (where the data is streamed one pattern at a time, rather than being available in a matrix as in Eq. 10) using a linear version of the delta learning rule, the Linear Delta Rule (LDR):

$$w_i = w_i + \alpha(f(\mathbf{x}) - \hat{f}(\mathbf{x})) \prod_{u \in Q_i} u \tag{12}$$

where $\alpha < 1$ is the learning rate. Experimental results have suggested that one divided by the number of weights in the network is a good value for α, i.e. $\alpha = \frac{1}{|W|}$. This allows the correction made in response to each prediction error to be spread across all of the weights.

The online learning algorithm is very similar to the perceptron (or MLP) learning algorithm. The iterative nature of the algorithm allows for early stopping to be used to control for overfitting with reference to an independent test set. Algorithm 3 describes the learning process.

Algorithm 3. On Line MOHN Learning with the Linear Delta Rule.

Let D_r be a subset of the available data to be used for training the network
Let D_s be a subset of the available data to be used for testing the network
for all $(\mathbf{x}, f(\mathbf{x})) \in D_r$ **do**
    Initialise the weights in the network using the parity rule of equation 7
end for
repeat
    for all $\mathbf{x} \in D_r$ **do**
        Update the weights using the delta learning rule of equation 12
    end for
    Let e be the root mean squared error from evaluating D_s with the model
until e is sufficiently low or starts to increase consistently

---

Note that the weights are initialised with the parity count learning rule, not to random values as with an MLP. This is because there are no local minima in the error function and so no need for random starting points. In cases where the entire input, output space of the function may be noiselessly sampled, the initialisation step will produce the correct weights immediately, without the need for additional error descent learning. The learning algorithm will work without the initialisation (the weights can be set to zero) but then requires more iterations of the learning cycle.

3.4 Capacity of a MOHN Content Addressable Memory

Used as content addressable memories, MOHNs have a certain capacity for storing and perfectly recalling a number of distinct patterns. The size of this capacity varies with the number of weights in the network and also depends on the patterns being stored. Used as optimisation tools, MOHNs have a capacity for storing a number of local and global maxima in a fitness function. Once the number of optima in a function exceeds the capacity of a network, the ability of the network to produce useful candidates for an evolutionary optimisation algorithm degrades. For these reasons, it is useful to understand the limits of the capacity of MOHNs for storing patterns and attractors. This paper addresses those limits.

[19] report a capacity for binary valued order d networks of the order of $n^d / \ln n$, a figure that is also reported by [7]. Their definition of order differs from that used here. It reflects the number of other inputs that are included in the correlation between each neuron, meaning that their value of d is one less than that used in this work. A Hopfield network, then, would be a first order model as each weight from a neuron connects to one other neuron.

This section investigates the capacity of MOHNs of various sizes for storing random patterns when the patterns are known and the learning is the simple Hebbian rule of Eq. 2. The traditional method for experimentally testing the capacity of a Hopfield network ([9] for example) is to load random patterns one at a time and then test whether the network still maintains all of the patterns learned so far as attractors. The process is given in Algorithm 4.

To conclude that a pattern is no longer an attractor state in a network, it is only necessary to find a single neuron that would change its value when the network is running. Rather than the usual method of running a network by updating the neurons in random order, the attractor test simply updates each neuron in turn in a fixed order (avoiding the overhead of randomisation) and stops as soon as a neuron changes value, indicating that the pattern is not an attractor. When all neurons have been tested and none have changed, the pattern is proved to be an attractor.

Figure 2 shows an example result for a network with up to 26 neurons, fully connected at orders from one to five. The squares and associated error bars show the mean and the range of the capacity for storing and perfectly recalling patterns. The theoretical lower bounds stated by [19] are shown as red (the weak lower bound) and green (the lower bound) lines.

3.5 Improving Capacity with Structure Discovery

To overcome the problem of exponential weight growth, a content addressable memory can be incrementally built using the structure discovery method of [12,15]. In this approach, weights are added to a network until it is able to store the patterns in the training set and removed if they do not contribute any improvement. The resulting network is sparsely connected, unlike those in Fig. 2, having some weights at several different orders (hence the name, Mixed Order Hyper Network).

Algorithm 4. Testing the capacity of a MOHN.

$P \leftarrow \emptyset$	▷ Start with an empty pattern set
$W \leftarrow 0$	▷ Set all the weights in the MOHN to zero
repeat	
Generate $\mathbf{p} \notin P$	▷ Generate a random pattern that is not in the pattern set
$W \leftarrow W + \mathbf{p}$	▷ Allow the MOHN to learn the pattern
$stop \leftarrow false$	
for all $\mathbf{p} \in P$ **do**	
$\mathbf{u} \leftarrow \mathbf{p}$	▷ Set the inputs to each pattern in the list in turn
$\mathbf{u} = \text{settle}()$	▷ Update the neurons in fixed order
if $\mathbf{u} \neq \mathbf{p}$ **then**	▷ If any neuron value changes
$stop \leftarrow true$	▷ Attractor is destroyed and capacity exceeded
end if	
end for	
if $!stop$ **then**	▷ No pattern was lost, so add new one to list
$P \leftarrow P \cup \mathbf{p}$	▷ Add the pattern to the set
end if	
until stop	▷ End when a pattern is lost

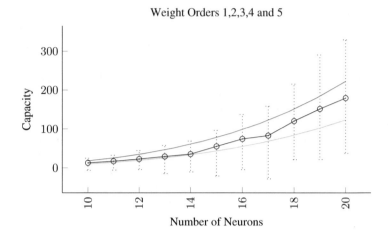

Fig. 2. Experimental mean and range of capacity of a MOHN fully connected at orders 1,2,3,4,5 (circles and error bars). The weak lower bound on capacity, $\frac{n^4}{240 \ln n}$ (red line) and the lower bound capacity, $\frac{n^4}{432 \ln n}$ (green line), both according to [19].

3.6   Experimental Example

To illustrate the stage capacity of a MOHN, a set of patterns representing the written digits from 0 to 9 were created over 25 neurons. A traditional fully connected Hopfield network with 300 s order connections can only store three or four such patterns. The patterns to store are shown in Fig. 3.

Fig. 3. The written digits from 0 to 9 as 25 bit patterns to be used to test the dynamic structure discovery algorithm applied to a CAM.

Firstly, static networks were tested to find the lowest order at which full connections were needed to store the patterns. Networks of 25 neurons, fully connected at all orders up to two, three and four all failed to store all 10 patterns as stable attractors. A network with all weights connected at all orders up to five was able to store the patterns. This network contained 53,131 weights. The next step is to try and discover a network that will store the same patterns in fewer weights.

The same patterns were then used to find a smaller structure that could store them all using the structure discovery algorithm used in [15]. The algorithm was able to find a network capable of correctly storing all of the patterns with a total of 362 weights, which is approximately the same number as found in a standard 25 neuron Hopfield network.

In principle, this approach gives MOHNs arbitrary storage capacity, as a MOHN can represent any function in $f : \{-1, 1\}^n \rightarrow \mathbb{R}$, and for any set of non-neighbouring patterns, $\mathbf{P}$ there exists a function in which each member of $\mathbf{P}$ is a local maximum. Of course, some functions may be difficult to discover the correct structure for, and some may require so many weights that a solution is impractical, but in principle, MOHNS can store arbitrary pattern sets. If $\mathbf{P}$ contains neighbouring patterns (two patterns are neighbours if there exists an input, u_i such that flipping its value, $u_i \leftarrow -u_i$, switches from one pattern to the other), then the neighbours will form a plateau where the output from the function is the same for all points. Whether these states can be considered stable attractors is a question of definition and implementation details of Algorithm 2.

If that algorithm only moves from a state to one with higher output, it will stay stable in the first state of a plateau that it finds. This would mean that seeding it with the target states would show them all to be attractors, but that some were not accessible from nearby states. As neurons are updated in random order, the same degraded pattern might produce a different pattern on the same plateau on repeated trials. An additional step can be added to Algorithm 2 in which neighbouring states of a first found attractor state are explored if they lie at the same height. This can be done by recursively making single steps from each point on the plateau until they have all been visited.

4 Analysis of Learning Rules

This section begins with a summary of the abilities and limitations of the different learning rules presented in this paper. It then goes on to analyse the rules and the resulting networks. Table 1 summarises some of the differences between the methods. Due to the structure of the MOHN, all of the learning rules are capable of reproducing the maximal turning points of the learned function, but the Hebbian based rules do not minimise the error elsewhere in the function space. The Hebbian rules learn in a single presentation of the data, so can operate in an on line mode without the need to iterate through the data set more than once. The others require either on line iterations or the entire data set to be present off line.

The weighted Hebbian rule is accurate only when a full sample of the input/output space is available, so is of limited practical use as the parity counting method gives more accurate estimates operating on weights independently with a single pass through the data. The parity counting method provides good starting weights for the linear delta rule. The following sections investigate different network structures in more detail.

4.1 Second Order Networks

When a MOHN has only second order connections, it is equivalent to a Hopfield Neural Network (HNN) [6] and the Hebbian learning rule of Eq. 2 is the standard

Table 1. Comparing four different MOHN learning rules in terms of the learning mode, any regularisation that is possible, whether or not the training error is minimised, and whether the training data is presented as input, output pairs (IO) or as patterns to store in a content addressable memory (CAM).

Method	Mode	Reg	Min. Err	Data
Hebbian	One shot	None	No	CAM
Weighted hebb	One shot	None	No	IO
LDR	On line	Early stop	Yes	IO
OLS	Off line	None	Yes	IO
LASSO	Off line	L^1-norm	Yes	IO

learning rule for a HNN. It is well known that HNNs are able to learn patterns as content addressable memories, but that they suffer from the presence of spurious attractors too. These spurious attractors may be reduced by defining an energy function for the network in which the patterns to be stored as memories are local maxima. This may be done using a Hamming distance based function and previous work [11,17] has shown that using the weighted Hebbian update rule of Eq. 5 and such a function on a second order MOHN (or equivalently, a HNN) is sufficient to produce a content addressable memory in which the turning points of the function are the memories to be stored. The capacity of a HNN for storing patterns is the same if the patterns are loaded directly with the Hebbian learning rule as it is when the patterns are learned from a Hamming distance based function.

To build the Hamming distance based function, denote the set of patterns to be stored as **T**:

$$\mathbf{T} = \{t_1, \ldots, t_s\} \tag{13}$$

and define a set of sub-functions, $f(\mathbf{x}|t_j)$ as a weighted Hamming distance between **x** and each target pattern t_j in **T** as

$$f(\mathbf{x}|t_j) = \sum \frac{\delta_{x_i,t_{ji}}}{n} \tag{14}$$

where t_{ji} is element i of target j and $\delta_{x_j,t_{ji}}$ is the Kronecker delta function between pattern element i in t_j and its equivalent in **x**. The function output given an input pattern, $f(\mathbf{x})$ is the maximal output across all the sub-functions given an input of $f(\mathbf{x})$.

$$f(\mathbf{x}|t) = max_{j=1\ldots s}(f(\mathbf{x}|t_j)) \tag{15}$$

By generating random input patterns, evaluating each using Eq. 15 and then using the LDR of Eq. 12 to learn each input, output pair sampled, a network with attractors at each member of **T** is learned. The network has the additional quality that as the number of samples learned increases, the number of spurious attractors in the network decreases.

To test this claim, experiments were run in which a 100 neuron MOHN was trained on a function that contained four true attractor states. Figure 4 shows the average results of running 100 trials in which the number of spurious attractors and the error of the network were measured for each iteration of the training data, which was a random sample of size 20,000 from the Hamming distance based function of Eq. 15.

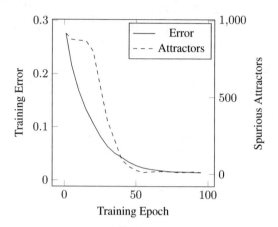

Fig. 4. As the number of learning iterations increases, the training error decreases as does the number of spurious attractors in the model.

More generally, the reduction in the number of spurious attractors depends on the ability of the structure of the network to represent the underlying function in which the only turning points are the desired attractor patterns. The better the network can represent the function, the fewer spurious attractors there are. Additionally, the linear delta rule allows a network to increase its capacity over the equivalent trained with the Hebb rule. Figure 5 shows the average and standard deviation of the capacity of networks from size 10 to 40, calculated from 50 trials of each learning rule at each network size.

Researchers have shown how the weights of a HNN can be designed to represent the travelling salesman problem [5, 21] and other problems such as graph colouring [2]. These approaches are limited by the fact that the weights must be chosen by hand to reflect the constraints of the problem to be solved. By training a HNN (or a MOHN) by sampling from a fitness function, it is now possible to build a network to represent any problem with a fitness function that can be evaluated, not just those that are amenable to having their weights set by hand.

4.2   Full Networks

When the data are noise free, a network is fully connected and the data sample is exhaustive (i.e. it covers every possible input pattern once), the weighted

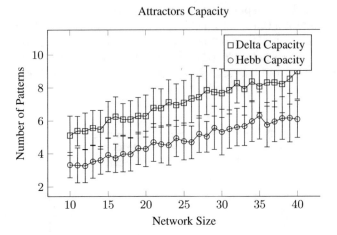

Fig. 5. The mean and standard deviation of the capacity of a fully connected Hopfield Network trained with the Hebb rule and the Linear Delta Rule.

Hebbian rule of Eq. 5 (with $|D| = 2^n$) will produce weights which reproduce the target function perfectly. In such cases, the product $\prod_{u \in Q_i} u$ provides a basis function for $f : \{-1, 1\}^n \to \mathbb{R}$. This basis function is very similar to the well know Walsh basis [1, 20].

A Walsh representation of a function $f(\mathbf{x})$ is defined by a vector of parameters, the Walsh coefficients, $\omega = \omega_0 \dots \omega_{2^n-1}$. Each ω_j is associated with the Walsh function ψ_j. The Walsh representation of $f(\mathbf{x})$ is constructed as a sum over all ω_j. In the sum, each ω_j is either added to or subtracted from the total, depending on the value of the Walsh function $\psi_j(x)$ which gives the function for the Walsh sum:

$$f(\mathbf{x}) = \sum_{j=0}^{2^n-1} \omega_j \psi_j(x) \tag{16}$$

A Walsh function, $\psi_j(x)$ returns $+1$ or -1 depending on the parity of the number of 1 bits in shared positions across $\mathbf{x}$ and $\mathbf{j}$ where $\mathbf{j}$ is the binary representation of the integer j. Using logical notation, a Walsh function is derived from the result of an XOR (parity count) of an AND (agreement of bits with a value of 1):

$$\psi_j(x) = \oplus_{i=1}^n (x_i \wedge j_i) \tag{17}$$

where $\oplus$ is a parity operator, which returns 1 if the argument list contains an even number of 1s and -1 otherwise. The Walsh transform of an n-bit function, $f(\mathbf{x})$, produces 2^n Walsh coefficients, ω_j, indexed by the 2^n combinations across $f(\mathbf{x})$. Each Walsh coefficient, ω_j is calculated by

$$\omega_j = \frac{1}{2^n} \sum_{x=0}^{2^n-1} f(x) \psi_j(x) \tag{18}$$

The weight values in a fully trained MOHN are equal in magnitude to the Walsh coefficients of the same index, but that they differ in sign when the weight order is an odd number. That is,

$$\omega_j = p(\omega_j)w_j \quad \forall w_j \in W \tag{19}$$

where $p(\omega_j)$ is the parity of the order of ω_j such that:

$$p(\omega_j) = \begin{cases} 1 & \text{if the order of } i \text{ is even} \\ -1 & \text{otherwise} \end{cases} \tag{20}$$

This is because the Walsh function returns a value based on a parity count of the number of variables set to one across the input variables that are connected to a given coefficient, as shown in Eq. 17. The parity function returns 1 if the number of variables with a value of one is even and -1 otherwise. The MOHN uses the product of those same values, which evaluates to -1 whenever there is an odd number of inputs set to -1. The MOHN indices match the Walsh coefficient indices because they both use the same method of deriving the index number from the binary representation of the connections described in Sect. 2.

As a fully connected MOHN provides a basis for all possible functions in $f : \{-1, 1\}^n \rightarrow \mathbb{R}$, then it follows that any function with coefficient values of zero may be perfectly represented by a less than fully connected MOHN so providing the correct structure can be found, a MOHN may represent any arbitrary function.

4.3   Discovering Network Structure

The structure of a MOHN is defined by W, which is a subset of all possible 2^n weights. As noted above, a fully connected second order network implements a HNN and a fully connected network at all orders forms a basis of all functions $f : \{-1, 1\}^n \rightarrow \mathbb{R}$. Any other pattern of connectivity is also possible, for example a first order only network is equivalent to a perceptron, or a multiple linear regression model. Adding higher order weights increases the power of the model to represent more complex functions.

Discovering the correct structure for the network is both challenging and instructive, compared to the same task when using an MLP, which is quite straight forward, but done in the dark. The question of discovering structure in functions from samples of data is of particular importance in the field of metaheuristic optimisation, where it is called linkage learning (see [4,8]).

The correct structure for a function may be discovered from the training data using an iterative approach of adding and removing weights as training progresses. The basics of the structure discovery algorithm are to train a partial network, test the significance of the weights it contains, remove those that are not significant, then add new weights according to some criteria. The weight picking criteria chosen for this work are based on maintaining a probability distribution over the possible weights, which is updated on each round of learning so that connection orders and neurons that have proved useful in previous rounds have a higher probability of being picked in subsequent rounds. The process is described briefly in [12] and in more detail in [15].

5   Experimental Results

In this section, the learning rules described in this paper are compared with each other and with a standard multi layer perceptron (MLP) for the speed at which they learn. The Hamming distance based function of Eq. 15 was used for these tests as it is possible to generate arbitrary functions containing a chosen number of turning points at random locations. This allows the different methods to be tested across thousands of different functions of varying degrees of complexity.

5.1   Speed Against Complexity

One way to vary the complexity of a function is to vary the number of turning points it contains. This section describes a set of experiments designed to measure the speed of learning of each of the MOHN learning rules and an MLP as the complexity of the function to be learned varies. Each single experiment involved training a MOHN and an MLP on a data set generated from a function with a random number of turning points. The same data was used to train three different MOHNs, one with each learning rule from OLS, LASSO and LDR. The function had 15 inputs and the MOHNs were fully connected up to order three, giving them 576 weights. The MLP has only 10 hidden units, giving it only 176 weights.

A sample of 580 random points was used for training each network. For the iterative learning methods (all except OLS) a target error of 0.01 was used as a stopping criteria, hence the measure of interest was time taken to reach a training error of 0.01. This process was repeated 1000 times, each with a new function with a number of turning points between 1 and 30.

Figure 6 shows the results. All methods except the MLP learned the function in a constant time, regardless of the degree of complexity. The MLP was able to learn the single turning point function (i.e. linear function) in less time than it was able to learn the more complex functions. The function with two turning points was also faster than those with more. After two turning points, the learning time for the MLP became constant. Regardless of the complexity of the function, the MLP always took considerably longer, followed by OLS. The LDR and LASSO algorithms took similar amounts of time and were the fastest.

5.2   Speed by Network Size

Another set of similar experiments related the training speed of each method to the size of the network. The number of inputs to a network was varied from 5 to 15 and 1000 trials were run. The mean squared error of the result of performing OLS was used as the stopping criteria for the MLP and the MOHN as it was

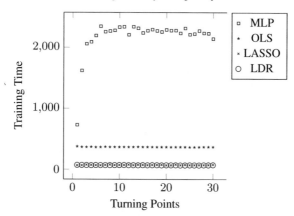

Fig. 6. Average learning time in milliseconds by function complexity for different MOHN learning rules and an MLP. The LASSO and LDR values are almost equal and can be seen along the bottom of the graph.

trained with LDR, ensuring that all models had the same level of accuracy. Figure 7 shows the results. OLS is known to have a time complexity of $O(np^2)$ where n is the number of data points and p is the number of variables. LASSO and LDR were of the same order, but the algorithms ran in less time. The MLP's training time grew exponentially with the number of variables in these particular experiments. As before the MOHN models all reached the target training error faster than the MLP.

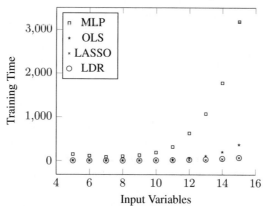

Fig. 7. Average learning time in milliseconds by number of inputs for different MOHN learning rules and an MLP.

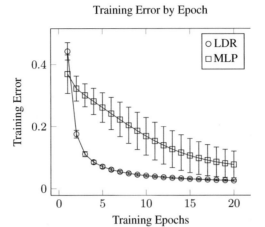

Fig. 8. The mean and standard deviation of training error as it descends over twenty training epochs, comparing an MLP with the Linear Delta Rule training a MOHN.

Error Descent Rate. The difference in training speed between the MOHN and an MLP was investigated further by recording the average error by training epoch for the first twenty passes through the training data. Figure 8 shows the average error on each pass of the training data from 1000 repeated trials on functions of varying complexity. The error bars show 1 standard deviation from the mean. Note that the MOHN error drops faster and that there is far less variation across trials (the error bars for the MOHN are sufficiently short that they sit inside the marks).

The improved learning speed of the MOHN and the slower, more varied speed of the MLP may be explained by the fact that the MLP combines fitting parameter values with feature selection. Recently, [14] provided an insight into the phases of MLP training, showing that early training cycles are taken up with fixing the role of the hidden units and later cycles then fit the parameters within the constraints of the features encoded by those hidden units. The MOHN does not have hidden units and so only needs to fit parameter values to its fixed structure. Of course, that structure needs to be discovered, but the task of structure discovery and parameter fitting are separated, unlike the case for the MLP.

Another consequence of the MLP's dual learning task of fitting both function structure and regression fit is that the error function contains local minima. These occur when the hidden units encode a suboptimal set of features and the network fits weight values to them. This is commonly solved by re-starting the training process from a different random set of initial weight values. The MOHN error function does not contain local minima, so the weights do not need to be randomised before learning, as shown in Algorithm 3. To illustrate this point, a final set of experiments compared an MLP trained with error back propagation to a MOHN trained with the LDR on a function designed to contain local minima

Training Error Descent for MLP

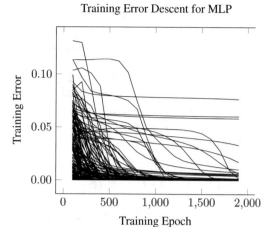

Fig. 9. Traces of training error over 200 different attempts at training an MLP on a concatenated XOR function. Note the variation in convergence time and the presence of a number of failed attempts after 2000 epochs.

Training Error Descent for MOHN

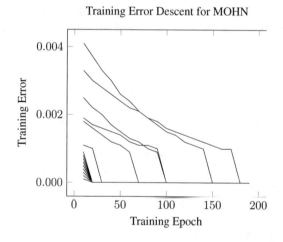

Fig. 10. Traces of training error over 200 different attempts at training a MOHN on a concatenated XOR function. Compare both the scale of the error and the number of training epochs involved with the same plot for the MLP in Fig. 9.

in its cost function. The function to be learned was a concatenation of XOR pairs such that each x_i where i is even is paired with x_{i+1} to form an XOR function. The function output is the normalised sum of the XOR of the pairs, so 101010 would produce an output of one and 110011 would produce zero. Figure 9 shows the traces of 200 MLPs started with random weights, each trained for 2000 cycles through the training data. The variation in error descent is clear, with some networks converging quickly, some taking many training epochs to converge, and some still stuck in local minima after 2000 epochs.

The MOHN networks were trained on random samples of input, output pairs. Figure 10 shows the trace of the training error during 200 attempts at learning the same XOR based function as that in Fig. 9 using a MOHN with the LDR. The variation is not due to random starting points—all networks start with weights at the same point—but is due to the fact that the training data is a small random subset of the full input space. Note that there are no traces that indicate a local minimum; all go to zero error.

6   Summary and Future Directions

In the space $f : \{-1, 1\}^n \to \mathbb{R}$, mixed order hyper networks are universal function models. They may be trained from a sample of data to act as either a regression function that attempts to fit the function that underlies the data across the entire function space or just to capture the function's turning points as energy minima. Learning may be off line, in which case all of the data needs to be available at one time, or on line in situations where data is streamed or the network structure is changing and existing weights need to be updated. This paper presented five learning rules designed to cover both on line and off line learning, and both regression and content addressable memory learning. Other learning methods might also be considered such as ridge regression or LARS, but that is left for future work.

This paper has only presented networks for function learning, but they may also be used for other machine learning tasks, all of which present interesting topics of research. By treating some of the neurons as inputs and some as outputs, a MOHN can implement a classifier. By introducing a link function that is the exponential of the network output, it is possible to model a Boltzmann distribution, turning the MOHN into a Markov random field. A MOHN used as a content addressable memory can also be trained so that the attractor states are centroids (or archetypes) in a clustering task. A MOHN can also be trained as a surrogate fitness function and there are methods for searching the resulting function for optimal points [13].

The issue of MOHN structure discovery was also raised. The experiments presented in this paper mostly worked on the assumption that the networks in question contained weights of sufficient order to capture the functions on which they were trained. This becomes increasingly difficult as the number of inputs grows. Problems with large numbers of inputs require a structure discovery phase to be carried out as part of the training process. See [15] for more details on structure discovery.

In conclusion, with a given network structure, training a MOHN is faster and has less error variance across trials than training with an MLP. Additionally, the training algorithm has no local minima when training a fixed structure MOHN, making training more reliable than that of an MLP.

References

1. Beauchamp, K.: Applications of Walsh and Related Functions. Academic Press, London (1984)
2. Caparrós, G.J., Ruiz, M.A.A., Hernández, F.S.: Hopfield neural networks for optimization: study of the different dynamics. Neurocomputing **43**(1–4), 219–237 (2002)
3. Hastie, T., Tibshirani, R., Friedman, J.: The Elements of Statistical Learning: Data Mining, Inference, and Prediction. Springer, New York (2009)
4. Heckendorn, R.B., Wright, A.H.: Efficient linkage discovery by limited probing. Evol. Comput. **12**(4), 517–545 (2004)
5. Hopfield, J.J., Tank, D.W.: Neural computation of decisions in optimization problems. Biol. Cybern. **52**, 141–152 (1985)
6. Hopfield, J.J.: Neural networks and physical systems with emergent collective computational abilities. Proc. Nat. Acad. Sci. USA **79**(8), 2554–2558 (1982)
7. Kubota, T.: A higher order associative memory with Mcculloch-Pitts neurons and plastic synapses. In: 2007 International Joint Conference on Neural Networks, IJCNN 2007, pp. 1982–1989, August 2007
8. Pelikan, M., Goldberg, D.E., Cantú-paz, E.E.: Linkage problem, distribution estimation, and Bayesian networks. Evol. Comput. **8**(3), 311–340 (2000)
9. Storkey, A.J., Valabregue, R.: The basins of attraction of a new hopfield learning rule. Neural Netw. **12**(6), 869–876 (1999)
10. Swingler, K.: Applying Neural Networks: A Practical Guide. Academic Press, London (1996)
11. Swingler, K.: On the capacity of Hopfield neural networks as EDAs for solving combinatorial optimisation problems. In: Proceedings of IJCCI (ECTA), pp. 152–157. SciTePress (2012)
12. Swingler, K.: A comparison of learning rules for mixed order hyper networks. In: Proceedings of IJCCI (NCTA) (2015)
13. Swingler, K.: Local optima suppression search in mixed order hyper networks. In: 2015 15th UK Workshop on Computational Intelligence (UKCI) (2015)
14. Swingler, K.: Opening the black box: analysing MLP functionality using walsh functions. In: Merelo, J.J., Rosa, A., Cadenas, J.M., Dourado, A., Madani, K., Filipe, J. (eds.) Computational Intelligence Studies in Computational Intelligence, vol. 620, pp. 303–323. Springer, Cham (2016)
15. Swingler, K.: Structure discovery in mixed order hyper networks. Big Data Analytics **1**(1), 8 (2016)
16. Swingler, K., Smith, L.: Training and making calculations with mixed order hypernetworks. Neurocomputing **141**, 65–75 (2014)
17. Swingler, K., Smith, L.: An analysis of the local optima storage capacity of hopfield network based fitness function models. In: Nguyen, N.T., Kowalczyk, R., Fred, A., Joaquim, F. (eds.) Transactions on Computational Collective Intelligence XVII. LNCS, vol. 8790, pp. 248–271. Springer, Heidelberg (2014). doi:10.1007/978-3-662-44994-3_13
18. Tibshirani, R.: Regression shrinkage and selection via the lasso. J. Roy. Stat. Soc. Series B (Methodological), 267–288 (1996)
19. Venkatesh, S.S., Baldi, P.: Programmed interactions in higher-order neural networks: maximal capacity. J. Complex. **7**(3), 316–337 (1991)

20. Walsh, J.: A closed set of normal orthogonal functions. Amer. J. Math. **45**, 5–24 (1923)
21. Wilson, G.V., Pawley, G.S.: On the stability of the travelling salesman problem algorithm of hopfield and tank. Biol. Cybern. **58**(1), 63–70 (1988)

Preindication Mining for Predicting Pedestrian Action Change

Kenji Nishida[1](✉), Takumi Kobayashi[1], Taro Iwamoto[2], and Shinya Yamasaki[2]

[1] National Institute of Advanced Industrial Science and Technology (AIST),
1-1-1 Umezono, Tsukuba, Ibaraki 305-8568, Japan
{kenji.nishida,takumi.kobayashi}@aist.go.jp
[2] Mazda Motor Co., 2-5 Moriya-cho Kanagawa-ku,
Yokohama, Kanagawa 221-0022, Japan
{iwamoto.tar,yamasaki.s}@mazda.co.jp

Abstract. The action prediction of pedestrians significantly contributes to an intelligent braking system in cars; knowing that the pedestrians will run in several seconds such as for crossing streets, the cars can start braking in advance, to effectively reduce the risk for crash accidents. In this paper, we propose a method to predict how the pedestrian act (run or walk) in the future based on preindication in video frames detected by only appearance-based image features. We empirically mine the distinctive frames that precede the target action, 'running' in this case, and are effective for predicting it in the framework of feature selection. By using the most effective frames, we can build the action prediction method by exploiting the image features extracted at those frames. As to the image feature extraction methods, we evaluate two types of features in our method, one is GLAC (Gradient Local AutoCorreration) and the other is HOG (Histogram of Oriented Gradient). In the experiments, the effective frames are successfully found around 0.37 s before running action by using GLAC feature; this is not the case of HOG. We also show that the results are closely related to human motion phases from walking to running via biomechanical analysis.

Keywords: Action prediction · Feature selection · Intelligent transport system · Image feature extraction

1 Introduction

According to Japanese traffic accident statistics [1], the number of pedestrian accidents are not decreasing while total number of accidents are decreasing. Moreover, the fatality rate in the pedestrian accidents are five times higher than the other accidents. Therefore, prevention of the pedestrian accidents is one of the most urgent issue in our society. The statistics [1] also reports that 70 % of the fatal pedestrian accidents occurred during crossing streets, and thus it is particularly important to safely detect/recognize those crossing pedestrians.

© Springer International Publishing AG 2017
J.J. Merelo et al. (eds.), *Computational Intelligence*, Studies in Computational Intelligence 669,
DOI 10.1007/978-3-319-48506-5_18

The fatality risk of pedestrian accidents is actually affected by the impact speed [2]: it is about 4 % at the impact speed of 40 km/h while it increases to about 10 % at 50 km/h and 20 % at 60 km/h. Thus, roughly speaking, the fatality risk decreases by 10 % as the impact speed decreases by 10 km/h. In the situation that automatic emergency braking (AEB) system works on 6 m/s^2 as defined by Euro-NCAP [3], it also means that if a car brakes 0.5 s earlier, the fatality risk in pedestrian accidents would be decreased by 10 %. For realizing early braking, it is not sufficient only to detect pedestrians, but it is highly required to recognize the pedestrian action of high risk, such as crossing street with running, as early as possible.

In the last decade, pedestrian detection is one of the most successful applications in the computer vision and pattern recognition fields. For example, Dalal and Triggs attained over 99 % detection rate by introducing HOG feature [4], and very recently it is further improved by deep CNN [5]. However, as described above, just detecting pedestrians is not sufficient for reducing the risk of pedestrian-car accidents. Keller and Gavrila detected crossing people by analyzing pedestrian movement which can be distinguished by the trajectory in the feature space [6]. Although they showed promising results such as the accuracy of 80 % in classifying the correct pedestrian action about 570 ms before the event, it is generally difficult to estimate the precise movement of pedestrians from on-board camera due to its self-motion (shaking). Reddy and Krishnaiah focused on a running pose to detect the pedestrian crossing streets [7]. These approaches detect the change of pedestrian action from walking to running with the accuracy of 92 %, but the detection is performed after the pedestrian already starts running, which is considered to be too late to contribute toward early braking.

We tackle a challenging problem to predict a high risk human action before it actually occurs. In the realistic situations, we have to pay careful attention to the pedestrians that cross a street with suddenly running and such (sudden) running is regarded as a high risk action to be treated by the AEB system with early braking. Therefore, in this paper, we address the problem to predict the (sudden) running action of pedestrians by mining the *sign* for that action which *preindicates* the running actions beforehand. In addition, we employ an appearance-based approach using only *static image features*, though motion features might be suitable for recognizing actions, since it is quite hard to extract reliable motion features from a moving on-board camera. There is a primary question how early we can predict the running action, or more basically, whether such sign (*preindicator*) exists or not, and the secondary question might be which image feature can extract the preindicator. We empirically answered these questions in the framework of feature selection and show the effective preindicator from the quantitative viewpoint. In our previous paper [8], we only examined gradient local auto-correration feature (GLAC) [9], and did not tested other image features. This paper gives a complementation of our previous paper by applying HOG feature for comparison to answer the second question.

2   Appearance Based Action Prediction

In this section, we detail the action prediction method using only *static image features*. This method is based on the assumption that the action preindicator can be sufficiently described by distinctive pedestrian shape, not motion itself.

2.1   Static Image Feature

To characterize the human shape in detail, we employ gradient local auto-correlation (GLAC) method [9]. The GLAC method extracts co-occurrence of gradient orientation as second-order statistics while HOG [4] is based only on first-order statistics of occurrence of gradient orientations. Suppose the pedestrian is detected by arbitrary methods and the bounding box enclosing the pedestrian is provided as shown in Fig. 1. As in the common approach such as of HOG [4], the bounding box is spatially partitioned into regular grids of 3×3 at each of which the GLAC features are extracted, then the final feature vector is constructed by concatenating those feature vectors; the setting of 9 orientation bins for gradients and 4 spatial co-occurrence patterns produces GLAC features of 324 dimensionality, and the final feature is formed as a $2916 = 324 \times 3 \times 3$ dimensional vector.

The spatial grids of 3×3 is much coarser compared to HOG-related methods. The GLAC method can characterize the human shape more discriminatively due to exploiting co-occurrence and thus even such coarser grids are enough for static image features. In addition, the coarser grids render robustness regarding spatial position of human shape; that is, the features are stably extracted even for miss-aligned bounding boxes. On the other hand, 3×3 grids are considered as the coarsest one for capturing the human shape; head, torso, two arms and two legs are roughly aligned to respective spatial grids.

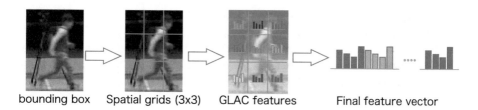

bounding box   Spatial grids (3x3)   GLAC features   Final feature vector

Fig. 1. Static image feature extraction by using GLAC method [9]. The bounding box is partitioned into 3×3 regular grids at each of which GLAC image feature is extracted and then they are concatenated into the final feature vector.

2.2   Action Prediction

Based on the time-series sequence of image features extracted in the bounding boxes, we predict the action which will occur in the near future.

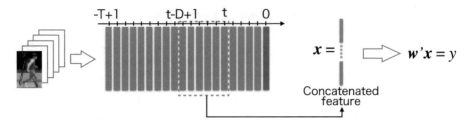

Fig. 2. Action prediction framework. We consider a T-frame subsequence as a unit. The action of running at $t = 0$ is predicted by using D-frame features preceding it.

We consider the subsequence of T frames as a unit which are represented by image feature vectors as described in the previous subsection. Then, we pick up D frames (feature vectors) from them, $[t - D + 1, t]$ $(-T + D <= t <= 0)$, to predict the action which will occur at the T-th frame indexed as time 0 (Fig. 2). Those D feature vectors are concatenated to single feature vector of relatively high dimension (Fig. 2) which is finally passed to a linear SVM classifier for predicting whether running will occur at time 0 or not. The concatenated feature indirectly encodes motion information of pedestrian during D frames. Because we can not know which timing $\{t, D\}$ produces better performance for predicting the running action, those parameters are empirically determined based on data from the quantitative viewpoint. It is obvious that the smaller t is preferable since it provides the earlier prediction; on the other hand, $t = 0$ means on time classification and does not give any prediction at all. And from the computational viewpoint, the smaller D is preferred.

3   Experiments

This section shows the experimental procedure for determining the parameters $\{t, D\}$ in the proposed method (Sect. 2.2) as well as evaluating it. For comparison, we also examined in the same procedure HOG feature of 2772 dimensional vector, which consists of 9 orientation bins on the block of 2×2 cells where the bounding box is partitioned into 8×12 cells.

3.1   Dataset

The dataset that we use contains 57 video sequences of 12 children captured by a (fixed) video camera with 30 fps in a gymnasium (Fig. 3)[1]. Children behave *unpredictably* in context and thus are regarded as the subjects to be carefully paid attention in a traffic scene. They first walk and then suddenly run in an arbitrary timing. The bounding boxes enclosing them are manually annotated since the pedestrian detection is out of our focus in this study. In addition, the

---

[1] This experiment is approved by the Ethical Review Board of Mazda Motor Corporation and the informed consent of all subjects were also obtained.

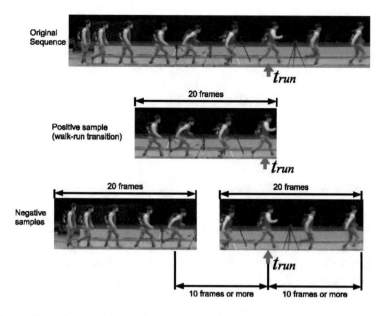

Fig. 3. Sampling positive and negative subsequences from a whole sequence.

frame when the subject starts running is also manually indicated; it is denoted as t_{run} (Fig. 3). The length of the subsequence is set to $T = 20$, since all the subjects of 57 sequences are *definitely* walking at the frame of $t_{run} - 19$; so the sign preindicating running is supposed to exist within this period from $t_{run} - 19$ to t_{run}.

The subsequence of $T = 20$ frames that ends at t_{run} is regarded as a *positive* sample, while we can regard as *negative* samples all the other subsequences except the ones overlapping the positive subsequence with over 10 frames. Note that we thereby obtain one positive sample and about 50 to 100 negative samples from each sequence.

3.2 Evaluation

The prediction performance is measured by *leave-one-sequence-out cross validation*, as follows. At the i-th iteration ($i = 1, .., 57$), we train the linear SVM classifier [10] over the samples excluding the ones drawn from the i-th sequence. Then, the samples from the i-th sequence are evaluated by applying the classifier. In this case, those evaluated samples are highly imbalanced due to containing only one positive sample. Therefore, we regard the i-th sequence as correctly classified only when all the sample from that sequence are successfully classified, which is a relatively hard criterion. In an overall evaluation, we measure the ratio of the correctly classified sequences out of 57 ones.

As to the prediction method (Sect. 2.2), we examined 210 pairs of $\{t, D\}$ parameters: considering $T = 20$, the prediction timing t varies from -19 to 0, and accordingly the period D can be changed from 1 to $t + 20$.

4 Experimental Results

The results by GLAC feature are firstly described and then those of HOG are shown.

4.1 Results by GLAC Feature

Figure 4 shows the classification performance for one frame duration ($D = 1$). We can see that the top accuracy was obtained at $t = -11$ and -7. This result suggests that the frames at $t = -11$ and -7 include distinctive features to preindicate running. It should be noted that though this task is to predict the running action at $t = 0$, the performance at $t = 0$ (on-time classification) is not high. This is because the pedestrians definitely run at $t = 0$ and some negative samples also contain the running action at $t = 0$, making hard to classify at $t = 0$.

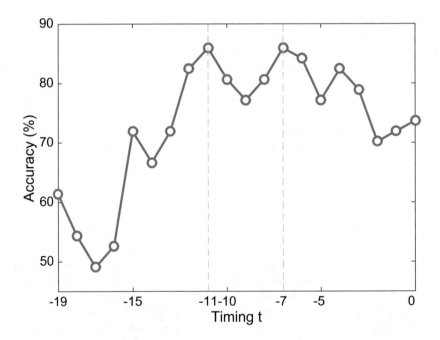

Fig. 4. Classification performance of one frame duration $D = 1$ by GLAC.

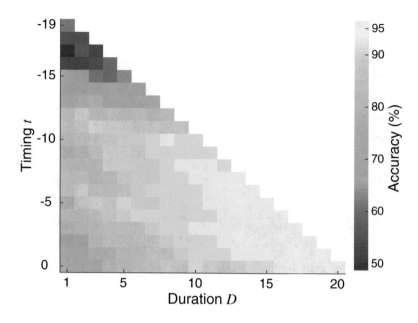

Fig. 5. Classification performance for all parameter pairs by GLAC.

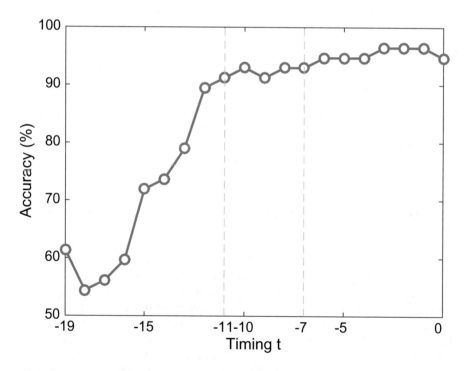

Fig. 6. Classification performance for timing t with maximizing over D by GLAC.

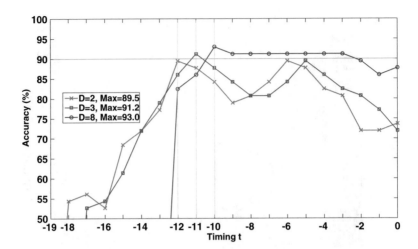

Fig. 7. Classification performance for duration $D = 2, 3, 8$ by GLAC.

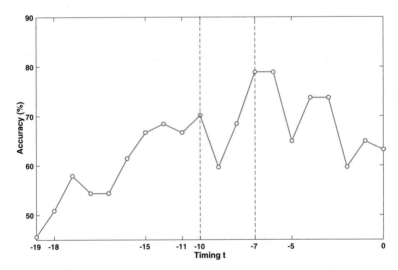

Fig. 8. Classification performance of one frame duration $D = 1$ by HOG.

Figure 5 shows the results for all parameter pairs of $\{t, D\}$. The best accuracy 96.5 % was attained at $t = -3$ with $D = 14, 16$ and $t = -2$ with $D = 18$; the whole sequence $(t = 0, D = 20)$ did not perform the best, exhibiting 94.7 %. However, $t = -2$ and $t = -3$ are not preferable for our purpose, early prediction.

As shown in Fig. 4, the distinctive features are found at $t = -7$ and -11, and thus we can push back the prediction earlier. For early prediction, the timing t is rather important than the duration D, and we show in Fig. 6 the best performance at each t by picking up the maximum accuracy over D. It apparently shows that the performance is saturated at $t = -11$, slightly increasing after

$t = -11$; for example, 93.0 % is attained at $t = -10$ with $D = 8$ which is close to the best 96.5 % at $t = -3$. However, a pedestrian has to be tracked through 8 frames for the duration $D = 8$, which is not preferable for on-board (moving) cameras. Figure 7 shows the classification accuracy and timing with the duration $D = 2, 3$ and 8. 91.2 % is attained at $t = -11$ with $D = 3$ which requires only 3 frame duration. Thus, we can conclude that it is possible to predict the action of running at about 0.37 s. earlier (corresponding to $t = -11$) with over 90 % accuracy. Moreover, if we can compromise with the classification accuracy of 89.5 %, the running action can be predicted at about 0.4 s. earlier ($t = -12$) by using only 2 frame duration.

4.2   Results by HOG Feature

Figure 8 shows the classification performance for one frame duration ($D = 1$). The highest accuracy was obtained at $t = -7$ and $t = -6$, however it is lower than 80 %, while GLAC feature obtained 85.9 %. We can also see a weak peak at $t = -10$, which is not as significant as the peak at $t = -11$ obtained by GLAC feature. This result indicates that the preindicator features at $t = -11$ could not find by using HOG feature, and $t = -7$ was the earliest prediction timing for HOG feature with duration $D = 1$.

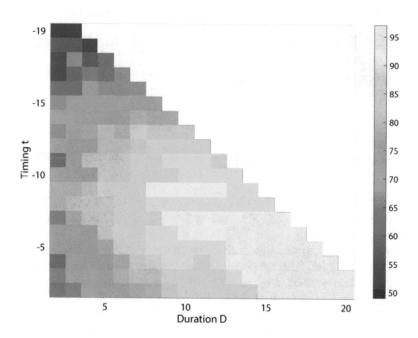

Fig. 9. Classification performance for all parameter pairs by HOG.

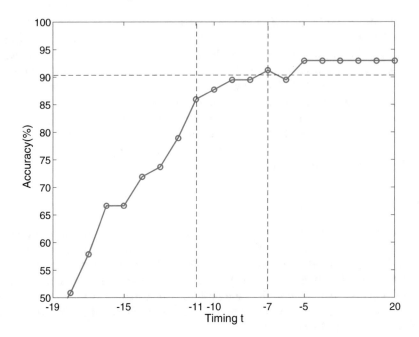

Fig. 10. Classification performance for timing t with maximizing over D by HOG.

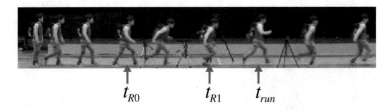

Fig. 11. Biomechanical analysis for transition from walking to running.

Figure 9 shows the results for all parameter pairs of $\{t, D\}$ with HOG feature. The best accuracy 93.0 % was attained at $t = -5$ with $D > 12$ and the whole sequence ($t = 0, D = 20$).

As shown in Fig. 8, the distinctive features are found at $t = -7$ and -8. Since the timing t is rather important than the duration D, and we show the best performance at each t by picking up the maximum accuracy over D in Fig. 10. It apparently shows that the performance is saturated at $t = -9$, slightly increasing after $t = -8$; for example, only 87.0 % is attained at $t = -10$, and the best accuracy 93 % was attained at $t = -5$. These results indicate that the preindicator could not be found in earlier timing by using HOG feature.

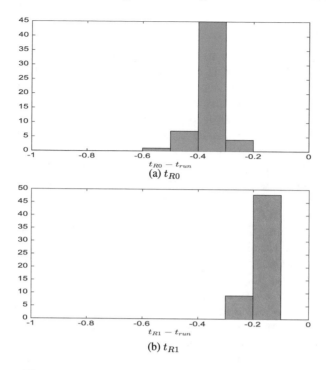

Fig. 12. Histogram for t_{R0} and t_{R1} compared to t_{run}.

5   Biomechanical Analysis

During the transition from walking to running, the visually most distinctive form is found when the head reaches the minimum height. After that, the pedestrian jumps up a little bit and subsequently the phase is completely changed into running. We call this point as t_{R1} (see Fig. 11). On the other hand, when a pedestrian starts running from walking, the form accordingly changes in order to facilitate its acceleration. At that point, the pedestrian's posture is leaning forward as well as stepping and shaking the arms more largely. This point is denoted as t_{R0} (see Fig. 11). The form at t_{R0} is less salient compared to that at t_{R1}, but t_{R0} precedes t_{R1}. For comparing the above results (Sect. 4) to these biomechanically distinct points, we manually annotated t_{R0} and t_{R1} in the sequences. The histograms for those timing points are shown in Fig. 12. Those timing points are not diverse across the pedestrians but relatively concentrated around the means. This result shows that those distinct points defined from the biomechanical viewpoint are also regarded as general measure for predicting running action. Those means are $\bar{t}_{R0} = t_{run} - 0.37$ s. and $\bar{t}_{R1} = t_{run} - 0.19$ s, corresponding to $t = -11$ and $t = -6$, respectively. These are surprisingly coincident with the points $t = -11$ and -7 which are qualitatively obtained in Fig. 4. Thus, we have shown that those quantitatively obtained timing points are also biomechanically meaningful.

6   Conclusion

We have proposed a method to predict the running action of pedestrians at earlier timing before the action actually occurs. The method is based on the appearance-based image features to extract distinctive forms of the pedestrian in transition from walking to running. In addition, the motion information is naively encoded via aggregating (concatenating) the consecutive frame-based features in a time-series sequence, with the two important parameters which indicate the timing and duration, respectively. In the experiments, we examined the performance of our method with two features, GLAC and HOG. By using GLAC, we could successfully determine those two parameters, showing favorable performance of prediction; the running action can be predicted at about 0.4 s before, while HOG enables the prediction at only 0.2 s before. By further analyzing the postures from the viewpoint of biomechanics, the prediction timing is shown to be closely related to the biomechanically distinct form. GLAC can effectively extract the distinct form from the image sequence in earlier timing, while HOG does not. The results indicate that extracting co-occurrence of gradient orientation as second-order statistics by GLAC is more effective than extracting only first-order statistics of occurrence of gradient orientation such as HOG. The experiments performed in this paper are limited due to such as indoor and fixed camera. Our future works include to apply the proposed method to the motion sequences which are captured in more realistic situations.

References

1. Ishikawa, T.: The analysis of pedestrian accidents. https://www.itarda.or.jp/ws/pdf/h22/13_01hokousyaziko.pdf. In Japanese
2. Rosen, E., Sander, U.: Pedestrian fatality risk as a function of car impact speed. Accid. Anal. Prev. **41**, 536–542 (2009)
3. Hulshof, W., Knight, I., Edwards, A., Avery, M., Grover, C.: Autonomous emergency braking test results. In: International Technical Conference on the Enhanced Safety of Vehicles (2013)
4. Dalal, N., Triggs, B.: Histograms of oriented gradients for human detection. In: IEEE Conference on Computer Vision and Pattern Recognition, pp. 886–893 (2005)
5. Ouyang, W., Wang, X., Zeng, X., Qiu, S., Luo, P., Tian, Y., Li, H., Yang, S., Wang, Z., Loy, C.C., Tang, X.: Deepid-net: deformable deep convolutional neural networks for object detection. In: IEEE Conference on Computer Vision and Pattern Recognition, pp. 2403–2412 (2015)
6. Keller, C.G., Gavrila, D.M.: Will the pedestrian cross? A study on pedestrian path prediction. IEEE Trans. Intell. Transp. Syst. **15**(2), 494–506 (2014)
7. Reddy, Y., Krishnaiah, R.: Driving assistance system for identification of sudden pedestrian crossings. Int. J. Res. Inf. Technol. **1**(12), 281–295 (2013)
8. Nishida, K., Kobayashi, T., Iwamoto, T., Yamasaki, S.: Pedestrian action prediction using static image feature. In: Proceedings of 7th International Joint Conference on Computational Intelligence, NCTA, vol. 3, pp. 99–105, Lisbon, Portugal, November 2015
9. Kobayashi, T., Otsu, N.: Image feature extraction using gradient local autocorrelations. In: European Conference on Computer Vision, pp. 346–358 (2008)
10. Vapnik, V.: Statistical Learning Theory. Wiley, New York (1998)

Weighted Nonlinear Line Attractor for Complex Manifold Learning

Theus H. Aspiras[1]([✉]), Vijayan K. Asari[1], and Wesam Sakla[2]

[1] University of Dayton, Dayton, OH 45496, USA
{aspirast1,vasari1}@udayton.edu
[2] Lawrence Livermore National Laboratory, Livermore, CA 94550, USA
wesam.sakla@us.af.mil

Abstract. An artificial neural network is modeled by weighting between different neurons to form synaptic connections. The nonlinear line attractor (NLA) models the weighting architecture by a polynomial weight set, which provides stronger connections between neurons. With the connections between neurons, we desired neuron weighting based on proximity using a Gaussian weighting strategy of the neurons that should reduce computational times significantly. Instead of using proximity to the neurons, it is found that utilizing the error found from estimating the output neurons to weight the connections between the neurons would provide the best results. The polynomial weights that are trained into the neural network will be reduced using a nonlinear dimensionality reduction which preserves the locality of the weights, since the weights are Gaussian weighted. A distance measure is then used to compare the test and training data. From testing the algorithm, it is observed that the proposed weighted NLA algorithm provides better recognition than both the GNLA algorithm and the original NLA algorithm.

Keywords: Nonlinear line attractor · Multidimensional data · Neural networks · Machine learning

1 Introduction

Artificial neural networks are a well-researched area, encompassing several different types architectures used to model biological neural networks in the brain. Most networks that have been researched are feed-forward neural networks, which propagate signals through several layers of neurons to create an output. The network that we will be focusing on is the attractor neural network, which are usually recurrent neural networks that connects neurons in the same layer for processing. There are different types of attractor networks that are available, like the fixed point attractor, also known as the Hopfield network [1]. These networks use the minimization of an energy function to get the network to converge to a desired state. Therefore given a specific data point with any number of distortions, the network should be able to attract towards a desired state, which is given by the fixed point. Other types of attractor networks include the cyclic

© Springer International Publishing AG 2017
J.J. Merelo et al. (eds.), *Computational Intelligence*, Studies in Computational Intelligence 669,
DOI 10.1007/978-3-319-48506-5_19

attractor network [2] and the line attractor network [3]. The attractor network that is of most interest is the line attractor network.

The line attractor, which differs from the fixed point attractor, attracts a given set of data points towards a line rather than a specific point. This type of attractor is useful in encompassing manifolds that are not contained spherically, but are elongated in shape. Due to the line attractor being created, estimation of data points is more useful than a fixed point attractor due to attraction to a specific point on the line rather than the only given point of a manifold. With the high dimensionality found in neural network, it is found that using a nonlinear line to encompass the manifold is better for modeling the manifold. We utilize the nonlinear line attractor as the basis for the research in this paper.

1.1   Biological Implications

Most recurrent autoassociative networks are interconnected together with all other nodes. In biological structures, neurons found in the brain are connected only to the surrounding neurons rather than every neuron found in the brain. These local connections are even weighted based on the proximity of the neurons towards each other. Tononi et al. [4] found that most networks are able to reduce the number of connections in the network while preserving the accuracy of the network. Guido et al. [5] found that even though part of the brain is damaged, due to modularity found in the structures of the visual cortex, the brain is able to retrain the systems to accommodate new data.

Even with the proximity of the neurons, there are connections that should be kept due to the least amount of error from the neuron. With proximity of the neurons, close neurons will be preserved due to the closeness of value, but due to the reduction of error in the neurons, the network will favor the least error in the network. This implication can be found in neural networks in the brain, since connections that have the least error should be weighted higher than connections that contain more error.

1.2   Modularity in Neural Networks

The local connectivity is able to create modularity in the network which still preserves the accuracy of the network. Happel et al. [6] found that different configurations in the network using modularity is able to improve the recognition of the network, since the redundancy, especially redundancy in error, found in the network is reduced. Gottumukkal et al. [7] used modularity in principal component analysis to create and recognize sub-images to recognize the whole image. Gomi et al. [8] found specific modules in the network which learns only portions of data to aid in the completion of the task. Auda et al. [9] used Cooperative Modular neural networks, which uses overlap between modules, to further improve the classification accuracy. Modularity is able to reduce the number of computations while improving the recognition capabilities of the network.

The base neural network that is researched in this paper is the nonlinear line attractor by [10], which has applications in pattern association [11], pose and expression invariant face recognition [12], and other machine learning tasks. Modularity and proximity weighting can be used to improve the capabilities of the network by reducing the redundancies found in the network and improving computational time. We will first look at the proximity weighting of the Gaussian weighting strategy and then implement the error-based weighting strategy, which should reduce the weights and improve the accuracy of the network.

The main contributions of this paper are:

- Gaussian weighting strategy to the Nonlinear Line Attractor Network to introduce modularity
- An error-based weighting strategy to the Nonlinear Line Attractor Network for comparison
- Reduction of the computational complexity to improve the convergence time of the NLA architecture
- An improved scenario for using the Nonlinear Dimensionality Reduction for object recognition.

2   Nonlinear Line Attractor

The nonlinear line attractor network trains patterns using a set of interconnected neurons with polynomial weighting. This network should provide better recognition than point and line attractor networks due to the stronger connection made through the polynomial weighting.

Let the response $x_{(i,s)}$ of the ith neuron for the sth pattern due to the excitations $x_{(j,s)}$ from other neurons in a fully connected recurrent neural network with N neurons be expressed as:

$$x_{(i,s)} = \frac{1}{N} \sum_{j=1}^{N} \Lambda_{ij}(x_{(j,s)}), \text{ for } 1 \leq i \leq N, \tag{1}$$

where Λ_{ij} is defined by a kth order nonlinear line as:

$$\Lambda_{ij}(x_{(j,s)}) = \sum_{m=0}^{k} w_{(m,ij)} x_{(j,s)}^{m} \text{ for } 1 \leq i,j \leq N \tag{2}$$

The equation provides the best fit line of the data points in a dataset for a given class as shown in Fig. 1. The mth order term of the resultant memory w_m can be expressed as:

$$W_m = \begin{pmatrix} w_{(m,11)} & \cdots & w_{(m,1N)} \\ \vdots & \ddots & \vdots \\ w_{(m,N1)} & \cdots & w_{(m,NN)} \end{pmatrix}, \text{ for } 0 \leq m \leq k, \tag{3}$$

To calculate the weights of the system, we can use error minimization of output. To minimize the least squares error in the weight matrix, we can formulate the following equation, which yields the optimum weight set.

$$E_{ij}[w_{(0,ij)}, w_{(1,ij)}, ..., w_{(k,ij)}] =$$
$$\sum_{s=1}^{P}[x_{(i,s)} - \Lambda_{ij}(x_{(j,s)})]^2, \text{ for } 1 \leq i, j \leq N, \tag{4}$$

To minimize the least squares error, we must equate the derivative of the error with respect to the weight to be zero, as shown in the following equation.

$$\frac{\delta E_{ij}}{\delta w_{(m,ij)}} = 0 \forall m = 0, 1, ..., k, \tag{5}$$

We can then find that the equation can be reduced to a set of linear equations based on the order of the polynomial, as shown below.

$$w_{(0,ij)} \sum_{s=1}^{P} x_{(j,s)}^0 + w_{(1,ij)} \sum_{s=1}^{P} x_{(j,s)}^1 + \cdots$$
$$+ w_{(k,ij)} \sum_{s=1}^{P} x_{(j,s)}^k = \sum_{s=1}^{P} x_{(i,s)} x_{(j,s)}^0$$

$$w_{(0,ij)} \sum_{s=1}^{P} x_{(j,s)}^1 + w_{(1,ij)} \sum_{s=1}^{P} x_{(j,s)}^2 + \cdots$$
$$+ w_{(k,ij)} \sum_{s=1}^{P} x_{(j,s)}^{k+1} = \sum_{s=1}^{P} x_{(i,s)} x_{(j,s)}^1 \tag{6}$$

$$\vdots$$

$$w_{(0,ij)} \sum_{s=1}^{P} x_{(j,s)}^k + w_{(1,ij)} \sum_{s=1}^{P} x_{(j,s)}^{k+1} + \cdots$$
$$+ w_{(k,ij)} \sum_{s=1}^{P} x_{(j,s)}^{2k} = \sum_{s=1}^{P} x_{(i,s)} x_{(j,s)}^k$$

Once the nonlinear best fit line is created to model the data points, we can then create limits to obtain the variance of the manifold, thus encompassing the entire dataset. This is done by creating an activation function, as shown in Eq. 7.

$$\Phi(\Lambda_{ij}[x_{(j,s)}(t)]) =$$
$$\begin{cases} x_{(i,s)}(t) \text{ if } \psi_{(ij,-)} \leq \{\Lambda_{ij}(x_{(j,s)}(t)) - x_{(i,s)}(t)\} \leq \psi_{(ij,+)} \\ \Lambda_{ij}(x_{(j,s)}(t)) \quad otherwise \end{cases} \tag{7}$$

where

$$\Lambda_{ij}(x_{(j,s)}(t)) = \sum_{m=0}^{k} w_{(m,ij)} x_{(j,s)}^m(t) \tag{8}$$

The activation function is used to define thresholds $[\psi_{(ij,-)}, \psi_{(ij,+)}]$ to ensure that data points do not update if the estimated data point lies within the manifold and that data points update if the estimated data point lies outside of the manifold. These threshold regions can be expressed as:

$$\psi_{(ij,-)} = \begin{cases} \psi_{(1,ij,-)} & \text{if } 0 \leq x_j < \frac{L}{\Omega} \\ \psi_{(2,ij,-)} & \text{if } \frac{L}{\Omega} \leq x_j < \frac{2L}{\Omega} \\ \quad \vdots \\ \psi_{(\Omega,ij,-)} & \text{if } (\Omega-1)\frac{L}{\Omega} \leq x_j < L \end{cases} \tag{9}$$

$$\psi_{(ij,+)} = \begin{cases} \psi_{(1,ij,+)} & \text{if } 0 \leq x_j < \frac{L}{\Omega} \\ \psi_{(2,ij,+)} & \text{if } \frac{L}{\Omega} \leq x_j < \frac{2L}{\Omega} \\ \quad \vdots \\ \psi_{(\Omega,ij,+)} & \text{if } (\Omega-1)\frac{L}{\Omega} \leq x_j < L \end{cases} \tag{10}$$

where Ω is the number of segments used for the piecewise threshold regions and L is the length of the manifold. Due to the usage of the nonlinear line attractor, we will no be using the activation and limits defined by the algorithm because of nonlinear dimensionality reduction. Figure 1 shows how the weights are interconnect through the inputs.

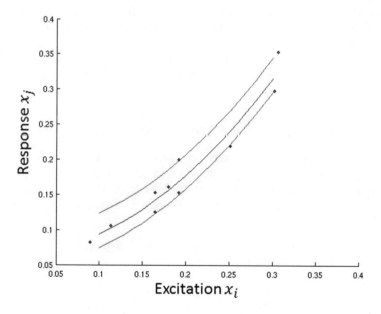

Fig. 1. Interconnection of weights. The blue line captures the nonlinear line modeling and the red lines capture the variances of the data.

Since this is modeling of a specific manifold, multiple manifolds may be needed to encompass more of a class. Most will require a specific manifold per class, so there will be at least one weight set per class.

Computational Strategy. We have devised an effective computational strategy for training data. Given Eq. 6, previous models of the computational strategy require computing powers for each interconnection, in which there are several calculations that are repeated in the equations while traversing through each interconnection. Instead of having redundant calculations, we can divide the weight calculation into different steps.

Stage 1 is the calculation of powers for the inputs. In Eq. 6, we see that every $x_{(j,s)}$ has an order associated with it. Calculation of these terms would be redundant for all different combinations of inputs and outputs, since there are multiple terms with the same order, hence the same value. Computing only the powers in this stage would tremendously reduce the computation time of the system.

Stage 2 is be the calculation of the weights, given the set of normal equations and using the values obtained from stage 1. The solving of the normal equations can be done using a linear solve algorithm. This stage will take a considerable amount of time due to the volume of data, specifically the number of inputs, since the weight matrix size is $N \times N \times k$ where N refer to the size of the image.

Stage 3 is the calculation of the activation function, which will also require a considerable amount of computation time. Since the computation of the orders for all of the input data are already known, the activation function can be formulated using that data.

Stage 4 is the calculation of the nonlinear dimensionality reduction. This step requires all of the weights and is dependent on the number of inputs and also the order of the weight system.

2.1   Gaussian Nonlinear Line Attractor (GNLA) Network

The Gaussian Nonlinear Line Attractor Network is a modified NLA network that incorporates proximity weighting which improves run times and recognition rates. When implementing a Gaussian neighborhood approach, we can change the coefficient in the front and add the distance equation. The equation can then be modified as:

$$x_{(i,s)} = \sum_{j=1}^{N} \alpha_{ij} \Lambda_{ij} x_{(j,s)} \text{ for } 1 \leq i \leq N, \tag{11}$$

where

$$\alpha_{ij} = exp\left(-\left(\frac{(q_i - q_j)^2}{2\sigma_q^2} + \frac{(r_i - r_j)^2}{2\sigma_r^2}\right)\right), \tag{12}$$

For these equations, q and r define the spatial coordinates of the input x. Instead of using the Gaussian function, we can use the Gaussian kernel, for

0	0	0	0	0	1	1	1	0	0	0	0	0
0	0	0	1	2	3	4	3	2	1	0	0	0
0	0	1	4	9	15	18	15	9	4	1	0	0
0	1	4	13	29	48	57	48	29	13	4	1	0
0	2	9	29	67	111	131	111	67	29	9	2	0
1	3	15	48	111	183	216	183	111	48	15	3	1
1	4	18	57	131	216	255	216	131	57	18	4	1
1	3	15	48	111	183	216	183	111	48	15	3	1
0	2	9	29	67	111	131	111	67	29	9	2	0
0	1	4	13	29	48	57	48	29	13	4	1	0
0	0	1	4	9	15	18	15	9	4	1	0	0
0	0	0	1	2	3	4	3	2	1	0	0	0
0	0	0	0	0	1	1	1	0	0	0	0	0

Fig. 2. An example of a 13×13 Gaussian kernel.

example a 13×13 Gaussian kernel as shown in Fig. 2. This will effectively reduce the computation time. We can then change the equation as

$$x_{(i,s)} = \sum_{j \in B} a_{ij} \Lambda_{ij} x_{(j,s)}, \text{ for } 1 < i < N, \tag{13}$$

where n is the size of the kernel and where

$$B = \{j | a_{ij} > 0\} \text{ for } 1 \leq i \leq N, \tag{14}$$

and

$$a_{ij} = exp\left(-\left(\frac{(\hat{x}_i - \hat{x}_j)^2}{2\sigma_{\hat{x}}^2} + \frac{(\hat{y}_i - \hat{y}_j)^2}{2\sigma_{\hat{y}}^2}\right)\right) \tag{15}$$

We can also effectively reduce the computation time of the kernel by not computing any portion that contains zeros. According to the Gaussian kernel above (which is a 13×13), roughly 28 % of the Gaussian kernel are zeros, as shown in Fig. 3, thus a reduction of the computation time can be accomplished by ignoring those computations.

					1	1	1					
			1	2	3	4	3	2	1			
		1	4	9	15	18	15	9	4	1		
	1	4	13	29	48	57	48	29	13	4	1	
	2	9	29	67	111	131	111	67	29	9	2	
1	3	15	48	111	183	216	183	111	48	15	3	1
1	4	18	57	131	216	255	216	131	57	18	4	1
1	3	15	48	111	183	216	183	111	48	15	3	1
	2	9	29	67	111	131	111	67	29	9	2	
	1	4	13	29	48	57	48	29	13	4	1	
		1	4	9	15	18	15	9	4	1		
			1	2	3	4	3	2	1			
					1	1	1					

Fig. 3. An example of a 13×13 Gaussian kernel with zeros removed.

Nonlinear Dimensionality Reduction. By using the weights in a nonlinear dimensionality reduction technique, we can leverage the nonlinear weights to give an optimum transform with less dimensions needed. Given that there are r different line attractor networks, there will be y different outputs, as shown in the following equation.

$$
\begin{aligned}
Y_1 &= W_{1,k}X^k + W_{1,k-1}X^{k-1} + \cdots + W_{1,0}X^0 \\
Y_2 &= W_{2,k}X^k + W_{2,k-1}X^{k-1} + \cdots + W_{2,0}X^0 \\
&\vdots \\
Y_r &= W_{r,k}X^k + W_{r,k-1}X^{k-1} + \cdots + W_{r,0}X^0
\end{aligned}
\tag{16}
$$

Singular value decomposition (SVD) [13] was used previously to reduce the weight set. We propose using the locally linear embedding (LLE) [14] algorithm which interconnected weight sets for different datapoints to reduce the weights of the NLA network.

Each mth term of the networks' memory is evaluated using the LLE algorithm. We first obtain a sparse matrix M using the following equation.

$$M_{(m,d)} = (I - W_{(m,d)})' * (I - W_{(m,d)}) \text{ for } 0 \le m \le k; 1 \le d \le r \tag{17}$$

We then take the smallest z eigenvectors from M and use them as the projection into the lower-dimensional subspace. The projection of the N-dimensional data to a z-dimensional subspace using a $z \times N$ sub-matrix obtained from the smallest z eigenvectors of the LLE yields a z-dimensional output Y'_m where $z \ll N$. The lower dimensional data can be used in a euclidean distance metric to evaluate the effectiveness of the algorithm.

The weight matrices do not contain the coefficients of the Gaussian weighting, thus we can incorporate the weighting inside the weight matrix. Given Eq. 17, we can embed the normalized coefficients to multiply the weights using the following equation.

$$A = \begin{pmatrix} \alpha_{11} & \cdots & \alpha_{1N} \\ \vdots & \ddots & \vdots \\ \alpha_{N1} & \cdots & \alpha_{NN} \end{pmatrix} \tag{18}$$

The resultant multiplication of the weight set is given by the equation below.

$$\bar{W}_{(m)} = \begin{pmatrix} \bar{w}_{(m,11)} & \cdots & \bar{w}_{(m,1N)} \\ \vdots & \ddots & \vdots \\ \bar{w}_{(m,N1)} & \cdots & \bar{w}_{(m,NN)} \end{pmatrix} \text{ for } 0 \le m \le k \tag{19}$$

For calculating the nonlinear dimensionality, each order must be calculated separately. Thus when computing the output, the orders still are computed in the same functionality as the original NLA architecture. Instead of losing some possible recognition ability due to the summation of orders, we can concatenate the orders to formulate a bigger vector, which will be tested later in this chapter.

2.2 Complexity

In Stage 1, we can find that it will be the same complexity as the previous algorithm since we are just computing the powers.

In Stage 2, the original complexity is $N \times N \times k^2$ due to the linear solve algorithm, but modified complexity is $N \times n \times k^2$, where # neighbors is significantly smaller than the size of the network. For example, given that we have a network of 60×80, which is 4800, and an order of 4 for the polynomial, the complexity of the algorithm becomes $4800 \times 4800 \times 4^2 = 368640000$ computations. If we create kernel of size 13×13, we would have a complexity of $4800 \times 169 \times 4^2 = 12979200$ computations, which is 3.52 % the computation time of the original. If we reduce the kernel by taking out all zeros in the function, we would have a complexity of $4800 \times 121 \times 4^2 = 9292800$ computation, which is only 2.52 % the computation time of the original and 71.6 % the computation time of the kernel.

In Stage 3, the complexity should be reduced just as stage 2. In Stage 4, the complexity should be the same as the previous algorithm. Table 1 shows the run

Table 1. Training times for each stage of the algorithms.

Runtime	Original	Current	Kernel	Kernel (No zeros)
Stage 1	N/A	15.2 min	15.2 min	15.2 min
Stage 2	1280 min	95.3 min	3.35 min	2.4 min
Stage 3	87.5 min	87.5 min	3.08 min	2.21 min
Stage 4	10.8 min	10.8 min	10.8 min	10.8 min
Total runtime	1378.3 min	208.8 min	32.43 min	30.61 min

times on a small subset of the EO Synthetic Vehicle dataset. It is found that the kernel algorithm without using zeros provides faster training times than all of the other algorithms.

2.3 Results

Datasets. The first dataset used to test the GNLA network is the EO Synthetic Vehicle Database, as shown in Fig. 4. This database contains several vehicles under different lighting conditions and viewing angles. The second dataset used is the Sheffield database, which contains face imagery under different viewing angles. For all datasets, a 13×13 kernel is used for training and testing the dataset.

Results 1. Table 4 shows the results on the EO Synthetic Vehicle database. A full 360 view of each vehicle and one lighting condition is used to train and test the validity of the technique. By concatenating all of the feautures found during the nonlinear dimensionality reduction, we can see better recognition results. Among all of the types of NLA architectures, it is found that the GNLA architecture with concatenation works the best (Table 2).

Results 2. Table 3 shows the results of the GNLA network and other classification algorithms on the Sheffield database. This database is split into two equal sized datasets, one for training and one for testing. It is found that the GNLA architecture with concatenation gives the best results amongst all of the other algorithms. Runtimes for the NLA and GNLA algorithms are 8.2 h and 0.8 h respectively.

Discussion. When training and testing with attractor networks, data that is orthogonal provides better recognition due to the separation of the data with the given input space. In-class variance in many of the dataset are very large while between-class variance is very small, thus providing the network less ability to separate the data in a classification task. The best case that the network should be able to handle is a completely orthogonal dataset for all classes.

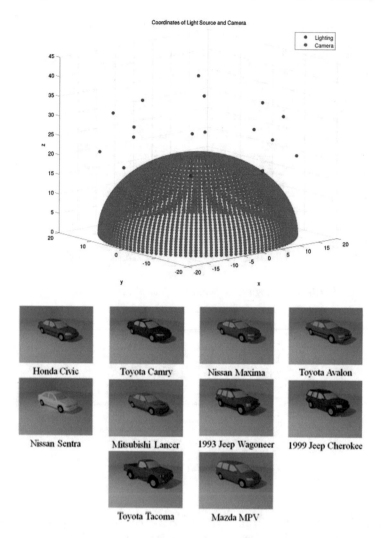

Fig. 4. EO Synthetic Vehicle database. The top image shows the different camera and lighting positions of the database and the images below show the different vehicles.

When using the GNLA architecture, orthogonality of the dataset increases due to the reduction of the number of inputs. When having data with many redundancies in the data will reduce the effectiveness of the data. Due to the weighting of the GNLA, less of the neurons are used in the recreation of the output, thus increasing orthogonality. ALso with the GNLA network, local information is used, thus increasing orthogonality due to regions of the data having only foreground or background data, thus reducing the influence of the background data.

Table 2. Recognition results on the EO Synthetic Vehicle database.

# prin. comp.	NLA reg.	NLA concat.	GNLA reg.	GNLA concat.
1	47.50	**60.56**	35.83	43.89
2	59.72	66.94	**82.78**	76.94
3	**86.67**	79.72	63.06	85.28
4	61.39	85.00	76.94	**88.61**
5	63.89	86.11	81.67	**88.89**
6	69.17	88.06	85.00	**91.11**
7	66.94	87.50	86.39	**91.39**
8	75.28	88.89	87.22	**92.22**
9	72.22	86.39	91.39	**95.00**
10	78.61	87.22	91.94	**94.72**
All	99.19	85.28	**99.44**	99.17

Table 3. Recognition results on the Sheffield database.

Algorithm	% Recognition
PCA	86.87
KPCA	87.64
LDA	90.87
DPCA	92.90
DCV	91.51
B2DPCA	93.38
GNLA	93.33
GNLA (concat)	94.07

With the GNLA algorithm, proximity of the neurons provides valuable insight as to improve the network. Since the influence of the convergence should mostly come from the foreground pixels, we should be able to leverage that data in the architecture, which will improve the recognition capabilities of the network.

3 Weighted Nonlinear Line Attractor (WNLA) Network

Using foreground information tends to reduce the error of the network due to the information being more valuable in discerning classes. When training the network with the weights, we are able to find the error that is created during training, which will be able to gives us insight as to which neurons to weight for the network. Neurons used in the reconstruction of the output that give less error should be weighted higher than other neurons.

The proposed weighting scheme finds neuron relationships based on the error of the reconstruction. With relationships that have least error, we can weight those relationships higher to improve the recognition capabilities. Therefore, we can expect that weighting relationships that contain less error, we will be able to reduce the overall error of the trained network. This is an improvement to the proposed GNLA architecture since the proximity weighting is spatial rather than based from least error. One benefit with this weighted NLA architecture is that creating more layers with the nonlinear dimensionality is possible since any spatial orientation is destroyed in the next layer.

3.1   Architecture

When calculating the weights for the weighted NLA, the polynomial weights must be calculated first for all of the training data to find the error of the data. This weight computation will be different than the GNLA architecture, which is able to reduce the training time. The weight NLA needs to compute all of the interconnections to find the best weight reduction for the network.

Once the weights are obtained, then the NLA architecture can be tested for all of the trained data points to find the error accrued in each relationship. Depending on the number of data points, the calculation may take longer for training. Once all of the nodal relationship error is found, a map can be created based on the order of the error that is produced by the network. The computation of the weights for each of the nodal relationships can be shown in Eq. 20. With the smallest error, the weighting will be the highest while the largest error will give a smaller weighting.

$$a_i = exp\left(-\left(\frac{i^2}{2\sigma^2}\right)\right) \qquad (20)$$

where i is the sorted index for all of the nodes. With the error obtained by the network, smallest error can be combined to remove collections of neurons for the entire network or for each neuron. For this paper, we will be using individual error to find the best neuron relationships for each neuron. Once the map for the weighting is obtained, we can perform nonlinear dimensionality reduction to reduce the number of dimensions for the classification task. This network should be able to reduce the global error due to the reduction of highly erroneous nodal relationships.

3.2   Results

The dataset used to test the weighted NLA architecture is the EO Synthetic Vehicle database. Table 4 shows the results of the algorithms on the database using a full 360 view of the vehicle and one lighting condition. It is found that the weighted NLA architecture provides the best recognition rates out of all of the proposed architectures.

Table 4. Recognition results on the EO Synthetic Vehicle database.

# of weights	NLA (R)	NLA (C)	GNLA (R)	GNLA (C)	WNLA (R)	WNLA (C)
1	47.50	**60.56**	35.83	43.89	38.89	42.22
2	59.72	66.94	**82.78**	76.94	49.44	66.94
3	**86.67**	79.72	63.06	85.28	61.11	76.67
4	61.39	85.00	76.94	**88.61**	70.83	86.11
5	63.89	86.11	81.67	88.89	86.94	**95.00**
6	69.17	88.06	85.00	91.11	86.39	**96.94**
7	66.94	87.50	86.39	91.39	90.00	**98.33**
8	75.28	88.89	87.22	92.22	94.44	**98.89**
9	72.22	86.39	91.39	95.00	94.72	**98.33**
10	78.61	87.22	91.94	94.72	93.89	**98.61**
All	99.19	85.28	99.44	99.17	**99.72**	**99.72**

The proposed weighted NLA architecture improves the classification rates and convergence of the network. Due to the averaging effect of the nodal relationships of the NLA architecture, the network seems to improve with any type of modification of the interconnectivity of the network. With the GNLA network, spatial connectivity takes precedence, but it is found insufficient due to the findings in the error-based weighting strategy. The weighted NLA architecture aims to reduce any error causing relationships in the network, thus providing the best recognition of the proposed architectures.

4   Conclusions

The proposed weighted nonlinear line attractor network has performed the best compared to the Gaussian NLA and the original NLA architectures. Due to the reduction of orthogonality based on the reduction of error, rather than reduction of nodes due to spatial proximity, the network is able to perform much better than the other proposed architectures and is projected to perform better in many other databases as well. We aim to leverage the knowledge found in the network to create a new network architecture, which encompasses the polynomial nature of the network, the modularity introduced by the GNLA and weighted NLA, and a multilayer architecture, that is not usually found in this network. These findings show promise in the development in a stronger, more robust neural network, capable of difficult recognition tasks and complex learning.

References

1. Hopfield, J.J.: Neural networks and physical systems with emergent collective computational abilities. Proc. Nat. Acad. sci. **79**, 2554–2558 (1982)
2. Lewis, J.E., Glass, L.: Steady states, limit cycles, and chaos in models of complex biological networks. Int. J. Bifurcat. Chaos **1**, 477–483 (1991)

3. Zhang, K.: Representation of spatial orientation by the intrinsic dynamics of the head-direction cell ensemble: a theory. J. Neurosci. **16**, 2112–2126 (1996)

4. Tononi, G., Sporns, O., Edelman, G.M.: Measures of degeneracy and redundancy in biological networks. Proc. Nat. Acad. Sci. **96**, 3257–3262 (1999)

5. Guido, W., Spear, P., Tong, L.: Functional compensation in the lateral suprasylvian visual area following bilateral visual cortex damage in kittens. Exp. Brain Res. **83**, 219–224 (1990)

6. Happel, B.L., Murre, J.M.: Design and evolution of modular neural network architectures. Neural Netw. **7**, 985–1004 (1994)

7. Gottumukkal, R., Asari, V.K.: An improved face recognition technique based on modular PCA approach. Pattern Recogn. Lett. **25**, 429–436 (2004)

8. Gomi, H., Kawato, M.: Recognition of manipulated objects by motor learning with modular architecture networks. Neural Netw. **6**, 485–497 (1993)

9. Auda, G., Kamel, M.: Cmnn: cooperative modular neural networks for pattern recognition. Pattern Recogn. Lett. **18**, 1391–1398 (1997)

10. Seow, M.-J., Asari, V.K.: Recurrent network as a nonlinear line attractor for skin color association. In: Yin, F.-L., Wang, J., Guo, C. (eds.) ISNN 2004. LNCS, vol. 3173, pp. 870–875. Springer, Heidelberg (2004). doi:10.1007/978-3-540-28647-9_143

11. Seow, M.J., Asari, V.K.: Recurrent neural network as a linear attractor for pattern association. IEEE Trans. Neural Netw. **17**, 246–250 (2006)

12. Seow, M.J., Alex, A.T., Asari, V.K.: Learning embedded lines of attraction by self organization for pose and expression invariant face recognition. Opt. Eng. **51**, 107201 (2012)

13. Golub, G.H., Reinsch, C.: Singular value decomposition and least squares solutions. Numer. Math. **14**, 403–420 (1970)

14. Roweis, S.T., Saul, L.K.: Nonlinear dimensionality reduction by locally linear embedding. Science **290**, 2323–2326 (2000)

Handling Selective Participation in Neuron Assembly Detection

Salatiel Ezennaya-Gomez[1,2]([✉]), Christian Borgelt[1], Christian Braune[2], Kristian Loewe[2,3], and Rudolf Kruse[2]

[1] Intelligent Data Analysis Research Unit, European Centre for Soft Computing, c/ Gonzalo Gutiérrez Quirós s/n, 33600 Mieres (Asturias), Spain
s.ezennaya@gmail.com
[2] Department of Knowledge and Language Processing, Otto-von-Guericke-University, Universitätsplatz 2, 39106 Magdeburg, Germany
[3] Department of Neurology, Experimental Neurology, Otto-von-Guericke-University, Leipziger Straße 44, 39120 Magdeburg, Germany

Abstract. With the objective to detect neuron assemblies in recorded parallel spike trains, we develop methods to find frequent parallel episodes in parallel point processes (or event sequences) that allow for imprecise synchrony of the events constituting occurrences (temporal imprecision) as well as incomplete occurrences (selective participation). The temporal imprecision problem is tackled by frequent pattern mining using two different notions of synchrony: a binary notion that captures only the number of instances of a pattern and a graded notion that captures both the number of instances as well as the precision of synchrony of its events. To cope with selective participation, which is the main focus of this paper, a reduction sequence of items (or event types) is formed based on found frequent patterns and guided by pattern overlap, for which we explore different concept. We demonstrate the performance of our methods on a large number of (artificially generated) data sets with injected parallel episodes, which mimic actually recorded parallel spike trains.

1 Introduction

We present methodology to identify meaningful frequent synchronous patterns in event sequences (see e.g. [16]), using principles of frequent item set mining (FIM) (see e.g. [2]). As is well known, the objective of FIM, which was originally developed for market basket analysis, is to find all item sets that are frequent in a transaction database. FIM uses the *support* (that is, the number of occurrences in the transactions) to define an item set as frequent, namely if its support reaches or exceeds a (user-specified) minimum support threshold. In standard FIM the support of an item set is a simple count of transactions. In our case, however, the event sequence data is continuous in nature, since it resides in the time domain, and thus no (natural) transactions exist. This continuous form causes several problems, especially w.r.t. the definition of a proper support measure.

© Springer International Publishing AG 2017
J.J. Merelo et al. (eds.), *Computational Intelligence*, Studies in Computational Intelligence 669,
DOI 10.1007/978-3-319-48506-5_20

Furthermore, frequent pattern mining in continuous time faces two main problems: *temporal imprecision* and *selective participation*. The former consists in the fact that events can be affected by temporal jitter, due to which the events underlying an occurrence of a pattern may not be perfectly aligned. In frequent pattern mining we tackle temporal imprecision by defining that items (or events) co-occur if they occur in a (user-specified) limited time span from each other. If a binary notion of synchrony is used (that is, a group of events is either considered to be synchronous or not synchronous—two values), the support of an item set can be defined as a maximum independent set (MIS) of its occurrences, which can be computed efficiently with a greedy algorithm (see [4, 19]).

Unfortunately, a greedy algorithm no longer guarantees an optimal solution to the MIS problem if a *graded notion* of synchrony is used, while a backtracking approach (as it would be used for a general MIS problem, which is NP-complete) takes exponential time in the worst case. As a consequence, an adaptation is necessary, which takes the form of an approximation procedure to compute the support, but maintaining the crucial property of support being *anti-monotone*. In this way, [7] defined a graded synchrony approach where the support computation takes the precision of synchrony into account. That is, a pattern that has fewer occurrences, but in each of these the items occur very closely together in time, is rated better than an item set, which has more instances, but in each of these the synchrony of the events is rather loose [7].

The second problem, that is, *selective participation*, is related to lack of occurrence of some items, which produces incomplete pattern instances. As a consequence, only subsets of the actual pattern are present in the instances underlying a pattern. This can be caused by imperfections of the measuring technology or by properties of the underlying process. [3] presented an approach to solve this problem in the binary synchrony setting we mentioned above.

Our motivating application area is *parallel spike train analysis* in neurobiology, where *spike trains* are sequences of points in time, one per neuron, that represent the times at which an electrical impulse (*action potential* or *spike*) is emitted. It is generally believed that biological neurons represent and transmit information by firing sequences of spikes in various temporal patterns [5]. However, in the research area of neural coding, many competing hypotheses have been proposed how groups of neurons represent and process information, and ongoing research tries to develop methods to confirm or reject (some of) these hypotheses by analyzing recordings of neuronal firing patterns. Here we focus on the *temporal coincidence coding hypothesis*, which assumes that neurons are arranged in *neuronal assemblies*, that is, groups of neurons that tend to exhibit synchronous spiking (such cell assemblies were proposed in [10]), and claims that the tighter the spikes are in time, the stronger the encoded stimulus is. In this setting, the *precision of synchrony* (which we tackle with our graded notion) is relevant, because (more tightly) synchronous spike input to receiving neurons is known to be more effective in generating output spikes [1, 13].

In this paper we investigate how selective participation can be handled with both binary and graded synchrony in an algorithm called CoCoNAD (for **Co**ntinuous-time **C**losed **N**euron **A**ssembly **D**etection), which was developed to detect significant synchronous patterns in event sequences.

As a first step to identify neuron assemblies, we look for *frequent neuronal patterns* (i.e. groups of neurons that exhibit *frequent synchronous spiking*). Here both temporal imprecision and selective participation are expected to be present and thus require proper treatment. Temporal imprecision is handled by binary and graded notions of synchrony that give rise to anti-monotone support measures, by which certain item (or event type) sets are characterized as frequent. Once frequent patterns are detected, statistical filtering is applied to remove those frequent patterns that are likely only chance events and thus not relevant. In a second step, selective participation is handled by analyzing the filtered patterns w.r.t. their overlap, forming a reduction sequence of items (or event types) from which a (candidate for) a neural assembly can finally be read.

The remainder of this paper is structured as follows: Sect. 2 covers basic terminology and notation and introduces the binary and graded notions of synchrony as well as the support computation for both methods. In Sect. 3 methods to mine and filter frequent synchronous patterns with pattern spectrum filtering are described. In Sect. 4 we present our methodologies to identify frequent parallel episodes with selective participation. Section 5 reports experimental results on (artificially generated) data sets with injected parallel episodes. Finally, in Sect. 6 we draw conclusions from our discussion.

2   Event Sequences

We adopt notation and terminology from [7,16,19]. The data are sequences of events $\mathcal{S} = \{\langle i_1, t_1 \rangle, \ldots, \langle i_m, t_m \rangle\}$, $m \in \mathbb{N}$, where i_k in the *event* $\langle i_k, t_k \rangle$ is the *event type* or *item* (taken from an item base B) and $t_k \in \mathbb{R}$ is the time of occurrence of i_k, $k \in \{1, \ldots, m\}$. Note that the fact that $\mathcal{S}$ is a set implies that there cannot be two events with the same item occurring at the same time: events with the same item must differ in their occurrence time and events occurring at the same time must have different types/items. Such data may as well be represented as *parallel point processes* $\mathcal{P} = \{\langle i_1, \{t_1^{(1)}, \ldots, t_{m_1}^{(1)}\}\rangle, \ldots, \langle i_n, \{t_1^{(n)}, \ldots, t_{m_n}^{(n)}\}\rangle\}$ by grouping events with the same item $i \in B$, $n = |B|$, and listing the times of their occurrences for each of them. Finally, note that in our motivating application (i.e. spike train analysis), the items are the neurons and the corresponding point processes list the times at which spikes were recorded for these neurons.

A *synchronous pattern* (in $\mathcal{S}$) is defined as a set of items $I \subseteq B$ that occur several times (approximately) synchronously in $\mathcal{S}$. Formally, an *occurrence* (or *instance*) of such a synchronous pattern (or a set of *synchronous events for I*) in an event sequence $\mathcal{S}$ with respect to a user-specified time span $w \in \mathbb{R}^+$ is defined as a subsequence $\mathcal{R} \subseteq \mathcal{S}$, which contains exactly one event per item $i \in I$ and which can be covered by a time window at most w wide. Let ϕ be an operator that yields the pattern underlying an instance, $\phi(\mathcal{R}) = \{i \mid \langle i, t \rangle \in \mathcal{R}\}$. Hence the set of all instances of a pattern $I \subseteq B$, $I \neq \emptyset$, in an event sequence $\mathcal{S}$ is

$$\mathcal{E}_{\mathcal{S},w}(I) = \{\mathcal{R} \subseteq \mathcal{S} \mid \phi(\mathcal{R}) = I \wedge |\mathcal{R}| = |I| \wedge \sigma_w(\mathcal{R}) > 0\},$$

where σ_w is the synchrony operator which measures the (degree of) synchrony of the events in $\mathcal{R}$. The following two sections describe two different synchrony operators: a binary and a graded one.

2.1   Binary Synchrony and Support Computation

For binary synchrony, the operator $\sigma_w^{(b)}$ captures the (approximate) synchrony of the events in $\mathcal{R}$ in a two-valued fashion:

$$\sigma_w^{(b)}(\mathcal{R}) = \begin{cases} 1 & \text{if } \max\{t \mid \langle i,t\rangle \in \mathcal{R}\} - \min\{t \mid \langle i,t\rangle \in \mathcal{R}\} \leq w, \\ 0 & \text{otherwise.} \end{cases}$$

That is, $\sigma_w^{(b)}(\mathcal{R}) = 1$ iff all events in the subsequence $\mathcal{R}$ can be covered by a (time) window at most w wide. Note that this allows for temporal imprecision.

Based on this notion of (imprecise) synchrony, we define the support of an item set $I \subseteq B$ as follows (see also [14,21] for a related, but still significantly different characterization that is based on covering windows rather than sets of underlying events as we employ them here):

$$s_{\mathcal{S},w}^{(b)}(I) = \max\big\{|\mathcal{U}| \mid \mathcal{U} \subseteq \mathcal{E}_{\mathcal{S},w}(I) \wedge \forall \mathcal{R}_1, \mathcal{R}_2 \in \mathcal{U}; \mathcal{R}_1 \neq \mathcal{R}_2 : \mathcal{R}_1 \cap \mathcal{R}_2 = \emptyset\big\}.$$

That is, we define the support (or total synchrony) of a pattern $I \subseteq B$ as the size of a *maximum independent set* (MIS) of its instances (where by *independent set* we mean a collection of instances that do not share any events, that is, the instances do not overlap). Such an approach has the advantage that the resulting support measure is guaranteed to be anti-monotone, as can be shown generally for maximum independent subset (or, in a graph interpretation, node set) approaches—see, for example, [8] or [24].

A parallel episode $I \subseteq B$ is called *frequent* (in $\mathcal{S}$) if its support $s_{\mathcal{S},w}^{(b)}(I)$ meets or exceeds a (user-specified) minimum support $s_{\min}$. The task of mining frequent parallel episodes consists in finding, for a given event sequence $\mathcal{S}$ and window width w, all parallel episodes $I \subseteq B$ that are frequent in $\mathcal{S}$. However, in order to reduce the output, it is common to report only the *closed frequent parallel episodes*, where a parallel episode I is called *closed* if no parallel episode that is a proper superset $J \supset I$ has the same support. We denote the set of all frequent parallel episodes that can be found in an event sequence $\mathcal{S}$ w.r.t. (user-specified) window width w and minimum support $s_{\min}$ by $\mathcal{C}_{\mathcal{S}}(w, s_{\min}) \subseteq 2^B$.

At least at first sight, a support measure based on (the size of) a maximum independent set (MIS) seems to suffer from the severe drawback that in the general case finding a maximum independent set is NP-complete [11] and even hard to approximate [9]. Intuitively speaking, this means that (unless $\mathsf{P} = \mathsf{NP}$) there is no (known) algorithm that does fundamentally better than an algorithm that tries all possibilities (here: enumerates all independent sets to find the maximum size). As a consequence, the algorithm has exponential time complexity (in the size of the set $\mathcal{E}_{\mathcal{S},w}^{(b)}(I)$, from which the maximum independent set is to be selected) and thus would take a prohibitively long time to find a solution.

Fortunately, though, the problem instances we are facing here are strongly constrained by the underlying one-dimensional time domain, which makes it possible to devise an efficient greedy algorithm that solves it exactly. For a given item set I, for which the support is to be determined, this algorithm starts with an empty selection of instances and proceeds by traversing the sequence $\mathcal{S}$ (or the parallel point processes $\mathcal{P}$) chronologically. It always selects as the next instance (i.e., the next element of $\mathcal{E}_{\mathcal{S},w}(I)^{(b)}$) that does not overlap any of the already selected instances and contains the earliest possible events for each of the items in I. For this, it does not even have to construct the set $\mathcal{E}_{\mathcal{S},w}^{(b)}(I)$ explicitly, but can work directly on the sequence $\mathcal{S}$ (or the parallel point processes $\mathcal{P}$). As a consequence, it has a time complexity of $m_I \cdot \log(|I|)$, where $m_I = \sum_{i \in I} m_i$ is the sum of the numbers of events of each item i (that is, the total number of events with items in I), since to compute the MIS size, m_I events have to be passed through a priority queue of size $|I|$. Details of this algorithm (including pseudo-code) can be found in [4], while a proof that it is guaranteed to find (the size of) a maximum independent set of $\mathcal{E}_{\mathcal{S},w}^{(b)}(I)$ can be found in [19].

Based on this support computation, frequent parallel episodes are then found with a standard divide-and-conquer scheme (or depth-first search scheme) as it is also known from standard frequent item set mining, particularly from the Eclat algorithm [2,25]. The algorithm proceeds as follows: for a chosen item i, the problem of finding all frequent parallel episodes is split into two subproblems: (1) find all frequent parallel episodes containing i and (2) find all frequent parallel episodes *not* containing i. Each subproblem is then further divided based on another item j: find all frequent patterns containing (1.1) both i and j, (1.2) i but not j, (2.1) j but not i, (2.2) neither i nor j etc.

The search is pruned with the so-called *apriori property*, which is a direct consequence of the fact that support is anti-monotone. A support measure s is called *anti-monotone* if it satisfies $\forall I, J \subseteq B : I \subseteq J \Rightarrow s(I) \geq s(J)$, that is, if an item set is extended, its support cannot increase, and the *apriori property* reads $\forall I, J \subseteq B : (J \supseteq I \wedge s(I) < s_{\min}) \Rightarrow s(J) < s_{\min}$, that is, *no superset of an infrequent parallel episode can be frequent*. Therefore the recursive division process can be terminated as soon as the support of the set of all included split items falls below the (user-specified) minimum support $s_{\min}$, since no frequent patterns can be found in deeper levels. Details of this approach in the context of FIM can be found, for example, in [2]. Details of this scheme for finding (closed) frequent parallel episodes (including pseudo-code) can be found in [4].

2.2 Graded Synchrony and Support Computation

A graded synchrony operator should coincide with binary synchrony for limiting cases as follows: if all events in $\mathcal{R}$ coincide (i.e., have exactly the same occurrence time, perfect synchrony), the degree of synchrony should be 1, while it should be 0 if the events are spread out farther than the window width w (no synchrony). However, if the (time) distance between the earliest and the latest event in $\mathcal{R}$ is between 0 and w, we want a degree of synchrony between 0 and 1.

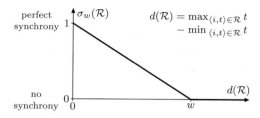

Fig. 1. Degree of synchrony as a function of the distance between the latest and the earliest event in an occurrence (or instance) of an item set.

Such a synchrony operator was described in [17] based on the notion of an influence map, which is placed at each event and describes the vicinity around an event in which synchrony with other events is defined. Such an influence map for an event occurring at time t is defined as the function

$$f_t(x) = \begin{cases} \frac{1}{w} & \text{if } x \in [t - \frac{w}{2}, t + \frac{w}{2}], \\ 0 & \text{otherwise.} \end{cases}$$

Note that an influence map is *not* a distribution function in the sense of probability theory, even though it shares its formal properties. In particular, it is *not* meant to describe uncertainty about the occurrence time of an event.

Based on influence maps, events are synchronous iff their influence maps overlap. The area of the overlap measures the degree of synchrony (Fig. 1):

$$\sigma_w(\mathcal{R}) = \int_0^\infty \min_{\langle i,t \rangle \in R} f_t(x, w) \, dx.$$

Alternatively, we may use the equivalent definition

$$\sigma_w^{(g)}(\mathcal{R}) = \max \left\{ 0, 1 - \frac{1}{w} \left(\max_{\langle i,t \rangle \in \mathcal{R}} t - \min_{\langle i,t \rangle \in \mathcal{R}} t \right) \right\}.$$

This synchrony operator underlies the definition of a graded support operator $s_{\mathcal{S},w}^{(g)}(I)$ that is used to mine synchronous patterns. Of course, such a support operator should (also) capture the number of occurrences of a pattern in a given event sequence $\mathcal{S}$. In addition, in order to be efficient, frequent pattern mining requires support to be *anti-monotone* so that it satisfies the *apriori property* in order to be able to prune the search effectively (cf. binary synchrony).

In principle, we can employ an approach that is analogous to the case of binary synchrony and define the support of an item set as the total degree of synchrony (i.e., the sum over the degrees of synchrony of the instances) of an independent set of instances that yields the maximum total synchrony. That is, we can define a maximum *weight* independent set support (where the weight of an event set is its degree of synchrony) instead of the maximum *size* independent set support for binary synchrony. Such an approach was considered in [17].

However, such an approach has severe drawbacks. Although the resulting support is intuitive and guaranteed to be anti-monotone, it cannot be computed efficiently with the same greedy algorithm that is used for binary synchrony support, since in the graded case this algorithm is no longer guaranteed to find the optimal (maximum weight) solution [7]. Replacing it with a general back-tracking algorithm that is guaranteed to find the optimal solution is not really a feasible alternative, because it has exponential time complexity in the worst case. Although the problem instances are still constrained by the underlying time domain, we have not been able up to now to find an efficient exact solution algorithm (for instance, a different greedy selection scheme). As a consequence, in order to avoid this problem, we opt for an approximation scheme.

As such an approximation, defined in [7], the integral over the maximum (union) of the minimum (intersection) of influence regions is chosen: the minimum represents the synchrony operator, the maximum takes care of a possible overlap between instances of synchronous event groups, and the integral finally aggregates over different instances. Formally:

$$s_{S,w}^{(g)}(I) = \int_{-\infty}^{\infty} \max_{\mathcal{R} \in \mathcal{E}_{S,w}(I)} \left(\min_{\langle i,t \rangle \in \mathcal{R}} f_t(x) \right) \, dx.$$

Note that, exploiting the properties of maxima and minima, this definition can conveniently be rewritten as

$$s_{S,w}^{(g)}(I) = \int_{-\infty}^{\infty} \min_{i \in I} \left(\max_{\langle j,t \rangle \in S; j=i} f_t(x) \right) \, dx.$$

The advantages of this support measure are mainly two: in the first place this support measure is anti-monotone due to the minimum over $i \in I$. Secondly, it allows to compute the support by a simple intersection of interval lists, since all occurring functions only take two values, namely 0 and $\frac{1}{w}$, and therefore it suffices to record where they are greater than 0. Thus, the list of intervals for each item $i \in B$ in which $\max_{\langle j,t \rangle \in S; j=i} f_t(x) > 0$ is computed. These intervals can then be intersected to account for the minimum. Summing the interval lengths and dividing by w (to account for the height of the influence maps) we obtain the area under the functions (cf. the example shown in Fig. 2).

Note that this computation scheme is very similar to the *Eclat algorithm* [25] (which intersects transaction identifier lists to compute support values), transferred to a continuous domain (and thus to effectively infinitely many transactions, one for each point in time). As a consequence, it can be applied with only few adaptations (concerning mainly the support computation) to obtain an efficient algorithm for mining frequent synchronous patterns (see e.g. [2]).

Given this support definition, the search for frequent item sets follows the same divide-and-conquer scheme described in Sect. 2.2. In order to reduce the output it is restricted to closed frequent patterns (as in Sect. 2.2). However, it should be noted that the restriction to closed patterns is less effective with graded synchrony than with binary synchrony, because adding an item can now reduce the support not only by losing instances, but also by worsening the precision of synchrony. Hence, most patterns are closed under graded synchrony [7].

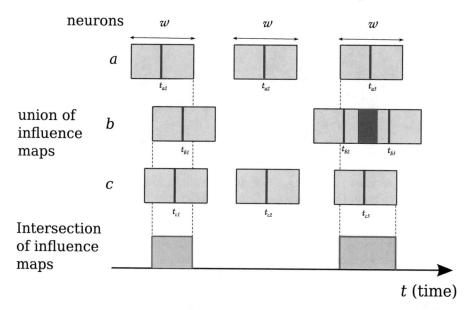

Fig. 2. Support computation for three items a, b, c. Each event has its influence map (represented as a rectangle). If two influence maps overlap, the resulting influence map is the maximum (union) of these influence maps. The intersection of influence maps is the minimum which defines the synchrony operator. In the diagram, item b has two events the influence regions of which overlap. The support results from the integral over the intersections.

3   Pattern Spectrum Filtering

The large number of patterns in the output of synchronous pattern mining method is a serious problem and thus further reduction is necessary. This is done by identifying statistically significant patterns and discarding all others. Previous work showed that statistical tests on individual patterns are not suitable [18,22]. The main problems are the lack of proper test statistics as well as *multiple testing*, that is, the huge number of patterns makes it very difficult to control the family-wise error rate, even with control methods like *Bonferroni correction*, the *Benjamini-Hochberg procedure* or the *false discovery rate* etc [6].

To overcome this problem, we rely here on the general approach suggested in [18] and refined in [22] for a time binning approach to event sequence analysis, namely *Pattern Spectrum Filtering* (PSF). This method is based on the following insight: even if it is highly unlikely that a *specific group* of z items co-occurs s times, it may still be likely that *some group* of z items co-occurs s times, even if items occur independently. The reason is simply that there are so many possible groups of z items (unless the item base B as well as the group size z are tiny) that even if each group has only a tiny probability of co-occurring s times, it may be almost certain that *one of them* co-occurs s times.

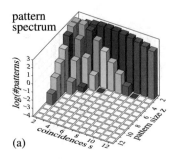
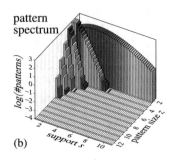

Fig. 3. Pattern spectrum generated from 10^4 surrogate data sets: (a) pattern spectrum for binary synchrony, (b) pattern spectrum for graded synchrony.

From this insight it was derived in [18] that patterns should rather be judged based on their *signature* $\langle z, s \rangle$, where $z = |I|$ is the size of a pattern I and s its support. A pattern is considered not significant if a counterpart (that has the same or larger pattern size z and same or higher support s) can be explained as a chance event under the null hypothesis of independent events.

In order to determine the likelihood of observing different pattern signatures $\langle z, s \rangle$ under the null hypothesis of independent items, a data randomization or surrogate data approach is employed. The general idea is to represent the null hypothesis implicitly by (surrogate) data sets that are generated from the original data in such a way that their occurrence probability is (approximately) equal to their occurrence probability under the null hypothesis. Such an approach has the advantage that it needs no explicit data model for the null hypothesis, which in many cases (including the one we are dealing with here) may be difficult to specify. Instead, the original data is modified in random ways to obtain data that are at least analogous to those that could be sampled under conditions in which the null hypothesis holds. An overview of several surrogate data methods in the context of neural spike train analysis can be found in [15].

In summary, the objective of PSF is to pool patterns with the same signatures $\langle z, s \rangle$ and to collect the occurrences of signatures over surrogate data sets. The result is called a *pattern spectrum* (examples are shown in Fig. 3). Given such a pattern spectrum, all patterns found in the original data are discarded, for which a counterpart (same or larger signature) is recorded in the pattern spectrum (that is, occurred in a surrogate data set). The reason is that patterns occurring in surrogate data are certainly chance events.

The only adaptation needed in comparison to [18,22] is that, due to the graded synchrony, support values are no longer integers, but can be arbitrary (non-negative) real numbers. As a consequence, the pattern spectrum changes from a bar chart with discrete values on both axes to a histogram, where support bins with a (user-specified) width are formed for the support axis. This does not change the general idea, though: patterns (found in the original data) are discarded if a counterpart (same or larger signature) is recorded in the pattern spectrum, as this indicates that it could merely be a chance event.

4   Handling Selective Participation

Our approach to identify parallel episodes in the presence of selective participation is based on the following insight: *although incomplete occurrences of a
pattern may make it impossible that the full pattern is reported by the mining
procedure, it is highly likely that several overlapping subsets will be reported.* An
example of such a situation is depicted in Fig. 4, which shows parallel spike
trains of six neurons labeled a to f with complete and incomplete instances of
parallel episodes comprising all six neurons (in blue; while background spikes are
shown in gray). Although the full set of neurons fires together only once (left-
most instance) and thus would not be detected (since its support is too low), the
other five incomplete occurrences give rise to five subsets of size 4, each of which
occurs twice, and many subsets of size 3, occurring 3 or more times. Since these
patterns overlap heavily, it should be possible to reconstruct the full pattern by
analyzing pattern overlap and combining patterns.

Our method views the set of patterns that were found in a given data set as a
hypergraph[1] on the set of items (which are the vertices of this hypergraph): each
pattern forms a hyperedge. Patterns that are affected by selective participation
thus give rise to densely connected sub-hypergraphs. Hence, we should be able
to identify such patterns by finding densely connected sub-hypergraphs [3].

Our detection method draws on the approach proposed in [23] for detecting
dense sub-hypergraphs. Although this approach was designed to find dense sub-
graphs in standard graphs, its basic idea is easily transferred and adapted: we
form a reduction sequence of items by removing, in each step, the item that is
least connected to the other items (that are still considered). Then we identify
from this sequence the set of items where the least connected item (i.e., the one
that was removed next) was most strongly connected (compared to other steps
of the sequence). This item set is the result of the procedure [3].

Although this method limits the basic procedure to the identification of a
single pattern, it is clear that multiple patterns can easily be found with the same
amendment as suggested in [23]: find a pattern and then remove the underlying
items (vertices) from the data. Repeat the procedure on the remaining items to
find a second pattern. Remove the items of this second pattern and so on. A
drawback of this approach is that it can find only disjoint patterns and thus fails

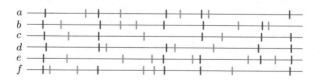

Fig. 4. Parallel episodes (indicating neuron assembly activity) with selective participation (blue) as well as background spikes (gray). (Color figure online)

---

[1] While in a standard graph any edge connects exactly two vertices, in a hypergraph
a single hyperedge can connect arbitrarily many vertices.

to identify overlapping patterns. However, given the general difficulty to handle selective participation, we believe that this is an acceptable shortcoming.

Formally, a reduction sequence of item sets is constructed, starting from the item base B (that is, the set of all considered items), as

$$J_n = B, \qquad \text{where } n = |B|,$$
$$J_k = J_{k+1} - \{\text{argmin}_{i \in J_{k+1}} \xi_{\mathcal{S},w,s_{\min}}(i, J_{k+1})\}, \qquad \text{for } k = n-1, n-2, \ldots, 0,$$

where $\xi_{\mathcal{S},w,s_{\min}}(i, J_k)$ denotes the strength of connection that item $i \in J_k$ has to the other items in the set J_k, as it is induced by the (closed) frequent patterns found when mining the sequence $\mathcal{S}$ with window width w and minimum support $s_{\min}$ (concrete functions $\xi_{\mathcal{S},w,s_{\min}}(i, J_k)$ are studied below). Then, we assign a quality measure to each element of this reduction sequence:

$$\forall k; 0 \leq k \leq n: \qquad \xi_{\mathcal{S},w,s_{\min}}(J_k) = \min_{i \in J_k} \xi_{\mathcal{S},w,s_{\min}}(i, J_k).$$

Finally, we select as the result of our procedure the pattern (item set)

$$I = \text{argmax}_{J_k;\ 0 \leq k \leq n} \xi_{\mathcal{S},w,s_{\min}}(J_k),$$

that is, the pattern with the highest quality (sub-hypergraph density).

To obtain concrete instances of the functions $\xi_{\mathcal{S},w,s_{\min}}(i, J_k)$, two different approaches are explored: a *pattern-based approach* that works with only patterns and their support (ignoring the specific instances of the patterns) and an *instance-based approach* that tries to remove instances to focus the evaluation on instances that likely resulted from the actual assembly activity (note that there can be chance instances, especially for small patterns). These methods can both be applied for binary and graded synchrony are described below.

4.1   Pattern-Based Approach

Let $\mathcal{C}_{\mathcal{S}}^*(w, s_{\min}) \subseteq 2^B$ be the set of closed frequent patterns that are identified by the CoCoNAD algorithm with a binary or a graded notion of synchrony (if executed with window width w and minimum support $s_{\min}$ on $\mathcal{S}$), for which no counterpart (no signature with the same size and greater or equal support constitutes a counterpart) was observed in any of the surrogate data sets (that is, the closed frequent patterns remaining after pattern spectrum filtering). Let $\mathcal{C}_{\mathcal{S},J}^*(w, s_{\min}) = \{I \in \mathcal{C}_{\mathcal{S}}^*(w, s_{\min}) \mid I \subseteq J\}$ be the subset of these patterns that are subsets of an item set J. Then we define the hypergraph connection strength of item $i \in J$ to the other items in J as

$$\xi_{\mathcal{S},w,s_{\min}}^{(\text{pat})}(i, J) = \sum_{I \in \mathcal{C}_{\mathcal{S},J}^*(w, s_{\min})} (|I| - r) \cdot s_{\mathcal{S},w}(I),$$

where $r \in \{0, 1\}$ is a parameter that determines whether the full pattern size (hyperedge size) should be considered ($r = 0$), or whether the item i itself should be disregarded ($r = 1$). The support of the item set I enters the definition because a larger support clearly means a stronger connection.

Intuitively, $\xi_{\mathcal{S},w,s_{\min}}^{(\text{pat})}(i,J)$ sums the (total) degrees of synchrony underlying each of the patterns that connect item i to the other items in J. Note that in this definition we assume (as is common practice and also intuitively plausible) that $\xi_{\mathcal{S},w,s_{\min}}^{(\text{pat})}(i,J) = 0$ if $\mathcal{C}_{\mathcal{S},J}^{*}(w,s_{\min}) = \emptyset$.

4.2   Instance-Based Approach

The pattern-based approach has the advantage that merely the filtered set of closed frequent patterns (together with their support values) is needed. However, it has the disadvantage that subset patterns which, by chance, occur again outside of the instances of the full pattern may deteriorate the detection quality. An example of such an occurrence can be seen in Fig. 4: the neurons a, b and e fire together between the second and third instance of the full set. However, this synchronous firing event is not an incomplete instance of the full set of neurons, but rather a chance coincidence resulting from the background spikes. This can lead to a subset being preferred to the full pattern, even though the sum in the above definition gives higher weight to events that support multiple instances (as these are counted multiple times). Hence removing such instances may be a good idea in order to improve the detection quality.

To achieve this, rely on the idea that we only want to consider instances that are not "isolated", but "overlap" some other instance (preferably of a different pattern). The reason is that isolated instances likely stem from chance coincidences, while instances that "overlap" other instances likely stem from the same (complete or incomplete) instance of the full pattern we try to identify.

Let $\mathcal{C}_{\mathcal{S}}^{*}(w,s_{\min})$ and $\mathcal{C}_{\mathcal{S},J}(w,s_{\min})$ be defined as above. Let $\mathcal{U}_{\mathcal{S},w}(I) \subseteq \mathcal{E}_{\mathcal{S},w}(I)$ be the set of all instances of I that was identified by the CoCoNAD algorithm in order to compute the support $s_{\mathcal{S},w}(I)$ (binary or graded, as desired). Furthermore, let $\mathcal{V}_{\mathcal{S},w,s_{\min}}(J) = \bigcup_{I \in \mathcal{C}_{\mathcal{S},J}^{*}(w,s_{\min})} \mathcal{U}_{\mathcal{S},w}(I)$. That is, $\mathcal{V}_{\mathcal{S},w,s_{\min}}(J)$ is the set of all instances underlying all patterns found in $\mathcal{S}$ that are subsets of J.

To implement our idea of keeping overlapping instances, we define

$$\mathcal{V}_{\mathcal{S},w,s_{\min}}^{*}(i,J) =$$
$$\{\mathcal{R} \in \mathcal{V}_{\mathcal{S},w,s_{\min}}(J) \mid \exists \mathcal{T} \in \mathcal{V}_{\mathcal{S},w,s_{\min}}(J)\colon \phi(\mathcal{T}) \neq \phi(\mathcal{R}) \wedge o(\mathcal{T},\mathcal{R}) = 1\},$$

where ϕ is the pattern operator defined in Sect. 2 and $o_i(\mathcal{R},\mathcal{T})$ is an operator that tests whether the instances $\mathcal{R}$ and $\mathcal{T}$ overlap. In words: $\mathcal{V}_{\mathcal{S},w,s_{\min}}^{*}(i,J)$ is the set of instances of patterns that contain the item $i \in J$ and are subsets of the set J, which overlap at least one other instance of a different pattern.

For the operator o we tried two different variants for binary synchrony:

$$o_i(\mathcal{R},\mathcal{T}) = \begin{cases} 1 & \text{if } \mathcal{R} \cap \mathcal{T} \neq \emptyset, \\ 0 & \text{otherwise,} \end{cases} \qquad \text{and}$$
$$o_s(\mathcal{R},\mathcal{T}) = \sigma_w(\mathcal{R} \cup \mathcal{T}),$$

where σ_w is the synchrony operator from Sect. 2.1. That is, o_i checks whether the instances have a non-empty intersection, while o_s only checks whether the events underlying the instances are synchronous.

Based on these definitions, we finally define

$$\xi_{\mathcal{S},w,s_{min}}^{(inst)}(i, J) = \left| \left\{ \langle j, t \rangle \in \bigcup_{\mathcal{R} \in \mathcal{V}_{\mathcal{S},w,s_{min}}^*(i,J)} \mathcal{R} \mid j \neq i \vee r = 0 \right\} \right|,$$

where the parameter $r \in \{0, 1\}$ determines whether events of the item i should be considered ($r = 0$) or disregarded ($r = 1$). That is, the parameter r has the same function as the parameter r in the pattern-based approach. Intuitively, $\xi_{\mathcal{S},w,s_{min}}^{(inst)}(i, J)$ is the total number of events (possibly ignoring events of the item i) underlying instances that connect it to other items in J.

Note that the operator o_s is not applicable for graded synchrony since there is no concept of the degree of synchrony of the events forming a single instance (which is a result of the approximation scheme we employ—there would be such a concept if we employed a maximum weight independent set support). To account for this, in our approach for graded synchrony, we use only the operator o_i and recompute the support from the reduced set of instances (using the formulas in Sect. 2.2), which is easy to do as the instances are known. That is, handling selective participation for graded synchrony consists in collecting only instances that overlap with other instances of different patterns, re-computing the support from these instances and finally applying the pattern-based approach. In contrast, for binary synchrony the instance-based approach does not require executing the pattern based approach as a second step.

Note that the instance-based approach has the advantage that chance coincidences are much less likely to deteriorate the detection quality. However, its disadvantage is that it is more costly to compute, because not just the patterns, but the individual instances of all relevant patterns have to be processed.

5   Experiments

We implemented our frequent synchronous pattern mining method in Python, using an efficient C-based Python extension module that implements the pattern mining and surrogate generation.[2] We generated event sequence data as independent Poisson processes with parameters chosen in reference to our application domain: 100 items (number of neurons that can be simultaneously recorded with current technology), 20 Hz event rates (typical average firing rate observed

---

[2] The selective participation handling is implemented in Python [20], while the CoCoNAD algorithm is implemented in C [12]. These implementations can be found at http://www.borgelt.net/coconad.html and http://www.borgelt.net/pycoco.html. A Java graphical user interface for the CoCoNAD algorithm is available at http://www.borgelt.net/cocogui.html. The scripts with which we executed our experiments as well as the complete result diagrams (all parameter combinations) will be made available at http://www.borgelt.net/hypernad.html.

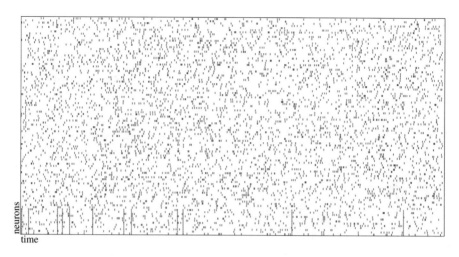

neurons

time

Fig. 5. Example of generated data sets that imitate parallel neural spike trains. Each row of blue dots represents the spike train of each neuron. In this example the injected patterns (here: full participation) are drawn in red. (Color figure online)

in spike train recordings), 3 s total time (typical recording times for spike trains range from a few seconds up to about an hour).

Into such independent data sets we injected a single synchronous pattern each, with sizes z ranging from 2 to 12 items and numbers c of occurrences (instances) ranging from 2 to 21. To simulate imprecise synchrony, the events of each pattern instance were jittered independently by drawing an offset from a uniform distribution on $[-1.5\,\text{ms}, +1.5\,\text{ms}]$ (for binary synchrony) and $[-1\,\text{ms}, +1\,\text{ms}]$ (for graded synchrony), which corresponds to typical bin lengths for time-binning of parallel neural spike trains (which are 1 to 7 ms). An example of such a data set is depicted in Fig. 5. To simulate selective participation, we deleted each item of a parallel episode from a number $\nu \in \{1, 2, 3, 4, 5\}$ of their instances (chosen randomly). This created data sets with instances similar to those shown in Fig. 4 (which corresponds to $z = 6$, $c = 6$ and $\nu = 1$, but has much fewer background spikes): a few instances may be complete, but most lack a varying number of items. For each signature $\langle z, c \rangle$ of a parallel episode and each value of ν we created 1000 such data sets. Then we tried to detect the injected synchronous patterns with the methods described in Sects. 3 and 4.

For mining closed frequent patterns with binary synchrony we used a window width of $w = 3\,\text{ms}$ (matching the jitter of the temporal imprecision), a minimum support $s_{\min} = 2$ and a minimum pattern size $z_{\min} = 2$. For graded synchrony, we chose $s_{\min} = 1.0$ and $z_{\min} = 2$ and, based on results presented in [7], the window width was set to $\frac{3}{2}j = 3\,\text{ms}$ where j is the temporal jitter width ($j = 2\,\text{ms}$, see above). Found patterns were filtered with pattern spectra derived from 100 and 1000 surrogate data sets with independent spike trains. Finally, the reduction sequence methods described in Sect. 4 are applied to the resulting patterns.

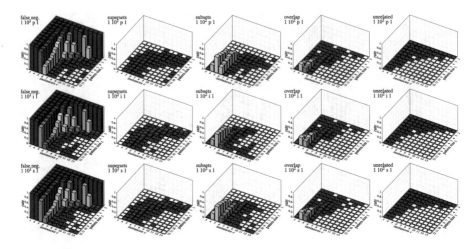

Fig. 6. Experimental results with $\nu = 1$ (each item missing from one instance) for binary synchrony, 100 surrogate data sets and $r = 1$.

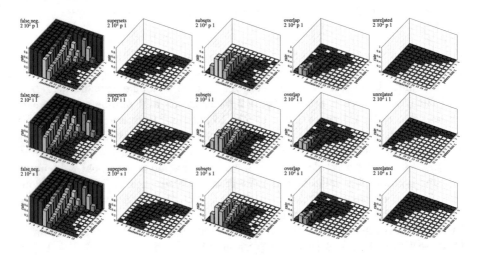

Fig. 7. Experimental results with $\nu = 2$ (each item missing from two instance) for binary synchrony, 100 surrogate data sets and $r = 1$.

Some of the results we obtained are shown in Figs. 6, 7, 8, 9, 10, 11 and 12. In each row of the figures, the first diagram shows the number of (strict) false negatives, that is, the fraction of runs (of 1000) in which either no pattern or some other pattern than exactly the injected pattern was found. In order to elucidate what happens in those runs in which the injected pattern was not (exactly) detected, the diagrams in columns 2 and 3 show the fraction of runs in which a superset or a subset, respectively, of the injected pattern was returned. Column 4 shows the fraction of runs with overlap patterns (the reported pattern contains some, but not all of the items of the injected pattern and at least one

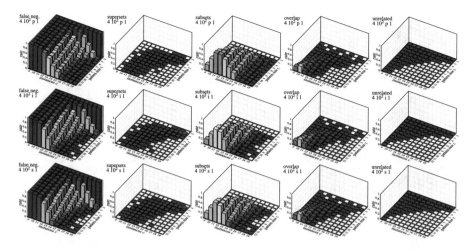

Fig. 8. Experimental results with $\nu = 4$ (each item missing from four instances) for binary synchrony, 100 surrogate data sets and $r = 1$.

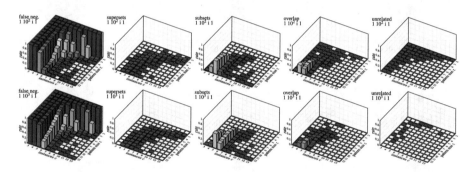

Fig. 9. Experimental results with $\nu = 1$, patterns filtered by 100, and 1000 surrogate data sets for binary synchrony.

other item), column 5 the fraction of runs with patterns that are unrelated to the injected parallel episode. At the top of each diagram the setup is specified by its parameters: the number ν of instances followed by the number of surrogate data sets, the reduction sequence approach applied, and the value of the parameter r.

If we compare the different rows of each of the Figs. 6, 7 and 8, we see that the instance-based approach performs slightly better than the pattern-based approach, and the more so, the more events are missing. The two instance-based approaches (distinguished by the overlap operator: o_i or o_s) are essentially tied, possibly with a very slight advantage for the overlap operator o_i.

If we compare the diagrams across the three Figs. 6, 7 and 8, we may also conjecture that each additional instance from which items are missing, requires about two additional instances to compensate the reduction in detection quality (note the different scales on the instance axis!). This is also plausible, since each item missing from one additional instance effectively removes an instance

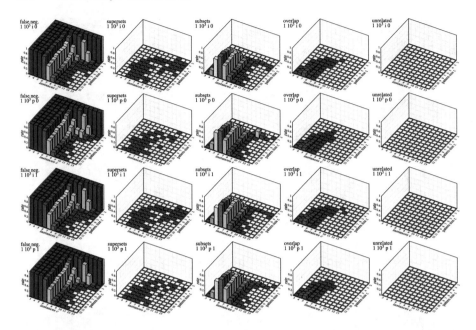

Fig. 10. Experimental results with instance- and and pattern-based approaches for graded synchrony for r values 0 and 1.

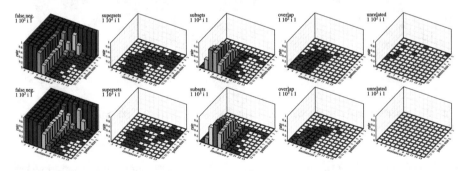

Fig. 11. Experimental results for graded synchrony, patterns filtered by 100, and 1000 surrogate data sets.

(as it removes as many events as an instance contains) and since the removals are distributed over multiple instances, additional compensation is needed. Furthermore, we see that we achieve reliable detection even if items are missing from up to about one quarter of the instances of a pattern.

W.r.t. graded synchrony, note that it has less problems with unrelated patterns: we observe that filtering with 100 and 1000 surrogate data sets performs better w.r.t. unrelated patterns compared to the binary approach. That is, in these cases every injected pattern is detected (at least partially as a subset or part of a superset or overlap pattern) if only something is detected at all.

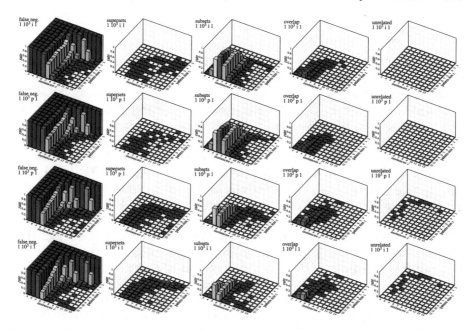

Fig. 12. Comparison between the instance- and pattern-based approaches for graded synchrony and the instance-based approach for binary synchrony.

Figure 10 shows a comparison between the instance- and pattern-based approaches using graded synchrony. The firsts two rows of diagrams show results for $r = 0$ and the second two rows correspond to $r = 1$. The pattern-based approach is slightly better than the instance-based approach in terms of false negatives (exact pattern detection). Concretely, for $r = 1$ the pattern-based approach has better ratios in supersets and overlaps, for which only a small price is paid terms in of a slightly worse ratio for subsets. We prefer the setup in which more subsets are detected, because subsets contain only items actually in the assembly, while superset and overlap patterns also contain unrelated items.

The first and second row of Fig. 12 correspond to the instance and the pattern-based approach for graded synchrony, while the third and fourth correspond to the pattern and instance-based approach for binary synchrony. Comparing the diagrams for unrelated patterns, the graded method detects all injected patterns (if something is detected at all, first and second row), while the binary method also produces unrelated patterns. It is demonstrated that the instance-based approach yields slightly better results than the pattern approach for the binary synchrony. However, this approach does not consider the precision of synchrony. Surprisingly, using only the pattern-based approach with a graded notion of synchrony yields a better ratio for overlap and superset patterns.

6   Conclusions

In this paper we presented a method to detect frequent synchronous patterns in event sequences for mining patterns in the presence of imprecise synchrony (temporal imprecision) of events constituting occurrences and incomplete occurrences (selective participation). We employed both a binary and a graded notion of synchrony, with the latter having the advantage that it takes not only the number of instances, but also the precision of synchrony into account. Selective participation, which was the main focus of this paper, is handled by exploiting that in its presence many overlapping patterns should be found. These are evaluated by interpreting them as a hypergraph on the set of items and constructing a reduction sequence to find a densely connected hypergraph. We presented different ways of defining the connecting strength of sub-hypergraphs, using either only the found patterns and their support or also analyzing the underlying instances and their overlap. We found that the computationally more expensive instance-based approach pays of in the case of a binary synchrony, while for graded synchrony the computationally cheaper pattern-based approach is preferable. We demonstrated in extensive experiments that selective participation can successfully be treated in this manner, obtaining fairly good detection rates, especially in the face of the scarcity of information that is available under such circumstances.

Acknowledgements. The work presented in this paper was partially supported by the Spanish Ministry for Economy and Competitiveness (MINECO Grant TIN2012-31372) and by the Principality of Asturias, through the 2013-2017 Science Technology and Innovation Plan (Programa Asturias, CT1405206), and the European Union, through FEDER funds.

References

1. Abeles, M.: Role of the cortical neuron: integrator or coincidence detector? Isr. J. Med. Sci. **18**(1), 83–92 (1982). Israel Medical Association, Ramat Gan, Israel (1982)
2. Borgelt, C.: Frequent item set mining. In: Wiley Interdisciplinary Reviews (WIREs): Data Mining and Knowledge Discovery, pp. 437–456. Wiley, Chichester (2012)
3. Borgelt, C., Braune, C., Loewe, K., Kruse, R.: Mining frequent parallel episodes with selective participation. In: Proceedings of 16th World Congress of the International Fuzzy Systems Association (IFSA) and 9th Conference of the European Society for Fuzzy Logic and Technology (EUSFLAT), IFSA-EUSFLAT2015, Gijon, Spain. Atlantis Press, Amsterdam, Netherlands (2015)
4. Borgelt, C., Picado-Muiño, D.: Finding frequent synchronous events in parallel point processes. In: Proceedings of 12th International Symposium on Intelligent Data Analysis, IDA, London, UK, pp. 116–126. Springer, Heidelberg (2013)
5. Dayan, P., Abbott, L.: Theoretical neuroscience: computational and mathematical modeling of neural systems. J. Cogn. Neurosci. **15**(1), 154–155 (2003). MIT Press, Cambridge

6. Dudoit, S., van der Laan, M.J.: Multiple Testing Procedures with Application to Genomics. Springer, New York (2008)
7. Ezennaya-Gómez, S., Borgelt, C.: Mining frequent synchronous patterns with a graded notion of synchrony. In: Proceedings of 16th World Congress Int. Fuzzy Systems Association (IFSA) and 9th Conference European Society for Fuzzy Logic and Technology (EUSFLAT), IFSA-EUSFLAT, Gijón, Spain. Atlantis Press, Amsterdam (2015)
8. Fiedler, M., Borgelt, C.: Subgraph support in a single graph. In: Proceedings of IEEE International Workshop on Mining Graphs and Complex Data, pp. 399–404. IEEE Press, Piscataway (2007)
9. Høastad, J.: Clique is hard to approximate within n^{1-e}. Acta Mathematica **182**, 105–142 (1999). Mittag-Leffler Institute, Stockholm
10. Hebb, D.O.: The Organization of Behavior. Wiley, New York (1949)
11. Karp, R.M.: Reducibility among combinatorial problems. In: Miller, R.E., Thatcher, J.W. (eds.) Complexity of Computer Computations, pp. 85–103. Plenum Press, New York (1972)
12. Kernighan, B.W., Ritchie, D.M.: The C Programming Language. Prentice Hall, Upper Saddle River (1978)
13. König, P., Engel, A.K., Singer, W.: Integrator or coincidence detector? the role of the cortical neuron revisited. Trends Neurosci. **19**(4), 130–137 (1996). Cell Press, Maryland Heights
14. Laxman, S., Sastry, P.S., Unnikrishnan, K.: Discovering frequent episodes and learning hidden Markov models: a formal connection. IEEE Trans. Knowl. Data Eng. **17**(11), 1505–1517 (2005). IEEE Press, Piscataway
15. Louis, S., Borgelt, C., Grün, S.: Generation and selection of surrogate methods for correlation analysis. In: Grün, S., Rotter, S. (eds.) Analysis of Parallel Spike Trains, pp. 359–382. Springer, Heidelberg (2010)
16. Mannila, H., Toivonen, H., Verkamo, A.: Discovery of frequent episodes in event sequences. Data Min. Knowl. Discovery **1**(3), 259–289 (1997). Springer, New York
17. Picado-Muiño, D., Castro-León, I., Borgelt, C.: Fuzzy characterization of spike synchrony in parallel spike trains. Soft Comput. **18**(1), 71–83 (2013). Springer, Heidelberg 2013 (online)/2014 (print)
18. Picado-Muiño, D., Borgelt, C., Berger, D., Gerstein, G.L., Grün, S.: finding neural assemblies with frequent item set mining. Front. Neuroinf.**7**(9). Frontiers Media, Lausanne, Switzerland (2013). doi:10.3389/fninf.2013.00009
19. Picado-Muiño, D., Borgelt, C.: Frequent itemset mining for sequential data: synchrony in neuronal spike trains. Intell. Data Anal. **18**(6), 997–1012 (2014). IOS Press, Amsterdam
20. van Rossum, G.: An Introduction to Python for Unix/C programmers. In: Proceedings of the NLUUG najaarsconferentie (Dutch UNIX users group) (1993)
21. Tatti, N.: Significance of episodes based on minimal windows. In: Proceedings of 9th IEEE International Conference on Data Mining (ICDM 2009, Miami, FL, USA), 513–522. IEEE Press, Piscataway (2009)
22. Torre, E., Picado-Muiño, D., Denker, M., Borgelt, C., Grün, S.: Statistical evaluation of synchronous spike patterns extracted by frequent itemset mining. Front. Comput. Neurosci. **7**, 132. Frontiers Media, Lausanne (2013)
23. Tsourakakis, C., Bonchi, F., Gionis, A., Gullo, F., Tsiarli, M.: Denser than the densest subgraph: extracting optimal quasi-cliques with quality guarantees. In: Proceedings of 19th ACM SIGMOD International Conference on Knowledge Discovery and Data Mining (KDD, Chicago, IL), pp. 104–112. ACM, New York (2013)

24. Vanetik, N., Gudes, E., Shimony, S.E.: Computing frequent graph patterns from semistructured data. In: Proceedings of IEEE International Conference on Data Mining, 458–465. IEEE Press, Piscataway (2002)
25. Zaki, M.J., Parthasarathy, S., Ogihara, M., Li, W.: New algorithms for fast discovery of association rules. In: Proceedings 3rd International Conference on Knowledge Discovery and Data Mining (KDD, Newport Beach, CA), pp. 283–296. AAAI Press, Menlo Park, CA, USA (1997)

A Modular Network Architecture Resolving Memory Interference Through Inhibition

Randa Kassab[1,2,3] and Frédéric Alexandre[1,2,3]([⊠])

[1] Inria Bordeaux Sud-Ouest, 200 Avenue de la Vieille Tour, 33405 Talence, France
{randa.kassab,frederic.alexandre}@inria.fr
[2] LaBRI, Université de Bordeaux, Bordeaux INP, CNRS, UMR 5800, Talence, France
[3] Institut des Maladies Neurodégénératives, Université de Bordeaux,
CNRS, UMR 5293, Bordeaux, France

Abstract. In real learning paradigms like pavlovian conditioning, several modes of learning are associated, including generalization from cues and integration of specific cases in their context. Associative memories have been shown to be interesting neuronal models to learn quickly specific cases but they are hardly used in realistic applications because of their limited storage capacities resulting in interference when too many examples are considered. Inspired by biological considerations, we propose a modular model of associative memory including mechanisms to manipulate properly multimodal inputs and to detect and manage interference. This paper reports experiments that demonstrate the good behavior of the model in a wide series of simulations and discusses its impact both in machine learning and in biological modeling.

Keywords: Associative memory · Interference · Inhibition · Biological systems

1 Introduction

In the domain of machine learning, models of neural networks are classified along their architecture and their mode of learning [1], specifically corresponding to supervised and unsupervised modes. In contrast, in the domain of cognitive science, a natural learning paradigm considered in a realistic behavioral and ecological environment often associates several neuronal architectures and learning modes. This is for example the case with pavlovian conditioning that has been shown to require learning a variety of invariants and to modify the neuronal circuitry in several brain regions including the amygdala, hippocampus and cortex [2]. Consequently, in addition to developing efficient models of neural networks designed for their specific characteristics, there is also a need for a more systemic view of learning, considered at the global cognitive level.

Such an approach was already proposed twenty years ago in [3] arguing that the brain exploits complementary learning systems, with a slow and procedural learning in the cortex, able to extract structures and regularities in the data and

© Springer International Publishing AG 2017
J.J. Merelo et al. (eds.), *Computational Intelligence*, Studies in Computational Intelligence 669,
DOI 10.1007/978-3-319-48506-5_21

to generalize, compared with a quick learning in the hippocampus able to retain the specifics of one's life experiences. This paper, with a very strong impact in both cognitive and machine learning communities, proposes that these systems might be respectively implemented with classical neural models of pattern matching like the multilayer perceptron for the slow learning and models of associative memory for the quick learning.

As an illustration, these models can be contrasted with the property of generalization. Generalization is often reported as a desirable property of artificial neural networks. This phenomenon occurs if, when a network is presented with an example it has never seen before, it is able to interpolate a satisfactory response from the combination of close previously learned examples. Such a response can be judged satisfactory not only because from a limited learning phase the network behaves well in a wider domain but also because in some sense learning goes beyond specific cases and is able to extract some general structures or regularities in the example space. In some cases, however, this property might be considered a flaw. This is the case for example when there is no useful topography in the example space or when the goal is to learn some arbitrary association. Consider for example learning to associate a phone number with a name: there is nothing to learn from the euclidean distance between two such numbers and you can in no way discover an association if it was not instructed to you before. This contrasts the case of learning a general rule from a set of examples, as it is for example studied with layered architectures like the multilayer perceptron, versus learning by heart specific cases like in associative memories.

Neural models of associative memories have been proposed with recurrent networks like the Hopfield model [4] and the Willshaw model [5]. Based on classical connectionist characteristics (like units with non linear activation functions and hebbian learning), the recurrent architecture of these networks indicates that learning is mainly focused on the inner characteristics of an example to be memorized and not on the elaboration of abstract representations in intermediate layers. Nevertheless, some problems can appear if too close examples are learned. In such a case, the network might elaborate an answer from the combination of several learned examples; what would be called generalization in other circumstances is called here interference.

As a consequence, models of associative memories are generally used as content addressable memories, where few prototypes are stored as stable states of the network and noisy or incomplete patterns are presented as inputs and reconstructed to the closest stored example. Beyond this use as an autoassociative memory (where initial input and final result have the same dimension), the adaptation to heteroassociative memory is straightforward: just virtually split the recurrent network in two sets of neurons A and B. The recurrent connectivity includes connections within A and within B (seen as two autoassociative memories) and between A and B (heteroassociative memory between the two sets of different dimension A and B). As configurations of A + B are learned as prototypes, proposing an incomplete pattern A (B neurons being set to 0) will result in the reconstruction of A + B, yielding the answer B. The main acknowledged weakness of these models is about their limited capacity of storage and the associated risk of catastrophic interference when this capacity is exceeded or when too close prototypes are stored [6,7]. The best solution to this problem is to require a sparse coding, which intrinsically also limits the maximum number of stored prototypes. An associated strategy is to orthogonalize the inputs and project their encoding in higher dimensions, which results in larger weight matrices to manipulate [8]. In both cases, this might prevent associative memories from being applied to large scale realistic problems and can accordingly explain why they didn't have the same expansion in the machine learning community than layered networks. It is consequently highly desirable to develop scalable models of associative memory.

In previous work, we have proposed a modular network model of associative memory [9] grounded on biological data [10,11]. These data report heterogeneities in the hippocampal structure that might support the coexistence of autoassociative and heteroassociative networks in this region. Specifically, the hippocampus is a neuronal structure known to be involved in episodic memory [12], corresponding to the storage of specific episodes including their context and their emotional or motivational significance. For example, the hippocampus is involved in contextual learning of pavlovian conditioning [13], linking neutral stimuli and their context to biologically significant events (reward and punishment). Though primarily oriented toward biological modeling, we have also explained in [9] the interest of such a segregation from an information processing point of view (cf. the concluding section for a summary). In addition, we have also postulated an additional mechanism for the association of autoassociative memories, that might result in a more robust system, particularly more resistant to interference. The goal of this paper is to evaluate more precisely the performances of this mechanism from an information processing point of view.

In the next section, we will present this model together with its formalism based on the associative memory initially proposed by Willshaw [5]. Then we will report the experiments that were conducted to evaluate its resistance to interference and the associated results. We will conclude by explaining the interest of such a mechanism both in neuroscience and in information processing domains.

2   Multiple Associative-Memory Model

The model is made up of two autoassociative networks that are heteroassociatively linked through a layer of intermediate cells (Fig. 1). The goal is to associate two multi-element patterns in such a way that when at least some elements of the first pattern are presented both patterns can be recalled as a whole. In the hippocampus, these two patterns are considered to represent two important dimensions of episodic memories: (1) The perceptual dimension arises from the integration of different kinds of signals coming from the perception of the outer world: exteroception. (2) The emotional dimension reflects the perception of internal cues of different valences related to pain and pleasure: interoception.

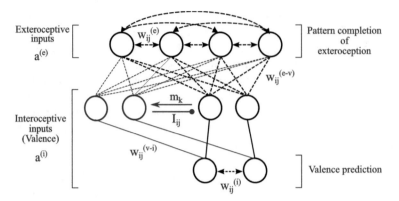

Fig. 1. The architecture of the hippocampal model. Black lines denote the basic circuit of the model while blue lines denote changes in circuitry mediated by one group of associated cells (blue) following the detection of valence-overload interference (red arrow). Autoassociative and heteroassociative connectivities between hippocampal cells are denoted respectively by bidirectional dashed lines and simple dashed lines without arrows. Inhibitory connections between valence cells are denoted by lines ended with circles. Stable non-plastic connections, both excitatory and inhibitory, are denoted by solid lines. (Color figure online)

Then, the two autoassociative networks considered in the model receive and store independently two types of input patterns, $a^{(e)}$ and $a^{(i)}$. The layer of intermediate cells is organized into a small number of ordered groups of valence cells that receive valence-related information from the same interoceptive pathways as the interoceptive autoassociative network. The cells in the first group can be directly activated by interoceptive inputs to the model and can therefore be thought of as the primary valence cells. Interoceptive inputs on the cells in the other groups, which are termed associated cells, are conditional, that is, they can not evoke postsynaptic activity within associated cells unless a concomitant signal, m_k, related to the activity pattern of a precedent group is applied.

The valence cells belonging to the same group of intermediate cells are not interconnected. By contrast, inhibitory connections, I_{ij}, exist between cells belonging to different groups. The inhibitory connections are not plastic. They are prewired such that an inhibitory connection from cell i to cell j exists ($I_{ij} = 1$) if the two cells belong respectively to different groups, k and l, and l precedes k ($l < k$). Thus, each group of associated cells, once activated, silences excitable cells in its preceding groups including the primary group of valence cells. This means that at most valence cells in one group can be active at a time.

The formation of extero-interoceptive associations is done at the level of heteroassociative links, $w_{ij}^{(e-v)}$, between the exteroceptive autoassociative network and the groups of intermediate valence cells. These latter provide direct excitatory input to the interoceptive autoassociative network through non-plastic connections, $w_{ij}^{(v-i)}$. These connections are prewired only between valence cells that are sensitive to the same kind of valence.

The classical binary version of the Willshaw network [5] is chosen as the basis for the implementation of both auto- and heteroassociative memory functions in the model. The neurons are simple McCulloch-Pitts binary threshold units and learning begins with all the synaptic weights set to zero. Synaptic plasticity is achieved according to a clipped version of Hebbian learning: a single coincidence of presynaptic and postsynaptic activity changes the synaptic weight w_{ij} from 0 to 1, while further co-activations do not induce further changes. The recall process is done by presenting a cue pattern $\tilde{x}$ and counting the dendritic sum for each cell j $(s_j = \sum_{i=1}^{n} w_{ij}\tilde{x}_i)$ in one-time step. The output cells that have a dendritic sum equal to or higher than the number of active inputs are activated. The quality of a recalled pattern can be assessed according to its Hamming distance (HD) from the originally stored pattern (i.e. the number of elements that differ between the two patterns. For example, if x = (0 1 1 1 0) and y = (1 1 0 1 0) then HD(x, y) = 2).

Similarly to cholinergic models of the hippocampus [14, 15], our model operates in transition between two modes, storage and recall, depending on a hyperparameter ACh. This mechanism is inspired from biological data describing mode switching under the dynamic regulation of the levels of acetylcholine (ACh) released from septal cholinergic projections to the hippocampus. During recall, a retrieval cue, $a^{(e)}$, is applied to the exteroceptive autoassociative network. The pattern of activity obtained at the output, $\hat{a}^{(e)}$, drives retrieval in the heteroassociative network. An intermediate valence cell, l, can fire only if the dendritic sum of its excitatory inputs exceeds the threshold value and if it does not receive inhibitory inputs from other valence cells that have already fired. The activity of the intermediate valence cells, $\tilde{a}^{(i)}$, triggers recall in the interoceptive autoassociative network yielding the valence prediction by the model, $\hat{a}^{(i)}$.

Just after delivery of the interoceptive information, two novelty-detection processes take place to compare the retrieved patterns to the actual patterns from extero- and interoception. The novelty condition occurs when the Hamming distance between two patterns exceeds pre-specified thresholds ($HD^{(e)} > e$ or $HD^{(i)} > v$). Novelty induces ACh dynamics that favor learning of new inputs, otherwise the model settles in recall mode.

During learning, excitatory intrinsic synaptic transmission along the recurrent connections is removed and activity in the model is purely driven by afferent extero- and interoceptive inputs, $a^{(e)}$ and $a^{(i)}$. In the model, two kinds of interference can occur due to a saturation, or overload of learning. The first kind of interference occurs within the autoassociative memories when too many or too close inputs are stored. It is called pattern overload and can be much mitigated using sparse patterns and low memory load conditions. The second kind of interference is called valence overload and is more likely to occur when elements making up the stored patterns become simultaneously associated to different valences. Consider for example learning AB+, AC− and BD−, where A, B, C and D are exteroceptive patterns and + and − are interoceptive valences. Since

A and B are simultaneously associated to $+$ and $-$ valences, the recall of AB would probably generate an interference (both responses produced). The model deals with valence-overload interference by monitoring activity of intermediate valence cells, $y^{(v)}$. If any activity is observed among intermediate valence cells ($\sum_i y_i^{(v)} > 0$) in response to exteroceptive inputs a matching process takes place to determine whether this activity matches interoceptive valence-specific inputs. A mismatch ($\text{HD}^{(v)} > v$) signals a potential interference to a successive group of associated valence cells that become able to respond to valence-related inputs and rapidly silence valence cells that were active in preceding groups.

3   Experiments

The validity of the proposed model is examined through a series of numerical experiments (cf. [9] for the description of other numerical experiments with this model). The simulated model is configured with 150 cells in the exteroceptive autoassociative network and 3 cells in the interoceptive autoassociative network. The intermediate valence cells are organized into 5 groups of 3 cells each.

Inputs are provided to the model as two independent patterns of activity. The exteroceptive inputs are generated as random 150-element binary patterns with 6 elements being active (set to 1). The interoceptive inputs are modeled by 3 binary cells to differentiate positive, negative and neutral valence states. One of these cells switches to its active state according to whether a pleasant (100), unpleasant (010), or neutral (001) stimulus is present.

The performance is evaluated by comparing the output patterns recalled by the model against the original representation of the input patterns that were presented to the model as new information to be stored. Specifically, two kinds of recall errors are considered when evaluating simulation results. Pattern completion errors which reflect the Hamming distance between the learned and retrieved activation for exteroceptive patterns, and valence prediction errors which reflect the Hamming distance between the correct and predicted valence. In both cases, errors are scored when Hamming distance is greater than zero.

Two types of simulations are set out to test the model for its ability to rapidly link exteroceptive patterns and their emotional valences while avoiding valence overload interference. The first set of simulations examines the effect of the number of stored patterns on the accuracy of valence prediction. The model is tested under full-cue and partial-cue recall conditions. The number of stored patterns is kept low enough that under full-cue conditions almost no pattern overload occurs at the level of autoassociative memories. This is important to ensure that any prediction errors might be detected arise directly from valence overload at the level of heteroassociative links between exteroceptive and interoceptive patterns. The second set of simulations focuses on how to quantify the ability of associated valence units to orthogonalize conflicting associations arising from a change in previously learned valence values.

In all of the simulations, the performance of the proposed model, also called the full model, is compared with that of a reduced model with the groups of associated cells removed. A third model with a single autoassociative memory in which both exteroceptive and interoceptive information are merged into a single pattern is also considered to further delineate benefits of the proposed architecture under partial-cue conditions. All results are averaged over 10 simulation runs and are displayed throughout the figures as mean $\pm$ standard error of the mean. The novelty-detection thresholds, e and v, are set to zero for all the simulations.

4   Results

4.1   Storage Capacity

The first set of simulations is run by varying the number of training patterns and observing how valence prediction is affected with and without the groups of associated cells included in the model (Fig. 2). Training patterns are presented randomly into blocks of N trials with N varying from 10 to 100 in steps of 10. At the different values of N, the full and reduced models were able to recognize exteroceptive patterns with pattern completion errors less than 0.3%. However, there was a noticeable difference between the two models in terms of valence prediction.

As illustrated in Fig. 2A, following the first presentation of training patterns, both models perform perfectly up to $N = 20$, after which point valence prediction errors begin to occur more frequently with increasing size of the blocks of training trials. But as expected, adding the associated cells decreases valence prediction errors at each value of N. For instance, at $N = 100$, the percentage of prediction errors is about 32% for the reduced model but falls to about 20% for the full model. This reduction results from the identification of about 7% of the stored associations as interfering associations (Fig. 2B). Interference effect is accordingly reduced through the recruitment of one group of associated cells (Fig. 2C). During the second presentation of training patterns, the full model detects all the interfering associations that remain and orthogonalizes them using the same group of associated cells (Fig. 2C). Therefore, the performance of valence prediction differs significantly between the two models after the second presentation of training patterns: the reduced model continues to commit the same prediction errors while the proposed model performs with no errors at all.

4.2   Pattern Completion

In the above simulations, it is pertinent to emphasize that no differences were observed between the autoassociative model and the heteroassociative model with the groups of associated cells removed. This is of no surprise because in both cases valence prediction is initiated by a complete set of exteroceptive cues. Under partial-cue conditions, the proposed model as well as its reduced version are expected to take advantage of the fact that pattern completion of exteroceptive cues is performed prior to valence prediction. To test this premise,

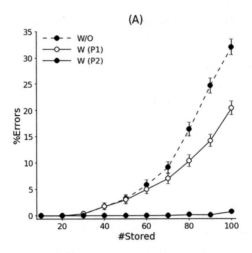

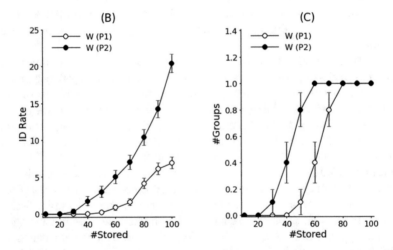

Fig. 2. Influence of the number of stored patterns on the accuracy of valence prediction. (A) Percentage of prediction errors of the model without associated cells (W/O) and with associated cells after one block (W (P1)) and two blocks (W (P2)) of training trials. (B) Rates of interference detection during the first (P1) and second (P2) training trials. (C) Number of groups of associated cells needed to resolve interference detected during training trials P1 and P2.

the three models are trained in the same manner as in the previous simulations except that recall is triggered by partial versions of the original trained patterns. Specifically, the block size is set to 100 training patterns and the model is cued with partial versions with either 1, 2 or 3 of the 6 active inputs turned off.

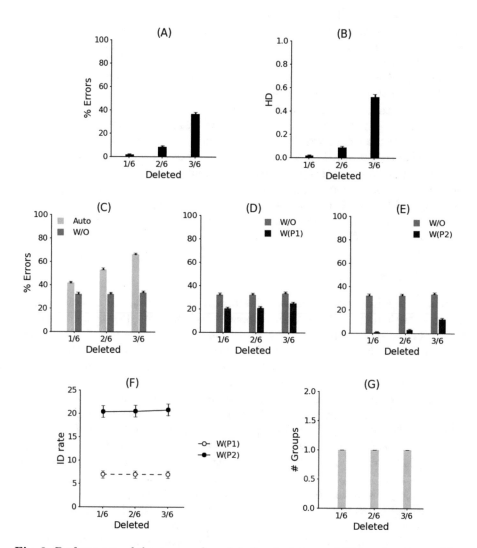

Fig. 3. Performance of the proposed model after training on 100 input patterns. The model is tested using partial cues in which 1, 2, or 3 out of 6 active elements in the original inputs are turned off. (A) Pattern completion performance, defined as the percentage of retrieved patterns that differ at least by one element from the originally stored patterns. (B) Pattern completion performance, defined in terms of Hamming distance between the stored and retrieved patterns. (C) Valence prediction performance of the simple auto- and heteroassociative models. (D and E) Valence prediction performance of the proposed model with (w) and without (w/o) associated cells after one and two blocks of training trials. (F) Rates of interference detection during the first (P1) and second (P2) training trials. (G) Maximal number of groups of associated cells needed to resolve interference detected during the training trials P1 and P2.

As shown in Fig. 3A and B, all the models perform similarly and reasonably well in terms of pattern completion of exteroceptive cues. The accuracy of valence prediction of the autoassociative model is much worse than that of the heteroassociative models and monotonously drops as the number of deleted elements increases (Fig. 3C). On the contrary, the heteroassociative models are much less sensitive to the percentage of deleted elements. The accuracy of valence prediction with the 1/6 partial-cue condition is the same as that obtained with the full-cue condition (Fig. 3D and E). This is because exteroceptive patterns are almost perfectly reconstructed as shown in Fig. 3B. The removal of two or three of the six active cues causes a proportional decrease in the accuracy of pattern completion of exteroceptive patterns. Consequently, the improvement in valence prediction by the proposed model is less pronounced but still highly significant as compared to the reduced model. For all the percentages of removal simulated, the model makes use of one group of associated cells to tackle valence-overload interference (Fig. 3G).

4.3   Discrimination

Here we investigate the functional significance of the groups of associated cells using numerical simulations with reversal learning tasks. The task in the first set of simulations involves two phases. In the first phase the model is presented repeatedly with 50 training patterns [e.g. A+, B−, C (neutral), etc.] over 4 blocks of trials and the percentage of prediction errors made at the beginning of each trial is measured and displayed in Fig. 4A.

This is a simple discrimination learning problem similar to those tested in the previous simulations. Thus as was observed before, valence-overload interference occurs at the early stages of learning and exhibits the recruitment of one group of associated cells to tackle it. When the groups of associated cells are removed the reduced model shows impaired performance that persists over the repeated trials. In the second phase, emotional valences of the training patterns are randomly changed to other value with a probability of 50 % [e.g. A−, B (neutral), C (neutral), etc.]. As shown in Fig. 4A the proposed model quickly learns to reverse its behavior as all the emotionally changed patterns are detected and learned on the first training trials after reversal. On the other hand, the reduced model fails to acquire the new associations since the old ones have not been unlearned.

4.4   Reversal Learning

Further analysis of the model behavior is based on a cue-context reversal learning task similar to that established by [16] to investigate reversal learning in patients with mild amnesic cognitive impairment. To simulate this task, three groups of 4 exteroceptive patterns each are formed such that one of the 6 active elements is used to encode the presence of a sensory cue and the others to encode contextual cues. No overlap is allowed between cells encoding for different cues or contexts (cf. Table 1).

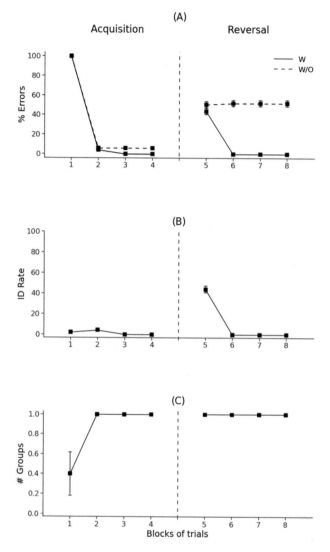

Fig. 4. Discrimination reversal learning. (A) Percentage of prediction errors of the model with (w) and without (w/o) associated cells. (B) Rates of interference detection over each block of trials. (C) Number of groups of associated cells needed to resolve interference across the different blocks of trials.

In the first phase of acquisition, the model is repeatedly presented with the training patterns in the first group and valence prediction is evaluated over four blocks of training trials. Figure 5 shows that both full and reduced models make correct valence prediction after a single exposure to the training patterns. Then, the reversal phase is immediately followed by exposing the models to new training patterns from the second and third groups, in addition to the old ones. The

Table 1. The experimental design of the task of [16]. Note. A–H refer to eight cue shapes, 1–8, eight contexts, + and − indicate respectively positive and negative valences.

Training patterns			Task	
Group 1 (original)	Group 2 (cue reversal)	Group 3 (context reversal)	Phase 1 (acquisition)	Phase 2 (retention & reversal)
A1+	E1−	A5−	Group1	Group1
B2+	F2−	B6−		Group2
C3−	G3+	C7+		Group3
D4−	H4+	D8+		

training patterns are also presented repeatedly four times in random order. The results show that, in the first block of trials, valence prediction errors are made for both new and old patterns. This reflects the fact that heteroassociative connections are irrelevantly strengthened between the original patterns and valences of new patterns. When interference is detected, one group of valence-associated cells is recruited and prediction errors fall to zero rapidly on the third block of trials after reversal. In contrast, the number of prediction errors the reduced model makes is still the same as the blocks progress for the same reason stated above.

5   Discussion

The primary goal of this paper was to propose a new framework to consider the storage of multimodal information in associative memories while making them efficient even in adverse conditions. Indeed, associative memories have powerful properties for learning by heart specific patterns and recalling them from partial information. They can learn quickly and recall patterns as they were initially presented, without modification nor generalization. It has been shown [3] that such properties are present and necessary in certain classes of cognitive functions and consequently it is of interest to build computational models able to emulate them. Nevertheless, associative memories are little exploited in classical machine learning because they suffer from limited storage capacities, particularly when patterns to be stored are close, resulting in interference and catastrophic forgetting [1].

It has also been shown [1] that associative memories can be simply used in autoassociation for pattern retrieval but also in heteroassociation between two different classes of inputs. In the model presented here, we exploit this heteroassociative view to propose a modular network. From an information processing point of view, we explain in the present paper that a heteroassociation between two data spaces of different size leads to more robust retrieval than a simple autoassociation with a flat vector concatenating both kinds of information

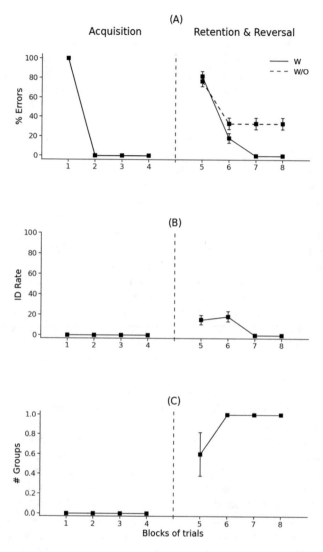

Fig. 5. Cue-context reversal learning. (A) Percentage of prediction errors of the model with (w) and without (w/o) associated cells. (B) Rates of interference detection over each block of trials. (C) Number of groups of associated cells needed to resolve interference across the different blocks of trials.

because the evaluation of the Hamming distance between stored and actual patterns would consider in this latter case that one error in any dimension yields the same penalty, which is obviously not the case. This is confirmed in the present paper, considering comparison of performances between similar autoassociative and heteroassociative models.

A modular view of associative memories is also exploited to implement another powerful property of our model, for managing interference, using another set of units called associated cells. When an association is learned between a high dimensional data space and a smaller space representing labels (valences in the present case), one central problem is about the association of close patterns with different labels or of different combinations of patterns with different labels. This classical problem has been termed configural learning [17]. With a fully automatic algorithm to insert associated cells between the heteroassociative modules, we have proposed in the present model a mechanism able to detect interference at the heteroassociative level and to trigger new learning accordingly. The experiments reported here, particularly comparing performances of reduced and full heteroassociative models, show that our model is very efficient at performing such a learning. In addition, this learning process is very quick, which preserves another important specificity of episodic learning.

In this paper, we also propose to relate the very good properties of modular heteroassociative memories to two different frameworks. In the framework of brain modeling, the model has been primarily built as a biologically informed model of the hippocampus [9]. In addition to proposing some evidences for the implementation of a modular network in this cerebral structure, we also propose that heteroassociation could take place between exteroception and interoception, corresponding to different kinds of hippocampal inputs. In further studies, this could be extended to other classes of hippocampal inputs, particularly related to the frontal cortex.

The focus was set here on interoception and exteroception because this study was related to other studies in the team [13] related to pavlovian conditioning. This learning paradigm is very interesting because it is an excellent basis for a systemic view of learning in the brain, with adaptive processing involving (at least) the amygdala, the hippocampus and the cortex [2]. Extending the duality between procedural learning in the cortex and specific cases learning in the hippocampus [3], we explain in [13] that the amygdala is designed to learn pavlovian associations from cues extracted by both structures with their own way of learning and also report, in accordance to other authors [18], a synergy between the three modes of learning, where an event in one learning module (an error of prediction, the occurrence or the storage of a specific case) can trigger or modify learning in another learning module. Considering the importance of such a distributed learning principle, better understanding its details deserves additional work.

Bio-inspiration was also a strong motivation for this work because, in addition to classical evaluation of performances, one of the experiments we made was also designed to reproduce behavioral and cognitive data in the medical domain for amnesic impairments [16]. Related medical data strongly suggest the central role of the hippocampus in this memory process, giving additional interest to the complementary learning system hypothesis [19]. The cognitive framework initiated in [3] postulates how procedural learning in the cortex, slowly learning and able of generalization, might be instructed by specific cases learned quickly in the hippocampus while avoiding interference. Adding emotional aspects with the dissociation between interoceptive and exteroceptive cues, extends this framework of mnemonic synergy in the brain, proposed for medical purposes.

In the framework of machine learning, we have also presented this work as a new model of associative memory and its main results have been described here mainly in the framework of information processing. This is also the reason why we use simple binary units in the hippocampal model, even if more complex functioning rules might be expected in the framework of a biologically inspired model: Even if more complex units might be considered in future works, particularly to fit with more precise biological data, the main goal of the present work was to settle the main computational principles of our modular model. Beyond the case for pavlovian conditioning with interoceptive and exteroceptive cues, we believe that it is not rare in the information processing domain to cope with such associations between data of different dimensions, as it is the case for example with labeled data (high-dimensional data associated with a symbolic label). In this case, we claim that combining autoassociation and heteroassociation as proposed here results in more robustness in the retrieval phase. More generally and beyond associative memories, heteroassociation using intermediate cells to reduce ambiguities is a general class of approaches in machine learning and the criteria proposed here to avoid interference and keep associations simple could be extended to other models of machine learning in further works. This could illustrate other cases where a primary biological inspiration yields efficient learning principles.

References

1. Hertz, J., Krogh, A., Palmer, R.: Introduction to the Theory of Neural Computation. Addison Wesley, Boston (1991)
2. Maren, S.: Building and burying fear memories in the brain. Neuroscientist **11**, 89–99 (2005)
3. McClelland, J.L., McNaughton, B.L., O'Reilly, R.C.: Why there are complementary learning systems in the hippocampus and neocortex: insights from the successes and failures of connectionist models of learning and memory. Psychol. Rev. **102**, 419–457 (1995)
4. Hopfield, J.J.: Neural networks and physical systems with emergent collective computational abilities. In: Proceedings of the National Academy of Sciences, USA, pp. 2554–2558(1982)
5. Willshaw, D.J., Buneman, O.P., Longuet-Higgins, H.C.: Non-holographic associative memory. Nature **222**, 960–962 (1969)
6. Graham, B., Willshaw, D.: Capacity and information efficiency of the associative net. Netw. Comput. Neural Syst. **8**, 35–54 (1997)
7. Knoblauch, A., Palm, G., Sommer, F.T.: Memory capacities for synaptic and structural plasticity. Neural Comput. **22**, 289–341 (2010)
8. McNaughton, B., Nadel, L.: Hebb-marr networks and the neurobiological representation of action in space. In: Neuroscience and Connectionist Theory, pp. 1–63. L. Erlbaum, Hillsdale (1990)

9. Kassab, R., Alexandre, F.: Integration of exteroceptive and interoceptive information within the hippocampus: a computational study. Front. Syst. Neurosci. **9** (2015)

10. Andersen, P.: The Hippocampus Book. Oxford Neuroscience Series. Oxford University Press, Oxford (2007)

11. Samura, T., Hattori, M., Ishizaki, S.: Sequence disambiguation and pattern completion by cooperation between autoassociative and heteroassociative memories of functionally divided hippocampal CA3. Neurocomputing **71**, 3176–183 (2008)

12. Tulving, E.: Episodic and semantic memory. In: Organization of Memory. Academic Press, New York (1972)

13. Carrere, M., Alexandre, F.: A pavlovian model of the amygdala and its influence within the medial temporal lobe. Front. Syst. Neurosci. **9** (2015)

14. Hasselmo, M.E., Wyble, B.P., Wallenstein, G.V.: Encoding and retrieval of episodic memories: role of cholinergic and gabaergic modulation in the hippocampus. Hippocampus **6**, 693–708 (1996)

15. Meeter, M., Murre, J.M., Talamini, L.M.: Mode shifting between storage and recall based on novelty detection in oscillating hippocampal circuits. Hippocampus **14**, 722–41 (2004)

16. LevyGigi, E., Kelemen, O., Gluck, M.A., Kéri, S.: Impaired context reversal learning, but not cue reversal learning, in patients with amnestic mild cognitive impairment. Neuropsychologia **49**, 3320–6 (2011)

17. Buhusi, C.V., Schmajuk, N.A.: Attention, configuration, and hippocampal function. Hippocampus **6**, 621–642 (1996)

18. Moustafa, A.A., Gilbertson, M.W., Orr, S.P., Herzallah, M.M., Servatius, R.J., Myers, C.E.: A model of amygdala-hippocampal-prefrontal interaction in fear conditioning and extinction in animals. Brain Cogn. **81**, 29–43 (2013)

19. O'Reilly, R.C., Bhattacharyya, R., Howard, M.D., Ketz, N.: Complementary Learning Systems. Cognit. Sci. **38**, 1229–1248 (2011)

An Emotional Robot

Masanao Obayashi[1(✉)], Shunsuke Uto[1], Takashi Kuremoto[1],
Shingo Mabu[1], and Kunikazu Kobayashi[2]

[1] Yamaguchi University, 2-16-1 Tokiwadai, Ube, Yamaguchi 755-8611, Japan
{m.obayas, t0055vk, wu, mabu}@yamaguchi-u.ac.jp
[2] Aichi Prefectural University, Nagakute, Aichi 480-1198, Japan
kobayashi@ist.aichi-pu.ac.jp

Abstract. The method to construct an emotional robot based on regulation of emotional responses with an emotion state embedded reinforcement learning system is proposed in this paper. Besides environmental states, the emotional robot has emotional states which are generated by stimulus received from sensor images. If the learning coefficient of emotions in Amygdala model is changed, generated emotional states in the robot are also different, even if the robot sees same sensor images. As a result, using the method, we can make kinds of robots with any emotions, having same structures. Through computer simulations, applying the proposed method to construct emotional robots, it is said the robots solve the path-finding problem including a variety of distinctive solutions. We find that each robot is able to have each individual solution depending on kinds of its emotions.

Keywords: Emotional state embedded reinforcement learning · Amygdala · Emotional model · Emotional robot

1 Introduction

Reinforcement learning (RL) for the behavior selection of robots has been proposed since 1950's. As a machine learning method, it uses trial-and-error search, and rewards are given by the environment as the results of exploration/exploitation behaviors of the robot to improve its policy of the action selection [1]. The architecture of RL system is shown in Fig. 1. However, when human makes a decision, he finally does it using the various functions in the brain, e.g., emotion. Even the environmental state is the same; many different selections of the behavior may be done depending on his emotional state then.

A computational emotion model has been proposed by J. Moren and C. Balkenius [2]. Their emotion model consists of four parts of the brain: "thalamus, sensory cortex, orbitofrontal cortex and amygdala" as shown in Fig. 2. Figure 2 represents the flow from receptors of sensory stimuli to assessing the value of it. So far, the emotion model has been applied to various fields, especially, the control field of something. For example, H. Rouhani et al. applied it to speed and position control of the switched reluctance motor [3] and micro heat exchanger control [4]. N. Goerke applied it to the robot control [5], E. Daglari et al. applied it to behavioral task processing for cognitive

© Springer International Publishing AG 2017
J.J. Merelo et al. (eds.), *Computational Intelligence*, Studies in Computational Intelligence 669,
DOI 10.1007/978-3-319-48506-5_22

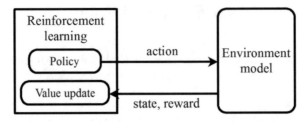

Fig. 1. Reinforcement learning system (Sutton et al. 1998) [1].

robot [6]. On the other hand, Obayashi et al. combined emotion model with rein-forcement Q learning to realize the robot with individuality [7]. F. Yang et al. also proposed the robot's behaviour decision-making system based on artificial emotion using cerebellar model arithmetic computer (CMAC) network [8]. H. Xue et al. pro-posed emotion expression method of robot with personality to enable robots have different personalities [9]. Kuremoto et al. applied it to a dynamic associative memory system [10]. All of these applications have good results.

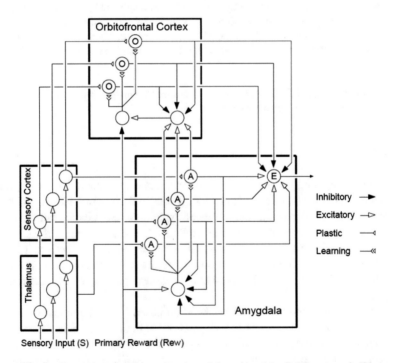

Fig. 2. A computational emotional model proposed by J. Moren et al. [2].

In this paper, we propose a method to construct an emotional robot based on regulation of emotional responses with an emotion state embedded reinforcement learning system.

The rest of this paper is organized as follows. In Sect. 2, a brief explanation about standard reinforcement Q learning is described, and in Sect. 3, a computational emotion model we used is provided. Our proposed method to construct an emotional robot based on regulation of emotional responses with an emotion state embedded reinforcement learning system is given in Sect. 4. Two computer simulations using two grid world environments are carried out to evaluate the proposed method in Sect. 5. This paper is concluded in Sect. 6.

2 Reinforcement Learning

In this section, a simple explanation of the commonly used reinforcement learning system, in particular, Q-learning, one of representative RL methods and used in this paper, is done. The framework of standard reinforcement learning system is shown in Fig. 1.

The algorithm of Q learning is as follows [1]:

```
Initialize Q(s,a) arbitrarily

Repeat (for each episode):
Initialize s
Repeat (for each step of episode9):
    Choose a among actions available in s using greedy policy de-
           rived from Q(s,a)

    Take action a , observe reward r and next state s'
```

$$Q(s,a) \leftarrow Q(s,a) + \alpha \left\{ r + \gamma \max_{a'} Q(s',a') - Q(s,a) \right\} \tag{1}$$

```
s ← s';
Until s becomes terminal state,
```

where, meaning of phrases and detail of this algorithm is referred to [1].

3 Computational Emotion Model

The computational emotional model is proposed by J. Moren and C. Balkenius [2] consists of 4 parts of the brain, "thalamus, sensory cortex, orbitofrontal cortex and amygdala" as shown in Fig. 2, it represents the flow from receptors of sensory stimuli to assessing the value of it. The dynamics of the computational emotional model are described as follows;

$$A_i = V_i S_i \tag{2}$$

$$O_i = W_i S_i \tag{3}$$

$$E = \sum_i A_i - \sum_i O_i \tag{4}$$

$$\Delta V_i = \alpha_{amy} \left(S_i \max \left(0, \; Rew - \sum_j A_j \right) \right) \tag{5}$$

$$\Delta W_i = \beta_{amy} \, S_i (E - Rew), \tag{6}$$

here, S_i denotes input stimuli from the sensory cortex and thalamus to the ith neuron in the amygdala, $i = 1, 2, \ldots, N_{amy}$, where N_{amy} corresponds to the number of neurons in the amygdala and A_i denotes the output of the ith neuron in the amygdala. Likewise, O_i denotes the output of ith neuron in the orbitofrontal cortex. E is the output of the amygdala after subtracting the input from the orbitofrontal cortex. α_{amy}, β_{amy} are emotion learning rates, V_i, W_i are synaptic weights of connections between the sensory cortex and amygdala, as well as the sensory cortex and orbitofrontal cortex, respectively. Primary reward Rew is the reinforcing signal.

To confirm the performance of Morén's emotional model, stimulus learning simulations were performed, and Fig. 3 shows the result. In the figure, S or $S1$, $S2$, $S3$ are sensory inputs, $Rew(=1.0)$ is a reward/reinforcing signal, E is the output of the model (Eq. 4). The emotion learning coefficients are $(b)\ \alpha_{amy} = 0.1$, $\beta_{amy} = 0.4$, $(c)\ \alpha_{amy} = 0.05$, $\beta_{amy} = 0.1$, respectively. From Fig. 3, we can conclude that the stimulus from the sensory cortex was associated with a reward, and disassociated when the reinforcing signal disappeared. Larger the α_{amy} is, larger the output E is, and the output E is close to the value of $Rew(=1.0)$ quickly. However, if the signal of Rew disappears, the output E also disappears. If there are any sensory inputs, smaller the β_{amy} is, slower the rate of the speed of disappearance is. As a result, it is said that we can control the output E of emotions by changing the emotion learning coefficients, α_{amy}, β_{amy}.

4   Emotion State Embedded Reinforcement Learning System

When a person saw an exciting landscape, he feels it pleasant or unpleasant. In this paper, we introduce the degree of - (pleasant-displeasant) impression of the image using the colour characteristics of the image as one of the emotional state to be defined in the internal robot. Figure 4 shows the proposed emotion state embedded reinforcement learning system. It has a hierarchical structure, the first layer is an image processing model, the second layer is a fuzzy inference model, the third layer is emotional models by Moren, the fourth layer is the integrated emotional state model by Russel and the fifth layer is the proposed extended reinforcement Q learning system [7]. In the next subsections short contents of them are described.

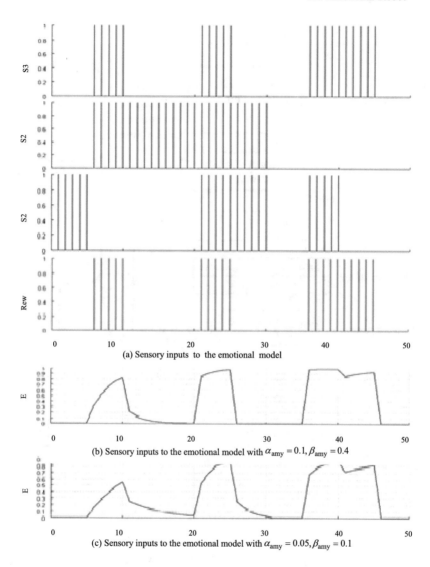

Fig. 3. Simulation results of the computational emotional model proposed by J. Moren et al. [2].

4.1 Image Processing Model: First Layer

In the first layer, RGB values of each pixel of the image acquired from the environment is converted to the HSV (Hue, Saturation and Value) values, using the following (7). These are transmitted to Fuzzy inference model of the second layer,

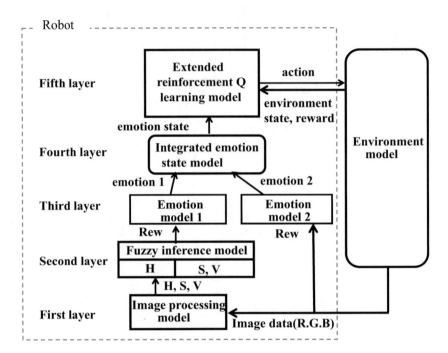

Fig. 4. Our proposed emotion state embedded hierarchical reinforcement learning system.

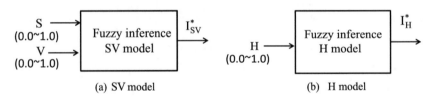

Fig. 5. Impression fuzzy inference model.

$$H = \begin{cases} \text{undefined} & \text{if MIN} = \text{MA} \\ 60 * \left(\frac{\text{G–R}}{\text{MAX–MIN}} + 1\right) & \text{if MIN} = \text{B} \\ 60 * \left(\frac{\text{G–R}}{\text{MAX–MIN}} + 3\right) & \text{if MIN} = \text{R} \\ 60 * \left(\frac{\text{G–R}}{\text{MAX–MIN}} + 5\right) & \text{if MIN} = \text{G} \end{cases} \tag{7}$$

$$V = \text{MAX}$$

$$S = \text{MAX} - \text{MIN}$$

where Max = max{R, G, B}, MIN = min{R, G, B}.

4.2   Fuzzy Inference Model: Second Layer

In the second layer, the colour features (Saturation, Value which represent modifier: dull thin, dark-bright-dark, and Hue which represents basic colour name: red, blue and green) provided from the first layer is converted to a degree of pleasure-displeasure using Mamdani type simplified singleton fuzzy inference.

The membership functions of Saturation, Value and Hue used in this paper are shown in Figs. 6, 7 and 8, respectively. They are set corresponding to their values. The fuzzy rules of Saturation and Value, Hue are shown in Tables 1 and 2. The impressions I_{sv} and I_H in these Tables are decided according to our human impression. In Table 2, the Impression (I_H) of red is set to high and that of blue is set to low. This represents to express the vitality impression with the colour.

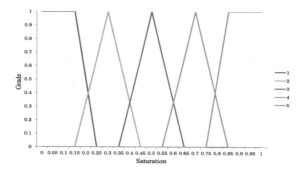

Fig. 6. Membership function for Saturation (S).

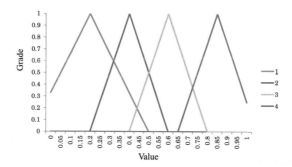

Fig. 7. Membership function for Value (V).

Concretely, we inference the impression (I_{SV}^*) from the Saturation and Value, taking the minimum value between the grade of S and V for each rule, and then taking fuzzy singleton inference for defuzzification (see Fig. 5(a)). The impression (I_H^*) from the Hue are calculated as same as I_{SV}^* (see Fig. 5(b)). Then, it is integrated to obtain an impression value (I_{HSV}) for a pixel by (8). This operation is applied to all the pixels. Then the emotion of the entire image (Image impression: Imi) is obtained by taking the

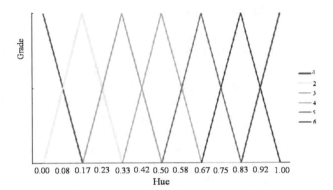

Fig. 8. Membership function for Hue (H).

Table 1. Fuzzy rule table for Saturation and Value.

Rule number	If		Then	Impression (I_{SV})
	Number of membership func. of S	Number of membership func. of V		
1	3	1	Very dark	0
2	3	2	Dark grayish	0.3
3	4	2	Dark	0.6
4	1	3	Grayish	0.9
5	5	3	Deep	1.2
6	1	4	Very pale	1.5
7	2	4	Pale	1.8
8	3	4	Light	2.1
9	5	4	Vivid	2.4

Table 2. Fuzzy rule table for Hue.

Rule number	If	Then	Impression (I_H)
	Number of membership func. of H		
1	1	Red	2.0
2	2	Yellow	1.5
3	3	Green	1.0
4	4	Light blue	−1
5	5	Blue	1.0
6	6	Purple	1.5

average of all of the impression values (9). Calculating Imi for each direction of the image, sum of them is input to emotion model 1 (the third layer) which is responsible for pleasure-displeasure as Rew.

$$\text{Impression } (I_{HSV}) = \text{Impression } (I_H^*) \bullet \text{Impression } (I_{SV}^*) \tag{8}$$

$$\text{Image impression (Imi)} = \frac{\sum\limits_{\text{pixel}} \text{Imression } (I_{HSV})}{\text{pixel length}} \tag{9}$$

4.3   Emotion Model: Third Layer

Figures 9 and 10 show the input and output for the pleasure-displeasure and activity-disactivity emotion models respectively. The structures of them are same and their learning method is explained in Sect. 3. The output of the emotion model for pleasure-displeasure is E_1, and E_2 is output of the activity-disactivity emotion model. These E_1 and E_2 are used for two axis for the integrated emotion state model in the fourth layer.

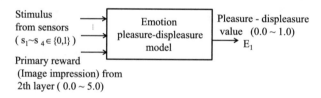

Fig. 9. Emotion pleasure-displeasure model.

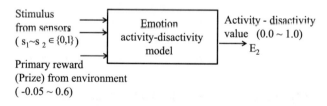

Fig. 10. Emotion activity-disactivity model.

Emotion Model 1. The function of the emotion model 1 whose structure is same as the computational emotion model in Sect. 3 is to produce the emotion of pleasure-displeasure by making use of characteristics of the image. Its input and output components are shown in Fig. 9.

Emotion Model 2. The function of the emotion model 2 whose structure is same as emotion model 1 is to produce the emotion of activity-disactivity by making use of the primary reward given by the environment. Its input and output components are shown in Fig. 10.

4.4   Integrated Emotion State Model: Fourth Layer

In this paper we use the circumplex emotion model [11] as the integrated emotion state model. The circumplex emotional model proposed by J.A. Russel consists of two axes that are pleasure-displeasure (horizontal axis) and activity-disactivity (vertical axis); it is shown in Fig. 11. The figure shows unidimensional scaling of 28 emotion words on the plane. Russel said that all the emotions of the living body can be dealt by this circumplex model. This model decides the current two dimensional emotional states of the robot using two inputs E_1 (displeasure–displeasure value) and E_2 (activity– disactivity value) from the third layer as shown in Fig. 4.

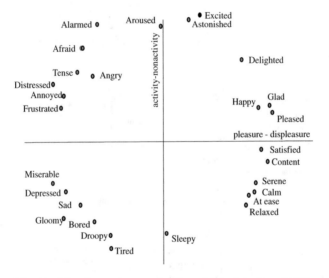

Fig. 11. The circumplex emotional model by J.A. Russel [11].

4.5   Extended Q Learning Model with Emotion State: Fifth Layer

The Emotion extended Q learning [7] is almost all of commonly used standard Q learning. The extended Q learning with emotion state has the emotion state of the robot in addition to environment state of standard Q learning. The value function of the state, emotion and action in the extended Q learning is represented as $Q(s, s_e, a)$. The update equation of $Q(s, s_e, a)$ is as follows;

$$Q(s, s_e, a) \leftarrow Q(s, s_e, a) + \alpha[r + \gamma \max_{a'} Q(s', s'_e, a') - Q(s, s_e, a)], \quad (10)$$

where s: current environment state, s_e: current emotion state with two dimensions from the fourth layer. a: current action, r: reward, s': next current environment state, s'_e: next current emotion state, α: learning rate, γ: discount rate. We use the greedy method as selection policy of behaviors of the robot.

5 Computer Simulation

5.1 Preparation

Problem Description. To evaluate our proposed method, we carried out two computer simulations using two grid world environments as shown in Figs. 13 and 14. The wall surrounds around them. There are meaningful plural paths from start to goal. Simulation 1 with environment A is easier path-finding problem than simulation 2 with environment B. There are two same red foods in aisles in environment A. We consider two types of robots constructed by changing emotion learning parameters. One robot takes a path from start to goal taking both foods, another robot takes another path to the goal, taking only one food. We found that each robot learned the different path from start to goal, forming the different emotions by use of the different parameter for learning of the emotion model.

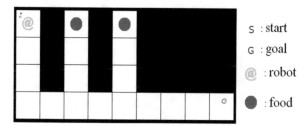

s : start
G : goal
@ : robot
● : food

Fig. 12. The environment A used in simulation 1.

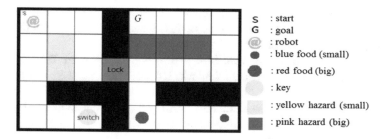

S : start
G : goal
@ : robot
● : blue food (small)
● : red food (big)
: key
: yellow hazard (small)
: pink hazard (big)

Fig. 13. The environment B used in simulation 2.

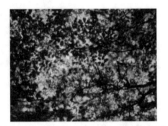

(a) Image given as the red big food   (b) Image given as the blue small food

Fig. 14. The image used as input to Image processing model in the simulation

Assumptions. In these simulations, next followings are assumed,

(1) The robot knows his own position.
(2) The action which the robot can take is "to move one cell to one direction among up, down, left and right".
(3) If the robot collides with the wall, the robot stays at the position before collision.
(4) Movable and touch area of the robot are as follows:

Movable areas by one step of the robot
(around one of the robot : up,
down, right and left)

Available areas of sensors of the robot
(around one of the robot : up, down,
right and left)

Environment Used in Simulations. In simulation 1 with the environment A shown in Fig. 12, two red foods are placed on the route. The autonomous robot is able to select the paths for "shortest to goal without collecting the food," "goal with only one of the feeds" and "longest to goal with two foods". Whether the robot choose which path depends on the individuality of the robot. In this simulation, we show that the path the robot select is decided according to the emotions.

In simulation 2 with environment B shown in Fig. 13, there are the cell which is locked and the switch cell to release the lock. It is necessary for the robot to visit the switch cell once to release the lock to get the goal. The robot has to take a circuitous route to get the red and blue foods and also has to take a hazard path to take the shortest path to the goal. So the robot has the dilemma, which route should be selected. It is verified the dilemma is solved by the individuality of the robot.

In both simulations, when the robot gets the food at the first visit, the food disappears and it keeps the food lost in the same episode.

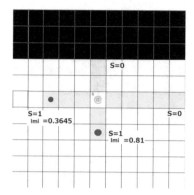

Fig. 15. Example of the sensory input s and primary reward input (Rew) of the emotion model for making the pleasure-displeasure value (E_1).

Emotion Formation in Simulations. In this simulation, the number n of the sensory inputs s_n is 4 in the computational emotion model shown in Fig. 2, toward the information about up, down, right and left. If there is a food within 5 cells from the robot, s_i is set to 1, otherwise 0 (see Fig. 15). According to the distance between the food and the robot, Rew is set as following equations;

$$\text{Prize} = 0.9^{\text{distance}} \text{ Image impression (Imi)} \tag{11}$$

$$Rew = \sum_{\text{image}} \text{Prize}, \tag{12}$$

The images used as input to Image processing model in the simulation are shown in Fig. 13. A calculation example of values is shown in Fig. 14. The emotion model of activity-disactivity is as to "activity of the robot itself". The number of sensory inputs S is 2, as to the information, one is always S = 1, the other is S = 1 if the robot is in hazardous yellow area, or pink area, S =0 for otherwise. The value of Rew changes step by step according to the rules of Table 3. Parameters used in the learning of the emotion models are shown in Table 4. In Table 4, the method "Q+AE" is our proposed extended Q learning with emotion state, however, the parameters used in the learning of the emotion model are fixed while in the simulations. The method "Q+AE+S" is also our proposed method. The bigger the learning coefficient parameter α_{amy} is, the bigger the output of the emotion model is. In reverse, the bigger the learning coefficient parameter β_{amy} is, the smaller the output of the emotion model is. In the emotional model 1, the learning parameters α_{amy} and β_{amy} are changed in order to reduce the level of the pleasure when the level is over 0.3. The emotion model 2 about the activity is almost same as the emotion model 1.

Integrated Emotion State Model. The object of the integrated emotion state model in the fourth layer is to decide the two dimensional emotion states $S_e(i)$, $(i = 1, \cdots, 4)$

Table 3. Primary rewards (Rew) for the emotion model 2 with activity-disactivity.

Initial value	0.4		
When after 1 step	−0.005	Blue food acquisition	+0.2
Hazardous area: yellow	−0.02	Red food acquisition	+0.6
Hazardous area: purple	−0.05	When release the yellow switch	+0.4

Table 4. Parameters used in emotion learning of the two emotion models.

	Pleasure - displeasure		Activity - disactivity	
	Learning rate α_{amy}	Learning rate β_{amy}	Learning rate α_{amy}	Learning rate β_{amy}
Q+AE	0.4	0.3	0.2	0.5
Q+AE+S	0.4 ($E_1 < 0.3$)	0.3 ($E_1 < 0.3$)	0.2 ($E_2 < 0.5$)	0.5 ($E_2 < 0.5$)
	0.01($E_1 \geq 0.3$)	0.8 ($E_1 \geq 0.3$)	0.01($E_2 \geq 0.5$)	0.8($E_2 \geq 0.5$)

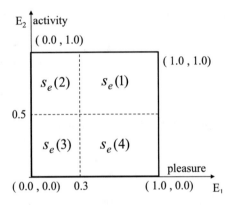

Fig. 16. The circumplex emotion model used in simulations.

(Fig. 16), using the output E_1 and E_2 of the emotion models 1 and 2, respectively. In the third layer and to transmit the state to extended Q learning system in the fifth layer.

Parameters and Rewards Used in the Emotion State Embedded Reinforcement Learning. Rewards given by the environment are shown at Table 5. The parameters used in the extended Q learning are given at Table 6.

Table 5. Reward r given by the environment for the extended Q learning in simulation 1 or 2.

Arrival to the goal	10.0	Red food acquisition (given as image of Fig. 14(a))	4.0
Collision to the wall	−2.0	Blue small food acquisition (given as image of Fig. 14(b))	1.5
Hazardous area: yellow	−0.5	When release the blue switch	5.0
Hazardous area: pink	−2.0	Others (when move 1 step)	−0.1

Table 6. Parameters used in extended Q learning.

Learning coefficient α	0.5	Discount rate γ	0.95
Policy	Greedy method		

5.2 Simulation 1 and Its Result

To confirm the performance of the proposed method, we compared with three methods:
(1) the conventional Q learning method named "Q", the other two methods are our
proposed methods, that is, (2) the method using extended Q learning with the learning
parameter fixed emotional model named "Q+AE", (3) the method using extended Q
learning with the learning parameter changed emotional model named "Q+AE+S".

The results of these three methods are shown at Table 7 and in Figs. 17, 18, 19, 20
and 21.

Table 7. Average convergence steps of 100 times in each method.

	Q	Q+AE	Q+AE+S
Convergence step	11	23	17

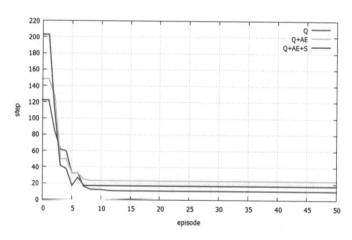

Fig. 17. The number of steps from start to goal for each method (average of 100 times).

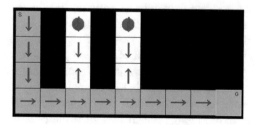

Fig. 18. The convergence path (arrow direction with green cells) for Q learning.

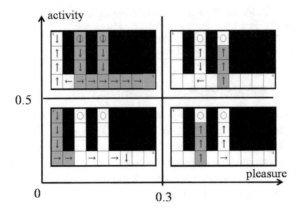

Fig. 19. The convergence path (arrow direction with green cells) in the four emotion states for the proposed method named "Q+AE".

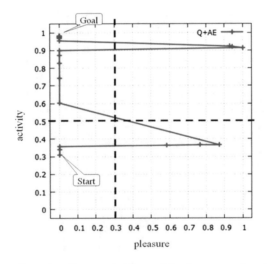

Fig. 20. The change of emotions from start to the goal for the proposed system named "Q+A E" after the convergence.

Table 7 shows the average convergence steps of 100 times in each method. The conventional Q learning method is the shortest steps from start to the goal without taking the food. In the Q+AE method, it took the longest steps because that the robot took the two foods. However in the Q+AE+S method, the robot took middle length of the steps because of taking only the first available food. This difference depends on the difference about emotion state with different emotional learning coefficient parameters.

Figure 18 shows the convergence path with green cells in the conventional Q learning. In the Q learning, though the robot visited the cells on which the foods is placed, it took the shortest path without taking foods finally.

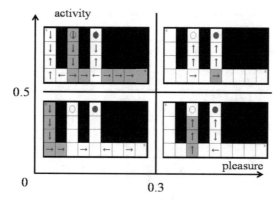

Fig. 21. The convergence path (arrow direction with green cells) in the four emotion states for the proposed method named "Q+AE+S".

Figures 19 and 20 show the simulation results of the Q+AE method. Figure 19 shows the convergence path along green cells in the four emotion states for the method. Figure 20 also shows the convergence path for the method. From Fig. 20, we can find that the robot starts with the emotion $S_e(3)$, passing through $S_e(4)$, $S_e(2)$ and $S_e(1)$, finally it got the goal with $S_e(2)$. Figure 19 shows the same situation with the green cells. For example, the robot visited the two foods cells with $S_e(4)$ and $S_e(2)$. Figures 21 and 22 show the simulation results of the Q+AE+S method. Figure 21 shows the convergence path along green cells in the four emotion states. In Table 7, the robot takes 17 steps from start to goal after getting only the food to be firstly discovered on the path. Comparing Fig. 20 with Fig. 22, it is found that although the result of the Q+AE method is the same as in the Q+AE method until the robot get the food to be

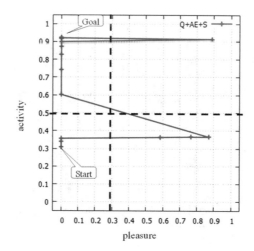

Fig. 22. The change of emotions from start to the goal for the proposed system named "Q+AE +S" after convergence.

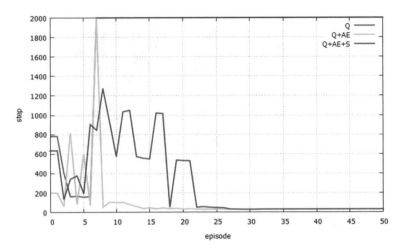

Fig. 23. The number of steps to the goal for each method (average of 100 times).

firstly discovered first on the path, however then in the Q+AE+S method the learning coefficient parameters of the emotion model was changed on the way to reduce the reaction for the stimulus from the environment. This is the reason why the robot didn't visit the cell the second food is placed.

5.3   Simulation 2 and Its Result

The difference between simulation 1 and simulation 2 is that simulation 2 is more difficult to solve than simulation 1. In simulation 2 as seen in Fig. 13, there is the

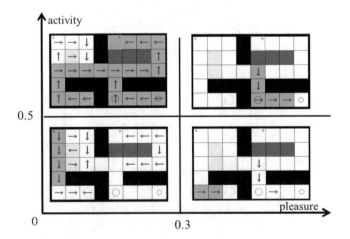

Fig. 24. The convergence path (arrow direction with green) in the each emotion state for the proposed method named "Q+AE".

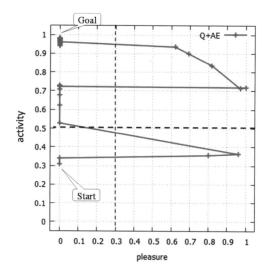

Fig. 25. The change of emotions from start to the goal for the proposed method named "Q+A E".

rocked door cell on the way to the goal, the robot has to visit the switch cell to release the rock of the door cell. Results of simulation 2 are shown in Figs. 23, 24, 25, 26 and 27. Explanations of them are similar with those of simulation 1.

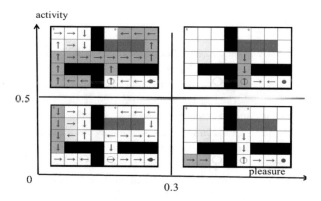

Fig. 26. The convergence path (arrow direction with green) in the each emotion state for the proposed method named "Q+AE+S"

The robot with the standard Q learning method failed to get goal, because it could not distinguish difference after or before visiting the switch cell. This is the reason simulation 2 are carried (Table 8).

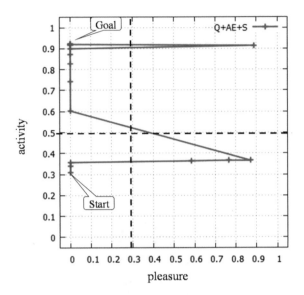

Fig. 27. The change of emotions from start to the goal for the proposed method "Q+AE+S".

Table 8. Average convergence steps to the goal of 100 times in each method.

	Q	Q+AE	Q+AE+S
Convergence step to the goal	–	32	28

6   Conclusions

The method to construct an emotional robot based on regulation of emotional responses with an emotion state embedded reinforcement learning system is proposed in this paper. Besides environmental states, the emotional robot has emotional states which are generated by stimulus received from sensor images. If the learning coefficient of emotions in Amygdala model is changed, generated emotional states in the robot are also different, even if the robot sees same sensor images. As a result, using the method, we can make any kinds of robots, having same structures. In the computer simulation for a path-finding problem with plural meaning paths, that is, having a dilemma, it was verified that the robot could get a variety of behavior patterns by setting the different learning parameters in emotional model learning. This means that by giving the different learning parameters to robots, it is able to make different robots, in spite of same internal structure.

In this study, we considered the single robot case. In the future, we would like to try to the multi robot case, i.e., to problems to be solved by multi-robots.

Acknowledgement. A part of this work was supported by Grant-in-Aid for Scientific Research (JSPS 25330287, and 26330254).

References

1. Sutton, R., Barto, A.: Reinforcement Learning: An Introduction. Bradford Book. The MIT Press, Cambridge (1998)
2. Moren, J., Balkenius, C.: A computational model of emotional learning in the amygdala. Cybern. Syst. **32**(6), 611–636 (2001)
3. Rouhani, H., Sadeghzadeh, A., Lucas, C., Araabi, B.N.: Emotional learning based intelligent speed and position control applied to neurofazzy model of switched reluctance motor. Control Cybern. **36**(1), 75–95 (2007)
4. Rouhani, H., Jalili, M., Araabi, B.N., Eppler, W., Lucas, C.: Brain emotional learning based intelligent controller applied to neurofuzzy model of micro heat exchanger. Expert Syst. Appl. **32**(3), 911–918 (2007)
5. Goerke, N.: EMOBOT: a robot control architecture based on emotional-like internal values, Chap. 4. In: J. Buchli (ed.) Mobile Robotics, Moving Intelligence. IntechOpen (2006). intechopen.com
6. Daglari, E., Temeltas, H., Yesiloglu, M.: Behavioral task processing for cognitive robots using artificial emotions. Neurocomputing **72**, 2835–2844 (2009)
7. Obayashi, M., Takuno, T., Kuremoto, T., Kobayashi, K.: An emotional model embedded reinforcement learning system. In: Proceedings of the IEEE International Conference on System, Man, and Cybernetics (IEEE SMC 2012), pp. 1058–1063 (2012)
8. Yang, F., Zhen, X.: Research on the agent's behavior decision-making based on artificial emotion. J. Inf. Comput. Sci. **11**(8), 2723–2733 (2014)
9. Hu, X., Xie, L., Liu, X., Wang, Z.: Emotion expression of robot with personality. Math. Problems Eng. **2013**, 1–10 (2013)
10. Kuremoto, T., Ohta, T., Kobayashi, K., Obayashi, M.: A dynamic associative memory system adopting amygdala model. Artif. Life Robot. **13**(2), 478–482 (2009)
11. Rusell, J.A.: A circumplex model of affect. J. Pers. Soc. Psychol. **39**(6), 1161–1178 (1980)

Recognition of Arm-and-Hand Visual Signals by Means of SVM to Increase Aircraft Security

Giovanni Saggio[1]([✉]), Francesco Cavrini[2], and Carlo Alberto Pinto[2]

[1] Department of Electronic Engineering, University of Rome "Tor Vergata",
Via del Politecnico 1, Rome, Italy
saggio@uniroma2.it
[2] Captiks S.r.l., Via Giacomo Peroni, 442-444 Rome, Italy

Abstract. In aircraft scenarios the proper interpretation of communication meanings is mandatory for security reasons. In particular some communications, occurring between the signalman and the pilot, rely on arm-and-hand visual signals, which can be prone to misunderstanding in some circumstances as it can be, for instance, because of low-visibility. This work intends to equip the signalman with wearable sensors, to collect data related to the signals and to interpret such data by means of a SVM classification. In such a way, the pilot can count on both his/her own evaluation and on the automatic interpretation of the visual signal (redundancy increase the safety), and all the communications can be stored for further querying (if necessary). Results indicate that the system performs with a classification accuracy as high as $94.11 \pm 5.54\%$ to $97.67 \pm 3.53\%$, depending on the type of gesture examined.

1 Introduction

Visual Signals (VS) can be referred as any means of communication depending on sight. In general, VS is useful to support communication, or to increase the communication security (for instance when silence is mandatory), or to realize rapid means. In particular, VS can be very effective when other forms of communications (written, voice, radio, tactile) are impossible, or unavailable, or inadequate. Emblematic is the real radio conversation occurred, on October 1995, between a US naval ship with Canadian authorities off the coast of Newfoundland (an island off the east coast of the North American mainland). Both the American than the Canadian asked the other to divert the respective course to avoid a collision, but none wanted to respect the request, realizing at the end, with a conceivable surprise, that the Canadians request come from a lighthouse and not from a ship.

VS can be realized by means of different visual aids, such as flags, pyro-techniques, flashlights, chemical light sticks, display panels, mirrors, strobes, smoke, etc. However, in the general frame of VS, our interest is specifically focused on arm-and-hand VS (A&HVS).

This special case of VS based on arm-and-hand posture/movements/gesture can be effective only when based on standardized signals, which have to be mandatorily familiar to all users. Unfortunately, a *universal* standard does not exist, but different standards are for different fields of application. In rugby, for instance, the referee has 53

J.J. Merelo et al. (eds.), *Computational Intelligence*, Studies in Computational Intelligence 669,
DOI 10.1007/978-3-319-48506-5_23

A&HVS to count on (even if the most commonly used are only 12) [1]. Although there are no official training A&HVS for dogs, several signals are anyway recognized by most professional trainers and specially used in obedience competition. An A&HVS guide exists, developed by Enform (the safety association for Canada's upstream oil and gas industry, www.enform.ca), for directing (driving and stopping) vehicles [2]. As an alternative way of communication to vocal and sound, standard A&HVS can be used in case of NBC (nuclear biological chemical) hazards [3]. Sign language necessary for deaf people to communicate is based on A&HVS. In the pilot's knowledge guide [4], A&HVS are reported in graphical view to be learned and used between the so-called *wingman* and the pilot.

This work deals with the special case of A&HVS related to specific vocabulary, receipting, acknowledging and identification procedures, adopted as code meanings by the US Department of the army ("Visual Signals, FM 21-60" manual).

The selection of this manual was made to take into account the requirements of the customer of this work, which was the Italian Ministry of Defence, General Air Armaments Directorate (Direzione Generale degli Armamenti Aeronautici "Armaereo", Ministero della Difesa).

We aim to take advantage from A&HVS, but intend also to try to overcome some of the severe limitations the A&HVS can suffer from. In particular these limitations come from the reduction (because of fog or other weather elements, or low light) or occlusion (because of an object between actors/units that disrupts the line of sight) of the visual range, from the possible misunderstanding or, in the special case of army applications, from the enemy interception.

One or more of those limitations can be so relevant to produce even severe consequences. We can mention, for instance, the airport disaster occurred, on October 2001, at Linate (Milan, Italy), when Scandinavian Airlines Flight 686 (McDonnell Douglas MD-87), carrying 110 people, collided on take-off with a Cessna Citation CJ2 business jet, carrying 4 people. All 114 people were killed plus 4 people on the ground. The following investigation revealed that the collision was caused by a number of non-functioning and non-conforming safety systems, standards, and procedures at the airport, but also because of the reduction of the visibility (thick fog, visibility less than 200 m) that highly limited the A&HVS communications between the *signalman* and the aircrafts.

All mentioned, for the purposes of this work, we dressed the signalman with wearable sensors to measure his/her posture/gesture, and used measured data to automatically interpret his/her communication meaning. Our system can offer the advantages of the redundancy of information, of the certainty of the signal meaning, and of the storing of the coded information. The redundancy is useful in increasing the security, the meaning certainty allows a double check to the pilot (with respect to what he/she understand by his/her own), and the information storage can allow retrieval of both rough and classified data, making it possible to realize a sort of "black box airport runway", similar to the "black box flight recorder".

2 Materials and Subjects

We intended to measure A&HVS. To do so, we could adopt the current gold-standard systems, which are based on optical technology, realized by means of markers which, dressed by the signalman, can be traced via cameras. In [5], for instance, 3D images were used to track body and hand together, in order to understand gestures for introducing a multi-signal gesture database for aircraft handling signals.

However, the optical technology is not here meaningfully suitable for different reasons: it is quite expensive (gold-standard apparatuses can cost as high as hundreds of Euros/Dollars), it is highly operator-dependent (its usage is not a trivial matter), it is high time-demanding for calibration procedure (which can last from several minutes up to hours), it suffer from occlusion/low-light problems, and finally it cannot be used everywhere (the cameras must be on-time arranged).

In order to overcome such limitations, we adopted wearable devices capable of measuring the movements/gestures/postures of the user, as detailed in the following. In particular, these wearable devices were intended for the measure of the kinematic of the fingers, of the hand, of the wrist, and of the forearm.

2.1 Sensors

In order to measure the flexion of the finger joints, we adopted flex sensors termed "Bend Sensors®" (by Flexpoint Sensor Systems Inc., Draper, Utah, USA). These sensors are low-cost, low-weight, unobtrusive and, for our purposes, offer suitable repeatability and reliability characteristics (already reported in [6]). The interested reader can find a comprehensive survey on their mechanical, physical and electrical properties in [7]. In particular, these flex sensors demonstrate a resistance versus bending angle curve suitably fitted, within some limits, by a second order polynomial curve.

Differently, in order to measure the movements of the wrist and of the forearm, we utilized two inertial measurement units (IMUs) termed Sparkfun Razor (by SparkFun Electronics, Niwot, Colorado, USA), each equipped with a 3-axis accelerometer and a 3-axis gyroscope.

2.2 Wearable Devices

Sensors integrated into a supporting glove realize a wearable device termed *sensory glove*. In the latest years, this kind of glove has been founding more and more applications in different contexts. For instance, in the medical field, sensory gloves realize tools for the evaluation of manual dexterity of the surgeons [8], for the objective assessment of hand rehabilitation after hand surgery of patients [9], for assessment of the hand motion control development in infants [10], and for arthritis rehabilitation applications [11]. In robotics, sensory gloves have been proposed in general for driving a mechanical hand under a master-slave configuration [12], and in particular for increasing the safety level in extravehicular manipulations [6]. In the social field, these gloves have been supported deaf people in automatic conversion of sign language to spoken sentences [13]. In the music field, sensory gloves have been used for instinctively controlling musical performances [14], and so ahead.

In all these occurrences, the sensory glove can be equipped with a different number and a different type of sensors.

In this work, ten flex sensors and two IMUs were integrated in a supporting glove made of Lycra®. In particular, ten flex sensors were housed into an equal number of pockets sewn in correspondence of the ' metacarpo-phalangeal (MCP) and proximal-inter-phalangeal (PIP) joints of each finger. None was utilized for the distal-inter-phalangeal (DIP) joints believed not relevant for our purposes (in any case, DIP flexion angles are normally correlated to the PIP ones in known percentages, as reported in [15]). The two IMUs were housed into two pockets in correspondence of the dorsal aspect of the hand and of the forearm respectively. In such a configuration, a total of 16 degrees of freedom (DOF) were measured (10 coming from the flex sensors and 6 from the inertial units).

The overall system was termed versatile-glove (V-Glove hereafter).

Figure 1 shows the arrangement of the sensors (here no glove is reported for the sake of clarity).

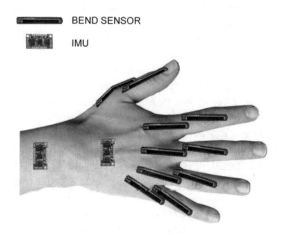

Fig. 1. Arrangement of the sensors. 10 flex sensors were adopted to measure flexion/extension of the fingers, 2 IMUs were adopted to measure movements of the hand and of the forearm. For the sake of clarity, here no glove is reported, but a supporting glove houses all the sensors.

2.3   Electronics

For the electronic circuitry we can distinguish one "source" and one "receiving" subsystems.

Let's start considering the source-subsystem (Fig. 2a), which is necessary to acquire signals from the sensors, to provide A/D conversions, and to wired/wireless transmit data to the receiving-subsystem. The wired transmission is intended for testing purposes, while the wireless one to be in-field adopted.

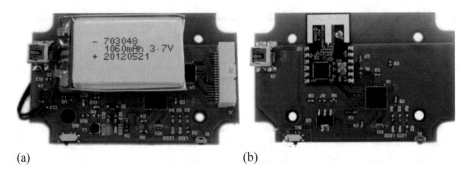

(a)   (b)

Fig. 2. Hardware of the (a) source and the (b) receiving subsystems. The source acquires data from the sensors and send data to the receiver which fed a personal computer. The source includes a battery, while the receiver is sourced directly by the personal computer to which it is connected.

Resistance values from the ten flex sensors were converted into voltage values by means of an equal number of voltage dividers, while voltage values from the IMUs fed directly the circuitry. The core of the source-subsystem was the integrated circuit (IC) PIC18F47J53 (by Microchip, Chandler, AZ, US), 48 MHz-clocked, capable of 12 bit A/D conversion in 21 µs. This IC offers only 10 analog inputs so that, in order to acquire all the 22 signals from the V-Glove, we used a 2×16 channel multiplexer, the ADG726 (by Analog Devices, Norwood, MA, USA) (its 10 spare input channels can be used for eventually additional requirements).

Requests for the wireless protocol included short (or medium) transmission range, low-medium transmission speed, low-power consumption, and scalability, so to handle data of up to four sensory gloves at a time, all in an auto user-independent configuration mode. To respond to these requests, we considered different protocols/standards, in particular the Bluetooth and Bluetooth smart/IEEE802.15.1, the IEC62591/WirelessHART, the ISA100.11a, the DASH7, the Z-Wave, the ANT, the Wavenis, and the IEEE802.15.1/ZigBee. The latter was our choice, since it better responded to the commitment requirements.

Here, the transmission security was not a mandatory parameter but, in any case, all the aforementioned protocols can be considered similar from a cryptography point of view. The interested reader can find a survey comparison in [16].

Our wireless transmission was then obtained with the IEEE 802.15.4 radio transceiver module MRF24J40MA (by Microchip, Chandler, AZ, USA), which allows a 0 dBm transmission within a 100 m range.

The DC power supply was realized with a 3.7 V/1060 mAh Li-Ion single-cell battery, charged and controlled by the IC BQ25015 (by Texas Instruments, Dallas, TX, USA), which includes a DC-DC buck converter capable of 300 mA @ 3.3 V.

The receiving-subsystem is the coordinator of the wireless network (Fig. 2b). Its functions are to create the network, to manage the communication with the source subsystem and to furnish received data to the personal computer. It was based on the same integrated circuit MRF24J40MA of the source-subsystem. The USB transmission was based on the inner full-speed module of a second PIC18F47J53. The DC power supply was obtained from the USB port, with its 5 V reduced to the necessary 3.3 V adopting an LDO voltage regulator.

2.4   Software

Software routines were developed to realize a virtual ambient to manage the calibration of the V-Glove, to train the users, to acquire and to classify data, and to graphically represent the recognized A&HVS. In order to make the overall system *user-friendly*, the ensemble was operated via an ad-hoc developed graphical user interface (GUI) (Fig. 3). Figure 4 shows the overall view of the system, including the sensory V-Glove, the transmission and receiving hardware and the software interface.

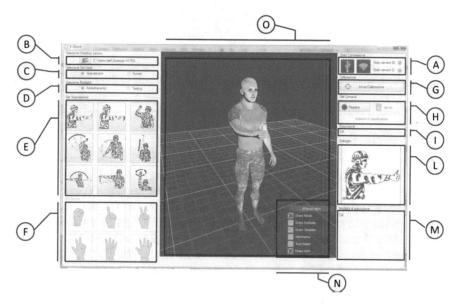

Fig. 3. Panel view. A: status connection, B: directory selection, C: Signals/Numbers selection, D: Training/Testing selection, E: Gesture selection, F: numbers selection, G: guided calibration, H: visualization of the Training/Testing phase, I/L/M: name/image/description of the selected gesture, N/O: 3D control-panel/window.

2.5   Participants

Ten subjects took part in training and testing procedures. They were six males and four females, 22–49 aged (average 29.4) right handed, with no motor or intellectual limitations. All participants signed a written consensus.

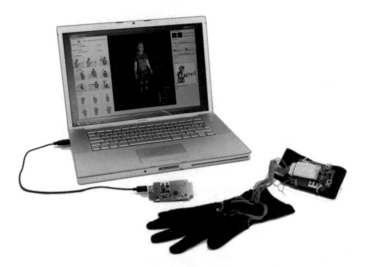

Fig. 4. The overall view of the system, made of a sensory glove, interfaced with a source circuitry, which treats and send data to a personal computer running an ad-hoc developed software.

2.6   Classifier

Among the most used classification approaches in the field of hand gesture recognition, we have Artificial Neural Networks, e.g. the widely popular Multilayer Perceptron (MLP), Hidden Markov Models (HMM) [17, 18], and Support Vector Machines (SVM) [19]. In particular, HMM is a time-aware model, i.e. the output at a given instant depends also on the previous history, while MLP and SVM are time-agnostic, i.e. the current classification does not take into account previous ones [21].

The input to the algorithms can be either an entire movement (firstly recorded, then classified) or a part of it by means, e.g., of the segmentation of data into overlapping windows of a short duration (classified while recorded) [22]. The first approach is typically simpler and allows the recognition process to have a global view of the gesture (data is processed after the movement is completed), while the second approach is appropriate in all those scenarios in which real-time constraints are important. Time-agnostic models such as SVM can be used in either mode, while time-aware algorithms such as HMM typically make more sense in the real-time scenario.

In the present project, we had a number of requirements and some degree of flexibility that guided our choice regarding the classification approach. In particular, with respect to flexibility, the option of entire movement registration followed by its recognition was acceptable. With respect to constraints, it was important to use a method that did not require the acquisition of too much training data, as the technical staff has to be operative in a few time. Thus we opted for SVM, given that:

1. It allows for the classification of both linearly and non-linearly separable datasets (by means of the "kernel trick", see [23] for further information).

2. Its training procedure is typically faster than that of MLP, since depends only on the number of example instances and not on their dimensionality [22].
3. It is less susceptible to *overfitting* with respect to MLP [24].
4. Good results can be obtained even with few training data, whereas MLP typically requires a lot of example instances in order to mitigate the effect of overfitting.

3   Methods

3.1   Selection of Gestures

Here we work with A&HVS divided into "signalling" and "numbering", both based on a single arm-and-hand (A&HVS based on both arms-and-hands will be part of a future

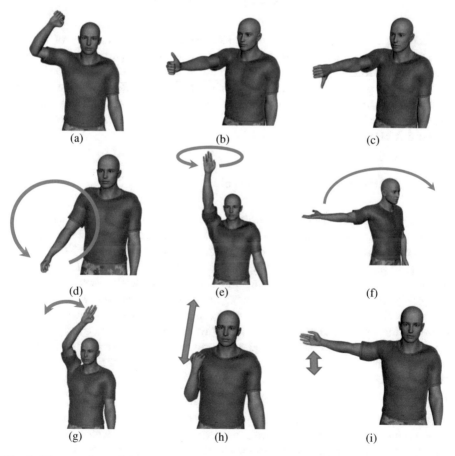

Fig. 5. The meanings of the gestures are: (a) freeze, (b) raise the boom, (c) lower the boom, (d) start engine or prepare to move, (e) assemble or rally, (f) advance or move out, (g) attention, (h) increase speed, double time or rush (i) slow down.

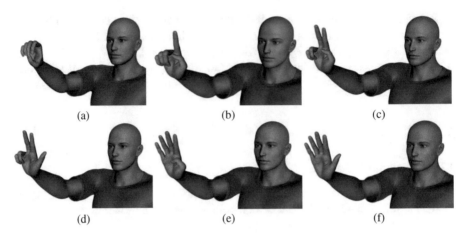

(a)   (b)   (c)

(d)   (e)   (f)

Fig. 6. Numbering from (a) zero to (f) five.

development). In particular, the signal codes, we refer to, are a subset of the standardized ones, reported in Fig. 5, according to the US Department of the Army (FM21-60 handbook, September 1987), while the number codes are the ones, made with fingers, reported in Fig. 6, as mostly universally known. Regarding the signalling showed in Fig. 5, a single arrowhead indicates a not continuously repeated (but eventually repeated at intervals) A&HVS; a double arrowhead indicates a continuously repeated A&HVS until acknowledged.

3.2   Signalling and Numbering Protocol

V-Glove Calibration. Before proceeding with the training/testing protocol, a calibration procedure was necessary for the V-Glove. This was to acquire the minimum and maximum electrical resistance values assumed by each flex sensor. Therefore, the user had to sit in front of a desk with his arm placed on the table, the arm-forearm forming an angle of 90°, and posing the hand in completely open (Fig. 7a) and completely closed (fist, Fig. 7b) position, for at least 2 s respectively. From these extreme positions, we obtained minimum and maximum resistance values, utilized to establish, for each flex sensor, its first order polynomial fit of the resistance vs. angular bending curve. We knew that a second order polynomial produced a better fitting with respect the first order one that we adopted, anyway the introduced error was not so effective, since it is meaningful only for angles <30° roughly, and low value angles were not so relevant for our purposes.

Experimental Protocol. In executing each A&HVS, the user starts always from the same posture: the body upright, facing forwards the pc screen, with shoulder blades back and the arms hanged loosely to the sides, palms facing sides of the body.

For repeated A&HVS, the repetition was executed three times, while static postures were requested to be maintained for 2 s as minimum.

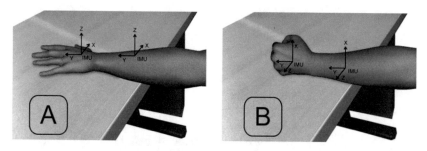

Fig. 7. The two hand poses (flat hand and fist) for the calibration steps. The sensory glove is omitted in the figure for clarity reasons.

The protocol was as follows:

1. Glove dressing and calibration;
2. Training: three replays of each arm-and-hand visual signal;
3. Glove removal and rest period;
4. Glove dressing and calibration;
5. Training: two replays of each arm-and-hand visual signal;
6. Testing: five replays of all the visual signals;
7. Glove removal and rest period;
8. Glove dressing and calibration;
9. Testing: five replays of all the visual signals.

For numbering we used a training/testing protocol identical to the one described above. We acquired 5 training repetitions and 10 testing repetitions of each signalling and numbering procedure. We felt it was worth to stress that both training and test data has been acquired in two distinct settings, i.e. after glove removal and re-dressing, so to improve the generalization capability of the classifier (during learning) and better estimate performance in a real-usage scenario (during testing).

3.3   Classification Framework

A rather general schema of a pattern recognition system is depicted in Fig. 8. The data pre-processing block represents all those procedures that one may apply in order to prepare data to enter the classification process. For example, noise reduction techniques may be used, a portion may be extracted from a continuous stream, and aggregation of different information sources may take place. Then, pre-processed data goes through the feature extraction step, which is in charge of mapping the input patterns to a number of relevant descriptors. The generated feature vector may include redundant, correlated or actually irrelevant components, and/or be of too high dimensionality. In all those cases, it may be worthwhile to consider a dimensionality reduction step, e.g. by means of Principal Components Analysis (PCA). The reduced feature vector is then given as input to the classifier, which assigns it to the class to which it is expected to belong, e.g. an A&HVS in the present project. Finally, the classifier response can be post-processed, e.g. to examine the related uncertainty and abstain if it is deemed to be too high.

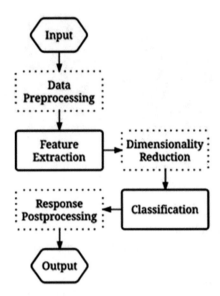

Fig. 8. Generic schema of a pattern recognition framework. Optional steps are indicated as dotted blocks.

Data Pre-processing. In the present system, to enhanced data quality, pre-processing involved the use of an infinite impulse response (IIR) filter based on an exponentially weighted moving average.

Feature Extraction. For the sake of simplicity and to discard irrelevant information, whenever possible we considered only data coming from a subset of the sensors. In particular, for the numbering gestures (Fig. 6), IMU information is not relevant (arm and hand spatial orientation is not meaningful as it should be rather fixed in a conventional, easily visible, configuration), so we analysed data coming from the flex sensors only. Differently, the IMUs furnished fundamental indicators in the case of the other arm-and-hand visual signals (Fig. 5).

Feature extraction for numbering gestures was based on the average value of each flex sensor resistance. In particular, that value was computed with respected to three adjacent signal windows positioned, respectively, at the beginning, around the center, and at the end of the gesture. After preliminary analysis, the value at the beginning was omitted since it resulted as non-relevant.

Feature extraction for the other A&HVS is more involved, as it requires considering all the sensors, and cannot be reasonably limited to a small portion of the registration; the entire gesture duration is meaningful. Various approaches are possible and they can be roughly divided into time-domain based, frequency-domain based, and time-frequency-domain based, or an ensemble of the previous. In this study, we opted for the time-frequency-domain strategy and used as features the first 32 *average coefficients* of the Discrete Wavelet Transform (DWT) based on the Daubechies-4 wavelet [25]. This choice originates from the desire to use a technique that is able to successfully highlight frequency-domain characteristics of the signals while maintaining temporal localization, which we regarded as an important discriminatory factor. Furthermore, the DWT can be implemented in an efficient manner, using e.g. the filter banks approach [26]. The specific type of wavelet and the number of coefficients were determined after preliminary evaluations. In particular, 32 coefficients were found to be sufficient for expressing the complexity of the gestures while discarding, at the same time, those detail differences that are not meaningful to the pattern recognition process.

Classification: Support Vector Machines [27]. Support Vector Machines (SVM) represents a powerful learning paradigm with solid statistical foundations [28]. In a geometrically intuitive way, we can approach SVM using the concept of *margin* in a two-class recognition problem. Consider the example in Fig. 9. The image shows two separating lines that we may interpret as the result of two algorithms that successfully learned to distinguish between green and red objects. However, the blue line seems to be preferable to the orange one, as it leads to a greater separation between the two opposite instances that are closer to the boundary. In other words, the blue line seems to be *safer*. This concept of safety margin is at the heart of SVM. Indeed, a SVM searches for the separating hyperplane that maximizes the margin on training data, in the hope that it will also provide better results on new instances. The learning problem can be put in terms of a *convex quadratic program* and thus solved efficiently using well-known techniques [29]. Moreover, the program complexity depends only on the number of training patterns and not on their dimensionality. In case the data is not linearly separable, it is possible to exploit the "kernel trick", i.e. operate like if non-linearly projecting data into another space and performing classification therein. A variety of kernels can be used, e.g. polynomials, radial-basis functions, sigmoidal functions. See [29] for further details.

In the present project, after preliminary evaluations, we opted for a Gaussian kernel. This resulted into two free SVM parameters we had to select: the penalty factor C and

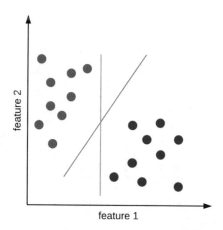

Fig. 9. Two discriminant line with different margin for the same (hypothetical) dataset. The blue line is preferable, as it leads to a greater separation between the two opposite instances that are closer to the boundary.

the kernel radius R. In particular, the C parameter weights the influence of misclassified points. A small value for C is likely to lead to a greater margin even if some points are misclassified. Conversely, a large value for C will make the SVM to search for a hyperplane that misclassifies few points, even if the resulting margin is small. The R parameter, instead, represents the standard deviation of the Gaussian-like kernel function. To estimate C and R we opted for a simple grid-search procedure, computing the corresponding model accuracy on a subset of the training data.

Finally, we introduced an abstention mechanism in order to avoid producing a possibly-wrong response in vague situations. In particular, if the difference in score between the two most probable classes, a given input feature vector may be belong to is below a given threshold, and then we prefer to abstain rather than outputting an uncertain decision. The threshold value was chosen after preliminary evaluations as a trade-off between responsiveness and safety (the higher the threshold, the higher the safety, but also the higher the number of correct classifications that do not pass the check, and thus the lower the responsiveness of the system).

Framework Structure and Implementation. Given the above described data pre-processing, feature extraction and classification steps, the resulting pattern recognition framework can be schematized as depicted in Fig. 10. The related software was implemented as a set of C++ dynamic libraries, to be used by the main application we realized in this project; the corresponding UML Component Diagram is shown in Fig. 11. As reported in the diagram, we used the two open-source libraries *wave++* [30] and *libsvm* [31] to implement, respectively, the DWT and the SVM classifier. In particular, to better suit our needs, we encapsulated the *wave++* complexities inside an *ad-hoc* developed library, VGloveWavelet, while *libsvm* was used directly inside the main module, VGlovePRS. Figure 12 shows the structure of VGlovePRS according to the UML Class Diagram formalism.

The class *VGloveFeatureExtractor* realizes feature extraction, differently for signalling and numbering as already mentioned. This class uses both the *VGloveWavelet*

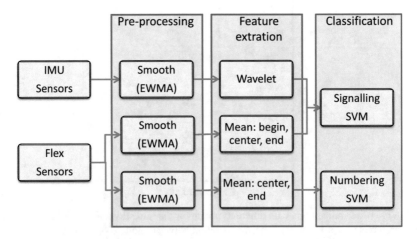

Fig. 10. Pattern recognition framework of the present project.

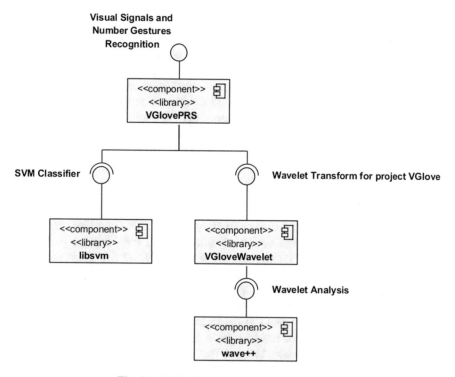

Fig. 11. UML Component Diagram (overall).

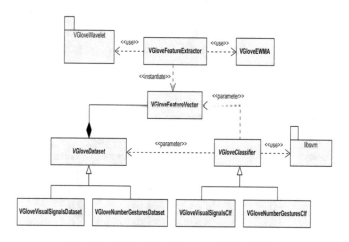

Fig. 12. UML Class Diagram VGlovePRS.

library and the *VGloveEWMA* class, which implements the exponentially-weighted moving average filter. Feature vectors are represented with the homonymous class.

The *VGloveDataSet* class holds an ensemble of feature vectors (possibly with associated labels) useful for training or testing purposes. The *VGloveVisualSignalsDataset* and the *VGloveNumberGesturesDataset* child classes represent, respectively, signalling and numbering datasets. The *VGloveClassifier* class stands for the SVM-based classifier, which relies on the *libsvm* library. Finally, the *VGloveVisualSignalsClf* e *VGloveNumberGesturesClf* specializations represent the classifiers for arm-and-hand signalling and numbering, respectively.

4 Results and Discussion

Table 1 reports the system performance in recognizing A&H signalling and numbering. As the results indicate, the system obtained a quite high average performance (94.11 ± 5.54 % of accuracy for signalling gestures and 97.67 ± 3.53 % of accuracy for numbering gestures) with remarkable peaks for some subjects, e.g. M1. It is also worth to point out that no significant difference in performance was experienced between males and females, even though, generally speaking, males' performance was slightly higher. This may be related to the fact that the V-Glove prototype was conceived for a medium-sized male hand, and thus may have not perfectly fit some of the females' hands. In such a context, also the IMUs may have experienced a few trembling because of the imperfect adherence of the glove to the skin. However, it is likely that such problems could be easily overcome by considering further glove sizes. Familiarity with the system seems to be important. Indeed, subjects M1, M2 and M3, the ones that practised with the system the most, obtained optimal or near-optimal performance, whereas greater variability characterized the accuracy achieved by the other subjects. For example, F5 reached a very high recognition rate, whereas F2 and

Table 1. System accuracy, error rate, and abstention rate for the 10 subjects that participated to the study (males indicated with M, females with F). Values are in percentage.

Subject	Signalling			Numbering		
	Accuracy	ErrRate	AbstRate	Accuracy	ErrRate	AbstRate
M1	100	0	0	100	0	0
M2	97.78	0	2.22	100	0	0
M3	95.56	0	4.44	100	0	0
M4	94.44	0	5.56	100	0	0
M5	96.67	1.11	2.22	96.67	1.67	1.67
F1	87.78	3.33	8.89	90.00	3.33	6.67
F2	88.89	5.56	5.56	93.33	3.33	3.33
F3	97.78	0	2.22	100	0	0
F4	83.33	6.67	10.00	96.67	0	3.33
F5	98.89	0	1.11	100	0	0
Average	94.11 ± 5.54	1.67 ± 2.58	4.22 ± 3.30	97.67 ± 3.53	0.83 ± 1.41	1.5 ± 2.28

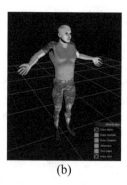

(a)   (b)

Fig. 13. The gesture (a) is recognized, (b) is not recognized (abstention).

F4 probably needed further experience with the system and the gestures. Finally, it is possible to observe that the introduction of the abstention mechanism was effective. In fact, the error rate of the system was quite low both for signalling and numbering gestures, which is important since a misclassification can be critical in the application context we refer to. Figure 13 reports the example of the graphical response of the system to the user for a recognized gesture and for a non-recognized one (abstention).

5   Conclusions

We propose a hardware/software solution for recognition of arm-and-hand visual signals (A&HVS). The system is intended to be camera-free, so grounded on the use of wearable sensors, in particular flex sensors to register finger bending and inertial measures units (IMU) to record arm and hand movements. All is packaged inside a glove, called Versatile-Glove (V-Glove), so to be of immediate use for the subject. Indeed, the system does not require a long calibration procedure, as for example camera-based solutions do, and it also overcomes the typical limitations of optic products, e.g. lighting, obstruction, need for complex setup of the environment (camera placing). Moreover, all the components are cost-effective.

We focused on A&HVS related to specific vocabulary, receipting, acknowledging and identification procedures, adopted as code meanings by the US Department of the army ("Visual Signals, FM 21-60" manual). Our system can offer the advantages of the redundancy of information, of the certainty of the signal meaning, of the storing of the coded information. The redundancy is useful in increasing the security, making it possible to realize a sort of "black box airport runway", similar to the "black box flight recorder".

System testing involved 10 subjects, who performed both 6 numbering and 9 signalling gestures. Numbering gestures were mainly static, whereas the dynamic evolution of movement was fundamental in signalling gestures. By means of a suitable pattern recognition framework (signal pre-processing via an IIR filter, feature extraction via both time-domain and time-frequency-domain characteristics, classification via Support Vector Machines), we were able to successfully recognize the above mentioned

A&HVS (94.11 ± 5.54 % of accuracy for signalling gestures and 97.67 ± 3.53 % of accuracy for numbering gestures). Moreover, given that misclassification is dangerous in the application field we refer to, we introduced an abstention mechanism to avoid producing a possibly wrong response in vague situations. Experimental tests confirmed that such a procedure is effective. Indeed the error rate was only 1.67 ± 2.58 % for signalling gestures and 0.83 ± 1.41 % for numbering gestures, with many probable-errors turned into abstentions (abstention rate: 4.22 ± 3.30 % for signalling gestures and 1.5 ± 2.28 % for numbering gestures).

Future work includes further tests with a greater number of participants, in order to have a statistically relevant sample and thus further confirm (or refute) the insights we obtained in the present study. Moreover, we would like to extend the gesture set and consider also A&HVS for which both hands are necessary.

Acknowledgements. This Work Was Funded by the "Armaereo" (Direzione Generale Degli Armamenti Aeronautici, Ministero Della Difesa), Contract #A2009.90, for Which We Would like to thank T.Col. Garn Aldo Spagnolini and T.Col. Garn Salvatore Vignola.

References

1. http://www.coca-cola.co.uk/stories/rugby-world-cup/instant-expert-6-deciphering-the-referees-hand-signals/
2. http://www.pistonwell.com/irp/irp12_hand_signals_for_controlling_vehicles.pdf
3. http://www.armystudyguide.com/content/army_board_study_guide_topics/cbrn/nbc-warning-signals.shtml
4. http://www.flyfast.org/sites/all/docs/FAST_FKG.pdf
5. Song, Y., Demirdjian, D., Davis, R.: Tracking body and hands for gesture recognition: Natops aircraft handling signals database. In: Proceedings of the 9th IEEE International Conference on Automatic Face and Gesture Recognition (2011)
6. Saggio, G., Bizzarri, M.: Feasibility of teleoperations with multi-fingered robotic hand for safe extravehicular manipulations. Aerosp. Sci. Technol. **39**, 666–674 (2014)
7. Saggio, G., Riillo, F., Sbernini, L., Quitadamo, L.R.: Resistive flex sensors: a survey. Smart Mater. Struct. **25**(1), 1–30 (2016)
8. Saggio, G., Lazzaro, A., Sbernini, L., Carrano, F.M., Passi, D., Corona, A., Panetta, V., Gaspari, A.L., Di Lorenzo, N.: Objective surgical skill assessment: an initial experience by means of a sensory glove paving the way to open surgery simulation? J. Surg. Educ. **72**(5), 910–917 (2015)
9. Saggio, G., Sbernini, L., De Leo, A., Awaid, M., Di Lorenzo, N., Gaspari, A.L.: Assessment of hand rehabilitation after hand surgery by means of a sensory glove. In: Proceedings of the 9th International Joint Conference on Biomedical Engineering Systems and Technologies (BIOSTEC 2016), co-located 9th International Conference on Biomedical Electronics and Devices (BIODEVICES 2016) (2016)
10. Tsagarakis, N. G., Kenward, B., Rosander, K., Caldwell, D.G., von Hofsten, C.: BabyGlove: a device to study hand motion control development in infants. In: EuroHaptics (2006)
11. O'Flynn, B., Sanchez, J. T., Angove, P., Connolly, J., Condell, J., Curran, K., Gardiner, P.: Novel smart sensor glove for arthritis rehabilitation. In: Proceedings of IEEE International Conference on Body Sensor Networks (BSN), pp. 1–6 (2013)

12. Zaid, A.M., Yaqub, M.A.: UTHM HAND: performance of complete system of dexterous anthropomorphic robotic hand. Procedia Eng. **41**, 777–783 (2012)
13. Cavallo, P., Saggio, G.: Conversion of sign language to spoken sentences by means of a sensory glove. J. Softw. **9**(8), 2002–2009 (2014)
14. Costantini, G., Saggio, G., Todisco, M.: A glove based adaptive sensor interface for live musical performances. In: Proceedings of the 1st International Conference on Sensor Device Technologies and Applications (SENSORDEVICES), pp. 18–25 (2010)
15. Ghosh, S.: Capturing human hand kinematics for object grasping and manipulation. Thesis Submitted to the Office of Graduate Studies of Texas and A&M University, for the Degree of Master of Science (2013)
16. Gomez, C., Paradells, J.: Wireless home automation networks: a survey of architectures and technologies. IEEE Commun. Mag. **48**(6), 92–101 (2010)
17. Mitra, S., Acharya, T.: Gesture recognition: a survey. IEEE Trans. Syst. Man Cybern. Part C Appl. Rev. **37**(3), 311–324 (2007)
18. LaViola, J.: A Survey of Hand Posture and Gesture Recognition Techniques and Technology, Brown University, Providence, Rhode Island, USA, CS-99-11 (1999)
19. Wu, J., Pan, G., Zhang, D., Qi, G., Li, S.: Gesture recognition with a 3-D accelerometer. In: Zhang, D., Portmann, M., Tan, A.-H., Indulska, J. (eds.) UIC 2009. LNCS, vol. 5585, pp. 25–38. Springer, Heidelberg (2009). doi:10.1007/978-3-642-02830-4_4
20. He, Z.: Accelerometer based gesture recognition using fusion features and SVM. J. Softw. **6**(6), 1042–1049 (2011)
21. Theodoridis, S., Koutroumbas, K.: Pattern Recognition, 4th edn. Academic Press, San Diego (2008)
22. Oskoei, M.A., Huosheng, H.: Hu, Myoelectric control systems—a survey. Biomed. Sig. Process. Control **2**(4), 275–294 (2007)
23. Alpaydin, E.: Introduction to Machine Learning. MIT press, Cambridge (2014)
24. Mitchell, T.: Machine Learning. McGraw Hill, New York (1997)
25. Walker, J.S.: A Primer on Wavelets and Their Scientific Applications, 2nd edn. Chapman and Hall/CRC, Boca Raton (2008)
26. Strang, G., Nguyen, T.: Wavelets and Filter Banks. SIAM, Wellesley (1996)
27. Cavrini, F.: (This section is based on) Hand gesture recognition for the benefit of transradial amputees. MSc Engineering thesis, University of Rome "La Sapienza" (2016)
28. Cherkassky, V., Mulier, F.M.: Learning from Data: Concepts, Theory, and Methods. Wiley–IEEE Press, New York (2007)
29. Alpaydin, E.: Introduction to Machine Learning, 2nd edn. The MIT Press, Cambridge (2009)
30. Ferrando, S.E., Kolasa, L.A, Kovacevic, N.: Wave++: a C++ wavelet library (2000). http://www.math.ryerson.ca/~lkolasa/CppWavelet.html
31. Chang, C.-C., Lin, C.-J.: LIBSVM: a library for support vector machines. ACM Trans. Intell. Syst. Technol. **2**(27), 1–27 (2011). Software, http://www.csie.ntu.edu.tw/~cjlin/libsvm

Machine-Learning-Based Visual Objects' Distances Evaluation: A Comparison of ANFIS, MLP, SVR and Bilinear Interpolation Models

Kurosh Madani$^{(\boxtimes)}$, Hossam Fraihat, and Christophe Sabourin

Images, Signals and Intelligence Systems Laboratory (LISSI/EA 3956),
Senart-FB Institute of Technology, University PARIS-EST Creteil (UPEC),
Bât.A, 36-37 Rue Charpak, 77127 Lieusaint, France
{madani, sabourin}@u-pec.fr, fraihathossam@yahoo.fr

Abstract. Spatial characterization of objects is a key-step for robots' awareness about the surrounding environment in which they are supposed to evolve and for their autonomy within that environment. Within this context, this chapter deals with visual evaluation of objects' distances using Soft-Computing based approaches and pseudo-3D standard low-cost sensor, namely the Kinect. However, although presenting appealing advantages for indoor environment's perception, the Kinect has not been designed for metrological aims. The investigated approach offers the possibility to use this low-cost pseudo-3D sensor for the aforementioned purpose by avoiding 3D feature extraction and thus exploiting the simplicity of the only 2D image' processing. Experimental results show the viability of the proposed approach and provide comparison between different Machine-Learning techniques as Adaptive-network-based fuzzy inference (ANFIS), Multi-layer Perceptron (MLP), Support vector regression (SVR) and Bilinear Interpolation (BLI).

1 Introduction and Problem Stating

Robots' visual perception of their surrounding environment and their ability of metrological information extraction from the perceived environment are the most important requirements for reaching or increasing robots' autonomy (for example for autonomous navigation or localization) within the environment in which they evolve [1]. However, the complexity of real-world environment and real-time processing constraints inherent to the robotics field make the above-mentioned tasks challenging. In fact, if the use of sophisticated vision systems (e.g. high-precision visual sensors, sophisticated stereovision apparatuses, etc....) combined with sophisticated processing techniques may offer an issue for overcoming a number of the above-mentioned requirements within the condition of quite slow dynamics, they remain either too expensive for every-day applications or out of real-time processing ability for prevailing dynamics inherent to the concerned field.

The recent decade has been a token of numerous progresses in computer vision techniques and visual sensors offering appealing potential to look at the above-mentioned

© Springer International Publishing AG 2017
J.J. Merelo et al. (eds.), *Computational Intelligence*, Studies in Computational Intelligence 669,
DOI 10.1007/978-3-319-48506-5_24

dilemma within innovative slants. In fact, on the one hand, numerous image processing techniques with reduced computational complexity have been designed ([2]) and on the other hand, a number of new combined visual sensors with appealing features and accessible prices have been presented as standard market products. Microsoft's "Kinect" [3], a Microsoft product which has been initially designed for Xbox play station in 2008, is a typical example of such combined low-priced standard-market visual sensor that allows a pseudo-3-D visual capture of the surrounding environment by providing the depth (in meters) using an infra-red device and an color image using a standard camera [4]. Even if its field of view is limited (about 60° vertical and 40° horizontal) and the data is noisy, its ever-increasing usage in many domains as medical, robotics, home automation, holograms' creation, has been appreciable during recent years ([5–9]). Kinect provides spatial depth coordinates between 0.6 m to 4 m and thus could swathe spatial features within the aforementioned area. It is pertinent to notice that the above-mentioned spatial coverage gap fits with the human's typical indoor living space making Kinect an appealing pseudo-3D sensor for our purpose.

It a recent previously realized work [10], we have investigated an Adaptive Neuro-Fuzzy Inference System (ANFIS) approach and its comparison with a geometric method using the Kinect for estimate distance between the objects. The depth and color image provided by Kinect have been subjected to a ANFIS-based approach hybridizing conventional image processing and the ANFIS model in order to extract the estimated distance between the objects. In another recent work [11], a first comparison, realized on the basis of simple objects' distances evaluation in laboratory environment, has been performed involving a number of most important popular Soft-computing issued techniques. Extending the above-mentioned works, the present chapter gathers the investigated concepts and extends the accomplished investigations to the real objects' distance estimation within realistic environment.

The chapter is organized in five sections. The next section introduces the investigated objects' distance estimation concept. Section 3 briefly describes the considered estimation models. In Sect. 4, validation's results, obtained from implementation of the investigated concept versus different considered estimation models are reported and discussed. Finally, the last section concludes the chapter.

2 Brief Overview of Considered Multi-level Conjecture Models

The investigated approach combines a conventional preprocessing stage and an estimation inference model (see Fig. 1) exploiting the preprocessed 2-D color and depth information images provided by the pseudo-3D sensor (namely Kinect). The estimation inference model may be a conventional estimator or a Soft-Computing-based technique. The next section provides the considered estimation inference models used in the present work. The computation chain consists of three phases:

1. Capturing 2-D color and depth images from Kinect.

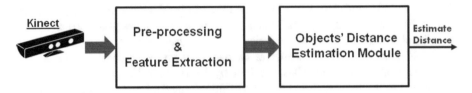

Fig. 1. General block diagram of the proposed approach.

2. Conventional processing of the Kinect issued images extracting appropriate features.
3. Inference model-based computation of distance between objects.

If the estimation inference model is a Soft-computing issued model, the computation of the estimated distance results from a Machine-Learning-based mapping involving inputs (e.g. features extracted from preprocessing of the above-mentioned 2D images) and corresponding examples of correct distances provided by a set of samples composing a "learning database". In this case, a learning phase is necessary in order to carry out the aforementioned mapping. Figure 2 illustrates block diagrams of learning and estimation (generalization) modes when the inference model is a Soft-computing-based model.

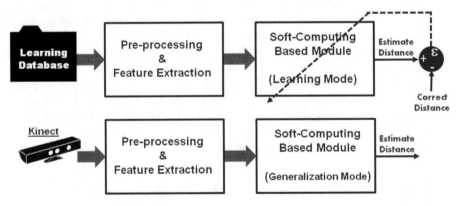

Fig. 2. General block diagrams of the proposed approach when estimation inference uses a Soft-computing-based model, showing learning mode (upper scheme) and generalization mode (lower scheme).

The first phase is devoted to the capturing of 2-D color and depth images from Kinect. The Kinect sensor can capture 2D color images at a resolution of up to 640 × 480 pixels at 30 frames per second cadency. The depth images contain the depth information providing directly the distance between the Kinect and visual patterns located within the depth sensor's field of view. The brightest fields in the depth images correspond to closer distances and the darkest ones to the farther distances. Kinect system provides directly the above-mentioned depth information in millimeters.

The Kinect can provide the depth image in 3 different resolutions: 640×480 (the default), 320×240, and 80×60. Figure 3 gives an example of color and depth images corresponding to two cubic objects captured by Kinect.

Fig. 3. Example of color (left-side picture) and depth (right-side picture) images of two cubic objects provided by Kinect.

The second phase is devoted to conventional processing of the Kinect issued images extracting appropriate features. It consists of several preprocessing tasks. The visual data (namely the color image) is segmented and a resulting binary image is constructed. The considered techniques are conventional segmentation techniques, described in [2], which have been chosen on the basis the low-computational complexity in order to fit real-time computation constraints. However, more sophisticated processing techniques may be used as those proposed by Moreno, Ramik, Graña and Madani in [12]. In the present work we used the Mean Shift Segmentation (MSS) method, described and applied by [13–15]. As indicates the name of the so-called segmentation, the segmentation task is based mainly on "mean shift" method which is a computational technique to estimate the most accurate mean location "$m(x)$" of the data (center of mass in Fig. 4). The estimation is performed by determining the "mean shift vector" within some initial mean region (called also region of interest in Fig. 4), which corresponds to the center of the region that represents maximum density of pixels. The iterative process is repeated until find the mean shift vector following the direction of the maximum increasing of the pixels' density.

To calculate the mean location $m(x)$ at the point x, we use the Eq. (1), where n represent the number of point in the kernel K of the region of interest, x_i is data point, x initial mean location and $K(x)$ stands for kernel function relative to the samples x contributing to the estimation of the mean location. The mean shift, defined as the difference between $m(x)$ and x, is computed iteratively for obtaining the maximum density in the local neighborhood. The process stops when $m(x) = x$. The corresponding mean shift vector has the direction of the gradient of the density estimate, and thus, the interruption condition corresponds to search of regions within the kernel where pixels are similar. Figure 5 shows an example of resulting segmented objects using MSS method exploiting images provided by Kinect for two objects located at a

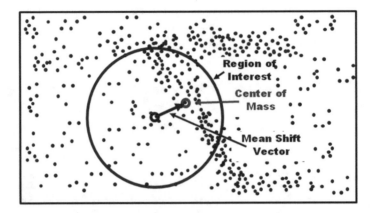

Fig. 4. The principle of the mean shift segmentation method.

(a)   (b)

Fig. 5. Example resulting objects' segmentation for two given objects located at a given distance from each other (a) and the definition of distance between those two objects as the distance between the closest pixels of each segmented region visible as line-1 (b).

given distance from each other. Once segmentation is performed, the minimum distance between the objects is calculated (number of pixels). Such distance is defined as the minimum distance between two horizontal pixels in each object (line 1 in Fig. 5-b).

$$m(x) = \frac{\sum_{i=1}^{n} K(x - x_i)\, x_i}{\sum_{i=1}^{n} K(x - x_i)} \tag{1}$$

The last phase of the computational chain deals with inference-based estimation of the objects' distance. The inputs are the features extracted from the preprocessing stage and output is the estimated distance between the concerned objects provided in cm. It is pertinent to note that the distance between two objects from the visual information of detected items usually requires an accurate acquaintance relating the location of a same detected item in both RGB and Depth images. In other words, before computing the spatial attributes of a detected object, it is necessary to calibrate depth and RGB cameras. Concerning the proposed approach, the estimation using Soft-computing-based model allows escaping the calibration task, because the estimation is performed by mapping the correspondence between the input visual features and the corresponding distances directly from representative examples (called learning patters), constructing the estimation model directly. However, it is also pertinent to note that the learning database (e.g. the used examples) has to contain enough examples in order to be representative of the matching function. In fact, the quality of the learning database plays a primary role in accuracy of estimated distances.

3 Brief Overview of Considered Multi-level Conjecture Models

As it has been mentioned-above, several main Soft-computing techniques have been considered for estimation inference model, namely Adaptive-network-based fuzzy inference (ANFIS), Multi-layer Perceptron (MLP), Support vector regression (SVR) and Bilinear Interpolation (BLI). This section is devoted to a brief reminder of each considered model.

3.1 Adaptive-Network-Based Fuzzy Inference

Adaptive-network-based fuzzy inference (ANFIS) is a Fuzzy Inference System (FIS) using Artificial Neural Network [16–18]. The rule base contains two fuzzy rules of Takagi and Sugeno's type, expressed here-bellow, where x and y are two input data, f_i is the fuzzy inference according to the desired output, A_i and B_i are labels of fuzzy sets characterized by appropriate membership functions.

$$\textbf{Rule 1}: if\ x\ is\ A_1\ and\ y\ is\ B_1\ then f_1 = p_1\, x + q_1\, y + r_1$$

$$\textbf{Rule 2}: if\ x\ is\ A_2\ and\ y\ is\ B_2\ then f_2 = p_2\, x + q_2\, y + r_2$$

The membership functions of A_i and B_i, denoted $\mu_{Ai}(x)$ and $\mu_{Bi}(y)$ respectively, are given by Eq. (2), where $\{a_i,\ b_i,\ c_i\}$ is the parameters set. Figure 6 depicts the functional block diagram of ANFIS.

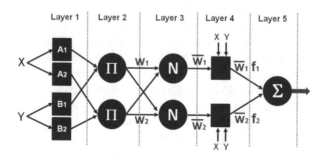

Fig. 6. ANFIS structure.

$$\mu_{Ai}(x) = e^{-\left(\frac{x-c_i}{a_i}\right)^2} \quad \text{and} \quad \mu_{Ai}(x) = e^{-\left(\frac{y-c_i}{b_i}\right)^2} \tag{2}$$

Layer1: Generating membership degree, where $O_{k,i}$ is the node function, where k is the number of the layer and i is the node position in the layer.

$$O_{1i} = \mu_{Ai}(x) \text{ with } i \in \{1, 2\}$$

Layer 2: Fuzzy intersection.

$$O_{2i} = w_i = \mu_{Ai}(x) \cdot \mu_{Bi}(y) \text{ with } i \in \{1, 2\}$$

Layer3: Normalization.

$$O_{3i} = \overline{w}_i = \frac{w_i}{w_1 + w_2} \text{ with } i \in \{1, 2\}$$

Layer4: Defuzzyfication, where $\{p_i, q_i, r_i\}$ is the parameters set (consequent parameters).

$$O_{4i} = \overline{w}_i f_i = \overline{w}_i (p_i x + q_i y + r_i)$$

Layer 5: The final output

$$O_{5i} = \sum_i \overline{w}_i f_i = \frac{\sum_i w_i f_i}{\sum_i w_i}$$

3.2   Multi-layer Perceptron

The Multi-Layer Perceptron (MLP) is a very well known artificial neural network organized in layers and where information travels in one direction, from the input layer to the output layer ([19, 20]). The input layer represents a virtual layer associated to the inputs of data. It contains no neuron. The following hidden layers are layers of neurons. The outputs of the neurons of the last layer always correspond to the desired data outputs. MLP structure may include any number of layers and each layer may include any number of neurons. Neurons are connected together by weighted connections. It is the weight $w_{i,j}$ of these connections that manages the operation of the network and ensures the transformation of inputs data to outputs data.

The back-propagation algorithm is used to minimize the quadratic error between the current output (computed by the network in response to a given input stimulus) and the desired value of the network's output expected for this input. The neural network's weights are updated accordingly to the output error gradient and back-propagated in order to minimize the output error. In our work we use a MLP with one hidden layer, where it have 304 input variables (inputs), 100 neurons on the hidden layer and 19 neurons on the output layer.

3.3 Support-Vector Machine

We will focus only the SVM regression basic principles. However, a detailed representation can be found in [21, 22].

Let suppose a dataset $D = \{(x_i, y_i) \mid 1 \leq i \leq N\}$ where $x_i \in \Re^N$ and $y_i \in \Re$. In the ε-SVM regression ([23]) the goal is to determine the function $f(x)$ which deviates by at most ε from the actual target y_i for all training data, and at the same time be as regular as possible. In other words, the errors that are less than ε be tolerated, while any greater deviation than ε be penalized. We begin by describing the case of the linear version (functions), given by Eq. (3), where $\langle \cdot, \cdot \rangle$ Denotes the dot product in $\Re^N$.

$$f(x_i) = \langle w, x_i \rangle + b \tag{3}$$

The problem could be formulated as an optimization process minimizing what is called "Flatness" w (an interval in the feature-space less sensitive to the perturbations) accordingly to the set of conditions expressed by Eq. (4).

$$\min\left(\frac{1}{2}\|w\|^2\right) \text{ subject to } \begin{cases} y_i - \langle w, x_i \rangle - b \leq \varepsilon \\ \langle w, x_i \rangle + b - y_i \leq \varepsilon \end{cases} \tag{4}$$

$$f(x) = \sum_{i=1}^{N} \left(a_i^+ - a_i^-\right) \langle x, x_i \rangle + b$$

$$w = \sum_{i=1}^{N} \left(a_i^+ - a_i^-\right) x_i \tag{5}$$

$$\begin{cases} \alpha_i^+ \left(\varepsilon + \xi_i^+ - y_i + \langle w, x_i \rangle + b\right) = 0 \\ \alpha_i^- \left(\varepsilon + \xi_i^- - y_i - \langle w, x_i \rangle + b\right) = 0 \\ \begin{cases} \mu_i^+ \xi_i^+ = \left(C - \alpha_i^+\right) \xi_i^+ = 0 \\ \mu_i^- \xi_i^- = \left(C - \alpha_i^-\right) \xi_i^- = 0 \end{cases} \end{cases} \tag{6}$$

$$\begin{aligned} &\max\left(y_i - \langle w, x_i \rangle + \varepsilon \mid \alpha_i^+ < C \text{ or } \alpha_i^- > 0\right) \\ &\min\left(y_i - \langle w, x_i \rangle - \varepsilon \mid \alpha_i^+ > 0 \text{ or } \alpha_i^- < C\right) \end{aligned} \tag{7}$$

f approximates all pairs (x_i, y_i) with ε precision. By associating a Lagrange multiplier to each constraint described above, the initial problem can be described by its dual problem, which is a quadratic optimization problem without constraints. Such dual formulation of the initial problem leads to express the function f as the set of Eqs. (5). This is called "Support Vector" in which w can be completely described as a linear combination of the training patterns x_i. The parameter b in the Eq. 5 can be computed by Karush-Kuhn-Tucker conditions expressed by the set of Eqs. (6). Then, within these

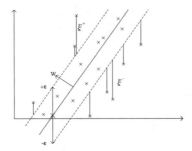

Fig. 7. Adjusting the loss function in the case of a linear SVM.

conditions, one can exploit the system given by the set of Eqs. (7). Figure 7 shows such a minimization process in a 2-D feature-space.

3.4   Bilinear Interpolation

The Bilinear Interpolation is based on a set of points in a given (considered) feature-space ([24–27]). An example of such interpolation involving the points P_1, P_2, P_3 and P_4 is shown in Fig. 8. Relating the investigated purpose, this example represents four points (representing data: here, the distance between two objects expressed in centimeters) in a 2D feature space that axes (features) are depth (e.g. distance of object regarding the Kinect, expressed in centimeters) and distance between objects between two objects (expressed in number of pixels). In other words, the goal is to search the intermediate bilinear distance between two classes (points), each class represents a distance between two objects in centimeters. This intermediate distance (P) is given by Eq. (8).

$$P = (1 - \lambda) \cdot [(1 - \mu) P_1 + \mu P_3] + \lambda [(1 - \mu) P_2 + \mu P_4] \tag{8}$$

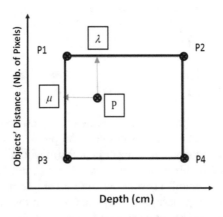

Fig. 8. Schematic diagram of Bilinear Interpolation algorithm.

4   Experimental Validation and Experimental Results

The reported results have been achieved on the basis of two databases collecting data relative to various positions (e.g. different distances of those objects from each other and different positions relative to the Kinect's position) of two kind of objects. The first one contains two simple (regular shape) objects and the second includes same kind of data for more complex objects (e.g. with irregular shapes). The considered objects have been placed on various positions regarding the Kinect's referential (e.g. 100 cm to 270 cm from Kinect). The first database (database 1) contains 495 color images of the regular objects and the second database (database 2) 304 pictures of irregular objects (trapezoidal). Different distances between the concerned objects have been considered: from 4 cm to 100 cm for the database 1 and from 1.7 cm to 91.7 cm for the database 2. On the other hand, different positions relative to the Kinect have been considered: 100 cm to 263 cm for the database 1 and 100 cm to 250 cm for the database 2. The capturing and segmenting processes have been developed using PYTHON. The distance prediction model has been realized using Matlab R2011environment.

The Table 1 resumes the different training and testing experiences.

Table 1. Summary.

Number of sample used for the experiment	Distance estimation's Max-Error rate
Database1 = 495	35 % (Fig. 13)

Accordingly to the above-mentioned protocol and the experimental setup of Fig. 9, the two aforementioned databases have been constructed following two policies:

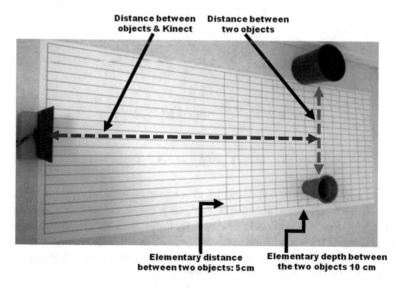

Fig. 9. Experimental setup.

- For a same distance between the two concerned objects, a set of RGB and depth images are collected from Kinect for different depths within the scale of 100 cm to 250 cm with a 10 cm step.
- The images' collection is repeated for various distances between the two concerned objects (e.g. increasing distance between the two objects) within the range of 5 cm to 100 cm by increasing step of 5 cm and following the same depth range whenever.

4.1 Objects' Distance Estimation Using Direct Geometrical Computation

Before assessing the investigated approach, let have a look at an objects' distance estimation technique which uses geometrical computation issued from Kinect's technological features and it's cameras' optical parameters. In fact, it is pertinent to remind that a number of geometrical estimation of rough distance between objects perceived by using Kinect have already been developed and described in [4]. However, they have been designed in the frame of a sketchy estimation of the perceived objects and within the spirit of a quite vague localization of objects by Xbox and the supported video games.

A quite simple and direct computation of objects' distance, taking into account Kinect's viewing fields exploits the fact that Kinect's color imaging sensor provides an image of 640-by-480 pixels with a 57° vertical viewing and 43 horizontal viewing angles respectively. Thus, considering a 4-by-2.5 square meters viewing area (due to Kinect camera's limitations), spatial coordinates (in meters) of a perceived object in Kinect's referential could be estimated by Eqs. (9) and (10), where z_m is the object's distance from the Kinect (depth information provided by infra-red sensor in meters), $p(x)$ and $p(x)$ are horizontal and vertical pixels' positions in image, respectively. In this conditions, the distance between two objects characterized by their respective coordinates (x_{m1}, y_{m1}, z_{m1}) and (x_{m2}, y_{m2}, z_{m2}) is estimated by Eq. (11).

$$x_m = 4 \cdot \left[\frac{640}{2} - p(x) \right] tg(57/2) \cdot \frac{z_m}{640} \tag{9}$$

$$y_m = 2.5 \cdot \left[\frac{480}{2} - p(y) \right] tg(43/2) \cdot \frac{z_m}{480} \tag{10}$$

$$d_m = \sqrt{(x_{m2} - x_{m1})^2 + (y_{m2} - y_{m1})^2 + (z_{m2} - z_{m1})^2} \tag{11}$$

Table 1 summarizes the experimental evaluation relating the geometric approach and Fig. 10 gives the obtained results providing estimation error versus depth for distance-between-objects included in database 1. As it is visible in this figure, the estimation error remains quite high revealing an average estimation error of 25 % while attaining its maximum value around 35 %. As it has already been mentioned, these unsatisfactory results could be explained taking into account the fact that the RGB camera and the depth camera are not calibrated and thus the location of a same object in

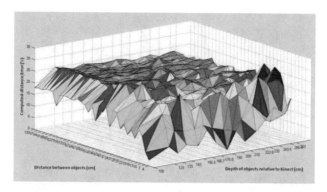

Fig. 10. Experimental setup. Estimation error versus Distance-between-objects and depth relative to geometric approach for database1.

RGB image doesn't fit the location of that object in depth image leading to quite significant estimation error.

4.2 Soft-Computing Based Distance Estimation and Comparative Results

The two above-mentioned databases have been used for evaluating the considered Soft-computing models in Machine-Learning-based distance evaluation. Table 2 resumes different training and testing configurations realized for this evaluation. We show the experimental results of different Machine-Learning models: ANFIS, MLP, SVR and Bilinear Interpolation. Figure 11 shows example of distance estimation results for ANFIS, indicating the estimation error for the case where learning has been performed using the second database and the test was performed using the first base data.

Table 2. Databases configuration in learning and testing phases.

Learning	Testing
Database2 (304 samples)	Database1 (495 samples)
Database2 + 50 % Database1 (552 samples)	50 % Database1 (247 samples)

Figures 12 and 13 show comparative results representing the distance estimation error's distribution in learning and testing modes, respectively. It is pertinent to remind that Bilinear Interpolation (BLI) isn't a learning-based technique and thus doesn't include a learning mode. Table 3 resumes mean-value and standard deviation of distance estimation error for each considered Soft-computing-based estimator. These results highlight an improved accuracy of ANFIS in objects' distances estimation: lower estimation error as well in learning mode ($\overline{Error} = 1 \pm 0.95\,\%$) as in testing mode ($\overline{Error} = 2.4 \pm 1.30\,\%$). This estimator leads also to lower standard deviation and thus leads to lower boundary (e.g. maximum) estimation error.

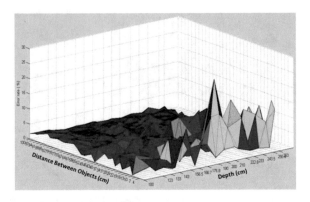

Fig. 11. Example of distance estimation error in testing Mode (ANFIS).

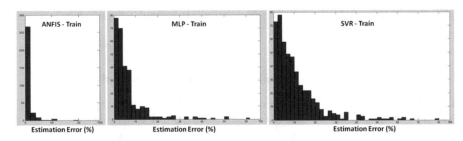

Fig. 12. Comparison of distance estimation error's distribution in "Learning" mode using Database 2 (304 samples).

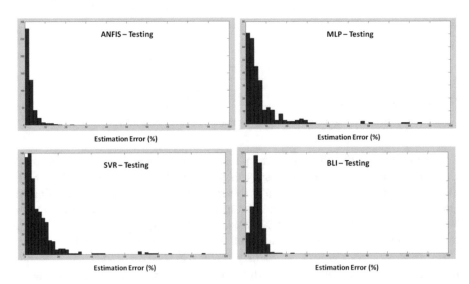

Fig. 13. Comparison of distance estimation error's distribution in "Testing" mode using Database1 (495 samples).

Table 3. Databases configuration in "Learning" and "Testing" phases.

Considered model	Learning		Testing	
	$\overline{Error}$	σ_{Error}	$\overline{Error}$	σ_{Error}
ANFIS	1.03 %	1.94 %	2.46 %	2.73 %
MLP	6.95 %	10.73 %	7.57 %	11.21 %
SVR	11.08 %	12.39 %	8.73 %	12.12 %
BLI			5.70 %	2.58 %

4.3 Generalization to Objects' Distance Evaluation in Realistic Environment

The accomplished comparative study highlights the supremacy of ANFIS versus the other considered Soft-computing-based approaches for estimating the objects' distances. This is obvious from its mean-error and standard deviation relating the distribution of estimation-error in learning (Fig. 12) and testing (Fig. 13) phases. In fact, as shown by Fig. 12, relating the learning phase, more than 270 patters (among 304 learned patterns) have been correctly learned (estimation-error less than 1 %), about 25 (among 304) estimated with less than 2 % estimation-error and the estimation of only a the estimation of a few of them (less than 15 among 304) surpasses 10 % estimation-error. In the same way, concerning the testing phase, more than 410 patters (among 495 tested patterns) have been correctly evaluated (estimation-error less than 2 %) and also the estimation of a few of them (less than 20 among 495) surpasses 10 % estimation-error.

Taking into account the aforementioned very encouraging result, the decision has been made to extend the evaluation of the ANFIS-based model considering realistic environment with every-day objects that it may contain. It is pertinent to note that the learning phase has been performed on the basis of the training databases mentioned in Table 2. In other words, the present assessment of the investigated approach deals with unknown (unlearned) objects. The system has been tested in a working space containing customary every-day objects relative to such spaces as well as accustomed users (humans) working in that locale. Figures 14 and 15 show two examples of sights acquired by the Kinect and the corresponding depth images, respectively.

Fig. 14. Two examples of sights acquired by the Kinect.

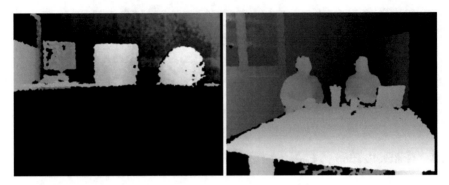

Fig. 15. Depth images providing by Kinect system corresponding to RGB images of Fig. 14.

Fig. 16. Segmented items corresponding to objects of images of Fig. 14.

The left-side image of the Fig. 14 corresponds to a typical working-corner where viewed from the meeting corner located within the same space close to that working-corner. The right-side picture corresponds to the above-mentioned meeting-corner including two users (individuals). Figure 16 shows the segmented items corresponding to objects of Fig. 14 and Table 4 labels the considered items of this figure. The last column in Table 4 gives nominal (e.g. measured) distances between the considered items. Finally, $d(Item_i, Item_j)$ denotes nominal distance between two given items, where $Item \in \{O, H\}$, $i \in \{1, \cdots, 6\}$ and $j \in \{1, \cdots, 6\}$ with $i \neq j$. Objects O_1 and O_2 are located at about 190 cm (e.g. $depth \cong 190$ cm) from Kinect and 20 cm from each other (e.g. $d(O_1, O_2) = 20$ cm). Concerning objects O_1, O_2, O_3 and O_4, they are located at about 210 cm (e.g. $depth \cong 210$ cm) from Kinect. The two individuals (e.g. humans H_1 and H_2), are located at about 260 cm (e.g. $depth \cong 260$ cm) from Kinect Their respective distances from each others are given in the last column of Table 4.

Table 5 resumes estimated distances between different considered items. $d^*(Item_i, Item_j)$ denotes estimated distance between two given items, where $Item \in \{O, H\}$, $i \in \{1, \cdots, 6\}$ and $j \in \{1, \cdots, 6\}$ with $i \neq j$. As it is visible from this table, distances between considered items (objects or humans) are accurately estimated revealing estimation-errors not exceeding the error interval corresponding to ANFIS-based estimator carried out in Subsect. 4.2. The obtained results for this unlearned (by the system) set of items and the magnitude of the related errors show the leeway of the

Table 4. Nature and label of considered items within the Fig. 14 and nominal distances between items.

Item	Label	Distance between items (cm)
Square Chair (left-side picture)	O_1	$d(O_1,O_2) = 20$
Round Chair (left-side picture)	O_2	$d(O_2,O_1) = 20$
Yellow Cap (right-side picture)	O_3	$d(O_3,O_4) = 16$; $d(O_3,O_5) = 38$; d $(O_3,O_6) = 64$
Black Boiler (right-side picture)	O_4	$d(O_4,O_3) = 16$; $d(O_4,O_5) = 10$; d $(O_4,O_6) = 36$
Orange Cap (right-side picture)	O_5	$d(O_5,O_3) = 38$; $d(O_5,O_4) = 10$; $d(O_5, O_6) = 19$
Laptop (right-side picture)	O_6	$d(O_6,O_5) = 19$; $d(O_6,O_4) = 36$; d $(O_6,O_3) = 64$
Left-side Human (right-side picture)	H_1	$d(H_1,H_2) = 35.5$
Right-side Human (right-side picture)	H_2	$d(H_2,H_1) = 35.5$

Table 5. Estimated distances between two items $d^*\left(Item_i, Item_j\right)$ and estimation-error.

Items	O_1	O_2	O_3	O_4	O_5	O_6	H_1	H_2
O_1		19.4 cm 3 %						
O_2	19.4 cm 3 %							
O_3				15.0 cm 6 %	35.9 cm 5.5 %	61.2 cm 4 %		
O_4			15.0 cm 6 %		9.3 cm 7 %	34.6 cm 3.8 %		
O_5			35.9 cm 5.5 %	9.3 cm 7 %		18.3 cm 4 %		
O_6			61.2 cm 4 %	34.6 cm 3.8 %	18.3 cm 4 %			
H_1								34.3 cm 3 %
H_2							34.3 cm 3 %	

investigated approach for handling vision-based robots' navigation within realistic and sufficiently complex indoor environments.

5   Conclusion

The obtained distance-estimation errors between two objects in generalization mode show ANFIS-based estimator's supremacy versus the three other considered Soft-computing-based models ($\overline{Error} = 2.4 \pm 1.30\,\%$ for ANFIS comparing to

$\overline{Error} = 7.6 \pm 5.1\%$ for MLP, $\overline{Error} = 8.7 \pm 6.0\%$ for SVR and $\overline{Error} = 5.7 \pm 1.30\%$ for BLI, respectively). Concerning MLP and SVR, they have been used within a classification-like paradigm and thus lead to generating a large number of classes. That is why the generalization remains quite far from expected accuracy. Concerning the Bilinear Interpolation, this method is based on the local approximation strategy. In fact, the disadvantage of this method is that the distance is calculated from the four neighborhood distance values and depends on the precision of these four distances values, without the possibility of a correction or adjustment. Although, out of sufficient accuracy for metrological applications (where an estimation with high precision is required), the ANFIS-based estimator presents appealing features relating robots' navigation oriented applications. This has been corroborated by estimating distances between a set of unlearned objects within the realistic environment of an everyday working space containing as well objects as humans. Concerning the usage of Kinect as sensory system, it is certain that this low-cost pseudo-3D sensor presents appealing features for Machine-Awareness within indoor environments, although initially not been designed for such range of applications. This appealing potential opens a wide range of promising prospect vision and robotics applications. Concerning outdoor environments, its limited visual field (covering a frontal sight comprised within 90 cm and 4 m), even though compatible with human's indoor living-space, remains inadequate.

Farther works relating the investigated technique will concern the enhancement of the estimation precision by using more sophisticated interpolation techniques.

References

1. Hoffmann, J., Jüngel, M., Lötzsch, M.: A vision based system for goal-directed obstacle avoidance. In: Nardi, D., Riedmiller, M., Sammut, C., Santos-Victor, J. (eds.) RoboCup 2004. LNCS, vol. 3276, pp. 418–425. Springer, Heidelberg (2005). doi:10.1007/978-3-540-32256-6_35
2. Gonzalez, R.C., Woods, R.E.: Digital Image Processing, 2nd edn. Prentice Hall, Upper Saddle River (2002)
3. Kinect camera. http://www.xbox.com/en-US/kinect/default.htm
4. Borenstein, G.: Making Things See: 3D Vision with Kinect, Processing, Arduino, and MakerBot. O'Reilly Media Inc., Sebastopol (2012)
5. Zhang, Z.: Microsoft Kinect sensor and its effect. IEEE Multimedia Mag. **19**(2), 4–10 (2012)
6. Meister, S., Izadi, S., Kohli, P., Haemmerle, M., Rother, C., Kondermann, D.: When can we use Kinect Fusion for ground truth acquisition? In: Proceedings of Workshop Color-Depth Camera Fusion Robot (2012)
7. Roth, H., Vona, M.: Moving volume Kinect Fusion. In: Proceedings of British Machinery Vision Conference, pp. 1–11 (2012)
8. Whelan, T., Kaess, M., Fallon, M., Johannsson, H., Leonard, J., McDonald, J.: Kintinuous: spatially extended Kinect Fusion. In: RSS Workshop on RGB-D: Advanced Reasoning with Depth Cameras, no. 4 (2012)
9. Han, J., Shao, L., Xu, D., Shotton, J.: Enhanced computer vision with microsoft kinect sensor: a review. IEEE Trans. Cybern. **43**(5), 1318–1334 (2013)

10. Fraihat, H., Sabourin, C., Madani, K.: Soft-computing based fast visual objects' distance evaluation for robots' vision. In: Proceedings of the 8th IEEE International Conference on Intelligent Data Acquisition and Advanced Computing Systems, (IEEE/IDAACS 2015), Warsaw, Poland, 24–26 September, vol. 1, pp. 81–86 (2015)

11. Fraihat, H., Sabourin, C., Madani, K.: Learning-based distance evaluation in robot vision: a comparison of ANFIS, MLP, SVR and bilinear interpolation models. In: Proceedings of the International Conference on Neural Computation Theory and Applications (NCTA 2015), Lisbon, Portugal, 12–14 November, pp. 168–173 (2015)

12. Moreno, R., Ramik, D.M., Graña, M., Madani, K.: Image segmentation on the spherical coordinate representation of the RGB color space. IET Image Procession J. 6(9), 1275–1283 (2012)

13. Comaniciu, D., Ramesh, V., Meer, P.: Real-time tracking of non-rigid objects using mean shift. In: Proceedings of IEEE Conference on Computer Vision and Pattern Recognition, vol. 2, pp. 142–149 (2000)

14. Comaniciu, D., Meer, P.: Mean shift: a robust approach toward feature space analysis. IEEE Trans. Pattern Anal. Mach. Intell. 24(5), 603–619 (2000)

15. Kheng, L.W.: Mean shift tracking, Technical report, National University of Singapore (2011). www.comp.nus.edu.sg/~cs6240/lecture/particle.pdf

16. Jang, J.R.: ANFIS: Adaptive-Network-based Fussy Inference System. IEEE Trans. Syst. Man Cybern. 23, 665–685 (1993)

17. Jang, J.R., Sun, C-T.: Neuro-fuzzy modeling and control. In: Proceedings of the IEEE, vol. 83, no. 3, pp. 378–406 (1995)

18. Jang, J.R., Sun, C.-T., Mizutani, E.: Neuro-fuzzy and Soft Computing: A Computational Approach to Learning and Machine Intelligence. Prentice Hall, Upper Saddle River (1997). ISBN 978-0132610667

19. Rumelhart, D., Hinton, G., Williams, R.: Learning internal representations by error propagation. In: Parallel Distributed Processing: Explorations in the Microstructure of Cognition. MIT Press, Cambridge (1986)

20. Lippman, R.P.: An introduction to computing with neural nets. IEEE ASSP Mag. 4(2) 4–22 (1987)

21. Smola, A.J., Scholkopf, B.: A tutorial on support vector regression. Stat. Comput. 14(3), 199–222 (2004)

22. Cristianini, N., Shawe-Taylor, J.: An Introduction to Support Vector Machines and Other Kernel-Based Learning Methods. Cambridge University Press, Cambridge (2000). ISBN 0521-78019-5

23. Cortes, C., Vapnik, V.: Support-vector networks. Mach. Learn. 20(3), 273–297 (1995). Kluwer Academic Publishers

24. Cok, D.R.: Signal Processing Method and Apparatus for Producing Interpolated Chrominance Values in a Sampled Color Image Signal, US Patent 4,642,678, (1987)

25. Intel, Using MMX™ Instructions to Implement Bilinear Interpolation of Video RGB Values (1996). https://software.intel.com/sites/landingpage/legacy/mmx/MMX_App_Bilinear_Inter polation_RGB.pdf

26. Lu, G.Y., Wong, D.W.: An adaptive inverse-distance weighting spatial interpolation technique. Comput. Geosci. 34(9), 1044–1055 (2008)

27. Chen, D., Ou, T., Gong, L.: Spatial interpolation of daily precipitation in China: 1951–2005. Adv. Atmos. Sci. 27(6), 1221–1232 (2010)

Author Index

© Springer International Publishing AG 2017   481
J.J. Merelo et al. (eds.), *Computational Intelligence*, Studies in Computational Intelligence 669,
DOI 10.1007/978-3-319-48506-5

Printed in the United States
By Bookmasters